Developments in Geotechnical Engineering

Series editors
Braja M. Das, Henderson, USA
Nagaratnam Sivakugan, Townsville, Australia

This series on Geotechnical Engineering has been initiated to provide an integrated platform for publishing in all areas of geotechnics and their applications to engineering problems. The series focuses on the topics that have garnered recent interest from a large research audience, including, but not limited to, saturated and unsaturated soils, expansive soils, geosynthetics, marine geotechnics, and soil improvement. This series includes monographs, textbooks, and references in the general area of Geotechnical Engineering. The books in this series will serve researchers, practitioners, and students alike.

More information about this series at http://www.springer.com/series/13410

Madhavi Latha G.
Editor

Frontiers in Geotechnical Engineering

Springer

Editor
Madhavi Latha G.
Department of Civil Engineering
Indian Institute of Science Bangalore
Bangalore, Karnataka, India

ISSN 2364-5156 ISSN 2364-5164 (electronic)
Developments in Geotechnical Engineering
ISBN 978-981-13-5870-8 ISBN 978-981-13-5871-5 (eBook)
https://doi.org/10.1007/978-981-13-5871-5

Library of Congress Control Number: 2018965901

This Springer imprint is published by the registered company Springer Nature Singapore Pte Ltd.
The registered company address is: 152 Beach Road, #21-01/04 Gateway East, Singapore 189721, Singapore

Foreword

The edited volume on *Frontiers in Geotechnical Engineering* comprises invited lecture contributions of the Indian Geotechnical Conference held at Indian Institute of Science, Bangalore, during 13–15 December 2018.

Various chapters of the book cover most of the important and emerging topics in geotechnical engineering. These chapters present an interesting blend of academic research and industrial case studies, sometimes bridging the gaps between the two. Some of the fundamental aspects of geotechnical engineering like study of pore spaces and their role in soil behaviour, characterization of soft clays, dredged fills and rocks, and seismic site characterization are covered well in some chapters, while the advancement of technological innovations with respect to geotechnical applications is demonstrated through studies on geoinfrastructure monitoring, soft ground improvement, and foundations on tall buildings. Sustainability issues in geotechnical engineering are covered through a holistic approach, considering bioengineering aspects and green techniques for soil improvement, energy in buried pipeline systems, and mitigation of landslides. Studies presented in these chapters include advanced laboratory testing, centrifuge and other model studies, field tests, and computational modelling. Some chapters presented a problem-solving approach for some of the contemporary geotechnical challenges, especially contaminant transport issues, seismic response of structures, and constructions in soft and problematic soils. Two of the chapters discuss professional vigilance in investigation, design and construction, and geotechnical education from practice.

All contributions of this book are highly valuable and provide insights into the contemporary understanding of the state of the art as well as future needs. The editors need to be congratulated for bringing out this important contribution and appreciations to authors for the diligent presentation of the knowledge and the quality in their contributions.

Bangalore, India

Prof. G. L. Siva Kumar Babu
President, Indian Geotechnical Society

Professor, Indian Institute of Science

Preface

It is indeed a great honour and pleasure to be the editor of this book. When I started inviting the contributing chapters for this book, my expectations were not very high, considering the limited time we had in our hands. At some point, I feared that the title *Frontiers in Geotechnical Engineering* is slightly broader and overambitious. Today, while I am making the line-up of the chapters, all my fears are calmed, and I am feeling elated to see the depth and breadth of knowledge capsuled into these 21 chapters. I thank all the contributors for their excellent contributions. Together, we made this valuable compilation possible.

My goal was to produce a comprehensive book on all important topics in geotechnical engineering, which is rigorous in terms of scientific content and, at the same time, interesting for a bigger cross section of readers. This book has some unique strengths, which make the learning experience interesting. Most of the investigations reported in the chapters used latest laboratory and field testing methods, showcasing the rich technological advances in the field of geotechnical engineering. Though the demonstration of new techniques is highlighted substantially, fundamental aspects of soil behaviour are given utmost importance while characterizing, testing, and improving soils. Thus, the content will keep the young as well as established researchers glued on alike. Concepts are established through theory and analysis, learnt in classrooms, tested in smaller scales through experiments, validated for real field situations through case studies, and extrapolated to many other possible situations through numerical studies. All these facets are equally important and indispensable in the context of geotechnical designs and constructions. This book contains studies on all these aspects and strikes a perfect balance between conceptual and applied geotechnics, which will interest both academia and practising industry.

Substantial portions of the book contain studies on fundamental aspects of soil behaviour and geotechnical characterization. In their chapter "Pores—The Almost Invisible Part of Soil" (Chap. 1), David Frost and his team demonstrated how the missing information on pore spaces in soil can be precisely captured through advanced digital imaging techniques and gained new insights into the role of pores in soil behaviour. The importance of internal pore fabric and distribution for characterizing expansive soils and the use of unmanned aerial vehicles (UAVs) for geostructural health monitoring is beautifully explained by Anand and his team in their chapter "Research Advancements in Expansive Soil Characterization, Stabilization and Geoinfrastructure Monitoring" (Chap. 2). A much-needed new perspective about the effects of testing method on basic soil properties like liquid limit, swell pressure, and friction angle is discussed by Robinson in his chapter "Testing Mode and Soil Parameters" (Chap. 3). Shin and his co-workers in their chapter "Estimation of Dredged Fill Volume Based on Self-weight Consolidation Characteristics of Dredged Soil" (Chap. 4) proposed sedimentation consolidation experiments in place of usual retention tests for the determination of dredged fill volume. Thakur and Degago illustrated the consequences of sample disturbances on the estimation of strength and deformation characteristics of soils in their chapter "Recommended Practice for Soft Clay Characterization with a Focus on Settlement and Stability Analysis" (Chap. 5). Bharat and his co-workers assessed the effectiveness of different laboratory diffusion studies and theoretical frameworks for accurately estimating mass transport parameters in their chapter "Insights into Contaminant Transport Modeling Through Compacted Bentonites" (Chap. 6). Vanapalli and Al-khazaali in their chapter "Axial Load-Displacement Behavior of Energy Pipeline Systems in Sand" (Chap. 7) demonstrated the role of matric suction on the mechanical behaviour of pipeline systems through experimental studies. The use of digital image correlation techniques for condition monitoring and crack growth assessment in pavements is described by Sireesh and Kumar in their chapter "Reflection Crack Assessment Using Digital Image Analysis" (Chap. 8).

Physical modelling is one of the powerful tools in geotechnical experimentation. The importance of high-*g* centrifuge modelling for studying the performance of reinforced soil models is presented by Viswanadham in his chapter "Centrifuge Model Studies on the Performance of Geosynthetic-Reinforced Soil Structures" (Chap. 9). The chapter "Soft Ground Improvement—Theoretical, Experimental, Numerical and Field Studies" (Chap. 10) by Indraratna and his co-workers presented a comprehensive approach to completely understand the soft soil improvement using prefabricated drains through large-scale testing and numerical studies.

Dynamic testing of soils and seismic site characterization gained new impetus in recent times due to growing infrastructural needs and awareness on earthquake-resistant designs. Some of the most important aspects of seismic site characterization are discussed in "Issues on Seismic Site Characterization" (Chap. 11) by Prasad and his co-authors. Analytical and numerical studies are presented by Deepankar and co-workers in their chapter "Dynamic Tunnel–Soil Interaction in Soft Soils Considering Site-Specific Seismic Ground Response" (Chap. 12) to understand the complex deformation modes and lining forces of tunnel during a seismic event, considering local site effects. Dammala and Krishna outlined various laboratory and field tests to measure the dynamic properties of soils along with the theoretical formulations involved in these tests.

The use of new materials to enhance the performance of soil structures is a topic of continuous innovation, and the research in this direction has no boundaries. Sreedeep and co-workers provided directives to adopt various bioengineering options in developing sustainable green infrastructure in their chapter "Sustainable Geotechnics: A Bio-geotechnical Perspective" (Chap. 15). Dasaka and Gade presented model tests on rigid retaining walls with geofoam inclusions in their chapter "Recent Developments in Earth Pressure Reduction Techniques" (Chap. 16), to bring out the beneficial isolation effects of geofoam in reducing the lateral thrust on retaining walls. The chapter by Vinod and his team on "Cyclic Behaviour of Scrap-Tyre Soil Mixtures" (Chap. 14) presented a detailed account of cyclic loading characteristics of soil mixed with tyre chips. In their chapter "Probabilistic Study on the Geotechnical Behavior of Fiber Reinforced Soil" (Chap. 17), Moghal and his team carried out a reliability-based study to determine the optimum fibre content required to stabilize expansive soils through laboratory experiments.

Case studies presented by seasoned industrialists always enhance the quality of any technical document. This book contains many such case studies, some of them from ongoing live projects. Ravi Sundaram and his team presented an excellent case study of a 38-storeyed building in the Indo-Gangetic alluvium in their chapter "Foundations for Tall Buildings on Alluvial Deposits—Geotechnical Aspects" (Chap. 18) to emphasize the importance of the quality of geotechnical investigations. "Sustainable Landslide Mitigation Measures-Demonstration Through a Live Case Study in Uttarakhand" (Chap. 19) is the chapter in which Minimol and her team narrated a live case study in Uttarakhand, India, where a comprehensive landslide mitigation scheme is implemented in complex geological conditions, giving a complete picture of this case, starting from geotechnical investigations to instrumentation.

"Geotechnical Education from Field" (Chap. 20) is an experience-based narrative of Anirudhan, which suggests that classroom learning without a feel for realistic field problems is futile and how a problem-solving approach can be developed through collaborative classroom and field learning. In similar lines, in the chapter by Ranganath and his team on "Role of Professional Vigilance in Design and Construction—A Case Study of Solar Projects" (Chap. 21), the authors brought out the importance of professional vigilance in executing geotechnical projects, with an example case study.

I would like to thank Springer for publishing this book, Swati Meherishi for encouraging me to produce the book, and TCA Avni for helping in publishing the book. My sincere thanks to my colleague and President of Indian Geotechnical Society, Prof. Siva Kumar Babu, for writing the foreword for this book.

Bangalore, India Prof. Madhavi Latha G.

Contents

1 **Pores—The Almost Invisible Part of Soil** 1
J. David Frost, Nimisha Roy, Mahdi Roozbahani, Ye Lu, Jie Cao and Prashanth Vangla

2 **Research Advancements in Expansive Soil Characterization, Stabilization and Geoinfrastructure Monitoring** 15
Anand J. Puppala, Surya S. C. Congress and Aritra Banerjee

3 **Testing Mode and Soil Parameters** 31
R. G. Robinson

4 **Estimation of Dredged Fill Volume Based on Self-weight Consolidation Characteristics of Dredged Soil** 51
Eun Chul Shin, Young Jin Park, Jeong Ku Kang and Jeong Jun Park

5 **Recommended Practice for Soft Clay Characterization with a Focus on Settlement and Stability Analysis** 75
Vikas Thakur and Samson Abate Degago

6 **Insights into Contaminant Transport Modeling Through Compacted Bentonites** 101
Tadikonda Venkata Bharat, Partha Das and Ankti Srivastava

7 **Axial Load-Displacement Behavior of Energy Pipeline Systems in Sand** 121
Sai K. Vanapalli and Mohammed Al-Khazaali

8 **Reflection Crack Assessment Using Digital Image Analysis** 139
Sireesh Saride and V. Vinay Kumar

9 **Centrifuge Model Studies on the Performance of Geosynthetic-Reinforced Soil Structures** 157
B. V. S. Viswanadham

10 Soft Ground Improvement—Theoretical, Experimental, Numerical and Field Studies 183
Buddhima Indraratna, Pankaj Baral, Cholachat Rujikiatkamjorn and Thanh Trung Nguyen

11 Issues on Seismic Site Characterization 217
S. K. Prasad, K. V. Vijayendra and Sitaram Nayak

12 Dynamic Tunnel–Soil Interaction in Soft Soils Considering Site-Specific Seismic Ground Response 249
Deepankar Choudhury, Milind Patil, P. G. Ranjith and Jian Zhao

13 Dynamic Characterization of Soils Using Various Methods for Seismic Site Response Studies 273
Pradeep Kumar Dammala and A. Murali Krishna

14 Cyclic Behaviour of Scrap-Tyre Soil Mixtures 303
J. S. Vinod, M. Neaz Sheikh and Soledad Mashiri

15 Sustainable Geotechnics: A Bio-geotechnical Perspective 313
Sreedeep Sekharan, Vinay Kumar Gadi, Sanandam Bordoloi, Abhisekh Saha, Himanshu Kumar, Budhaditya Hazra and Ankit Garg

16 Recent Developments in Earth Pressure Reduction Techniques 333
S. M. Dasaka and V. K. Gade

17 Probabilistic Study on the Geotechnical Behavior of Fiber Reinforced Soil 345
Arif Ali Baig Moghal, B. Munwar Basha and Mohammed Ashfaq

18 Foundations for Tall Buildings on Alluvial Deposits—Geotechnical Aspects 369
Ravi Sundaram, Sanjay Gupta and Sorabh Gupa

19 Sustainable Landslide Mitigation Measures-Demonstration Through a Live Case Study in Uttarakhand 395
Minimol Korulla, Rudra Budhbhatti, P. S. Meenu and Vikalp Kamal

20 Geotechnical Education from Field 417
I. V. Anirudhan

21 Role of Professional Vigilance in Design and Construction—A Case Study of Solar Projects 441
N. Ranganath, Debasis Sarkar, Surendra Singh Kachuwaha, Vinaykumar S. Mathad and Saurav Kumar Ghosh

About the Editor

Madhavi Latha G. is a Professor in the Department of Civil Engineering, Indian Institute of Science (IISc) Bangalore, India. She has completed her Ph.D. from Indian Institute of Technology Madras, and has previously worked as a post-doctoral fellow and assistant professor at IISc and IIT Guwahati respectively. Professor Latha is a member of various professional bodies including IGS, ISSMGE and ISRM, and is the Editor-in-Chief of the *Indian Geotechnical Journal*, and an Editorial board member in many reputed journals. Her research work looks at fundamental aspects of soil and ground reinforcement, and she has authored 70 journal articles, 4 book chapters and has developed a web-course on Geotechnical Earthquake Engineering on the NPTEL platform, sponsored by the Ministry of Human Resources Development, Government of India.

Chapter 1
Pores—The Almost Invisible Part of Soil

J. David Frost, Nimisha Roy, Mahdi Roozbahani, Ye Lu, Jie Cao and Prashanth Vangla

1.1 Introduction

Soil mechanics has been practiced as a field of endeavor for about a century [1] although some of the theories utilized in the application of soil mechanics date back two centuries and more [2, 3]. The practice of soil mechanics today involves soil characterization (laboratory and field methods), analytical design methods, numerical simulation methods, construction control, and performance monitoring, among others. In each of these stages, techniques are used which predominantly focus on the soil as a multiphase material consisting of particles (solids) and pores (either fluid-filled, gas-filled, or filled with a combination of fluid and gas). In the development of soil mechanics, the fact that particles were physical objects whose individual sizes and size distributions could be readily quantified, even if the soil was disturbed, led to them being routinely dominant in the description of the soil and their subsequent association with the engineering properties of the soil such as strength, deformation, and hydraulic conductivity. While recognized to be important in the response of the soil also, the inability to quantify pores individually led to them being effectively ignored and only the total volume of pores is typically quantified and referenced in subsequent engineering analyses.

J. David Frost (✉) · N. Roy · M. Roozbahani
Georgia Institute of Technology, Atlanta, USA
e-mail: david.frost@ce.gatech.edu

Y. Lu
Shanghai University, Shanghai, China

J. Cao
Golder Associates, Seattle, USA

P. Vangla
Indian Institute of Technology Delhi, New Delhi, India

M. Latha G. (ed.), *Frontiers in Geotechnical Engineering*, Developments in Geotechnical Engineering, https://doi.org/10.1007/978-981-13-5871-5_1

1.2 Digital Reconstruction

The emergence of the digital age has opened up new paradigms for the study of soils. The historical vestiges can effectively be eliminated if the soil is first transformed into a digital equivalent of the physical soil. Emerging techniques including optical microscopy combined with image montage, serial sectioning and image stacking, and X-ray computer tomography have opened the door for such techniques to be used in soil mechanics. A brief summary of the former method applied to a sand specimen is provided below.

Reconstruction of a digital soil specimen using optical microscopy involves a number of sequential steps (Fig. 1.1).

Once a specimen is ready to be reconstructed, its structure is preserved using a resin impregnation technique. This ideally involves filling the pore space with a low viscosity resin that cures at room temperature with minimal volume change upon curing. Once the specimen has been cured, coupons of an appropriate size are cut from the larger specimen and one surface of each of the coupons is subjected to a series of grinding and polishing steps to yield a surface suitable for optical microscopy. A series of overlapping images are captured and stitched together to create a mosaic image of that surface (Fig. 1.2). This allows a large field of view to be captured at high resolution.

When the images on a specific surface have been captured, the specimen is subjected to an additional step of grinding and polishing to remove an additional layer

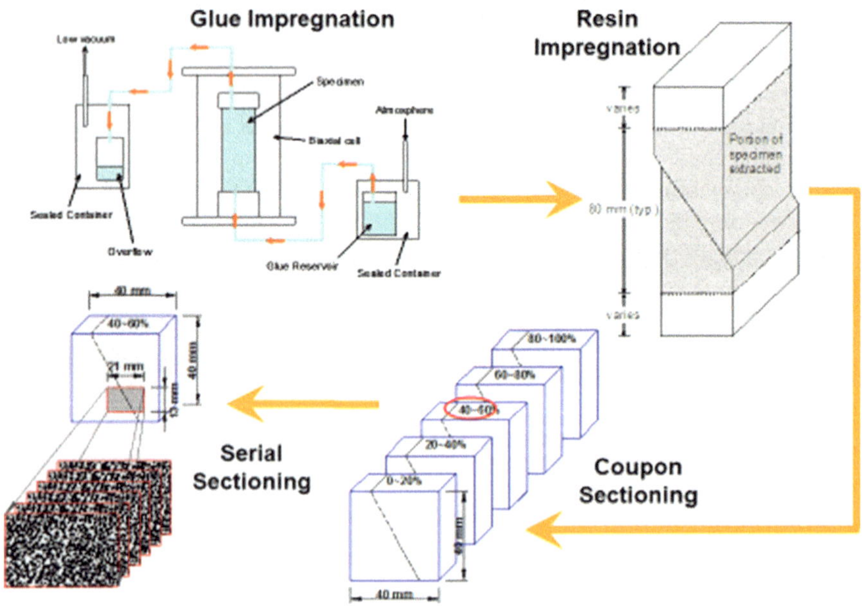

Fig. 1.1 Sample Impregnation technique to prepare soil specimens for serial sectioning

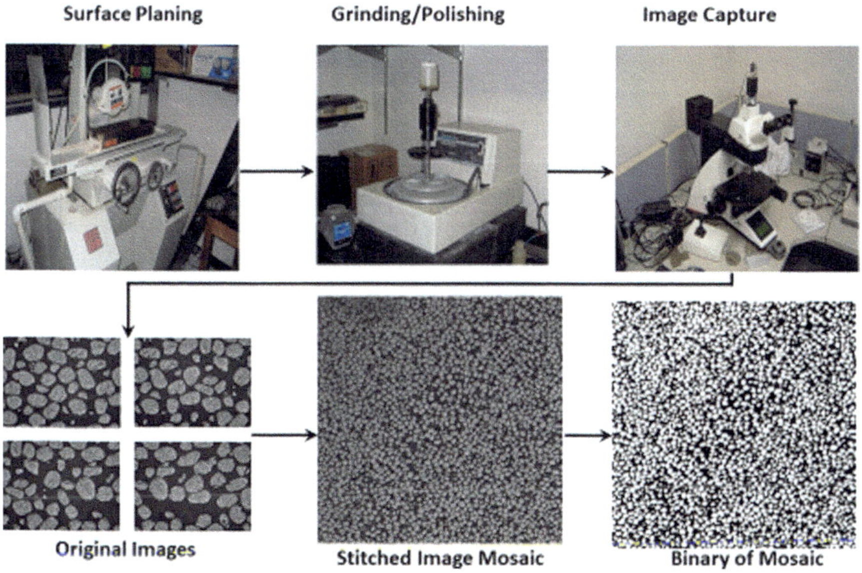

Fig. 1.2 Surface Preparation and image capture of specimens to create a mosaic image

of resin-impregnated sand. A layer of thickness equal to the size of a pixel in the image is typically removed.

For sands, this is typically a thickness of between 10 and 20 μm to ensure that each sand grain and each pore is dissected by about 40 to 50 slices to ensure high-fidelity reconstruction of the desired volume. The new surface is imaged as noted above to create a montage parallel to the previous one but separated by a distance equal to the pixel dimensions. This process yields equidimensional voxels and simplifies any future volumetric calculations. The process of mosaic generation and serial sectioning is repeated until a stack of montage images from about 600 slices has been generated (Fig. 1.3).

Upon completion of the image generation and stacking phase, volume rendering of the soil coupon is accomplished using Avizo which is a general purpose commercial software for scientific and industrial visualization and analysis. An image of a volume rendered specimen with a partial cutaway is shown in Fig. 1.4. This volume was generated using about 12,000 images and contains about 20,000 sand particles and a comparable number of pores. The volume contains about 1.2 billion voxels and with a voxel side of about 8 μm represents an actual volume of soil of about 14 mm by 9 mm by 5 mm.

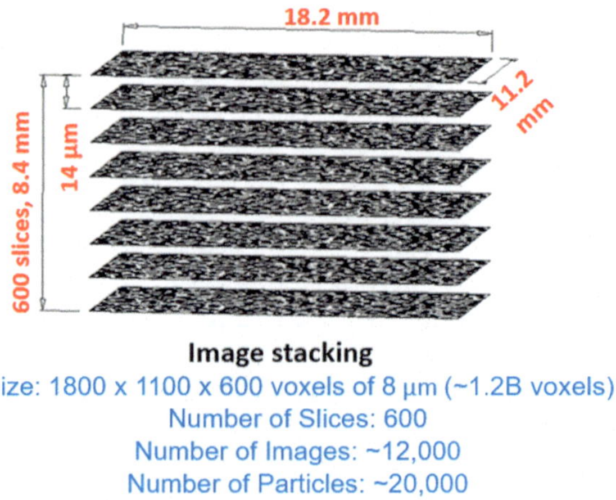

Fig. 1.3 Demonstration of stacking montage images to create 3D reconstructed structure

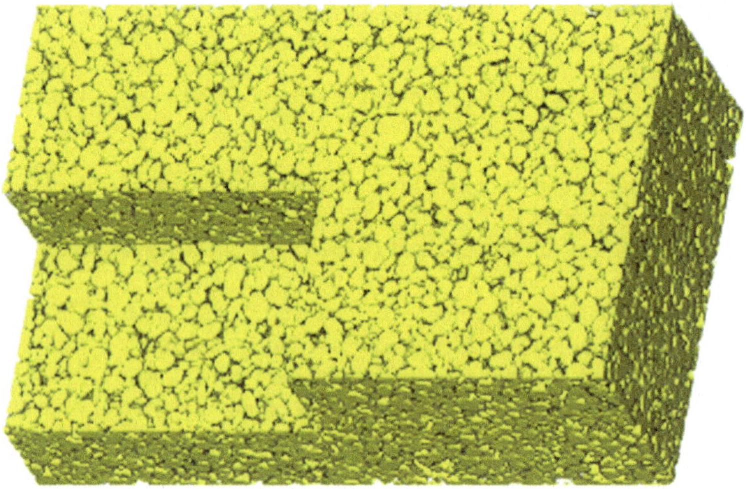

Fig. 1.4 Image of a volume-rendered soil specimen

1.3 Digital Dissection—Meso-scale Analysis

Notwithstanding that this volume of soil was reconstructed from 600 slices that had an area of about 14 mm by 9 mm and the resulting digital volume can be dissected in any orientation since it is now a voxelized representation of the soil. A schematic illustrating this process is shown in Fig. 1.5. This redissection process serves to illustrate the power of conducting studies using a digital representation of the soil rather than being restricted to using only a physical specimen. Apart from the fact that the soil can be analyzed from multiple perspectives, it is important to note that

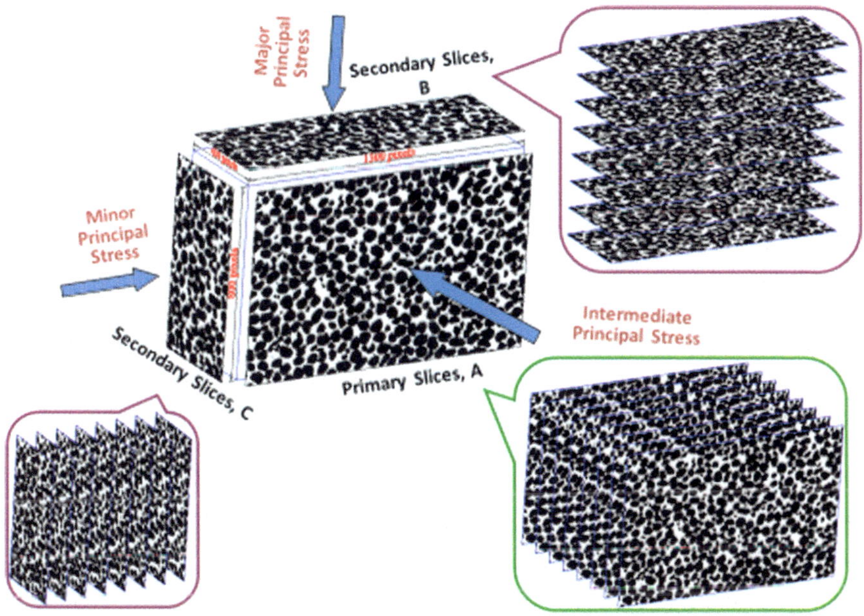

Fig. 1.5 Illustration of dissection of reconstructed digital volume of a soil specimen

such actions are repeatable and that the identical digital structure can be examined in multiple different ways.

An interesting illustration of this concept is shown in Fig. 1.6 which schematically shows a specimen that has a shear band traversing diagonally across the A plane as illustrated in the sketch. A series of slices were cut through this volume of sand and are called the A slices. Similarly, slices cut perpendicular to this plane are noted in the figure as B slices and C slices. It can be envisaged that irrespective of the depth into the specimen that one of the A slices is recovered from, the porosity of that plane (area of voids/area of plane) should be relatively constant since each plane will have the same amount of shear band propagating through it. Similarly, for the B slices, it can be seen that while the location of the shear band will be different in every slice, the porosity of that plane should be relatively constant since each B slice should have about the same amount of shear band propagating through it. However, as a result of the orientation of the C slices, a different porosity can be expected as the slices transition from the left of the volume (zero shear band in the slices) to the center of the specimen (entire slice is contained within the shear band) to the right of the specimen (zero shear band in the slices again). In that this was a dilatant specimen, one would expect to see lower void ratio values at the extremities of the specimen and larger values in the center of the specimen where the slice was fully contained within the shear zone.

These meso-scale variations in void ratio can be readily seen in Fig. 1.7, which shows the variations in void ratio of each slice as a function of distance from the edge

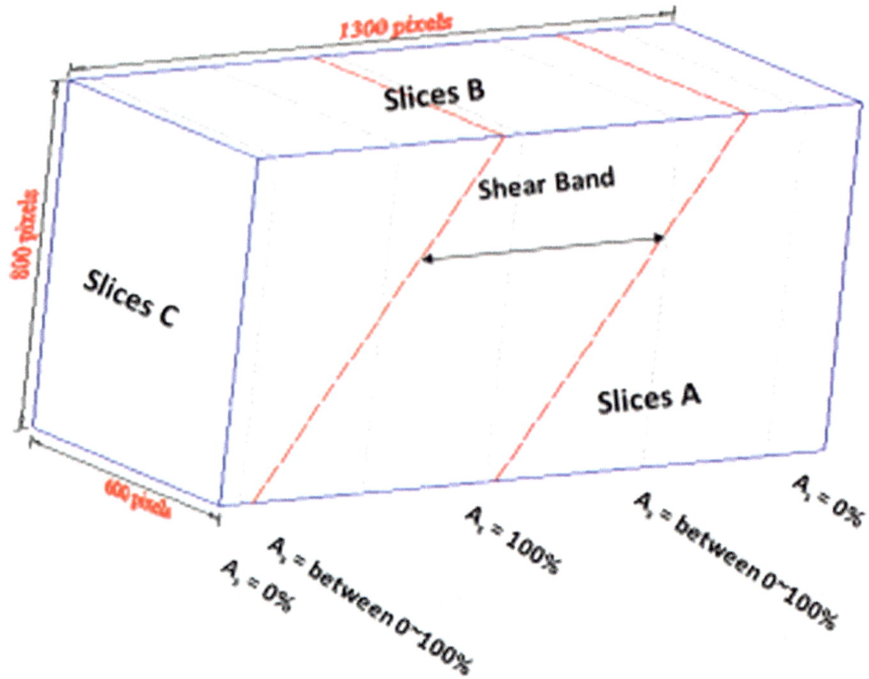

$$A_s = \frac{\text{Area of Shear Band}}{\text{Total Area}} \times 100\%$$

Fig. 1.6 Demonstration of dissection of a sheared soil specimen

of the specimen. It can be seen that, as expected, the void ratio values for the A and B slices fluctuate about the mean void ratio value of the total specimen. In contrast, the value of void ratio for the C slices shows a distinct trend that, as expected, shows lowest void ratio values at the edges of the specimen and highest values for the slices from the center of the specimen, consistent with the location of the dilatant shear band. The minor fluctuations about the mean lines for all slices are reflective of statistical variations controlled by the mean particle size of the sand used in the specimens and are termed the "packing signature effect".

The validity and importance of the void ratio variations or "packing signature effect" introduced in the preceding discussion are verified by a similar analysis conducted on two idealized particle packing arrangements shown in Fig. 1.8. The schematic and plots shown in the left-hand side are for a simple cubic-packed structure. The plot of void ratio versus slice location for the three orthogonal orientations parallel to the face of the cubic structure is identical. Equally importantly, they illustrate that for the planes that pass through the contact points of the particles, the void

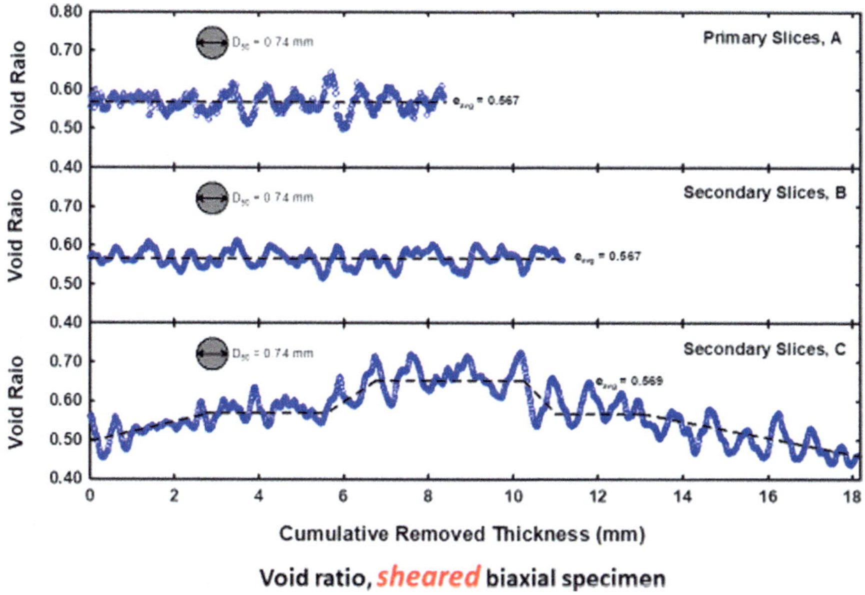

Fig. 1.7 Meso-scale void ratio variations along three mutually perpendicular directions of a sheared soil specimen

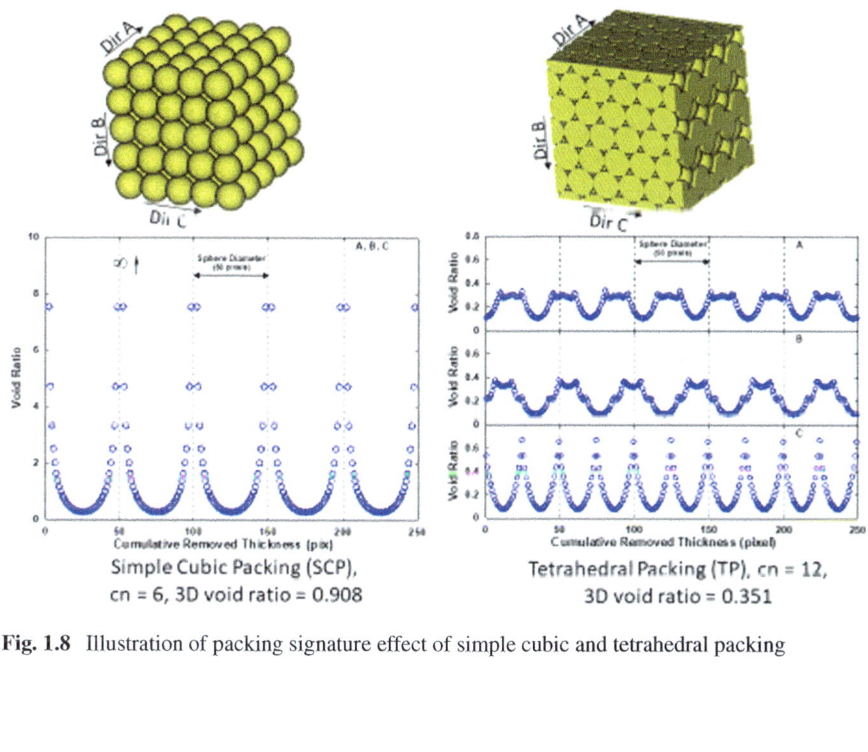

Fig. 1.8 Illustration of packing signature effect of simple cubic and tetrahedral packing

ratio goes to infinity while for planes that pass through the centers of the particles, it approaches a low but nonzero value of void ratio.

In contrast, for the tetrahedral packing, the three void ratio profiles are different and do not reflect the same singularity seen with the simple cubic packing. The fact that these meso-scale measurements show distinct void ratio patterns as a function of orientation provides critical new insight into anisotropy in solid and other granular materials. Equally well, it also shows the limitations of representing the void space by a single value.

1.4 Sub-volume Visualization

The preceding section of the paper presented new insights into pore structures from a meso-scale perspective. This analysis was performed on the complete reconstructed specimen. As previously noted, the fact that the soil exists in a digital format allows for multiple analyses on the same specimen including sub-specimens extracted from the full specimen. An additional abstraction that can be performed on the digital specimen or sub-specimen is the separation of the particle and pore phases and thus comparison of various properties. This type of subsampling and phase separation is illustrated in Fig. 1.9.

This figure shows three images of the same volume extracted from a digital soil. The size of the three volumes is about 300 by 300 by 300 μm. The volume on the left is what might be termed the continuum representation of the soil that includes both particles and pores. In contrast, the volume in the center is a representation of the particles only where the pore filling material has been eliminated. This representation is similar to what is generated and used in particulate modeling software such as those for Discrete Element Modeling (DEM). Finally, the volume on the right is a representation of the pore volume only where the particles have been eliminated. The ability to consider the soil from these different perspectives has created significant new opportunities to analyze and visualize soils from different scales and perspectives.

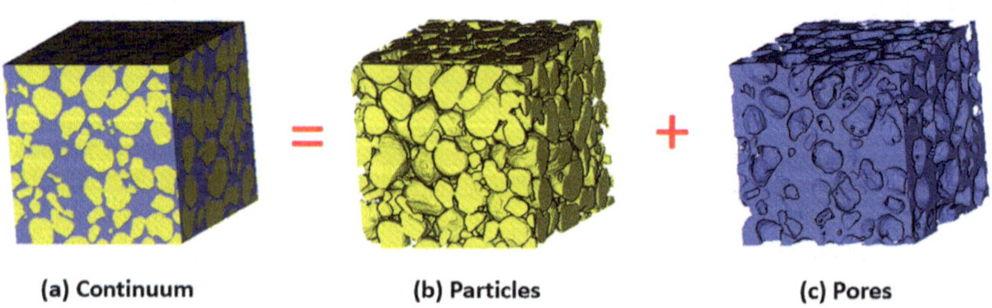

Fig. 1.9 Visualization of subsampling and phase separation of a digital specimen

1.5 Single Particle and Pore Visualization

In the same manner that digital structures allow for new and unique insights into soil characterization and behavior, they also provide opportunities for particle and pore-scale studies. While other techniques such as laser scanning can be used to generate 3-D high-fidelity representations of individual particles, the generation of a digital structure can allow for even greater examination of both individual particles and pores that can be readily extracted from a reconstructed volume. For example, Fig. 1.10 shows a set of sand grains that were part of a reconstructed volume and then were subsequently extracted from the digital structure.

It can be seen that despite the fact that the particles were extracted from a digital specimen that contained about 20,000 particles, the shape and characteristics of the individual particles are very realistic. In this example, the volume was reconstructed from a stack of serial montage images that were 8 μm apart and thus given that the particles had a D_{50} of about 350 μm, each particle was intersected by about 40 to 50 slices.

In a similar manner, Fig. 1.11 shows a set of pores that were part of the same reconstructed volume and then were subsequently extracted from the digital structure. A number of interesting attributes can be seen from the individual pore images. First, the shapes of the pores reveal a set of scalloped surfaces, reflective of the particles which surrounded them. In general, the pore bodies have primary dimensions that are smaller than the surrounding particles and equally importantly, are interconnected by even narrower pore throats. It is of interest to note that these attributes of the pores are rarely considered in engineering characterization and analyses of soils.

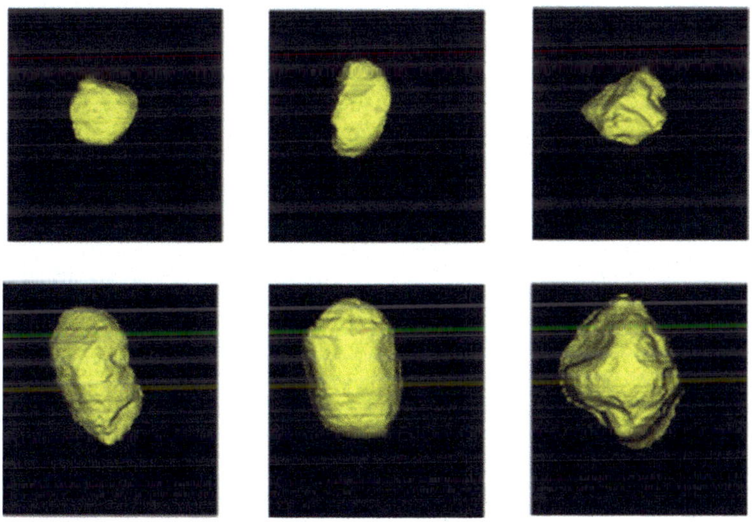

Fig. 1.10 3D particle visualization of digital sand grains

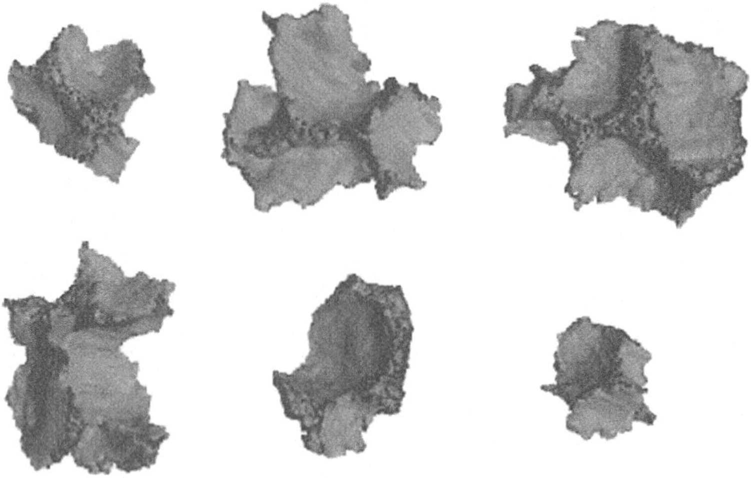

Fig. 1.11 3D pore visualization between digital sand grains

1.6 Pore Size Distributions

The engineering behavior of soils is often correlated to the mean particle size. At the same time, other measures of the particle size distribution are better predictors of engineering behavior. For example, correlation between hydraulic conductivity and D_{10} [5] or shear modulus and D_5 [6] are often more reliable. The historical lack of robust measures of pore size distribution including both pore throat and connectivity has resulted in, for example, conservative empirical correlations for filter designs in major earth structures. To illustrate the potential merits of determining pore size distributions in soils, an initial study was performed using four different mono-sized sphere packing models that were adapted from the work of Dullien [4]. The descriptions of the models are summarized in Table 1.1 along with the theoretical packing density.

Table 1.1 Four different mono- sized packing models adopted from Dullien (2012)

Model	Description	Packing density
Very loose random packing	Spheres slowly settled	0.56
Loose random packing	Dropped into bed or packed by hand	0.59–0.60
Poured random packing	Spheres poured into bed	0.609–0.625
Dense random packing	The bed vibrated	0.625–0.641

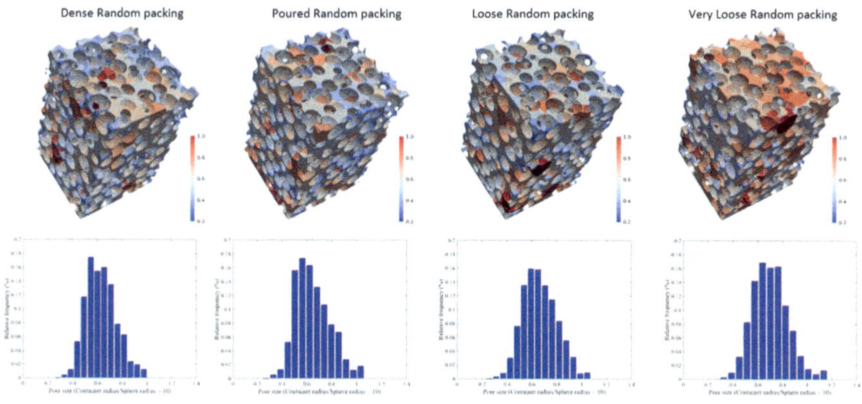

Fig. 1.12 Packing structures with the histograms of their pore size distributions

Figure 1.12 shows images for each of these packing structures along with the histograms for their pore size distributions [7]. As expected, the mean pore size and the distributions change as the packing density increases. It is noted that without particle breakage, the particle size distribution for such a set of packing models would not change even as the density changed; thus the particle size distribution does not yield the same insight that the pore size distribution does.

A similar comparison is illustrated for two specimens of real sands as shown in Fig. 1.13. In the left-hand images, the digital reconstruction of an unsheared specimen of sand is shown and the pore size distribution is uniformly variable across the entire volume. The histogram shows the mean and distribution of the pore sizes [7].

In contrast, the images on the right-hand side of the figure clearly show the location of the shear band traversing diagonally across the specimen. Equally importantly, the associated histogram of pore size distribution shows a higher mean value, reflective of the dilation in the shear zone as well as a skewing of the histogram towards a higher void distribution.

While both the size and distribution of pores in soil are of importance, other measures of the pore network structure can be equally important in predicting the engineering behavior of soil. For example, the tortuosity of the pore network paths has important implications on the genesis and dissipation of pore water pressure in soils subjected to dynamic loading such as earthquakes and machine vibrations or in predicting the hydraulic conductivity of a soil. Despite this potential role, tortuosity has not been considered a critical index property in characterizing soils despite its obvious relationship to phenomena such as fluid flow and pressure dissipation. With the availability of digital soils, quantification of such important properties becomes readily achievable.

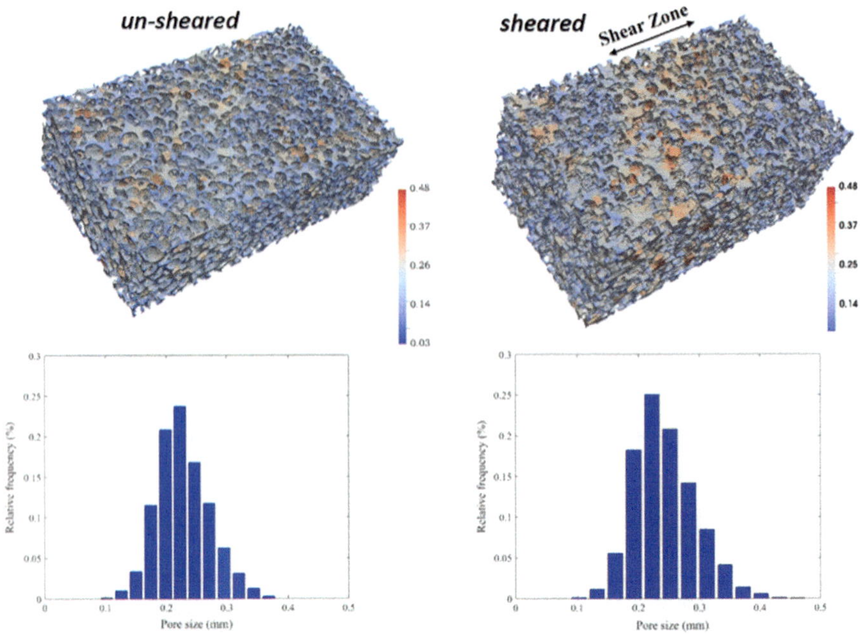

Fig. 1.13 Pore size distribution of unsheared and sheared real sands

Figure 1.14 shows the results of a study conducted on a real soil specimen to measure the tortuosity in three orthogonal orientations. It can be seen that the distribution of tortuosity for each orientation is different, providing unique insight into the anisotropy of tortuosity and thus engineering properties that are dependent on fluid or gas flow in these paths. A similar analysis conducted on sub-volumes extracted from the sheared specimen of sand is summarized in Fig. 1.15. As illustrated in the figure, three sub-volumes from inside the shear band, outside the shear band, and in the transition zone between the shear band and the non-shearing soil are examined. The histograms of tortuosity for these three sub-volumes show the lowest tortuosity for the volume from within the shear zone as expected since the particles dilating would reduce the path tortuosity. Likewise, the sub-volume outside the shear zone had the intermediate tortuosity that remained essentially unchanged from the initial specimen structure. The path tortuosity for the sub-volume in the transition zone had the highest tortuosity value since in that region adjacent to the dilating shear zone, it is known that the soil behavior contracts thereby bringing the particles closer together and increasing the path tortuosity.

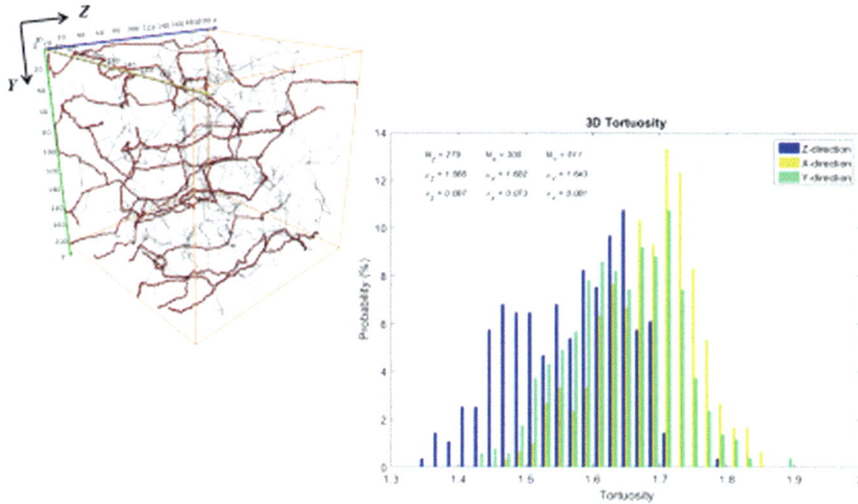

Fig. 1.14 3D tortuosity distribution of a real soil specimen

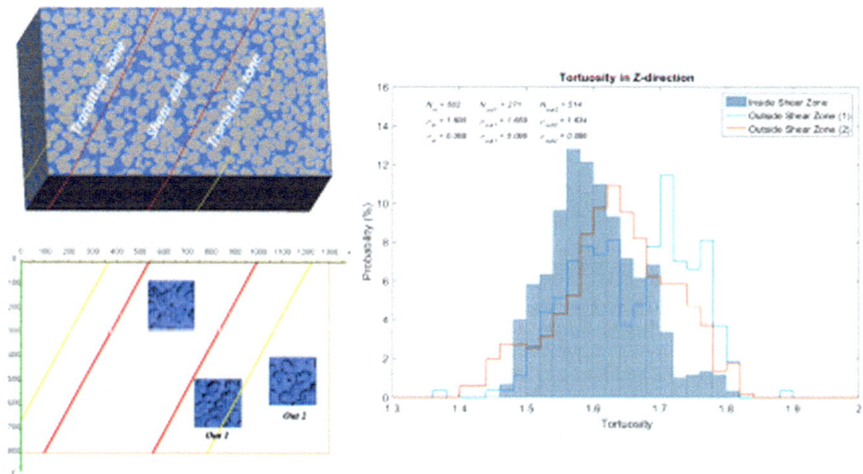

Fig. 1.15 3D pore structure analysis of a sheared sand specimen

1.7 Conclusions

Through a series of examples, this paper has attempted to show how digital reconstruction of soils and their subsequent study is opening up new approaches to understanding soil behavior and more importantly, the important role that pores and pore networks have in influencing soil behavior. A number of salient conclusions can be made:

- There is much to be learned from studying and understanding pores and pore networks.
- New transformational technologies are enablers for numerical modeling and vice versa—both work in "digital" environments.
- We should not continue to ignore pores as a critical factor in our understanding of the engineering behavior of soils—we can see them now!!
- More importantly, they have a contributing role that we need to understand and appreciate.
- Do we get anything from studying pores that we cannot get by studying particles only? The answer to that may vary as function of phenomena we are trying to study. We should begin by asking ourselves a few simple questions:
  - Which requires more energy—crushing particles or collapsing pores? (note we might want to call this "crushing pores" since they are equally tangible in digital world!)
  - What is more important for liquefaction triggering—collapsing pores or crushing grains?
  - Are there always unique relationships between force chains and path tortuosity?

Acknowledgements The contributions of a number of former and current graduate students at Georgia Tech including Deh-Jeng Jang, Jin-Young Park, Chien-Chang Chen, Xuan Yang, and Rodrigo Borela in conducting some of the experiments and simulations presented in this manuscript are gratefully acknowledged.

References

1. Terzaghi, K.: Erdbaumechanik auf bodenphysikalischer Grundlage. Leipzig and Vienna: Franz Deuticke, 399 pp. (1925)
2. Coulomb, C.A.: Essai sur une application des règles de maximis et minimis à quelques problèmes de statique relatifs à l'architecture, in Mémoires de la mathématique et de physique, présentés à l'Académie Royale des Sciences, par divers Savants, et lÛs dans sés Assemblées. Paris: De l'Imprimerie Royale, Année. **7**, 343–384 (1773)
3. Rankine, W.J.M.: On the stability of loose earth. Roy. Soc. Lond. Philos. Trans. **147**(pt. I), 9–27 (1857)
4. Dullien, F.A.L.: Structure of porous media. In: Bear, J., Corapcioglu, M.Y. (eds.) Transport Processes in Porous Media. NATO ASI Series (Series E: Applied Sciences), vol. 202. Springer, Dordrecht (1991)
5. Darcy, H.: Les fontaines publiques de la ville de Dijon. Dalmont, Paris (1856)
6. Hardin, B.O., Kalinski, M.E.: Estimating the shear modulus of gravelly soils. ASCE J. Geotech. Geo-environ. Eng. **131**(7), 867–875 (2005)
7. Roozbahani, M.M., Borela, R., Frost, J.D.: Pore size distribution in granular material microstructure. J. Mater. **10**(11), 1237–1258 (2017)

Chapter 2
Research Advancements in Expansive Soil Characterization, Stabilization and Geoinfrastructure Monitoring

Anand J. Puppala, Surya S. C. Congress and Aritra Banerjee

2.1 Introduction and Background

Expansive soils undergo significant volume changes due to variation in precipitation, temperature, and ground-water table causing daily and seasonal fluctuations in environmental conditions. These volume changes can inflict moderate to severe damages to civil infrastructure with shallow foundations such as pavements and one or two storey buildings. Many locations in India and elsewhere around the world have expansive soils which are either used as subgrade for pavement sections or are within the influence zone of a structure. Hence, volume changes in these soils caused primarily by change in moisture regime result in excessive damages to the geoinfrastructure assets. Mostly researchers and designers resort to simplistic design and analysis which utilizes basic parameters, such as soil indices, for characterizing the behavior of expansive soils [1–5].

For instance, it is well-known that two-clayey soils having same index properties subjected to moisture changes may have different volumetric response due to difference in clay mineralogies [6]. Hence, using these crude simplification results in problematic characterization are leading to erroneous pavement or foundation design. Additionally, the environmental conditions at each site will lead to different volume change pattern [4, 7–9].

Over the years, it has been observed that there is no decrease in the maintenance cost involving swell-shrink-induced damage repairs [6]. This has been the motive behind this paper which identifies some of the recent developments in characteriza-

A. J. Puppala (✉) · S. S. C. Congress · A. Banerjee
Civil Engineering Department, The University of Texas at Arlington, Arlington, USA
e-mail: anand@uta.edu

S. S. C. Congress
e-mail: surya.congress@uta.edu

A. Banerjee
e-mail: aritra@uta.edu

M. Latha G. (ed.), *Frontiers in Geotechnical Engineering*, Developments in Geotechnical Engineering, https://doi.org/10.1007/978-981-13-5871-5_2

tion of expansive soils for developing models which are capable of accurately predicting the behavior of the soil in field conditions. Other contributions in the paper focus on new technologies using photogrammetry principles for health monitoring of transportation infrastructure using unmanned aerial vehicle or UAV platforms.

The characterizations that do not consider the compositional and environmental conditions for analyzing the behavior of soils generally do not perform well in predicting soil response. Mineralogy and soil suction are the essential parameters which are often ignored in contemporary design. However, when these parameters are considered for design, the effect of environmental conditions on the response of soils especially during swell-shrink cycles are well predicted [10]. Studies have shown that the engineering behavior of soils can be better interpreted if the influence of matric suction or suction stress is taken into account [11, 12]. Pore-size distribution is also an important parameter for studying the hydraulic conductivity and moisture absorption characteristics [13]. However, the contemporary methodology used do not consider the pore-size distribution. Advanced technique to determine the pore-size distribution of soils is also discussed in brevity in this paper.

Additionally, poor characterization of soils may also lead to inefficient stabilization as variation of clay mineralogy could result in significant difference in the behavior of stabilized soil under similar conditions [14]. Most of the current stabilization approaches primarily depend on the plasticity index (PI) and gradation characteristics of the soil. The dosage for cement stabilization is based on the unconfined compressive strength (UCS) value obtained after moisture conditioning of the soil specimen. While, the dosage for lime stabilization is based on the pH methods similar to the Eades–Grim test (ASTM D6276), the UCS values are indirect parameters to identify the behavior of stabilized soils.

The mineralogy of a soil governs its physical and chemical properties. The most common clay minerals found in soils are Kaolinite, Illite, and Montmorillonite, which have significantly different properties. Among these, montmorillonite has the highest specific surface area and is primarily responsible for inducing the expansive behavior in soils. The procedures such as X-ray diffraction, scanning electron microscopy (SEM), and thermal gravimetric analysis (TGA) are predominantly used to identify clay minerals, but these test procedures are time-consuming and expensive [15]. Chittoori [16] had proposed a novel technique to quantify the presence of various types of clay minerals using a regression analysis of chemical soil properties like cation exchange capacity (CEC), specific surface area (SSA) and total potassium (TP). This technique is fast and relatively much cheaper than current procedures used.

Photogrammetry is a science of making measurements using two or more images. Photogrammetry can be conducted at wide range of distances using varied sensors ranging from smart phone camera to sensor mounted satellites. Close range photogrammetry (CRP) is a subsidiary of photogrammetry with sensor operating within a 1000 ft distance from the inspecting object. Unmanned aerial vehicles (UAV) offer mobile and stable platforms to mount the visible range camera, operating at localized distances, to conduct CRP inspection. UAV-CRP technology data can be processed to obtain various mapping outputs at accuracy levels necessary to make engineering judgements [17]. Visualization of these 3-dimensional data outputs provides an in-depth idea about the condition of the infrastructure [18, 19]. This innovative technology is used in this research to monitor the geoinfrastructure laid over problematic sulfate soils. A pavement site that was rehabilitated after exhibiting sulfate induced heave have been inspected to find out the effectiveness of the treatment.

This paper covers the characterization methods and novel ground improvement technologies for expansive soils in the first two sections of the paper. The final section introduces new monitoring technology using unmanned aerial vehicles coupled with close range photogrammetry (UAV-CRP) to collect the performance data of geo-infrastructure.

2.2 Advancements in Soil Characterization and Stabilization

In this section, the influence of presence of montmorillonite in soils is discussed along with the relationship between plasticity index (PI) and percentage montmorillonite for the effective stabilization of clayey soils. The durability studies exhibited that soils having similar plasticity index but different montmorillonite content demonstrated dissimilar response when treated with standard stabilization procedures. To illustrate this theory, soils samples from Fort Worth and Bryan (in North Texas, USA) having the same PI of 30 are collected. Both Fort Worth and Bryan soils with montmorillonite content of 60% and 30%, respectively were stabilized using standard stabilization design procedures and their performances were monitored under durability and swell tests.

These studies have demonstrated the need for incorporating both clay mineralogy and PI properties within the standard design procedure for stabilization of clayey soils. Additional durability tests on different clayey soils to study the volumetric changes after wetting and drying cycles on various soils had led to the conclusion that for soils having PI less than 35% and montmorillonite content less than 40%, the conventional treatment methods would mostly prove to be efficient. However, soils with higher PI and montmorillonite content need to be stabilized with higher concentration of traditional chemical stabilizers, such as lime (greater than 8%) and cement (greater than 6%). Based on the observations of this study, a stabilization design chart (Fig. 2.1) was prepared, which requires both PI values and montmorillonite content for effective stabilization.

The recommendations made in this paper are as a result of studies performed on six clayey soils, but to comprehensively conclude on the importance of characterization of soils prior to selection of stabilizer, further independent studies are required. Such studies would improve the performance of designed infrastructure by mitigating the issues related to expansive clayey soils.

Additional advancement in soil characterization have been made by studying the internal distribution of pores within the soil specimen by mercury intrusion porosime-

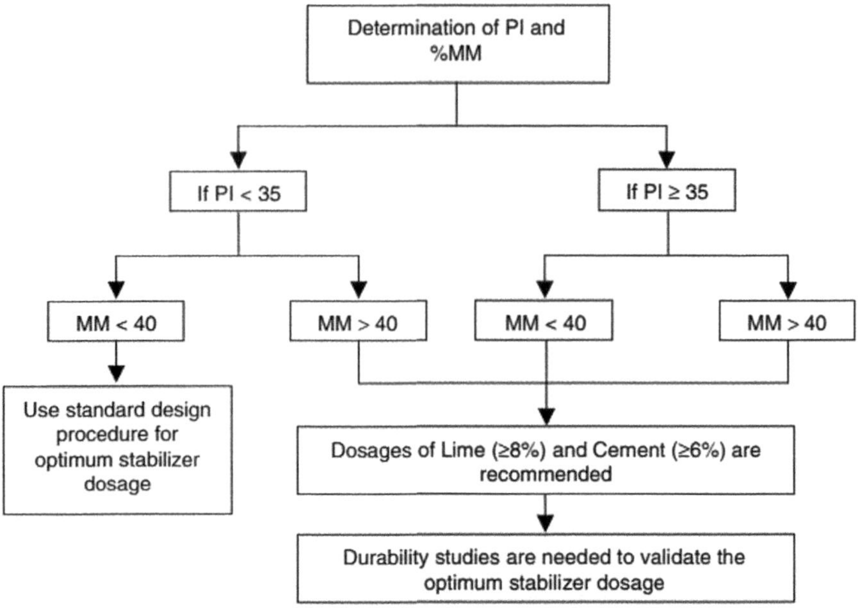

Fig. 2.1 Proposed stabilization design chart modified to include montmorillonite content [15]

try (MIP) technique. In this technique, the different sizes of pores are differentiated on the basis of MIP test results. From MIP testing on various soils at different void ratio, it was observed that at maximum dry density (MDD) condition, the high-particle density resulted in lower volume of pore voids, which mostly consisted on micro and medium sized pores. Whereas, at 95% MDD condition (lower density), there was abundance of macro pore voids within the soil specimen [10].

A model was developed by Pedarla [10] for predicting the one-dimensional swell pressure in soils during an increase in degree of saturation, which was based on the ratio of total surface area estimated from clay mineralogy data and mercury intrusion porosimetry results. The model showed good validation results with prediction matching the measured swell test results. This demonstrates the accuracy of using advanced characterization techniques in predicting complex phenomena such as swelling in soils.

2.3 Soil Stabilization Advances

The establishment of transportation infrastructure using native materials is often the most desirable solution. However, the presence of soft, highly compressible or expansive soils in various parts of the world, including India, often hinder the use of native soils in its original form. Stabilization of these soils with cementitious materials, such as lime and cement, generally result in enhanced mechanical properties due to formation of pozzolanic compounds which may satisfy the design requirements [20–22]. Apart from traffic loading, additional pavement distress occurs due to the diurnal and seasonal changes in weather, which alters the moisture regime within the subgrade soil and cause significant damages to pavement [23].

When sulfate-rich native soils are encountered, traditional calcium-based stabilizers, such as lime, often become counter-productive, due to the formation of highly expansive minerals ettringite and thaumasite. Multiple cases of severe pavement heaving and distress were reported when sulfate-rich expansive soils treated with traditional calcium based stabilizers were used as subgrade soils [24–33]. This phenomena is referred to as 'sulfate-induced heave' and the cost of repairs of such pavements exceed the cost of stabilization [31].

Recently, a new class of alumino-silicate-polymers, commonly known as Geopolymers have received much attention for its eco-friendly and sustainable nature and its cementitious properties. Additionally, Geopolymers are often manufactured using raw materials which lack calcium, hence it has the potential to mitigate 'sulfate-induced heave'. Other types of treatment methods which have been recently used include bio-polymers and chemical injections to mitigate the shrinkage characteristics [34–36]. As such, there is a need for new and improved ground improvement solutions that are sustainable, durable and enhance the engineering properties of expansive soils.

The production of traditional stabilizer like cement is energy intensive and emits large quantity of CO_2. Geopolymers as alternative to Ordinary Portland Cement (OPC) [37, 38] constitute a family of materials consisting of covalently bonded alumino-silicates, non-crystalline networks and are generally substituted for many engineering applications due to their high strength. They have received much attention as an eco-friendly and sustainable alternative to OPC because they can be processed at room temperature and can be inexpensive as they are prepared from waste materials such as fly ash or natural sources (e.g. clay), thus providing the plentiful worldwide raw material supply [39]. More importantly, the use of Geopolymer rather than OPC can reduce CO_2 emissions by 44–64% [40].

Bell et al. [41] defined Geopolymers as "a class of totally inorganic, alumino-silicate based ceramics that are charge balanced by group I oxides. They are rigid gels, which are made under relatively ambient conditions of temperature and pressure into near-net dimension bodies, and which can subsequently be converted to crystalline or glass-ceramic materials". In the last 40 years, Geopolymers have been extensively studied as a source of cementitious materials in concrete structures [37, 38, 42] and others. However, its application for soil stabilization has been limited.

Since Geopolymers represent complex materials system, many factors influence the formation and properties of Geopolymer including aluminosilicate source, Silica and alumina (Si/Al) ratio in precursor materials, water to solids ratio used during synthesis, type and amount of alkali activator, mixing and curing conditions. Influence of all these factors on structure and properties of the cured Geopolymer specimen have been extensively studied over the last decades [37, 42, 43]. This study evaluates the effectiveness of metakaolin-based Geopolymer stabilizer in mitigating the volume changes in the native North Texas soil during wetting or drying periods.

Although, significant progress has been achieved in understanding the effects of different factors in processing Geopolymer (GP) on their properties, the study of its effectiveness for expansive soil stabilization and in presence of sulfates in soils hasn't been elaborately investigated. In this research, the native soil from Dallas-Fort Worth (DFW) metroplex in Texas (Fig. 2.2) was obtained to evaluate the use of geopolymers as a soil stabilizer to mitigate its volume change properties to acceptable levels. Tests were conducted on treated and untreated soils to study the volume change behavior of the soil samples. The results were analyzed to evaluate the utility of Geopolymers in mitigating volume change during change in moisture content.

A preliminary study was attempted to assess the mixability of GP and an expansive soil classified as clay of high-plasticity (CH). Since the GP has multiple interdependent parameters, such as chemical composition, curing time and temperature, which defines its behavior, this study explores the behavior of a potassium-based GP with Si/Al = 2 and Al/K = 1, to reduce swell-shrink characteristics. Subse-

Fig. 2.2 Typical desiccation cracks observed in slopes consisting of expansive soils in the Dallas-Fort worth region

quently, three different GP compositions with the already-stated parameters and varying water/solids ratio from 2 to 4 were made and compacted into moulds having a diameter and height of 3.2 cm (1.25 in.) and 1.9 cm (0.75 in.), respectively. It was observed that both K431 (K = Potassium, $SiO_2/Al_2O_3 = 4$, Water/Solids = 3, Al/K = 1) and K441 mixed well with CH, but K431 was chosen for the first set of testing as it was observed to yield higher strength (from UCS testing).

Linear shrinkage tests were conducted on treated CH specimens, which indicated that GP treatment resulted in significant mitigation of shrinkage as compared to that for untreated specimens, which could be observed in Fig. 2.3. Treated CH specimens exhibited a decrease in shrinkage from 16.6 to 13.1% (Fig. 2.4).

Similarly, when Atterberg's limit tests were conducted on treated CH specimens, it showed that GP treatment of CH specimens resulted in about 72% decrease in its PI, from 21 to 6. Swell tests on treated CH specimens indicated a significant reduction in vertical swell strain from about 0.9 to 0.04% (Fig. 2.5).

This study explores the use of metakaolin-based geopolymer for evaluating the swell-shrink behavior of a native North Texas soil. The soil was mixed with the

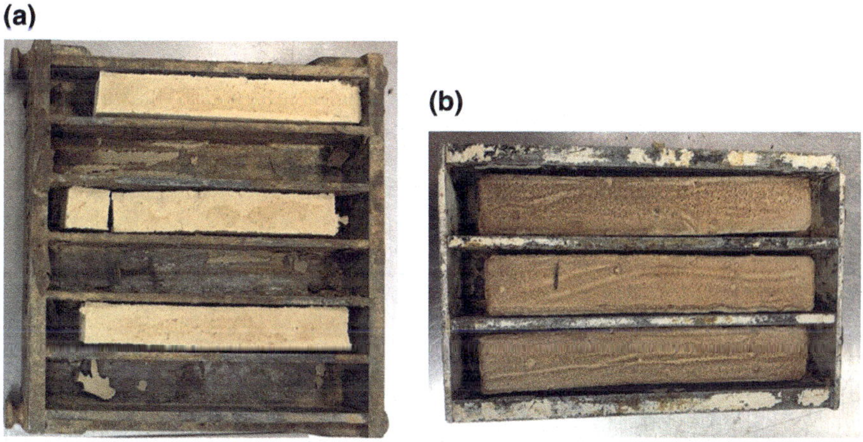

Fig. 2.3 Linear shrinkage strains of **a** untreated CH, and **b** GP-treated CH

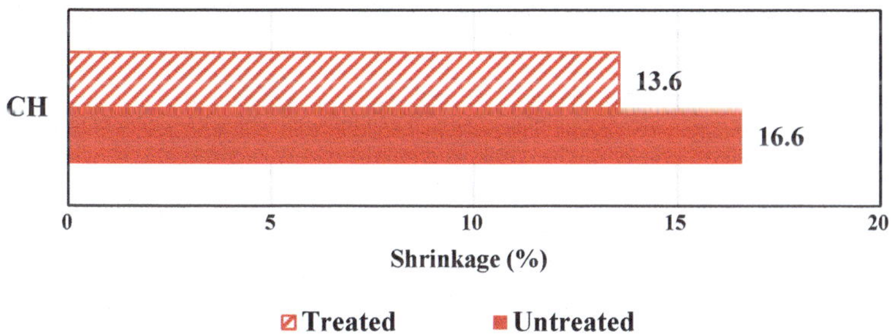

Fig. 2.4 Shrinkage strain (%) plot of untreated and GP treated CH

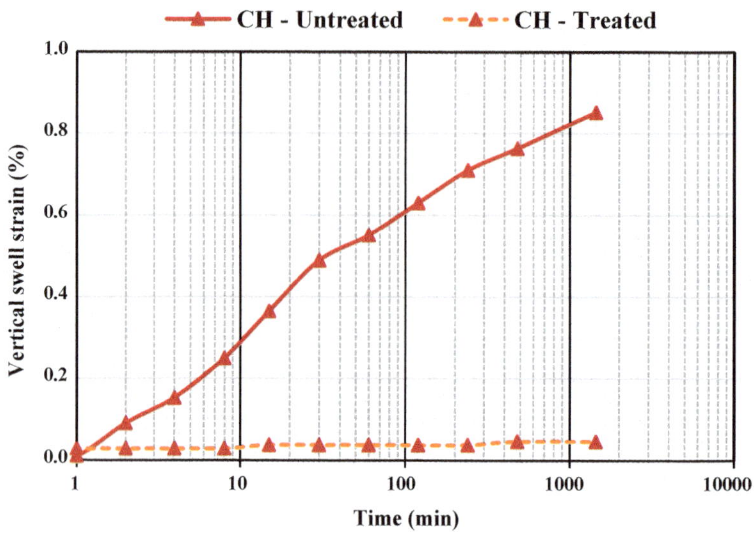

Fig. 2.5 Swell strain plot of untreated and GP treated fat clay (CH)

K431 GP mix at a ratio of 8% (by weight) dry GP to dry soil. Shrinkage tests showed that geopolymer-treated soils are efficient in reducing shrinkage, without developing cracks. Swell tests show that the swell potential of the soil is mitigated within acceptable limits, on treating with GP. To summarize, GP treatment of soils reduces the swell-shrink potential of soils significantly, which is a major concern for expansive soils.

There is a potential for wide-scale application of GP as a sustainable soil stabilizer for expansive soils around the world. Based on this study, it is recommended that parameters affecting GP strength, such as dry GP to dry soil ratio, GP composition, processing methods, and alkali-activator, be varied, to observe its influence on the engineering characteristics of GP treated subgrade soils. Furthermore, durability studies as well as sustainability metrics and life cycle cost analysis studies would be useful in practical implementation of this soil stabilization method.

2.4 Geoinfrastructure Monitoring Using UAV

UAV-CRP technology have been extensively used for monitoring geoinfrastructure assets including pavements, railways, bridges, and tall structures [44]. In this research, pavement data was collected on a nearby road connecting FM 1752 and west bound US 82 highway located at 33.61388°N, 96.34675°W. This infrastructure site had heaving problems due to the presence of rich sulfates in the soils reacting with lime and cement stabilizers.

Fig. 2.6 Cautionary sign for placing ahead of drone operating area

Researchers from UT Arlington (UTA) formulated comprehensive stabilizing strategies for this sulfate-heaving pavement site as part of their research work for Texas Department of Transportation (TxDOT) [30, 45]. After treating the soils overlain by the pavement, technologies involving innovative equipment like terrestrial LiDAR and unmanned aerial vehicle equipped with camera are used to detect any trace of sulfate induced heaving. The contours and elevation maps obtained from these technologies can offer a better idea about heaving of the pavement. There will be localized abrupt elevation changes in the elevation maps and contours due to the presence of heaving. Cautionary signs, as shown in Fig. 2.6, following the TxDOT guidelines for a specific allowable road speed were placed during UAV data collection.

An autonomous flight plan was prepared and followed to collect the infrastructure images at 80% longitudinal overlap and 60% lateral overlap. Due to the vegetation on the shoulder of the pavement, a large wooden plank was placed to serve for both landing and takeoff. Data collection was only conducted on one side of the pavement due to the research stage of this project. A multiple flight path was adopted with 6 m (20 ft) and 23 m (75 ft) as operating flight altitudes. At lower flight altitude the camera was obliquely facing the road and at higher flight altitude the camera was placed in nadir direction. This procedure was adopted to obtain a clear picture of the road and its surroundings.

Images can be geotagged using high accurate global navigation satellite system (GNSS) or less accurate global positioning system (GPS) systems. Although the GNSS equipment is costly, it was purchased and used to obtain the required accuracy. There are two methods of using the GNSS data: real-time kinematics (RTK) or post processing kinematics (PPK). As the names suggest RTK can be used immediately after collecting data whereas PPK accesses the data from the base station which takes couple of days to update the data [19]. In this research RTK is used to geotag the images. The geotagged images are processed in image analysis software to obtain dense point cloud models, orthomosaics, digital elevation models (DEM), contours, and other mapping outputs. Orthomosaic, digital elevation model, and contour map of the pavement infrastructure laid over high sulfate soils are shown in Figs. 2.7, 2.8 and 2.9, respectively.

Fig. 2.7 Orthomosaic of the road connecting FM 1752 and West Bound US 82 Highway, Paris District, Texas

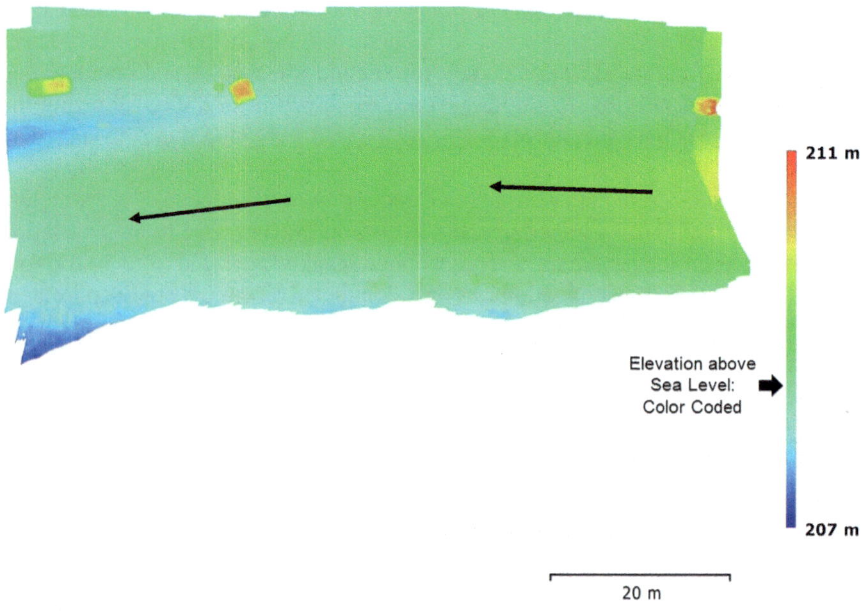

Fig. 2.8 DEM of road connecting FM 1752 and West Bound US 82 Highway, Paris District, Texas

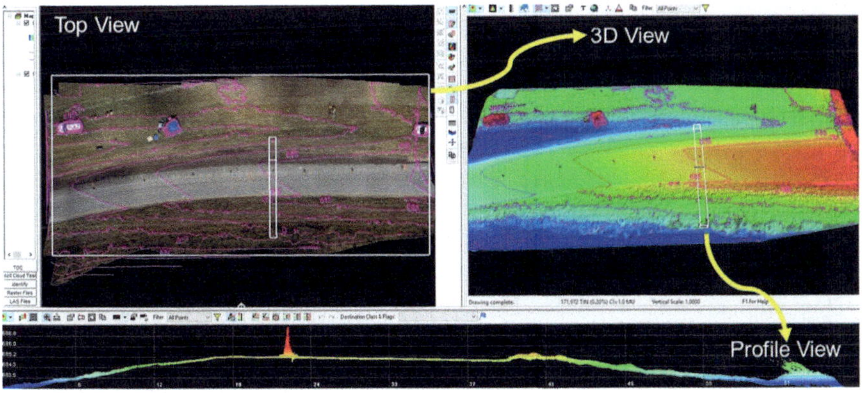

Fig. 2.9 Contour road map connecting FM 1752 and West Bound US 82 Highway, Paris District, Texas

Ground control points (GCP) coordinates are used to develop the model and check point (CK) coordinates are used to estimate the quality of the model in terms of RMSE values in X, Y, and Z directions. The known coordinates of the CK are imported into the developed 3-dimensional models and are manually located to calculate the difference in the estimated and actual location of the check points in the developed model. The RMSE values in X, Y, and Z directions are obtained as 1.8 cm (0.06 ft), 2.1 cm (0.07 ft), and 0.6 cm (0.02 ft), respectively. According to the New ASPRS Positional Accuracy Standards for Digital Geospatial Data [46] for non-vegetated areas, the horizontal accuracy is classified by the square root of sum of the squares of RMSE values in X and Y direction. Vertical accuracy is classified based upon the RMSE in Z direction.

Accuracy of horizontal data obtained is better than horizontal accuracy class of 2.5-cm and accuracy of the vertical data obtained is very close to vertical accuracy class of 1-cm. Considering the safety, man power and cost of inspection, the accuracy obtained for this pavement site is acceptable for most infrastructure sites. This also gives the confidence in making engineering judgements.

DEM and contour map of the nearby road connecting FM 1752 and west bound US 82 Highway in the Paris district provided an idea of the heaving as well as the drainage flow patterns at the site (as shown in Figs. 2.8 and 2.9). DEM provides the color coded elevation of the data collected. The orthomosaic shown in Fig. 2.7 and DEM shown in Fig. 2.8 are oriented in the same direction, which helps in understanding the relative elevations of the pavement and its surroundings. The highest elevation point in the DEM, roof of the van and the tent, are represented by orange and the lowest elevation point, adjacent areas of the pavement shoulder, are represented by dark blue. This is in line with the observations that were made during the field data collection. The elevation of road is sloping down along the direction of vehicular traffic indicated by two black arrows in orthomosaic and DEM shown in Figs. 2.7 and 2.8, respectively.

The DEM does not indicate any abrupt changes in elevation that confirm the absence of sulfate induced heaving at this site.

The UAV studies were performed along the edge of the shoulder and data was collected within 15 min. Processing and data analyses required several hours and the final output results as shown in Figs. 2.7 and 2.8 indicate that this technology is viable one and can provide reliable data in a safe and efficient manner. This tool can provide health monitoring of other structures and can be used to evaluate the performance of new additives or materials in strengthening pavement system including underlying subsoil layers.

Contour maps are also obtained for the pavement site. There are three views: top, 3D, and profile view of the pavement as shown in Fig. 2.9. The 3-dimensional view of the big white rectangle covering the whole top view is provided at the right corner of Fig. 2.9. The profile view of the small white rectangle drawn over the pavement in the top view is provided at the bottom of Fig. 2.9. Contours are laid over the top and 3D view of the pavement and a detailed profile view also gives an idea of the elevation profile of points across the pavement. The contours also confirm that there are no irregular elevation points, which indicate that the stabilization of sulfate soils adopted at this site is effective in mitigating heave distress.

2.5 Summary and Conclusions

Several important findings of various research studies related to expansive soils are briefly summarized and covered. Both natural expansive soils and sulfate rich expansive soils are major focus of some of these research studies. Some of the salient findings are presented below.

Expansive soil characterizations witnessed some major advancements partly due to the abilities to measure clay mineralogy fractions, internal pore distributions and unsaturated soil properties that govern their shrink-swell behaviors. Many of the new models incorporate these properties in the expansive soil characterization. Though these property measurements can be expensive and time consuming, the models are still superior to existing expansive soil characterization practices and can assist in better design of stabilization of these soils.

Different chemical additives are being evaluated for effective stabilization of expansive soils. Recently, a new class of alumino-silicate-polymers, commonly known as Geopolymers (GP) have received much attention for its eco-friendly and sustainable nature and its cementitious properties. This method was investigated in stabilizing an expansive soil. Atterberg's limit tests were conducted on GP treated CH specimens, which showed that GP treatments resulted in about a considerable decrease in the plasticity properties (PI). Swell tests on treated CH specimens also indicated a significant reduction in vertical swell strains from about 0.9 to 0.04%.

Innovative way of quantifying the presence of pavement heaving using UAV-CRP technology has been successfully attempted and developed in this study. The aerial data collected from UAV-CRP studies indicated that there is no sulfate induced heaving on the pavement when the underlying soils are treated with lime-fly ash additives with an extended mellowing period. The multi-faceted data obtained from this technology can be used to identify other infrastructure distress patterns including cracking on pavement surface layers. Overall, the research studies showed that the UAVs can be used and considered for monitoring geoinfrastructure assets laid over problematic soils.

Acknowledgements Authors would like to acknowledge National Science Foundation (NSF), Texas Department of Transportation, US Army Corps of Engineers, US Department of Transportation Funded Tier 1 Center on Center for Transportation Equity, Decisions and Dollars (CTEDD) and Transportation Consortium of South-Central States (Transet) for funding various projects which provided data to the present keynote paper. Several folks including Joe Adams, Jonathan Martin, Arturo Perez, Cody Lundberg, John Vasquez, Richard Williammee, and Wade Blackmon provided their assistance during data collection for TxDOT projects. Several UTA former doctoral students, Bhaskar Chittoori, Tejo Bheemasetti, Aravind Pedarla, and Ujwal Patil who did research studies that have contributed to the keynote paper.

References

1. Chen, F.H.: Foundations on Expansive Soils. Elsevier (2012)
2. Abduljauwad, S.N.: Study on the performance of calcareous expansive clays. Bull. Assoc. Eng. Geol. **30**, 481–498 (1993)
3. Al-Rawas, A.A.: The factors controlling the expansive nature of the soils and rocks of northern Oman. Eng. Geol. **53**, 327–350 (1999)
4. Puppala, A.J., Katha, B., Hoyos, L.R.: Volumetric shrinkage strain measurements in expansive soils using digital imaging technology. Geotech. Test. J. **27**, 547–556 (2004)
5. Puppala, A.J., Manosuthikij, T., Chittoori, B.C.S.: Swell and shrinkage characterizations of unsaturated expansive clays from Texas. Eng. Geol. **164**, 187–194 (2013)
6. Puppala, A.J., Pedarla, A., Hoyos, L.R., Zapata, C., Bheemasetti, T.V.: A semi-empirical swell prediction model formulated from 'clay mineralogy and unsaturated soil' properties. Eng. Geol. **200**, 114–121 (2016)
7. López-Lara, T., Zaragoza, J.B.H., Lopez-Cajun, C.: Mineralogical characterization of stabilized soils. Electron. J. Geotech. Eng. **9**, 1 (2004)
8. Puppala, A.J., Cerato, A.: Heave distress problems in chemically-treated sulfate-laden materials. Geo-Strata Geo Inst. ASCE **10**, 28 (2009)
9. Chittoori, B., Puppala, A.J.: Quantitative estimation of clay mineralogy in fine-grained soils. J. Geotech. Geoenviron. Eng. **137**, 997–1008 (2011)
10. Pedarla, A.: SWCC and Clay Mineralogy Based Models for Realistic Simulation of Swell Behavior of Expansive Soils (2013)
11. Alonso, E.E., Vaunat, J., Gens, A.: Modelling the mechanical behaviour of expansive clays. Eng. Geol. **54**, 173–183 (1999)
12. Banerjee, A., Puppala, A.J., Patil, U.D., Hoyos, L.R., Bhaskar, P.: A simplified approach to determine the response of unsaturated soils using multistage triaxial test. IFCEE **2018**, 332–342 (2018)
13. Mitchell, J.K., Soga, K.: Fundamentals of Soil Behavior. Wiley, New York (2005)
14. Puppala, A.J., Hoyos, L., Viyanant, C., Musenda, C.: Fiber and fly ash stabilization methods to treat soft expansive soils. In: Soft Ground Technology, pp. 136–145 (2001)
15. Pedarla, A., Chittoori, S., Puppala, A.J.: Influence of mineralogy and plasticity index on the stabilization effectiveness of expansive clays. Transp. Res. Rec. **2212**, 91–99 (2011)

16. Chittoori, B.C.S.: Clay Mineralogy Effects on Long-Term Performance of Chemically Treated Expansive Clays (2008)
17. Congress, S.S.C., Puppala, A.J., Lundberg, C.L.: Total system error analysis of UAV-CRP technology for monitoring transportation infrastructure assets. Eng. Geol. **247** (2018). https://doi.org/10.1016/j.enggeo.2018.11.002
18. Puppala, A.J., Congress, S.S.C., Bheemasetti, T.V., Caballero, S.: Geotechnical data visualization and modeling of civil infrastructure projects. In: GeoShanghai International Conference, pp. 1–12. Springer (2018)
19. Congress, S.S.C.: Novel infrastructure monitoring using multifaceted unmanned aerial vehicle systems - close range photogrammetry (UAV - CRP) data analysis (2018)
20. Hausmann, M.R.: Engineering principles of ground modification. McGraw-Hill New York (1990)
21. Petry, T.M., Little, D.N.: Review of stabilization of clays and expansive soils in pavements and lightly loaded structures—history, practice, and future. J. Mater. Civ. Eng. **14**, 447–460 (2002). https://doi.org/10.1061/(asce)0899-1561(2002)14:6(447)
22. Chakraborty, S., Nair, S.: Impact of different hydrated cementitious phases on moisture-induced damage in lime-stabilised subgrade soils. Road Mater. Pavement Des. **19**, 1389–1405 (2018)
23. Nelson, J.D., Miller, D.J.: Expansive Soils: Problems and Practice in Foundation and Pavement Engineering. Wiley, New York (1992)
24. Mitchell, J.K.: Practical problems from surprising soil behavior. J. Geotech. Eng. **112**, 255–289 (1986). https://doi.org/10.1061/(asce)0733-9410(1986)112:3(255)
25. Hunter, D.: Lime induced heave in sulfate bearing clay soils. J. Geotech. Eng. **114**, 150–167 (1988). https://doi.org/10.1061/(asce)0733-9410(1988)114:2(150)
26. Perrin, L.: Expansion of lime-treated clays containing sulfates. In: Proceedings of the 7th International Conference on Expansive Soils, pp. 409–414. ASCE Expansive Soils Research Council New York (1992)
27. McCallister, L.D., Petry, T.M.: Leach tests on lime-treated clays. Geotech. Test. J. **15**, 106–114 (1992)
28. Puppala, A.J., Hanchanloet, S., Jadeja, M., Burkart, B.: Sulfate induced heave distress: a case study. In: Proceedings, Transportation Research Board Annual Meeting, Washington DC, USA (1999)
29. Puppala, A.J., Wattanasanticharoen, E., Punthutaecha, K.: Experimental evaluations of stabilisation methods for sulphate-rich expansive soils. Gr. Improv. **9**, 89–90 (2005). https://doi.org/10.1680/grim.9.2.89.63641
30. Puppala, A.J., Talluri, N.S., Chittoori, B.S., Gaily, A.: Lessons learned from sulfate induced heaving studies in chemically treated soils. In: Proceedings of the International Conference on Ground Improvement and Ground Control. Research Publishing, pp. 85–98 (2012)
31. Talluri, N., Puppala, A.J., Chittoori, B., Gaily, A., Harris, P.: Stabilization of high-sulfate soils by extended mellowing. Transp. Res. Rec. J. Transp. Res. Board. 96–104 (2013)
32. Puppala, A.J., Talluri, N., Chittoori, B.C.S.: Calcium-based stabiliser treatment of sulfate-bearing soils. Proc. Inst. Civ. Eng. Improv. **167**, 162–172 (2014)
33. Kota, P., Hazlett, D., Perrin, L.: Sulfate-bearing soils: problems with calcium-based stabilizers. Transp. Res. Rec. J. Transp. Res. Board. 62–69 (1996)
34. He, S., Yu, X., Banerjee, A., Puppala, A.J.: Expansive soil treatment with liquid ionic soil stabilizer. Transp. Res. Rec. 0361198118792996 (2018)
35. Acharya, R., Pedarla, A., Bheemasetti, T.V., Puppala, A.J.: Assessment of guar gum biopolymer treatment toward mitigation of desiccation cracking on slopes built with expansive soils. Transp. Res. Rec. J. Transp. Res. Board. **2657**, 78–88 (2017). https://doi.org/10.3141/2657-09
36. Caballero, S., Acharya, R., Banerjee, A., Bheemasetti, T.V., Puppala, A.J., Patil, U.: Sustainable slope stabilization using biopolymer-reinforced soil. In: Geo-Chicago 2016, pp. 116–126. American Society of Civil Engineers, Reston, VA (2016)
37. Davidovits, J.: Geopolymer, green chemistry and sustainable development solutions. In: Proceedings of the World Congress Geopolymer 2005. Geopolymer Institute (2005)

38. Provis, J.L., Van Deventer, J.S.J.: Geopolymers: Structures, Processing, Properties and Industrial Applications. Elsevier (2009)
39. Van Jaarsveld, J.G.S., Van Deventer, J.S.J., Schwartzman, A.: The potential use of geopolymeric materials to immobilise toxic metals: Part II. Material and leaching characteristics. Miner. Eng. **12**, 75–91 (1999). https://doi.org/10.1016/S0892-6875(98)00121-6
40. McLellan, B.C., Williams, R.P., Lay, J., van Riessen, A., Corder, G.D.: Costs and carbon emissions for geopolymer pastes in comparison to ordinary portland cement. J. Clean. Prod. **19**, 1080–1090 (2011). https://doi.org/10.1016/j.jclepro.2011.02.010
41. Bell, J.L., Driemeyer, P.E., Kriven, W.M.: Formation of ceramics from Metakaolin-based geopolymers. Part II: K-based geopolymer. J. Am. Ceram. Soc. **92**, 607–615 (2009). https://doi.org/10.1111/j.1551-2916.2008.02922.x
42. Duxson, P., Fernández-Jiménez, A., Provis, J.L., Lukey, G.C., Palomo, A., van Deventer, J.S.J.: Geopolymer technology: the current state of the art. J. Mater. Sci. **42**, 2917–2933 (2007). https://doi.org/10.1007/s10853-006-0637-z
43. Shayan, A., Tennakoon, C., Xu, A.: Specification and Use of Geopolymer Concrete in the Manufacture of Structural and Non-structural Components: Review of Literature, AP-T318-16 (2016)
44. Puppala, A.J., Congress, S.S.C., Bheemasetti, T.V., Caballero, S.R.: Visualization of civil infrastructure emphasizing geomaterial characterization and performance. J. Mater. Civ. Eng. **30**, 4018236 (2018). https://doi.org/10.1061/(ASCE)MT.1943-5533.0002434
45. Puppala, A.J., Talluri, N., Congress, S.S.C., Gaily, A.: Ettringite induced heaving in stabilized high sulfate soils. Innov. Infrastruct. Solut. **3**, 72 (2018). https://doi.org/10.1007/s41062-018-0179-7
46. Smith, D.L., Abdullah, Q.A., Maune, D., Heidemann, K.H.: New ASPRS positional accuracy standards for digital geospatial data (2014). https://doi.org/10.14358/PERS.81.3.A1-A26

Chapter 3
Testing Mode and Soil Parameters

R. G. Robinson

3.1 Introduction

Soil properties are commonly determined by performing element tests in the laboratory. The properties can be either index property, commonly used for soil classification and correlations, or the engineering properties for design purpose. The important index properties of soils are Atterberg's limits, grain size distribution, relative density, etc. For fine grained soils, the important index properties are the Atterberg limits such as liquid limit (w_L) and plastic limit (w_P). The major engineering properties of soils are shear strength, consolidation and permeability. For expansive soils, swell pressure is an important property.

In the laboratory, these properties are commonly determined using a variety of equipment. It was observed that the same property, when determined using different equipment often yields different results. In this paper, the variation of soil properties with type of equipment used is discussed. The properties discussed are:

(i) Liquid limit
(ii) Swelling pressure
(iii) Drained angle of internal friction of fine grained soils
(iv) Radial coefficient of consolidation and
(v) Interface friction angle

The types of apparatus used for each property and the values obtained are discussed. The possible reasons for the differences observed are highlighted.

R. G. Robinson (✉)
Department of Civil Engineering, Indian Institute of Technology Madras, Chennai 600036, India
e-mail: robinson@iitm.ac.in

M. Latha G. (ed.), *Frontiers in Geotechnical Engineering*, Developments in Geotechnical Engineering, https://doi.org/10.1007/978-981-13-5871-5_3

3.2 Liquid Limit

Liquid limit is one of the most important index properties of fine grained soils. It is an essential parameter for soil classification. In addition, it is also often used to estimate the consolidation parameters through empirical correlations. Presently, two standard methods are in use for the determination of liquid limit of soils: the percussion method, using Casagrande's apparatus, and fall cone method. The details of these tests are covered in IS 2720 [1] and many other codes. Several studies are reported in the literature that compares the values of liquid limit obtained from Casagrande's apparatus and the fall cone apparatus. A review of the results in the literature and the mechanisms controlling the liquid limit of soils can be found in Sridharan and Prakash [2]. It can be seen that for soils with lower liquid limit values, the cone penetration method gives slightly higher liquid limit compared to the Casagrande's method [3]. However, the trend changes at higher liquid limits. The possible reason for this difference is attributed to the fundamental mechanism of the test itself, which is briefly brought out below.

The tests conducted using Casagrande's apparatus are typically a slope failure test (Fig. 3.1) in which a miniature slope is made to fail by repeatedly mobilising the shear strength by the drops/blows. Therefore, the stability number (S_n), given in Eq. 3.1 governs the slope failure in the test [4].

$$S_n = \frac{c_u}{\gamma H} \tag{3.1}$$

where c_u, γ and H are the undrained shear strength, unit weight and height of the slope, respectively. The dynamic effects are ignored in the analysis. As per Eq. (3.1), the shear strength of the soil at liquid limit water content cannot be expected to be constant for all soils. However, the stability number is expected to be a constant. As the liquid limit increases, the value of γ decreases. Therefore, in order to maintain constant value of stability number, the shear strength decreases.

The factors involved in a fall cone test (Fig. 3.2) are the mass of the cone (m), the cone angle (α), the undrained shear strength (c_u) and depth of penetration (d). These factors are related to the cone factor k_α (which depends on α), as per Eq. (3.2) [3].

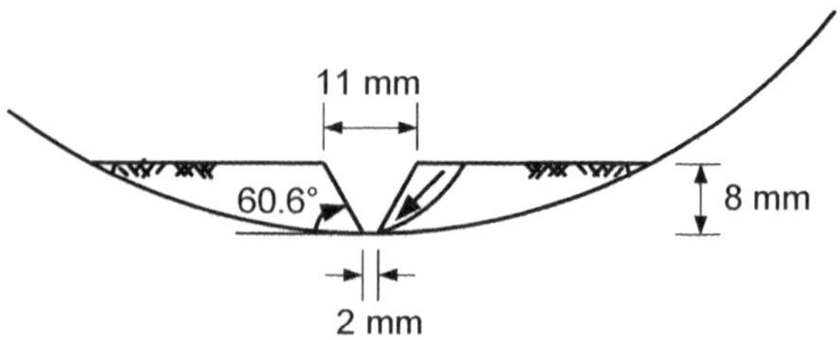

Fig. 3.1 The soil slope in Casagrande's liquid limit apparatus

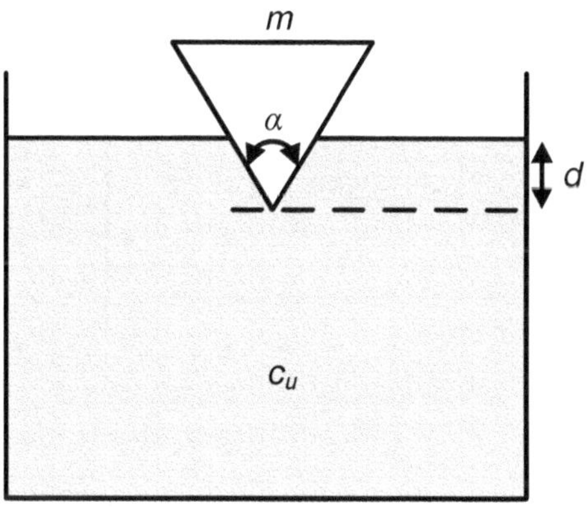

Fig. 3.2 Factors in a fall cone test

$$\frac{c_u d^2}{mg} = k_\alpha \tag{3.2}$$

Clearly, as per Eq. (3.2), c_u is expected to be a constant at liquid limit for all soils (as $d = 20$ mm). As the mechanisms of the tests are different, the liquid limit values are expected to be different. The data reinterpreted from Das et al. [5] (Fig. 3.3) validates the mechanisms discussed above that c_u decreases with increase in liquid limit by Casagrande's apparatus, whereas it is nearly constant in fall cone method.

3.3 Swelling Pressure

Swelling pressure is one of the important variables required in the prediction of heave of swelling soils. Swelling pressure is the pressure required to maintain constant volume on saturating an expansive soil from its unsaturated state. Several laboratory methods are available to evaluate the swelling pressure in the laboratory. Typically, oedometer swelling test, suction measurements, triaxial methods, etc. are commonly used [6–10] out of which oedometer testing method is popular due to its simplicity. Three different oedometer methods such as Method A, Method B and Method C [6] can be used for the determination of swelling pressure as explained below.

Method A—This method is also often called as swell-consolidation method. As the name indicates, the specimen in the oedometer cell is soaked with water and allowed to swell under a nominal seating pressure of about 5–6.25 kPa. Once the primary swell is completed, the specimen is one-dimensionally consolidated back to reach its initial thickness. The consolidation pressure required to consolidate the specimen to its initial thickness is taken as the swelling pressure. This method is similar to Method A of ASTM D4546-03 [11].

Method B—This method is also called as different pressure method in which three identical specimens are assembled in the oedometer and loaded with different

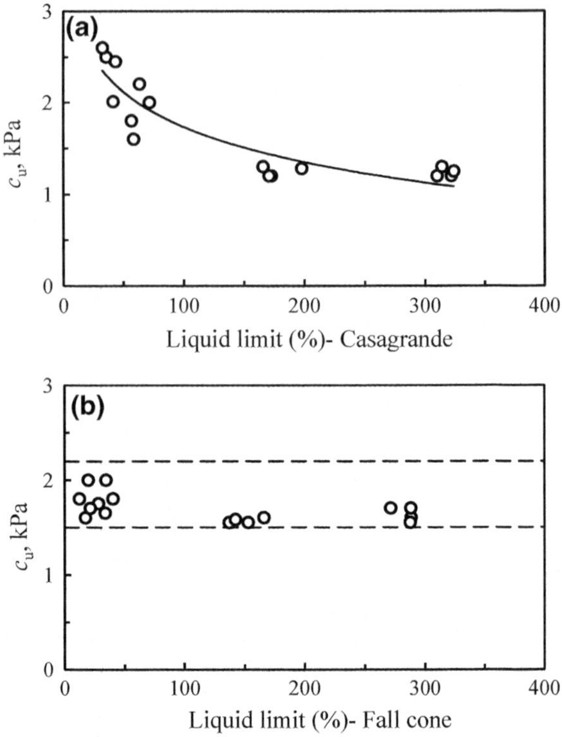

Fig. 3.3 Variation of undrained shear strength at liquid limit with liquid limit determined by **a** Casagrande's method and **b** fall cone method (data from Das et al. [5])

pressures close to the expected swelling pressure. The specimens are submerged in water and the volume change is monitored. The pressure corresponding to zero volume change is taken as swelling pressure. The limitation of the method is that three or more identical specimens are required unlike Method A, which requires only one specimen.

Method C—In this method, the specimen is maintained at constant volume by controlling the vertical pressure after the specimen is inundated in free water. The vertical pressure that maintains constant volume is taken as the swelling pressure. This method is also called as constant volume method and guidelines are given in BS 1377 [12] and IS 2720 [13].

In principle, all the three methods are equally valid. However, they often yield entirely different swelling pressure values. Johnson and Snethen [14] observed that the magnitude of swelling pressure depends on the degree of confinement. Ali and Elturabi [15] reported the swelling pressure values from Method A to be higher than that from method Sridharan et al. [16] carried out oedometer swelling tests using the Methods A, B and C on soils with different initial void ratio (e_i) and water content (w). They reported that Method A gives the highest swelling pressure value and Method B gives the lowest value. The results from Method C lie in between as summarised in Table 3.1. Definite relation between the three methods was not observed.

Table 3.1 Typical swelling pressure values by the three methods [16]

Soil sample	Initial conditions		Swelling pressure (kPa) by method		
	e_i	w (%)	A	B	C
BC1	0.893	0	300	160	190
BC4	1.002	0	390	160	220
BC5	0.742	0	510	140	310
BC8	0.656	20.8	150	–	70

Table 3.2 Index properties and compaction characteristics of the soil used

Properties	Gradation A	Gradation B
Atterberg limits		
(a) Liquid limit (%)	105	69
(b) Plastic limit (%)	40	30
(c) Shrinkage limit (%)	7	8
Compaction test results		
(a) Maximum dry unit weight (kN/m3)	14.55	14.96
(b) Optimum moisture content (%)	26.4	24
Differential free swell (%)	100	80

Thus, the values of swelling pressure depend on test method. However, the possible reason for the dependence of swelling pressure on the test method is not brought out in the literature. Soundara and Robinson [10] addressed this issue by carrying out oedometer tests on identical specimens of an expansive soil.

The soil used was taken from Siruseri, near Chennai. Two types of gradation (Gradation A and Gradation B) were prepared. Sample with Gradation A contains soil fraction passing through 75μ whereas Gradation B was prepared by sieving the soil through 1 mm sieve. The index properties and the compaction characteristics of the specimens are given in Table 3.2. The clay samples can be classified as clay of high to very high expansion potential [17]. The specimens with required water content and dry unit weight were prepared by static compaction. The Proctor compaction curve was used as a reference to fix the placement condition.

Table 3.3 reports the dry unit weight (γ_d), water content (w) and the specimen condition. Measured quantities of dry soil and water were mixed thoroughly and kept in airtight bags, overnight, for moisture equilibrium. Specimens of 14 mm thickness with the required unit weight were prepared by statically compacting in oedometer rings of 60 mm diameter and 20 mm height. The side wall friction between the ring and specimen was reduced by smearing the inner sides of the consolidation rings with silicon grease.

In addition to performing swelling pressure tests, falling head permeability tests were conducted by connecting a burette to the bottom of the consolidation cell.

Table 3.3 Placement conditions adopted for the experiments

Specimen	γ_d (kN/m^3)	w (%)	Specimen condition
A_1	14.55	26.4	Optimum
A_2	13.82	23.6	Dry side
B_1	14.96	24	Optimum
B_2	14.21	21	Dry side
B_3	14.21	28	Wet side

Table 3.4 Swelling Pressure values by different methods

Specimen	Swelling pressure (kPa) by method		
	A	B	C
A_1	325	160	185
A_2	270	150	175
B_1	240	160	175
B_2	210	115	140
B_3	60	–	50

Specimens A_1 and A_2 were prepared for the scanning electron microscope (SEM) images once the test was over. The specimens for SEM studies were prepared as per the guidelines given in Mitchel [18]. Specimens of 20 mm × 10 mm × 10 mm was cut and were subjected to freeze drying [19]. The broken surface was observed under the scanning electron microscope for fabric studies.

The swelling pressure values obtained for all the specimens by the three methods are listed in Table 3.4. The swelling pressure by Method A is the highest while that from Method B is the lowest. For the specimens compacted on the dry side of optimum, the swelling pressure by Method A is about 1.5–2 times higher than that from Method B. However, the swelling pressure values from Methods B and C are nearly the same, similar to those reported in the literature [15, 16]. Samples in the wet side displayed very low values of swelling pressure similar to Seed and Chan [20]. In addition, the swelling pressure values are also not very sensitive to test methods (Sample B_3).

Typical falling head permeability test results for sample B_1 in the form of void ratio (e) versus permeability (k) are shown in Fig. 3.4. Unlike in Methods A and B, the permeability values were obtained at only one void ratio in Method C as volume change and hence the void ratio change was not permitted. The results clearly show that the k values are significantly different, at the same void ratio corresponding to the swelling pressure, when the samples were subjected to different test modes. The permeability values corresponding to the initial void ratio (under the swelling pressure) for all the specimens are listed in Table 3.5. While permeability values of Methods B and C are nearly the same, the permeability values of samples in Method A are about 4–8 times less than those by Method C, when the samples are at optimum

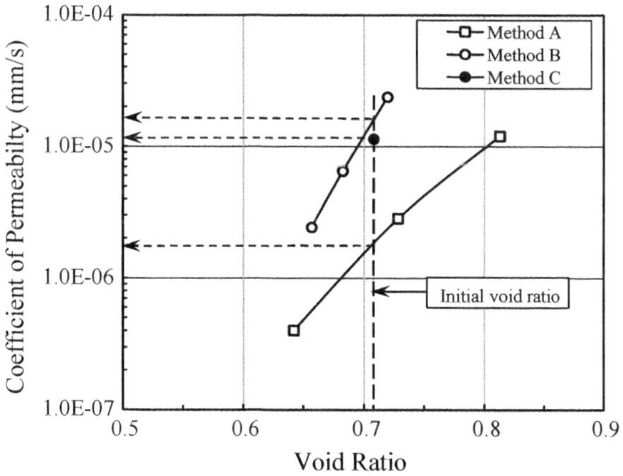

Fig. 3.4 Coefficient of permeability plots for Sample B_1 for all methods

Table 3.5 Permeability values at initial void ratio for different methods

Specimen number	Coefficient of permeability, k (mm/s)		
	Method A	Method B	Method C
A_1	4.9×10^{-7}	1.8×10^{-6}	2.4×10^{-6}
A_2	7.9×10^{-7}	2.2×10^{-6}	2.9×10^{-6}
B_1	1.8×10^{-6}	1.0×10^{-5}	1.1×10^{-5}
B_2	1.2×10^{-6}	1.3×10^{-5}	1.0×10^{-5}
B_3	4.2×10^{-7}	–	5.2×10^{-7}

or dry side of optimum. The permeability value of sample in the wet side of optimum (sample B_3) is practically the same for Methods A and C.

Lambe [21] reported that compacted soils at the dry side of optimum display flocculated structure. As the moisture content increases beyond optimum towards the wet side of optimum, the double layer of the clay particles expands more, resulting in an increased repulsion between the particles during compaction leading to a dispersed structure [22]. Therefore, specimens A_1, A_2, B_1 and B_2 are expected to be in a flocculated state before soaking in water. Samples in Method A underwent significant swelling under a nominal surcharge of 6.25 kPa on saturation. The volumetric strains experienced by the samples A_1, A_2, B_1 and B_2 are 26.30%, 34.44%, 16.81% and 26.82%, respectively. As a consequence of this swelling, the structure of the clay might have changed from flocculated to a more dispersed state. Further one-dimensional loading, to bring the soil specimen to its initial void ratio alters the fabric towards a more dispersed fabric. In Methods B and C, the flocculated structure is likely to be preserved as the specimens are not allowed to swell. The permeability values reported in Table 3.5 supports this possibility.

As sample B_3 was prepared in the wet side of optimum, its structure is expected to be in a dispersed state [21]. In Method A, the swelling experienced by the sample is only 1.38%. Therefore, the fabric alteration between Methods A and C may not be

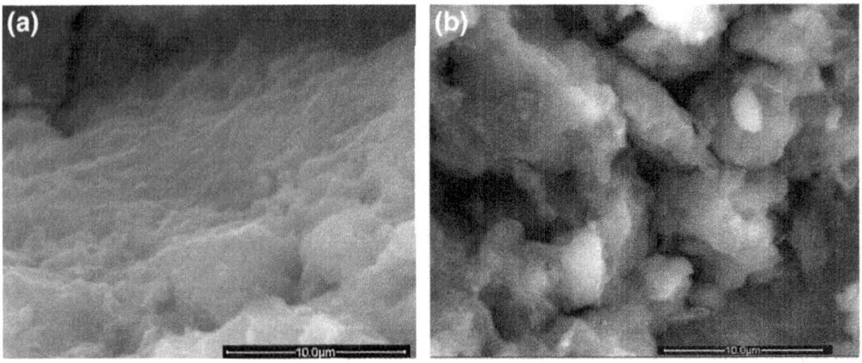

Fig. 3.5 SEM images in **a** Method A and **b** Method C

significant during the swelling pressure test. This can be corroborated from Table 3.5 that the coefficient of permeability is almost the same between Methods A and C. As the structure is not changed significantly between the methods, the difference in swelling pressure values between tests is only marginal in sample B_3.

Typical SEM images of samples tested in Method A and C for sample A_2 are shown in Fig. 3.5a and b, respectively. Parallel orientation of clay particles (dispersed structure) is seen on specimens tested using Method A. Flocculated structure is seen in the case of Method C.

The results of permeability tests and SEM images confirm that the flocculated structure of the compacted soil in the dry side of optimum gets altered to dispersed structure when Method A is adopted for measuring the swelling pressure, as a consequence of swelling. As volume change is not permitted during saturation in Methods B and C alteration of fabric is expected to be less. Therefore, the difference in swelling pressure value between methods is attributed to direct consequence of fabric change that occurs during the swelling pressure test.

3.4 Angle of Internal Friction

Among the types of apparatus, the direct shear apparatus and the triaxial apparatus are very popular to determine the shear strength parameters. Direct shear test is generally preferred for granular soils such as sands and gravels. However, the test is also often used for the determination of drained angle of internal friction of clays. It is widely accepted that the angle of internal friction from plane strain apparatus $\left(\phi'_{\text{ps}}\right)$ is higher than that obtained the triaxial angle of internal friction (ϕ'_{tri}) from triaxial compression test [23, 24]. Comparison of angle of internal friction values from direct shear test (ϕ'_{ds}) and triaxial compression tests for granular soils was also made in the literature [25]. Generally, ϕ'_{ds} is about 2–8° higher than ϕ'_{tri} for the case of sands.

Studies that compare the effective angle of internal friction from triaxial compression tests and direct shear tests for fine grained soils are scarce in the literature.

Lini Dev et al. [26] carried out direct shear tests and triaxial tests under drained conditions on fine grained soils. Three types of soils such as kaolinite, IIT lake clay and a silty soil were used as fine grained soils. Uniformly graded sand sample (particle size in the range of 0.425–1.0 mm) was also used for the study. The clay size particle present in the silty soil is only 12%. The other two soils contain clay sized particles of about 32%. The maximum and minimum dry unit weights of the sand are 17.2 kN/m^3 and 14.4 kN/m^3, respectively.

Reconstituted soil specimens, consolidated from slurry with water content of 1.5 times the liquid limit, were used for the study. Specimens of 50 mm diameter and 100 mm height were used for carrying out the triaxial tests. Effective confining pressures of 50, 100 and 200 kPa were used. In order to have better drainage performance double filter paper side drains suggested by Sivakumar et al. [27] were adopted. Back pressure saturation technique was adopted so as to achieve a minimum B value of 0.95. The tests were conducted at sufficiently slow rate to keep the specimens under drained condition. Time–volume change curve, during the consolidation stage, was used to calculate the deformation rate required to keep the specimens under drained condition as suggested by Head [28].

For direct shear tests, the specimens were directly consolidated in the in the shear box itself from slurry state to the required normal stress. Normal stress values of 50, 100 and 200 kPa were used. The rate of deformation was calculated as per the procedure given in Head [28] to keep the specimens under drained condition during shear. Area correction was applied on the shear stress calculations as per Petley [29]. Relative density of 65% was used for the sand specimen. Moisture tamping technique was used for the specimen preparation. Drained tests were conducted after saturating the specimens.

Table 3.6 compares the test results. Contrary to the generally reported results for the case of dense sands, the ϕ'_{tri} values for the fine grained soils are higher than ϕ'_{ds}. The difference in angle of internal friction for Kaolinite and IIT lake clay are about 8°. However, the difference is only 1° for the case of the silty soil. For the sand specimens, the values of ϕ'_{ds} and ϕ'_{tri} are 44.5° and 42°, respectively, which is in line with the reported trend for sands.

Table 3.6 Values of angle of internal friction from direct shear and triaxial tests

Sl. no.	Soil	Angle of internal friction from		$(\phi'_{tri} - \phi'_{ds})$
		Triaxial test	Direct shear test	
1	Kaolinite	29.5°	21°	8.5°
2	IIT clay	28°	20°	8°
3	Silty soil	26.5°	25.5°	1°
4	Sand	42°	44.5°	−2.5°
5	Sand + plastic disc	45.5°	35°	10.5°

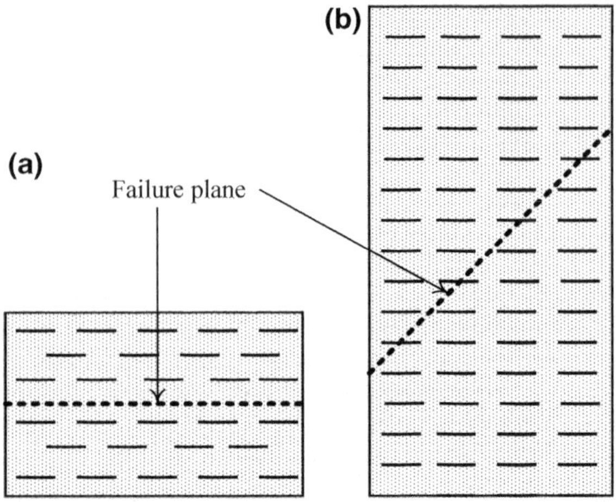

Fig. 3.6 Likely condition in **a** direct shear and **b** triaxial compression tests

The value of $\phi'_{tri} - \phi'_{ds}$ is not significant for the silty soil unlike the other two soils. One of the differences in the soils is the composition of clay sized particles, which is 12% for silty soil and 32% for the other two soils. Therefore, there is a possibility that the difference may be due to the presence of clay sized particles in the soil. It is known that the clay particles are like plates [30]. Therefore, the clay particles may orient into parallel array due to the consolidation pressure as shown in Fig. 3.6a, b. The particles slide one over the other during shearing in the direct shear mode unlike the triaxial specimen, resulting in higher shear strength in the triaxial test.

The above hypothesis was validated by mixing plate shaped discs with sand. The transparency film of 0.1 mm thickness, commonly used for overhead projectors, was cut using a punching machine to 4.5 mm diameter (Fig. 3.7) so that the diameter to thickness ratio (aspect ratio) is 45. The discs and sand were mixed in the proportion of 10% by weight of disc and 90% sand. Direct shear and triaxial tests were conducted under drained conditions on specimens prepared at unit weight of 15 kN/m^3.

The difference between the ϕ'_{tri} and ϕ'_{ds}, when the plastic discs were mixed with sand is 10.5° (Table 3.6). Therefore, the possible reason for higher values of angle of internal friction from triaxial compression test than the direct shear test for fine grained soils containing clay sized particles may be attributed to the presence of plate shaped clay particles.

3.5 Radial Coefficient of Consolidation

Design of vertical drain system is conventionally designed based on Barron's theory [31] because of its simplicity and ease of use. The solution for the condition of equal vertical strain without smear is given as

Fig. 3.7 The transparency sheet of 4.5 mm diameter cut for the test

$$U = 1 - \exp\left(\frac{8T_r}{F(n)}\right) \tag{3.3}$$

where,

$$F(n) = \frac{n^2}{(n^2 - 1)}\ln(n) - \frac{(3n^2 - 1)}{4n^2} \tag{3.4}$$

in which n is the drain spacing ratio given by

$$n = \frac{\text{Diameter of influence}\,(d_e)}{\text{Diameter of drain well}\,(d_w)} \tag{3.5}$$

and

$$T_r = \frac{c_h t}{d_e^2} \tag{3.6}$$

where T_r is the time factor and c_h is the coefficient of horizontal consolidation. The important design parameter required in Barron's theory is c_h.

Several types of laboratory apparatus with radial drainage were used in the literature for determining c_h [32–42]. However, Rowe consolidation cell [39–42] with radial drainage is popular. In the Rowe cell, radial consolidation tests can be conducted by allowing drainage in radially-in (Fig. 3.8a) or radially-out (Fig. 3.8b) directions. Modified conventional consolidation was also used to determine the radial coefficient of consolidation [33–35, 43]. A comparison of c_h values obtained from these two types of drainage conditions is scarce in the literature. Sridhar et al. [44]

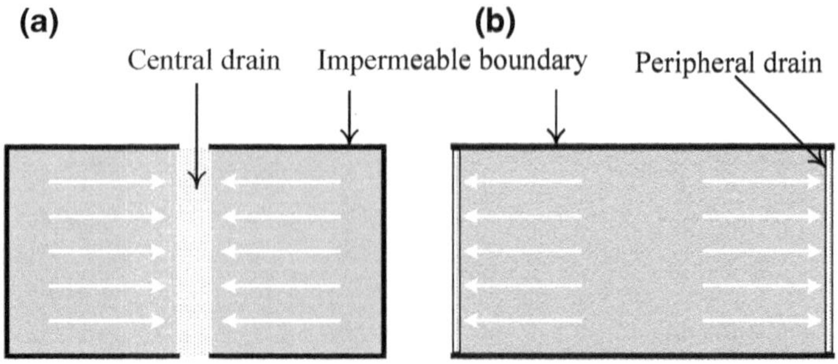

Fig. 3.8 Radial consolidation test with **a** central drain and **b** peripheral drain

addressed this issue by performing tests using a 100 mm diameter Rowe cell [45] with drain spacing ratio of 2–10.

The soil used is the commercially available kaolinite with liquid and plastic limits of 43% and 23%, respectively. For the radially inward test, a sand drain was placed at the middle. The diameters of the central drain used are 10, 15 and 50 mm. This gives drain spacing ratio of 10, 6.67 and 2. A peripheral drain was provided for conducting the radially outward flow consolidation.

Reconstituted soil specimens that were preconsolidated to a pressure of 12.5 kPa were used for the study. A central hole was made and filled with clean sand in loose condition. The sand used satisfies the requirement suggested by Shields and Rowe [46]. Radially outward drainage condition was created by placing a saturated porous plastic drain (1.5 mm thickness) at the periphery. Rubber membrane of 0.5 mm thickness was introduced at the top and bottom boundaries of the clay specimen so as to avoid vertical drainage. A 10-mm-thick rigid acrylic platen was placed on the top of the specimen so that conditions of equal strain hold good. Load increment ratio of 1.0 was used to carry out the consolidation test. The time-settlement data were continuously recorded during the consolidation test. It was reported in the literature that the water content is non-uniform along the radial direction in radial consolidation tests [47–49]. Therefore, the variation of water content along the radial direction was determined at the end of the test.

The time-settlement data obtained from the radially inward consolidation test was analysed using the log t method [35] and the c_h values were calculated. The square root of time fitting method was used to obtain the c_{h} values for the case of radially outward consolidation test [50].

The variation of c_{h} with effective consolidation pressure obtained from both types of radial consolidation test is shown in Fig. 3.9. It can be seen from the figure that the c_{h} values increase with increase in consolidation pressure similar to the variation of coefficient of consolidation c_{v} with consolidation pressure for kaolinite [51]. It can be inferred from Fig. 3.9 that the values of c_{h} increase with n. In other words, c_{h} increases as the diameter of the drain decreases for a given consolidation pressure. Shields and Rowe [46] also reported similar observations. They attributed this behaviour to the length of drainage path, which increases as the drain size decreases. The c_{h} values

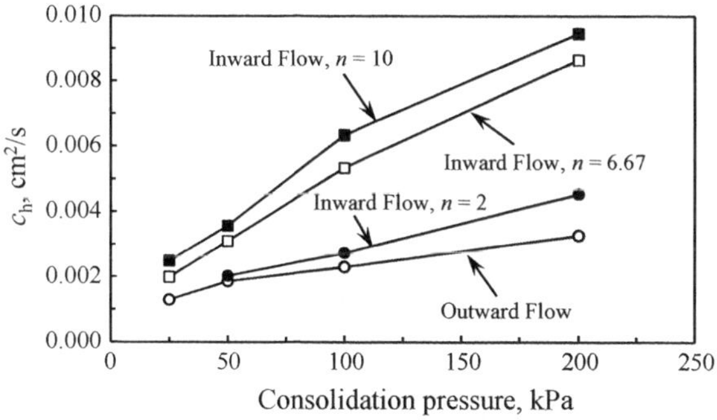

Fig. 3.9 Variation of c_h with consolidation pressure

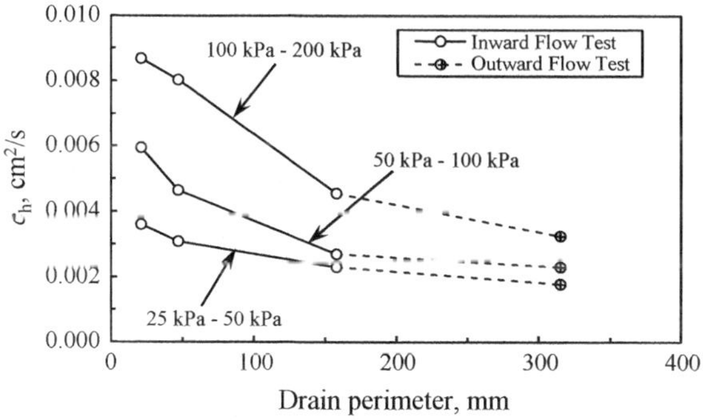

Fig. 3.10 Variation of c_h with drain perimeter

from outward flow consolidation tests are less compared to those obtained from inward flow consolidation test. In order to evaluate this trend further, the variation of c_h with perimeter of the drain is plotted in Fig. 3.10. It is interesting to note that the value of c_h decreases as the perimeter of the drain increases.

Further, the variation of water content along the radial direction is shown in Fig. 3.11. Clearly, the water content near the drain is less compared to other locations irrespective of direction of flow due to non-uniform stress distribution in the soil in radial consolidation tests [48, 49]. As a consequence, the permeability near the draining boundary is less compared to other locations. Therefore, the flow of water towards the drain gets reduced resulting to slower rate of consolidation. Higher the perimeter of the drainage boundary, lesser is the rate of consolidation and hence the magnitude of c_h.

3.6 Interface Friction Angle

Evaluation of interface friction between soils and construction materials (solid surfaces) is essential for the design of several structures like pile foundations, retaining structures, reinforced earth structures, etc. The interface friction is generally quanti-

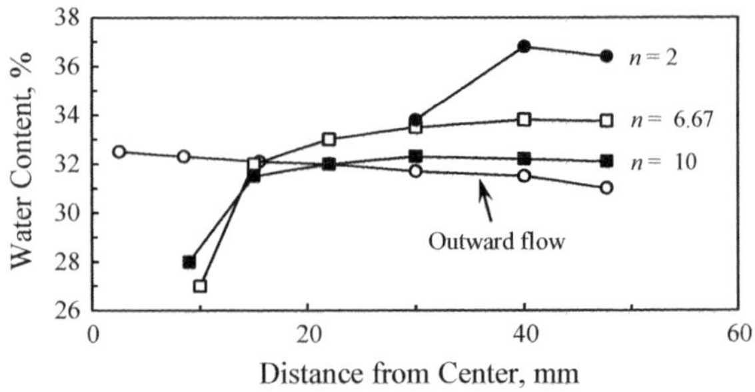

Fig. 3.11 Variation of water content within the specimen after radial consolidation test

fied in terms of friction angle (δ) or coefficient of friction (μ), based on Coulomb's law [52]. Meyerhof [53] carried out strain controlled direct shear tests to determine the interfacial friction angle between sand and brass sections, probably one of the earliest studies in this area. Subsequently, Kézdi [54] used a large direct shear box of 300×300 mm to determine δ. Potyondy [55] determined the interface friction angle between soils and solid surfaces by keeping the solid material in the lower portion of the box and the soil in the upper box. Rowe [56] replaced the entire bottom half of the shear box with test material to maintain constant test area. Silberman [57] reversed the configuration by keeping the test material above the soil.

Apart from direct shear apparatus, other types of apparatus were also developed in the literature. Detailed review of apparatus types available in the literature can be found in Subba Rao et al. [58]. In this section, the influence of mode shear on the interface friction between sand and solid surface is discussed.

The dependence of angle of internal friction (ϕ') of sand with dry unit weight is well documented in the literature. Studies are also available in the literature to understand the influence of dry unit weight on δ between sands and solid surfaces. These studies reported contradicting conclusions. Some studies [59–61] reported that the friction angle δ remains the same irrespective of the dry unit weight. However, several other studies observed that the angle of internal friction increases with the unit weight [62].

The interface friction angle depends on surface roughness of the solid surface [55, 60, 63, 64, to name a few]. The values of δ increase with roughness and for very rough surfaces it tend to a constant value. Potyondy [55] and Panchanathan and Ramaswamy [65] observed that for very rough surfaces, the limiting maximum friction angle δ_{lim} is close to the peak angle of internal friction (ϕ) of sand. Several other studies using simple shear apparatus also confirm that δ_{lim} is the peak angle of internal friction of the sand [63, 64, 66, 67]. Everton [68] also reported that $\delta_{\mathrm{lim}} = \phi$ [69]. However, the studies of Yoshimi and Kishida [60] contradict this observation. They carried out interface tests using ring torsion apparatus and the results showed that δ_{lim} is the critical state angle of internal friction (ϕ_{cv}) of sand.

Some of the results reported in the literature are summarised in Table 3.7, from which the following inferences can be drawn:

Table 3.7 Summary of literature

Author(s)	Type of testing apparatus	Results of investigation	Remarks
Potyondy [55]	Direct shear apparatus with the sand on the top of test material	δ increases with unit weight and $\delta = \phi$ for very rough surfaces	Sand was placed above the test material
Broms [61]	Direct shear mode with the test material over the sand	A δ value of 23° was obtained irrespective of unit weight of sand	Test material was placed above the sand
Yoshimi and Kishida [60]	Ring shear with the test material on top of sand	Unit weight has no influence and $\delta = \phi_{cv}$ for very rough surfaces	Test material was placed above the sand
Acar et al. [62]	Direct shear test similar to Potyondy [55]	δ increases with unit weight	Sand was placed above the test material
Desai et al. [70]	Translational test box	δ increases with unit weight	Sand was placed above the test material
Noorany [59]	Similar to Broms [61]	Influence of unit weight is negligible	Test material was placed above the sand
O'Rourke et al. [71]	Similar to Potyondy [55]	δ increases with unit weight	Sand was placed above the test material
Uesugi et al. [67]	Simple shear with the sand on top of the test material	δ increases with unit weight and $\delta_{lim} = \phi$	Sand was placed above the test material

(a) Surface roughness of the solid material has a strong influence on δ
(b) The value of δ_{lim} is either ϕ or ϕ_{cv}
(c) Contradicting observations are reported with regard to unit weight.

Subba Rao et al. [73] and Robinson [72] analysed the results reported in the literature (Table 3.7) and identified the possible reason for the discrepancy observed with reference to the limiting value of interface friction angle and the effect of unit weight on δ. From careful analysis of the results in Table 3.7, the following can be observed:

(a) Whenever δ_{lim} is equal to ϕ_{cv} and that δ was independent of unit weight, the interface in the test was prepared by placing the test material above the prepared soil. The normal stress required was then applied through the test material to the interface. This situation can be described as Type A.

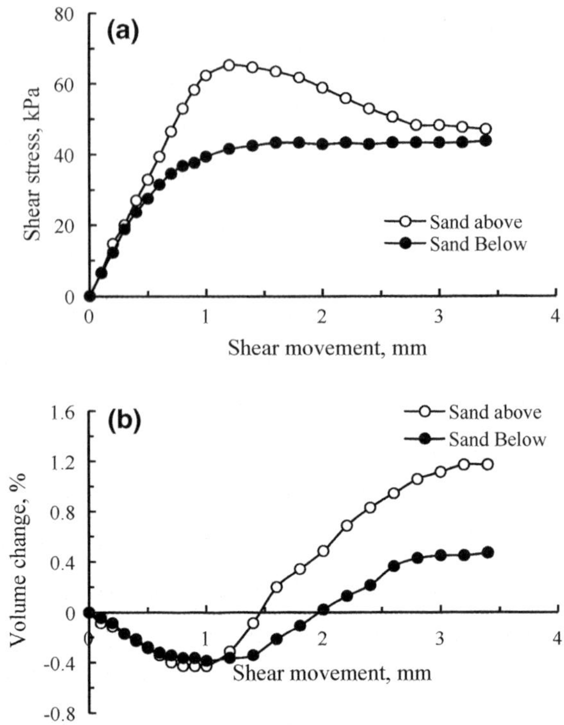

Fig. 3.12 Typical **a** shear stress–movement and **b** volume change–movement curves in Type A and Type B mode of shear

(b) Where ever it was observed that the limiting value of δ is equal to ϕ or δ increases with unit weight, the interface was prepared by preparing the soil above the test material. The normal stress was then applied through the soil to the interface.

Apart from the magnitude of the friction angle being affected by the mode of shear, as discussed above, the shape of shear stress–movement curve is also affected by the mode of shear. Typical results from a modified direct shear test for the sand–concrete interface are shown in Fig. 3.12a, b. The relative density of the sand used was 85%. It can be seen that for the Type B case, a well-defined peak is observed and at large movements, the shear stress for both Type B and Type A mode of shear is practically the same. There is also a notable difference in the volume change behaviour during shearing. The dilation for Type B case is more than that from Type A case.

3.7 Summary and Conclusions

The influence of apparatus type on soil properties was discussed in this paper. The properties discussed are the liquid limit, swelling pressure, angle of internal friction of fine grained soils, horizontal coefficient of consolidation and the interface friction angle. The following conclusions are drawn:

(i) The value of liquid limit obtained from fall cone method and Casagrande's apparatus is different. The reason for this difference is attributed to the difference in the mechanisms. The mechanism of fall cone test is a strength test whereas in Casagrande's apparatus, it is the undrained failure of slope, for which the stability number governs the slope failure.

(ii) The swell pressure values obtained from the different pressure method is the lowest while the swell-consolidation method yields the highest swell pressure. The reason for this difference in results is attributed to the structure change that occurs during swelling of the specimen in the swell-consolidation method.

(iii) The values of horizontal coefficient of consolidation determined from radial consolidation test depend whether the test is conducted with drainage direction towards the central drain or peripheral drain. When the drainage is towards outward direction the values of c_{h} is less compared to that obtained from radially inward test. The water content at the drainage boundary for both types of tests was found to be less than that at other locations. The likely reason for lower value of c_{h} is attributed to the larger drainage boundary with lesser void ratio.

(iv) The drained angle of internal friction of fine grained soils obtained from triaxial compression test was found to be higher than that obtained from direct shear test. The reason for this difference may be due to the shape of the clay sized particles present in the fine grained soils.

(v) Depending on the mode of shear in interface direct shear test, the interface conditions can be classified into two types such as Type A and Type B. In Type A mode of shear, the solid material is placed above the sand whereas the solid material is placed below the sand in Type B mode of shear. In Type A mode, the interface friction angle is independent of unit weight and its maximum limiting value for very rough interface is the critical state angle of internal friction of sand. However, in Type B, mode of shear the interface friction angle depends on the unit weight for a given material. For very rough interfaces, the maximum limiting value of interface friction angle is the peak angle of internal friction of sand.

References

1. IS 2720-5: Method for Test for Soils: Determination of Liquid and Plastic Limit. Bureau of Indian Standards, New Delhi (1985)
2. Sridharan, A., Prakash, K.: Percussion and cone methods of determining the liquid limit of soils. Geotech. Test. J. ASTM **23**(2), 242–250 (2000)
3. Wasti, Y., Bezirci, M.H.: Determination of the consistency limits of soils by the fall cone test. Can. Geotech. J. **23**(2), 241–246 (1986)
4. Wood, D.M.: Soil Behaviour and Critical State Soil Mechanics. Cambridge University Press (1990)
5. Das, N., Sarma, B., Singh, S., Sutradhar, B.B.: Comparison in undrained shear strength between low and high liquid limit soils. Int. J. Eng. Res. Technol. **2**(1), 1–6 (2013)
6. Brackley, J.J.A.: Swell pressure and free swell in compacted clay. In: Proceedings of the 3rd International Conference on Expansive Soils, Haifa, vol. 1, pp. 169–176 (1973)

7. Abduljauwad, S.N., Al-Sulaimani, G.J.: Determination of swell potential of Al-Qatif clay. Geotech. Test. J. ASTM **16**(4), 469–484 (1993)
8. Al-Shamrani, M.A., Al-Mhaidib, A.I.: Swelling behavior under oedometric and triaxial loading conditions. Geotech. Spec. Publ. **99**, 344–360 (2000)
9. Thompson, R.W., Perko, H.A., Rethamel, W.D.: Comparison of constant volume swell pressure and oedometer load-back pressure. In: Proceedings of the 4th International Conference on Unsaturated Soils, Carefree, AZ, United States, pp. 1787–1798 (2006)
10. Soundara, B., Robinson, R.G.: Testing method and swelling pressure of clays. Int. J. Geotech. Eng. USA **3**(3), 439–444 (2009)
11. ASTM D4546-03: Standard Test Methods for One-Dimensional Swell or Settlement Potential of Cohesive Soils. ASTM International, West Conshohocken, PA, USA (2003)
12. BS 1377-5: Methods of Test for Soils for Civil Engineering Purposes: Measurement of Swelling Pressure. British Standard Institution, London (1990)
13. IS 2720-41: Measurement of Swelling Pressure of Soils. Bureau of Indian Standards, New Delhi (1977)
14. Johnson, L.D., Snethen, D.R.: Prediction of potential heave of swelling soils. Geotech. Test. J. ASTM **1**, 117–124 (1978)
15. Ali, E.F.M., Elturabi, M.A.D.: Comparison of two methods for the measurement of swelling pressure. In: Proceedings of the 5th International Conference on Expansive Soils, Adelaide, Australia, vol. 84, no. 3, pp. 72–74 (1984)
16. Sridharan, A., Rao, S.A., Sivapullaiah, V.: Swelling pressure of clays. Geotech. Test. J. ASTM **9**(1), 24–33 (1986)
17. Holtz, W.G.: Expansive Clays—Properties and Problems, vol. 54, no. 4, pp. 89–125. Colorado School of Mines (1959)
18. Mitchel, J.K.: Fundamentals of Soil Behavior, 2nd edn. Wiley, New York (1993)
19. Tovey, N.K., Wong, K.Y.: Preparation of soils and other geological materials for the electron microscope. In: Proceedings of the International Symposium on Soil Structure, vol. 1, pp. 58-66. Swedish Geotechnical Institute, Stockholm (1973)
20. Seed, H.B., Chan, C.K.: Compacted clay: structure and strength characteristics. ASCE Trans. **126**, 1344–1385 (1961)
21. Lambe, T.W.: Compacted clay: engineering behavior. ASCE Trans. **125**(1), 718–741 (1960)
22. Das, B.M.: Principles of Geotechnical Engineering. PWS Publishing Company, Boston (1998)
23. Lee, K.L.: Comparison of plane strain and triaxial tests on sand. J. Soil Mech. Found. Eng. Div. ASCE **96**, 901–923 (1970)
24. Cornforth, D.H.: Some experiments on the influence of strain conditions on the strength of sand. Geotechnique **14**, 143–167 (1964)
25. Das, B.M.: Advanced Soil Mechanics. McGraw-Hill Book Publishing Co (1983)
26. Lini Dev, K., Pillai, R.J., Robinson, R.G.: Drained angle of internal friction from direct shear and triaxial compression tests. Int. J. Geotech. Eng. **10**(3), 283–287 (2016)
27. Sivakumar, V., Mackinnon, P., Zaini, J., Cairns, P.: Effectiveness of filters in reducing consolidation time in routine laboratory testing. Geotechnique **60**, 949–956 (2010)
28. Head, K.H.: Manual of Soil Laboratory Testing, Volume 3: Effective Stress Tests. Wiley, Singapore (1998)
29. Petley, D.J.: The shear strength of soils at large strains. Ph.D. Thesis, University of London, UK (1966)
30. Lambe, T.W., Whitman, R.V.: Soil Mechanics. Wiley, USA (1979)
31. Barron, R.A.: Consolidation of fine grained soils by drain-wells. ASCE Trans. **113**, 718–724 (1948)
32. Rowe, P.W.: Measurements of the coefficient of consolidation of lacustrine clay. Geotechnique **9**(3), 107–118 (1959)
33. Sridharan, A., Prakash, K., Asha, S.R.: Consolidation behavior of clayey soils under radial drainage. Geotech. Test. J. ASTM **19**(4), 421–431 (1996)
34. Robinson, R.G.: Analysis of radial consolidation test data using a log-log method. Geotech. Test. J. ASTM **32**(2), 119–125 (2009)

35. Sridhar, G., Robinson, R.G.: Determination of radial coefficient of consolidation using log T method. Int. J. Geotech. Eng. **5**(4), 373–381 (2011)
36. Ganesalingam, D., Sivakugan, N., Read, W.: Inflection point method to estimate ch from radial consolidation tests with peripheral drain. Geotech. Test. J. ASTM **36**(5), 1–6 (2013)
37. Rowe, P.W., Barden, L.: A new consolidation cell. Geotechnique **16**(2), 162–170 (1966)
38. Juirnarongrit, T.: Constant rate of strain consolidation test with radial drainage. Master thesis, Asian Institute of Technology, Bangkok, Thailand (1996)
39. Singh, G., Hattab, T.N.: Laboratory study of efficiency of sand drains in relation to methods of installation and spacing. Geotechnique **29**(4), 395–422 (1979)
40. Al Tabbaa, A.: Consolidation with radial drainage: observed and predicted behaviour. In: Proceedings of the 13th International Conference on Soil Mechanics and Foundation Engineering, New Delhi, India, 5–10 January 1994, pp. 75–78 (1994)
41. Cao, L.F., Chang, M.F., The, C.I., Na, Y.M.: Back-calculation of consolidation parameters from field measurements at a reclamation site. Can. Geotech. J. **38**(4), 755–769 (2001)
42. Jang, I.S., Chung, C.K., Kim, M.M., Cho, S.M.: Numerical assessment on the consolidation characteristics of clays from strain holding, self-boring pressure meter test. Comput. Geotech. **30**(2), 121–140 (2003)
43. Sridhar, G., Robinson, R.G., Rajagopal, K., Radhakrishnan, R.: Comparative study on horizontal coefficient of consolidation determined using rowe and conventional consolidation cell. Int. J. Geotech. Eng. **9**(4), 388–402 (2015)
44. Sridhar, G., Robinson, R.G., Rajagopal, K.: Horizontal coefficient of consolidation from inward-and outward-flow tests. J. Proc. Inst. Civ. Eng. Ground Improv. 1–8 (2018)
45. BS1377-6: Methods of Test for Soils for Civil Engineering Purposes, Consolidation and Permeability Tests in Hydraulic Cells and with Pore Pressure Measurement. British Standard Institution, London (1990)
46. Shields, D.H., Rowe, P.W.: A radial drainage oedometer for laminated clays. J. Soil Mech. Found. Div. ASCE **19**(4), 21–431 (1965)
47. Atkinson, J.H., Evans, J.S., Ho, E.W.L.: Non-uniformity of triaxial samples due to consolidation with radial drainage. Géotechnique **35**(3), 353–355 (1985)
48. Pyrah, I.C., Smith, I.G.N., Hull, D., Tanaka, Y.: Non-uniform consolidation around vertical drains installed in soft ground. Geotechnical engineering for transportation infrastructure. In: Proceedings of the 12th European Conference on Soil Mechanics and Geotechincal Engineering, Amsterdam, Netherlands, 7–10 June 1999, pp. 1563–1569 (1999)
49. Robinson, R.G., Shilpa, D.: Equal strain consolidation of clays under radial drainage. Indian Geotech. J. **38**(2), 204–220 (2008)
50. Head, K.H.: Manual of Soil Laboratory Testing: Volume 3 Effective Stress Tests, 3rd ed. Wiley, Chichester (2006)
51. Robinson, R.G., Allam, M.M.: Effect of clay mineralogy on coefficient of consolidation. Clays Clay Miner **46**(5), 596–600 (1998)
52. Coulomb, C.A.: Essai sur une application des regles de maximis et minimis quelques problemes de statique, relatits a l'architecture. Memoires de Mathematique de l'Academie Royale de Science 7, Paris (1776)
53. Meyerhof, G.G.: An Investigation of the Bearing Capacity of the Minimis a Quelques Problems de Statique Relatif a L'architecture. Mem. Acad. Royoal Pres. Divers Say, Paris (1948)
54. Kezdi, A.: Bearing capacity of piles and pile groups. In: Proceedings 4th International Conference on Soil Mechanics and Foundation Engineering, London, vol. 2, pp. 46–51 (1957)
55. Potyondy, J.G.: Skin friction between various soils and construction materials. Geotechnique **11**(4), 39–353 (1961)
56. Rowe, P.W.: The stress-dilatancy relations for static equilibrium of an assembly of particles in contact. Proc. R. Soc. Lond. Ser. A **269**, 500–527 (1962)
57. Silberman, J.O.: Some factors affecting the frictional resistance of piles driven in cohesionless soils. Thesis presented to Cornell University, Itaca, NY (1961); in partial fulfillment of requirements for the degree of Master of Science, vide Broms (1963)

58. Subba Rao, K.S., Allam, M.M., Robinson, R.G.: Interfacial friction between cohesionless soils and solid surfaces- a review. Indian Geotech. J. **31**(2), 107–137 (2001)
59. Noorany, I.: Side friction of piles in calcareous sands. In: Proceedings of the 11th International Conference on Soil Mechanics and Foundation Engineering, San Fransisco, vol. 3, pp. 1611–1614 (1985)
60. Yoshimi, Y., Kishida, H.: Friction between sand and metal surface. In: Proceedings of the 10th International Conference on Soil Mechanics and Foundation Engineering, Stockholm, vol. 1, pp. 831–834 (1981)
61. Broms, B.B.: Discussion on bearing capacity of piles in cohesionless soils. J. Soil Mech. Found. Div. ASCE **89**(6), 125–126 (1963)
62. Acar, Y.B., Durguroglu, H.T., Tumay, M.T.: Interface properties of sand. J. Geotech. Eng. ASCE **108**(4), 648–654 (1982)
63. Uesugi, M., Kishida, H.: Influential factors of friction between steel and dry sands. Soils Found. **26**(2), 33–46 (1986)
64. Uesugi, M., Kishida, H.: Frictional resistance at yield between dry sand and mild steel. Soils Found. **26**(4), 139–149 (1986)
65. Panchanathan, S., Ramaswamy, S.V.: Skin friction between sand and construction materials. J. Indian Natl. Soc. Soil Mech. Found. Eng. **3**(4), 325–336 (1964)
66. Kisheda, H., Uesugi, M.: Tests of interface between sand and steel in the simple shear apparatus. Geotechnique **37**(1), 45–52 (1987)
67. Uesugi, M., Kishida, H., Uchikawa, Y.: Friction between dry sand and concrete under monotonic and repeated loading. Soils Found. **30**(1), 125–128 (1990)
68. Everton, S.J.: Experimental study of frictional shearing resistance between non-cohesive soils and construction materials. M.Sc. (Engg) Dissertation, University of London (Imperial College) (1991)
69. Jardine, R.J., Lehane, B.M., Everston, S.J.: Friction coefficients for piles in sands and silts. In: Proceedings, Conference on Offshore Site Investigation and Foundation Behaviour, London, vol. 28, pp. 661–677 (1993)
70. Desai, C., Drumm, C., Zaman, M.: Cyclic testing and modeling of interfaces. J. Geotech. Eng. ASCE **111**(6), 793–815 (1985)
71. O'Rourke, T.D., Druschel, S.J., Netravalli, A.N.: Shear strength characteristics of sand-polymer interfaces. J. Geotech. Eng. ASCE **116**(3), 451–469 (1990)
72. Subba Rao, K.S., Allam, M.M., Robinson, R.G.: A note on the choice of interfacial friction angle. Geotech. Eng. Lond. **119**(2), 123–128 (1996)
73. Robinson, R.G.: Some studies on the interfacial friction between soils and solid surfaces. A thesis submitted for the degree of Doctor of Philosophy in Indian Institute of Science, Bangalore (1998)

Chapter 4
Estimation of Dredged Fill Volume Based on Self-weight Consolidation Characteristics of Dredged Soil

Eun Chul Shin, Young Jin Park, Jeong Ku Kang and Jeong Jun Park

4.1 Introduction

Dredged soil is widely distributed in the coastal area adjacent to land, so it is easy to get the volume. However, it does not settle immediately, because it flows into the state of slurry, which is a mixture of fine granular soil and water, and is sent out in a high water content state. After settling and sedimentation, it then shows self-weight consolidation settlement behavior. As a result, this makes it difficult to accurately predict the amount of landfilled dredge landfill, and sometimes there are problems in design and construction.

According to the Korean Geotechnical Society [1], the most important design factor is the proper calculation of dredging volume in landfill, and this can be estimated by the volume and loss rate of the landfill. According to the port design standards of the Ministry of Oceans and Fisheries [2], it is most desirable to evaluate the loss rate in dredging landfill based on experiments. If not, the retention ratio is proposed to calculate the dredging volume according to particle size. However, it is difficult to calculate accurate dredging quantities with a wide range of settings; therefore, it is necessary to apply a reasonable method to estimate the dredging volume properly.

In the case of dredged clay with high water content, consolidation occurs due to self-weight, and the length of drainage is too long, so that it is difficult to analyze by the theory of Terzaghi [3].

E. C. Shin (✉) · J. K. Kang · J. J. Park
Department of Civil and Environmental Engineering, Incheon National University, Academy-ro 119, Incheon 22012, Republic of Korea
e-mail: ecshin@inu.ac.kr

Y. J. Park
Kyonggi University, Gwanggyosan-ro 154-42, Yeongtong-gu, Suwon, Gyeonggi 16227, Republic of Korea

M. Latha G. (ed.), *Frontiers in Geotechnical Engineering*, Developments in Geotechnical Engineering, https://doi.org/10.1007/978-981-13-5871-5_4

The precipitation and self-weight consolidation behavior of dredged soil were suggested by Mikasa [4], Gibson et al. [5], and Shin [6] concluded that the laboratory results based on the finite deformation theory are in good agreement with the field measurements.

Kynch [7] proposed the characteristics of interference and interfacial settling by mathematically identifying the settling of particles. Imai [8] conducted precipitation experiments on Japanese dredged clay, and studied the interaction between settling and consolidation. Been and Sills [9] identified the initial void ratio of dredged soil through a series of experiments.

Comprehensive theory on the sedimentation and consolidation process of soil particles was developed by Pane and Schiffman [10]. In particular, Yano [11] proposed the empirical property relation, which is widely used in Korea.

In Korea, Jeong and Yoo [12] investigated the settlement consolidation behavior of dredged soil through Mikasa theory and laboratory tests, and experimental studies on the settling characteristics of marine clay were performed mainly by Kim [13], Song et al. [14], Lee et al. [15, 16]. Jee and Lee [17] compared the results of numerical analysis using the PSDDF program and model experiment.

Yoo et al. [18] conducted a numerical analysis using a centrifugal model test, analyzing the constitutional relationship of the dredged landfill with high water content. Kim et al. [19] proposed a method for estimating the amount of dredged landfill by conducting a study on the determination of the loss rate of dredged soil.

Recently, Lee and Lee [20] and Lee and Choi [21] analyzed the characteristics of the settling consolidation behavior of dredged soil in landfill area.

The purpose of this study is to investigate the method of calculating the amount of dredged landfill by self-weight consolidation experiment on the dredged soil of Incheon harbor used for the Songdo 11 district landfill construction project. The amount of dredged landfill was estimated through the basic physical property test of dredged soil that occurred when dredging the port area. The relationship between the loss rate and the amount of dredged landfill generated in the process of pump transportation to Songdo area was analyzed.

4.2 Consolidation Theory Considering Self-weight Consolidation

4.2.1 Estimation of Landfill Volume

The methods of landfilling are a method of pumping dredged soil to a landfill by pump type dredger, and a method of dredging by using bucket dredger, dipper, backhoe dredger, grab dredger, etc., and landfilling with soil from inland.

When dredging with pump dredging, in the process of excavating the dredged soil and transporting it to the landfill through the discharge pipe, the soil may float, and some of it may be lost outside of the planned landfill. At this time, the accumulation in the planned area without loss is called the retention amount. The percentage of retention amount is called the retention ratio. Estimating the dredging volume based on the retention ratio is based on Eq. 4.1:

$$V = \frac{V_0}{P} \tag{4.1}$$

where, V is the volume of the landfilled soil (m^3), V_0 is the volume of the total landfill (m^3) considering the settlement rate and the shrinkage rate, and P is the average retention rate (%) of the landfill by the pump dredger.

When designing a landfill site, the retention ratio is calculated by the average after self-weight consolidation test or calculating the retention ratio according to the grain size and species of the dredged soil. Table 4.1 shows the shrinkage rates according to soil types.

The settlement rate is expressed by the settlement amount for the ground layer thickness and settlement value by Terzaghi's one-dimensional consolidation theory.

The retention rate of the dredged landfill means the weight ratio of the landfill to the dredged soil. There are various factors for the retention rate, such as the grain size, location and height of extra banking, distance from the drainage hole, and the landfill area.

According to the design standards of harbor and fishing ports [2] when the dredging rate cannot be determined experimentally, as shown in Table 4.2, it is stipulated that the retention ratio can be applied to each soil type. But there are limitations to calculating the appropriate amount of dredging, because of the wide range of applications.

Table 4.1 Shrinkage ratio with soil type

Soil type	Shrinkage ratio (% with depth of layer)
Clay soil layer	Less than 20
Clay and sandy soil layer (mixed soil layer)	10–15
Granular soil layer	5

Table 4.2 Retention ratio with soil type

Soil type	Retention ratio (%)
Clay or clayey silt	Less than 70
Sand or sandy silt	70–95
Gravel	95–100

4.2.2 Settlement Consolidation Characteristics of Dredged Soils

Self-weight consolidation settling curves are classified into sections of flocculation, settling with increasing settling velocity, and consolidation settling. At this time, the starting point of consolidation settlement is generally based on the inflection point at which the settling velocity slows down.

The time of self-weight consolidation is the time corresponding to the intersection point of the tangent line in the settlement section and the consolidation settling step. In the time-settling curve, the point that meets the tangent line at the bottom curve becomes the end point of settling, and this end point can be estimated as the self-weight consolidation time.

Figure 4.1 shows the method for determining the starting point of self-weight consolidation through the separation of the settling and consolidation step.

When a sample having the same initial water content and different contents of fine particles has settled and consolidated, the dredged soil, which has the form of section settlement, enters the self-weight consolidation step according to the passage of time through the sedimentation and accumulation steps. In the relation between the interface height and sedimentation time, a linear relationship Eq. 4.2 can be obtained, and the interface can be obtained as Eq. 4.3:

$$\log H = \log h_1 - C_S \cdot \log t \tag{4.2}$$

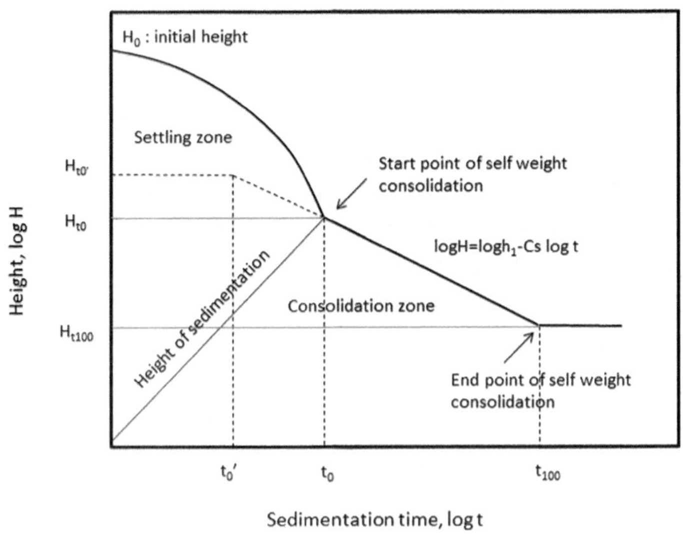

Fig. 4.1 Determination of the starting point of self-weight consolidation

$$H = h_1 \cdot t^{-C_S} \tag{4.3}$$

where, h_1 is a constant indicating the height at which $H = h_1$ at $t = 1$ (min), and C_S is the settling consolidation coefficient.

The beginning point of the straight line is the start point of self-weight consolidation, while the end of the straight line is defined as the self-weighted consolidation end point. The coefficient of sedimentation self-weight consolidation can be expressed as the slope between the two points, as expressed in Eq. 4.4:

$$C_s = \frac{\log(\mathrm{H}_{t0}/\mathrm{H})}{\log(t/t_0)} \tag{4.4}$$

where, H_{t0} is the interface height at time t_0.

In general, the relation between the real soil height and the interface height is given by:

$$H_s = \frac{H}{1 + G_s/G_w \cdot \bar{w}} = \frac{H}{1 + \bar{e}} \tag{4.5}$$

where, H_0 is the initial interfacial height, G_s is the specific gravity of the soil, G_w is the specific gravity of water, $\bar{w}$ is the average water content of the dredged soil, and $\bar{e}$ is the average void ratio of the dredged soil.

The dredged soil with the same initial water content and different heights were settled, then assuming that the consolidation settling start time shown in each settling curve is expressed as t_0, and the time when entering the creep behavior is t_{100}, the self-weight consolidation formulae in the consolidation settling zone between the time ($t = t_0$–t_{100}) and the interface height (H) can be obtained.

Figure 4.2 shows the height of the real soil height with the interfacial height at time (t_0, t_{100}):

When the self-weight consolidation start and end point according to the water content ratio are indicated by the real soil height of the logarithmic scale, a linear relationship is established. From this, it is possible to obtain the real soil height for the initial surface height of the landfill, as shown in Eq. 4.6:

$$\log H = \log h_2 + C_k \log H_s \tag{4.6}$$

where, h_2 is the surface height, and is the intercept at the real soil height $H_s = 1$, and C_k represents the slope of the surface high-real soil height graph.

The landfill height is estimated using the real soil height calculated from the sedimentation consolidation test. It is possible to finally calculate the volume change ratio by using the relationship between the void ratio and the water content ratio.

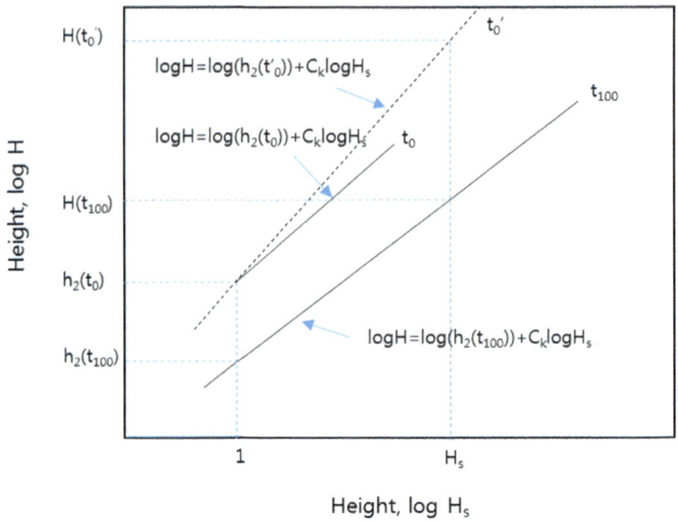

Fig. 4.2 Height of dredged soil with H_s at $t = 0$ and $t = 100$

4.3 Laboratory Model Experiments

4.3.1 Physical Properties Test of Dredged Soil and Residence Soil

Samples of the dredged area were collected in the first and third container wharfs, and entry route extension areas. The natural water content was (51.1–97.9)%, liquid limit was (25.6–28.1)%, and liquid index was 4.9 or more. This shows the characteristics of the unconsolidated clay.

The particle size distribution was in the range of (69.3–87)% on the #200 sieving percentage, and it was classified as clay silt or silt belonging to fine-grained soil. Table 4.3 shows the experimental results of the samples taken from each area.

The physical properties of soil samples of the landfill site were obtained from the three sites. The natural water content was (31.5–36.4)%, the liquid limit was (34.5–48.3)%, and the liquidity index was (0.7–1.0). These values show normal consolidation soil characteristics. The particle size distribution shows that the #200 sieve passage percentage is 95% or more, since the PI value exceeds the plasticity A-line. All of them were classified as clay (CL). Table 4.4 shows the test results of the landfill soil.

The 60% passing diameter (D_{60}) of the dredged sample was 0.05 mm, the 30% passing 155 mm diameter (D_{30}) was 0.015 mm, and the 10% passing diameter (D_{10}) was 0.002 mm. The coefficient of uniformity was 6 or more, while the curvature coefficient was about 2.25, which show that the landfill soil was composed of silt soil with relatively good grain size.

Table 4.3 Physical property of the dredged soil

Sample no.	w_n (%)	G_s	Atterberg limits (%)		Grain size distribution %, finer than				USCS
			LL	PI	0.002 (mm)	0.075 (mm)	0.425 (mm)	2.00 (mm)	
A	51.1	2.574	25.6	6.4	10.3	66.9	100	100	CL-ML
B	78.3	2.562	28.1	5.5	9.2	73.9	100	100	ML
C	97.9	2.564	26.4	4.1	8.3	50.85	99.4	100	ML

Table 4.4 Test results of the landfill soil

Sample no.	w_n (%)	G_s	Atterberg limits (%)		Grain size distribution %, finer than				USCS
			LL	PI	0.002 (mm)	0.075 (mm)	0.425 (mm)	2.00 (mm)	
1	34.3	2.68	40	20	12.3	92.4	99.3	100	CL
2	31.5	2.671	48.3	25.4	17.6	98.3	100	100	CL
3	36.4	2.681	34.5	25.2	18.3	95.8	100	100	CL

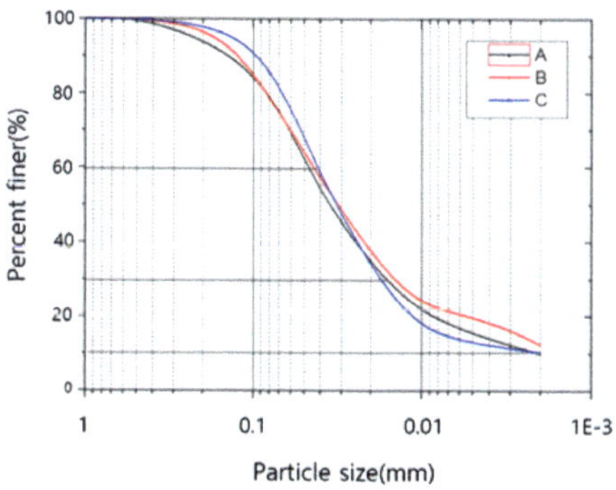

(a) Dredged soil

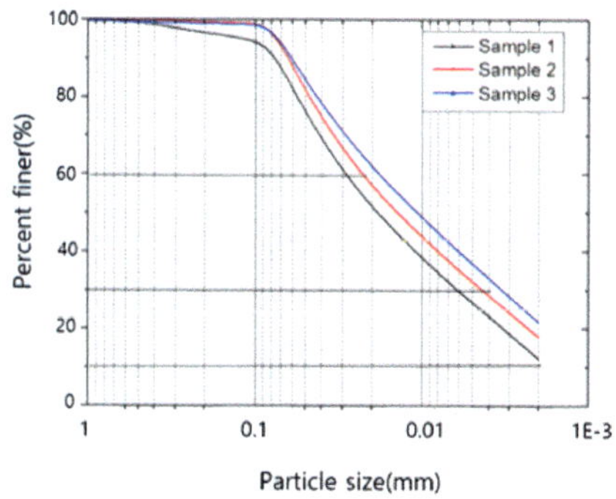

(b) Ground soil

Fig. 4.3 Cumulative particle distribution curve

In addition, the characteristics of the soil constituting the land fill area were 0.02 mm for D_{60}, 0.004 mm for D_{30}, 0.001 mm for D_{10}, with uniformity coefficient of 6 or more, and curvature coefficient of 1 or less. Figure 4.3 shows each soil characteristic:

4.3.2 Sedimentation Consolidation Test of Dredged Soil

The sedimentation and self-weight consolidation characteristics of the dredged soil used in the Songdo 11 district landfill in Incheon were analyzed.

For the sedimentation consolidation test, a cylindrical mold with a diameter of 150 mm and a height of 1,000 mm was produced according to the Yano test method, as shown in Fig. 4.4.

The coarse sand of #40 or more was excluded based on the grain size analysis of each dredged area sampled, and the mold was charged with each water content condition. Air was injected from the bottom of the mold, and sufficiently agitated, before performing the sedimentation experiments. The sedimentation consolidation characteristics were analyzed according to water content, fine particle content, and depth of sedimentation.

The sedimentation self-weight consolidation tests of dredged soils were carried out at (400, 700, and 1,000) % of the water content of samples. After the samples were prepared according to the water content ratio, the initial concentration of the sample was measured using a measuring cylinder to calculate the initial void ratio.

The experimental conditions were set by remeasuring the water content of the specimens dumped into the mold. In order to analyze the sedimentation-self weight consolidation characteristics according to the content of fine-grained soil, samples of (50, 65, and 80)%, respectively, of sediment of #200 or more were sedimented, and the consolidation coefficient C_s was calculated and compared.

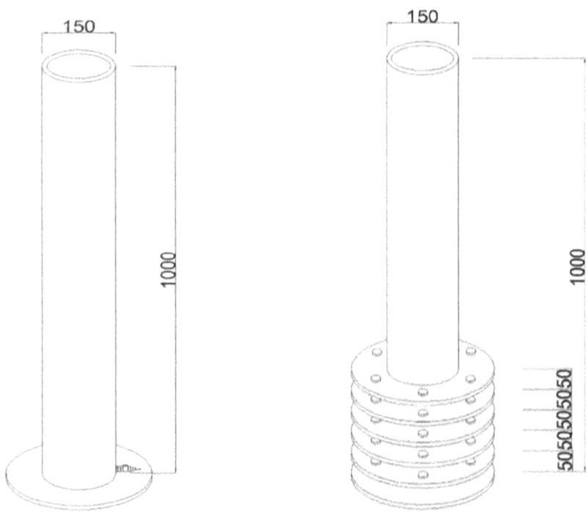

(a) Test for high water content (b) Test for low water content

Fig. 4.4 Instrument for self-weight consolidation test

Finally, the sample with the same water content was settled at the height of (60 and 80) cm, and the height of boundary surface and the height of real soil volume correlation at the self-weight consolidation start time t_0 and end time t_{100} were shown. Then, the linear relationship between the start and the end point was made. Figure 4.5 shows the procedure of the sedimentation consolidation test according to the depth of sedimentation:

4.4 Analysis for the Results of the Sedimentation-Self Weight Consolidation Test

4.4.1 Calculation of Sedimentation Consolidation Coefficient

In order to predict the amount of self-weight consolidation settlement and volume change after completion of dumping, the slope of the self-weight consolidation behavior should be checked by connecting the start and end points of the linear section. The starting point is the inflection point at which the sedimentation—consolidation curve slows down. According to the Yano method, the point where the straight line starts from the sedimentation—consolidation curve is taken as the starting point. After the consolidation is completed, the end point is set as the end point of self-weight consolidation, and the slope and intercept are calculated.

(a) Height of dredged soil, 80 cm (b) Height of dredged soil, 60 cm

Fig. 4.5 Self weight consolidation test with different height

For dredged soil samples of four areas of Incheon port, the water content was adjusted to (400, 700, and 1,000)%. Under the condition that the content of the fine grained soil is the same, as the water content increases, the sedimentation-consolidation factor value decreases. The higher the water content ratio, the shorter the self-weight consolidation time. Under the same conditions of water content, as the content of the clay is increased to (50, 65, and 80)%, the initiation of sedimentation -consolidation tends to be delayed.

Figure 4.6 shows the analysis results of the start and end points on the self-weight consolidation test for each section sample:

The content of the clay was adjusted to (50, 65, and 80)%, and as a result of estimating the sedimentation-consolidation coefficient Cs, the sedimentation consolidation coefficients for (50, 65, and 80)% were (0.071–0.044), (0.055–0.041), and (0.042–0.033), respectively. The consolidation coefficient was the largest at 50%. In the 400% water content, on increasing the amount of fine-grained soil, the sedimentation-consolidation coefficient decreased linearly. In the case of (700 and 1,000)%, as the content of fine-grained soil increased, the sedimentation-consolidation coefficient shows a decreasing tendency in the curve shape.

Figure 4.7 shows the sedimentation-consolidation coefficient, C_s variation for each water content.

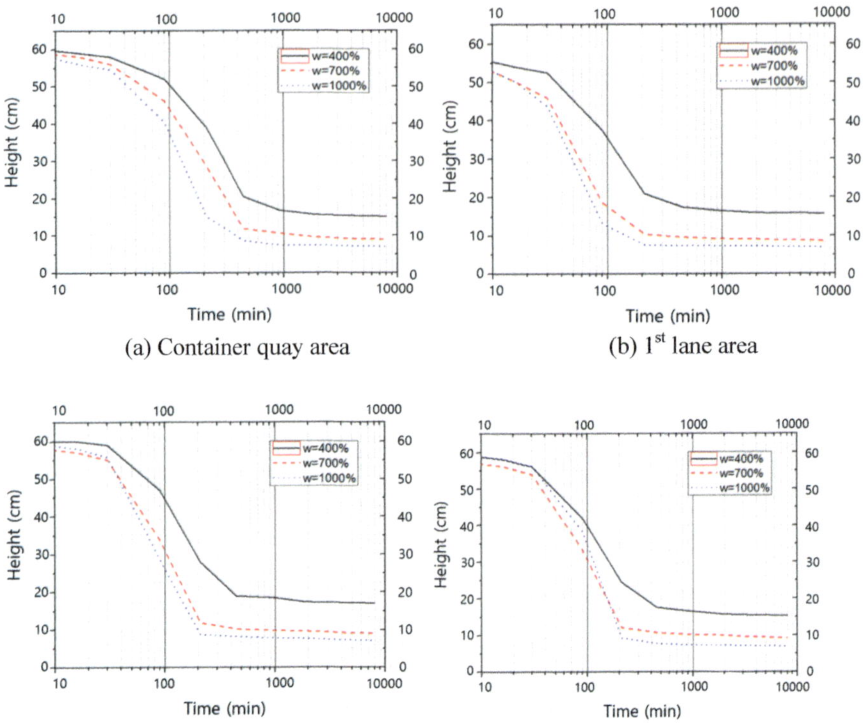

(a) Container quay area (b) 1st lane area (c) 3rd lane area (d) Expansion area for entrance

Fig. 4.6 Consolidation of dredged soil with water content

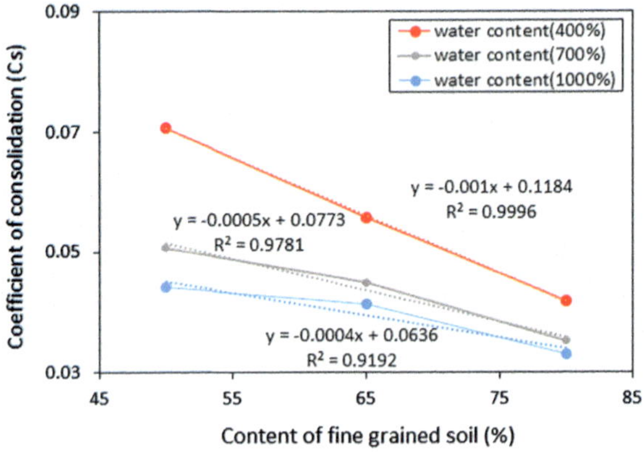

Fig. 4.7 Coefficient of consolidation with fine-grained soil content

4.4.2 *Relation Between the Height of Boundary Surface and the Height of Real Soil Volume*

The relationship ($\log H_s - \log H$) between the actual soil volume height and the surface height is linear, and the actual soil volume height can be calculated using the surface height between the start point (t_0) and the end point (t_{100}) of the self-weight consolidation test and the average void ratio.

If the water content is 400% at the start point of self-weight consolidation, A was 0.8041 and B was 0.9459, while at the end point, A was 0.6884 and B was 1.0026.

When the water content ratio is 1,000%, at the start point of self-weight consolidation, A was 0.8395 and B was 0.9476, while at the end point, A was 0.7419 and B was 0.9725.

Applying these values in Eq. 4.6, it is possible to derive the equation between the height of real soil volume and the height of boundary surface.

Figure 4.8 graphs the relationship between the height of real soil volume and the height of boundary surface to calculate the coefficient.

In case of water content of (400 and 80)% of fine-grained soil content, the relationship between the start and the end point of consolidation is as shown in Eq. 4.7:

$$\begin{aligned} t_0 &: \log H = 0.8041 + 0.9459 \log H_s \\ t_{100} &: \log H = 0.6884 + 1.0026 \log H_s \end{aligned} \tag{4.7}$$

Depending on the water content and fine-grained soil content of dredged soil, Jang [22] presented the relationship between the surface height and the actual soil height through various sedimentation experiments.

According to the empirical formulae in this study, when the initial water content was 400% and the fine-grained soil content was 80%, the sedimentation-consolidation coefficient value was 0.03297 cm/min.

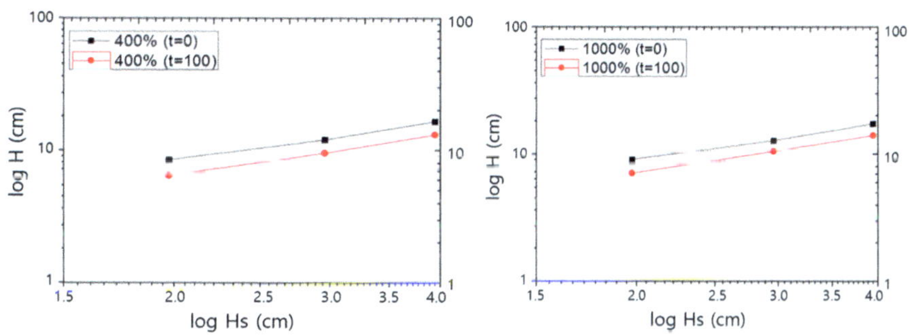

(a) Water content 400 % (b) Water content 1,000 %

Fig. 4.8 Relationship of H versus H_s with water content

4.4.3 Prediction of Volume Change Ratio

Using the surface height and A and B coefficients, we predicted the ratio of change in landfill height and volume over time. The volume change of dredged soil with time can be obtained by using the specific gravity and water content. The dredged soil was applied with the initial specific gravity of 2.7, the initial void ratio of 4.485 after completion of landfill, and the initial landfill height of 10.194 m.

H1, the initial value of the surface height at the time when the dredging dumping is completed ($t = 1$ min), is calculated by Eq. 4.2. After dredging and landfill, log H was calculated over time, and the landfill height change is estimated by Eq. 4.7.

Table 4.5 shows the actual soil volume height, landfill height, void ratio, water content, and volume change ratio after settlement. As a result, it can be seen that it takes about 2,920 days or more to change from the initial volumetric change rate of 1.50 to the volumetric change rate of 1.06. In addition, at the end of two years after the landfill has been completed, the volume change rate is 1.2; at this point, the volume of all the dredged landfill was reduced to 83%.

Figure 4.9 shows the predicted behavior for water content, void ratio, and volume change with elapsed time. According to the chart, it takes about 3,000 days or more to reach a soil condition with a water content of 110%, a void ratio of 3.1, and a volumetric change rate of 1.1.

Table 4.5 Height, void ratio, water content and volume change with elapsed time

Elapsed time (day)	log H	Height of reclamation (m)	Void ratio	Water content (%)	Volume change ratio
1	1.008	10.194	4.485	168.61	1.50
30	0.995	9.883	4.318	162.33	1.45
60	0.985	9.653	4.194	157.67	1.42
90	0.976	9.472	4.097	154.02	1.39
120	0.970	9.325	4.018	151.05	1.37
150	0.964	9.201	3.951	148.53	1.35
180	0.959	9.093	3.893	146.35	1.34
210	0.954	8.999	3.842	144.44	1.32
240	0.950	8.915	3.797	142.74	1.31
365	0.936	8.639	3.649	137.18	1.27
730	0.912	8.163	3.393	127.56	1.20
1,460	0.886	7.683	3.135	117.86	1.13
2,920	0.858	7.216	2.883	108.38	1.06

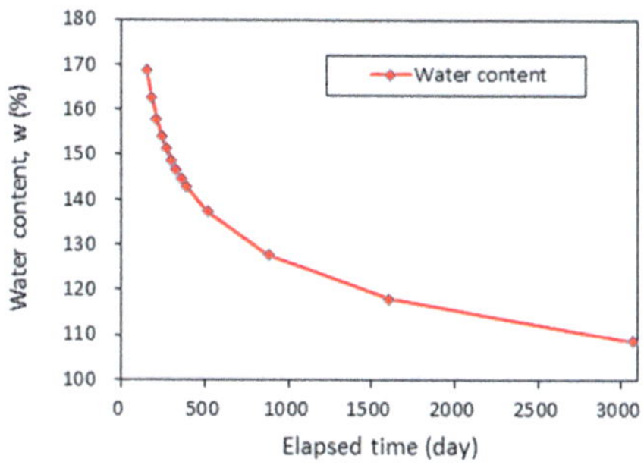

(a) Variation of water content

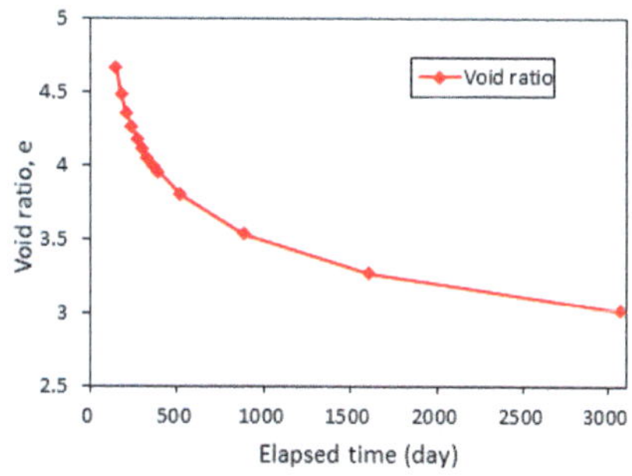

(b) Variation of void ratio

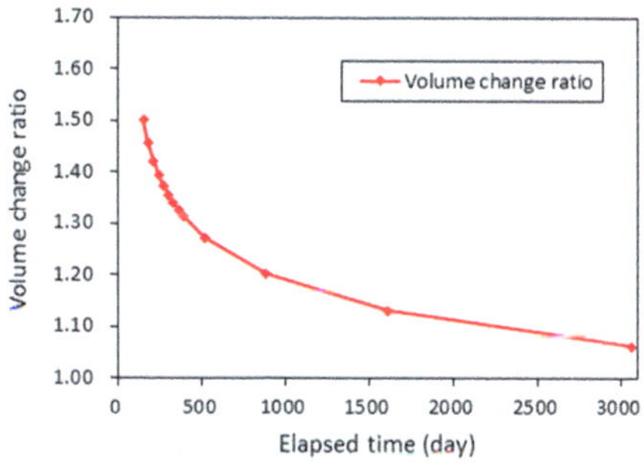

(c) Variation of volume change

Fig. 4.9 Prediction of water content, void ratio, and volume change with elapsed time

4.5 Prediction of Dredging Landfill Volume Using the Numerical Method

The program applied in this study is a finite difference analysis program for the settlement of dredged soil, and the primary consolidation, secondary compression, and desiccation of dredged fill were programmed.

Based on the mathematical model proposed by Gibson et al. [5], the concept of c_α/c_c proposed by Mesri and Godlewski [23] is introduced. The ratio of the compression index (c_α) in the second consolidation process and the compression index (c_c) in the primary consolidation process are constant under a certain pressure.

Figure 4.10 shows the relationship of c_α/c_c between (0.01 and 0.05) in the clay soil of the dredging area:

Fig. 4.10 Second consolidation coefficient of dredged soil in seaport harbor area of the west coast

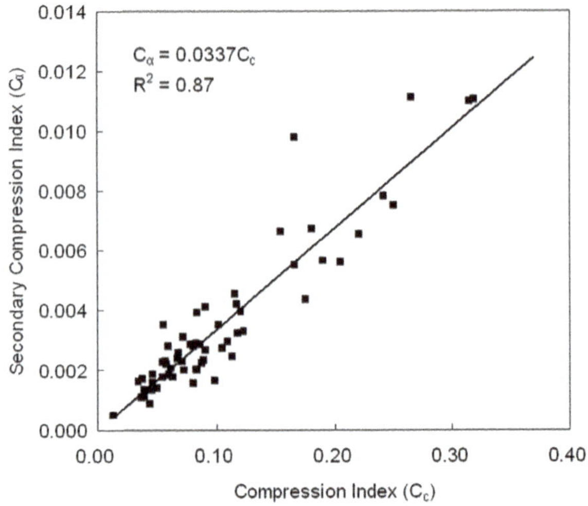

Fig. 4.11 Permeability according to void ratio

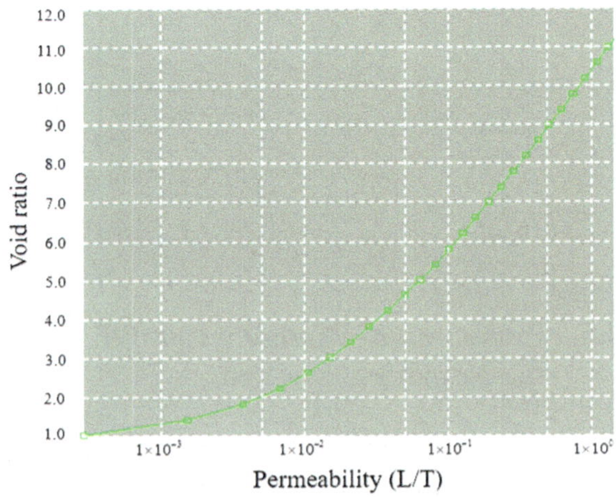

4.5.1 Permeability and Stress Characteristics According to Void Ratio

For self-weight consolidation analysis of dredged landfill, the important factors are the water content, specific gravity, elapsed time, and initial void ratio. Figure 4.11 shows the permeability characteristics according to the void ratio of dredged soil.

The plastic index, desiccation limit, shrinkage index, saturation degree of desiccation limit, desiccation effect, and maximum drying depth are necessary factors for effective stress calculation. Compression index and secondary consolidation coefficient ratios should be considered for long-term settlement.

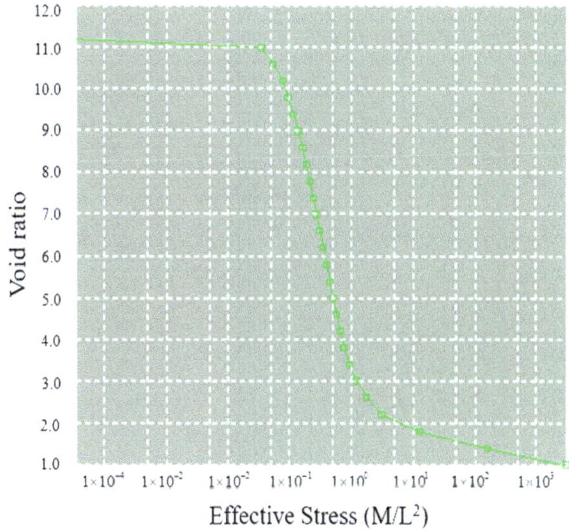

Fig. 4.12 Stress characteristics according to void ratio

Figure 4.12 shows the effective stress characteristics according to the void ratio of dredged soil.

4.5.2 Dredging Landfill Height and Void Ratio Change with Time

The dredged landfill shows consolidation characteristics in which large deformation of 50% or more occurs by self weight consolidation. Therefore, the differential analysis of the second differential equation is applied, rather than the micro strain analysis.

Figure 4.13 shows the result according to the step-by-step dumping of dredged soil in the port area of the west coast. The change of the void ratio is the main influence on the change of the dredging volume. Figure 4.14 shows the change in void ratio over time of the dredged soil.

The initial void ratio decreased from (4.485 to 2.0) after 4,900 days in the self-weight consolidation step.

4.5.3 Change of Void Ratio and Effective Stress with Time

The pore water pressure at the beginning of dredging forms a triangular distribution like water pressure and it can be confirmed that the pore water pressure is gradually dissipated over time.

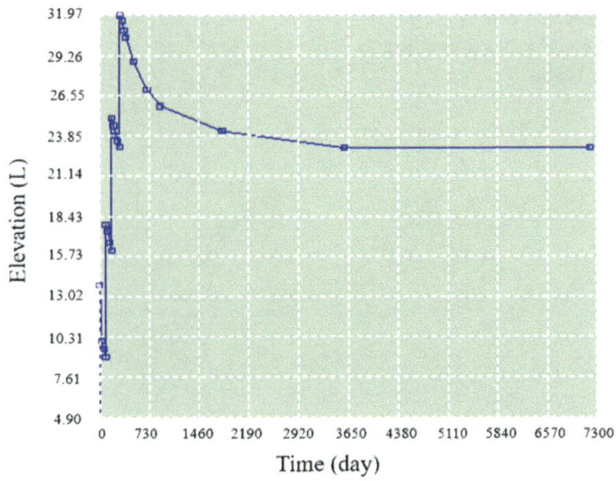

Fig. 4.13 Change of dredged landfill height with time

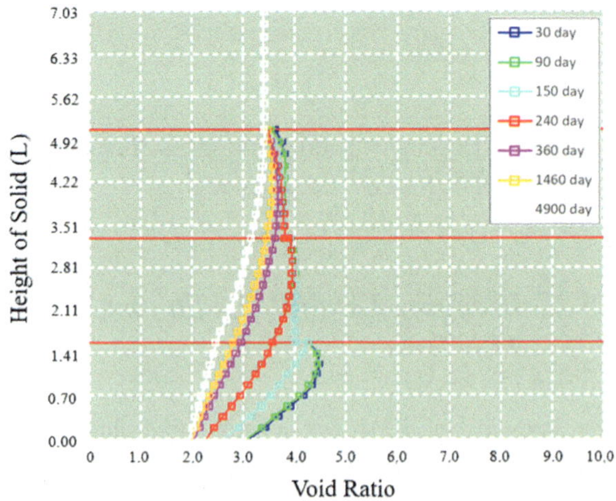

Fig. 4.14 Void ratio change with Time

Figure 4.15 shows the change in pore water pressure over time. The initial excess pore water pressure tends to increase toward the deepest part depending on the surface layer drainage, and as time passes, the dissipation process of the excess pore water pressure by the lower drainage can be confirmed.

Effective stress is not initially expressed. But the effective stress tends to gradually increase as the drainage of pore water is made by upper drainage, self-weight consolidation, consolidation settlement, etc. Figure 4.16 shows the increase trend of effective stress over time after the dredging of 4 stages.

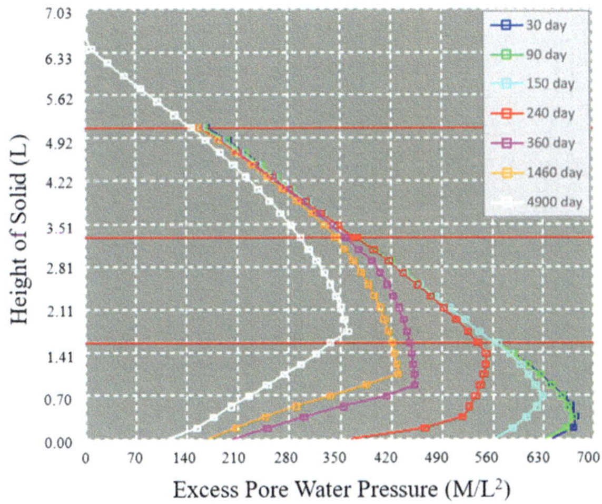

Fig. 4.15 Change of pore pressure with time

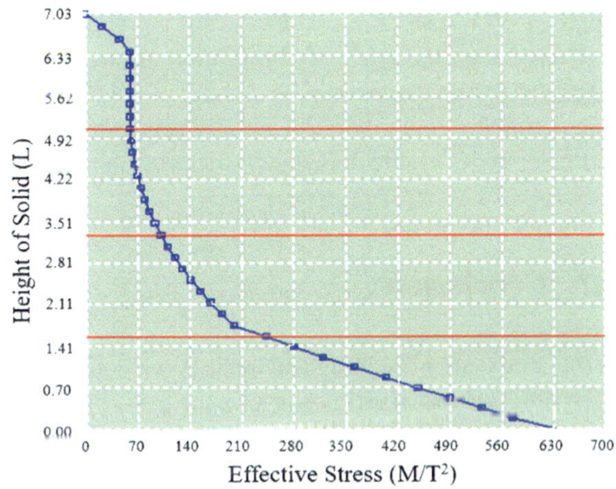

Fig. 4.16 Change of effective stress with time

4.6 Propose Calculation Method of the Dredged Landfill Volume

Regarding dredging landfill, factors of soil investigation of ground, soil penetration, negative pressure and desiccation, self-weight consolidation mechanism, numerical analysis modeling, dumping method, etc. should be considered.

In this study, for analysis of the mechanism of self-weight consolidation, estimation of the start & end point and volume ratio of self-weight consolidation, and experimental studies and numerical methods for the retention rate were conducted. The settlement rate of foundation ground was considered for the estimation of dredged soil volume.

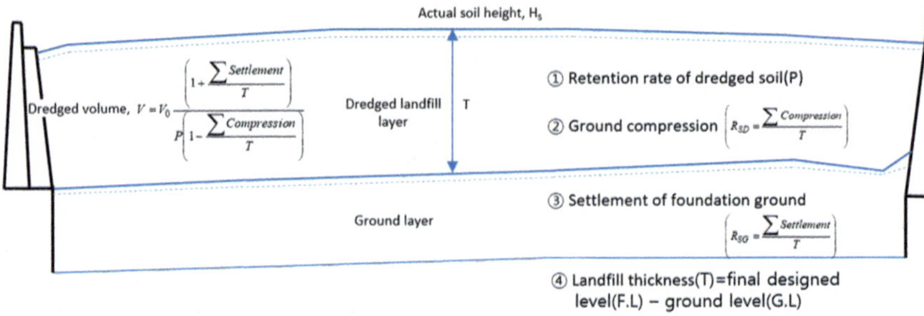

Fig. 4.17 Determination of landfill volume using the retention rate, compression rate, and settling rate

In case analysis, the improvement for calculation of dredging quantity is suggested. The amount of dredged soil is calculated by calculating the retention rate of dredged landfill, calculating the compression rate, and calculating the settlement rate of the foundation ground.

Considering the effects of infiltration, desiccation, and dumping of dredged soil, we proceed with estimating the amount of landfill. Figure 4.17 shows the concept of calculating the amount of dredged soil using the retention rate, compression rate and settlement rate. Figure 4.18 shows the procedure for predicting the dredged amount.

4.7 Conclusion

In this study, after dredging and landfill, the volume change was analyzed in the west coast dredged soil. For this purpose, the basic physical properties of the dredged and clay samples collected from the dredging area, and sedimentation and self-weight consolidation tests were carried out according to the contents of fine-grained soils.

Based on the results of the laboratory tests, and prediction of the future consolidation characteristics of the landfill, we revised the existing retention rate calculation standard.

The main conclusions of this study are as follows:

1. When calculating the landfill amount, according to the standard of construction, the range of retention ratio on sand or sandy silt is wide, from (70 to 95)%. It is not possible to provide a proper calculation of the dredged amount. In order to compensate for the fine changes of the landfill soil and comprehensive range of the existing, it was proposed that the retention rate be calculated according to the content of fine-grained soil.
2. As a result of the basic physical property tests of dredged samples and ground clay samples collected at the site, most dredged samples were classified as ML. After sedimentation of dredged soils containing (50, 65, and 80)% of fine-grained soil with water contents of (400, 700, and 1,000)%, the volumetric changes of dredged

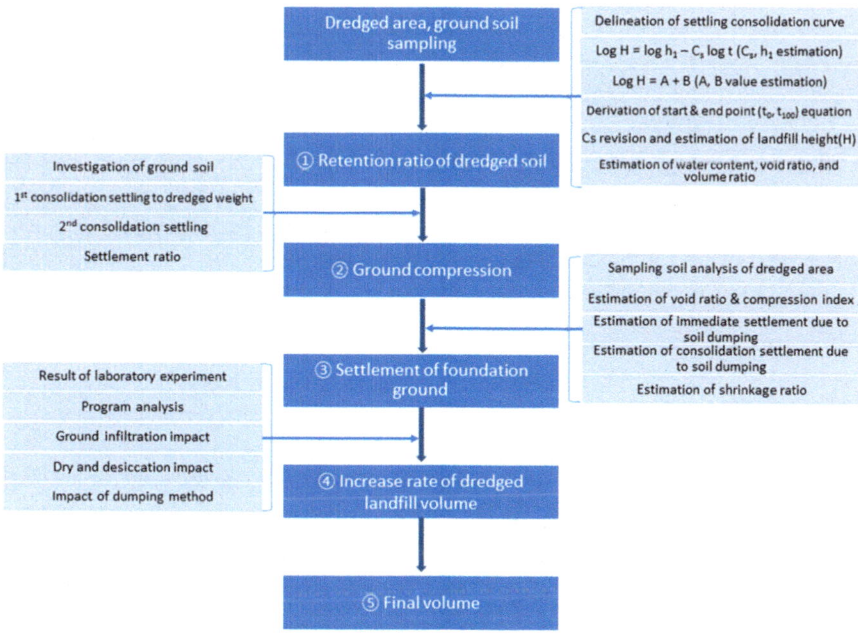

Fig. 4.18 Procedures for estimating the dredged soil volume

soil with time were analyzed. Analyzing the actual soil volume of dredged landfill soil, at the present time, two years after the dredging landfill, the retention ratio is lower than that of the existing design method.

3. In the sedimentation test using dredged soil, as the water content was lower, the sedimentation consolidation coefficient was higher; and as the water content was higher, the sedimentation consolidation time was faster. In contrast, when the content of fine-grained soil was increased to (50, 65, and 80)% under the condition of the same water content ratio, the starting point of sedimentation-self weight consolidation tended to be late. The reason for this is firstly, as the amount of coarse dredged soils is increased, consolidation sedimentation is terminated early, and secondly, as the fine-grained soil increases, the sedimentation consolidation coefficient C_0 becomes smaller, due to floating phenomena.
4. From the relationship between the sedimentation—self weight consolidation coefficient C_S and the surface height and the actual soil volume, the coefficients A and B were calculated by the total dumping method. After the completion of dumping, the water content, void ratio, and volumetric change ratio with time were predicted. It is possible to calculate the reasonable dredged amount by applying the retention ratio according to the content of fine-grained soil in various conditions.
5. Numerical analysis on the volume change of the dredged landfill based on the finite deformation theory shows that it behaves similarly to the volume change

obtained by self-weight consolidation experiment. Therefore, it is possible to design an economical and effective dredging landfill by carrying out sedimentation consolidation experiments under various conditions, and setting up the influential factors.

Although a reasonable method of estimating the retention ratio by the sedimentation-consolidation test method has already been proposed, various applications of the retention ratio estimation method are lacking. In order to improve estimation, it is expected that reliable data accumulation should be continued through additional experimental study on the water content and fine-grained soil content.

References

1. Korean Geotechnical Society: Dredging Landfill: Chapter 4 Dredged Landfill, Gumi West Building, Seoul, pp. 186–295 (2005)
2. Ministry of Oceans and Fisheries: Commentary on Design Criteria for Harbor and Fishing Ports (1st), 11-1192000-000184-14, pp. 725–752 (2014)
3. Terzaghi, K.: Erdbaumechanik auf Bodenphysikalischer Grundlager. Deuticke, Vienna (1925)
4. Mikasa, M.: The Consolidation of Soft Clay—A New Consolidation Theory and Its Application, Kajima Institution Publishing, pp. 21–26 (1963)
5. Gibson, R.E., England, G.L., Hussey, M.J.L.: The theory of one-dimensional consolidation of saturated clays, I: finite non-linear consolidation of thin homo-geneous layers. Geotechnique **17**, 261–273 (1967)
6. Shin, E.C.: Consolidation of Soft Clay—A New Theory and Application, pp. 55–67. Future Technology, Seoul (2009)
7. Kynch, G.J.: A theory of sedimentation. Trans. Faraday Soc. **48**, 166–176 (1952)
8. Imai, G.: Setting behaviour of clay suspension. Soil Found. **20**(2), 7–20 (1980)
9. Been, K., Sills, G.C.: Self-weight consolidation of soft soil: an experimental and theoretical study. Geotechnique **31**(4), 519–535 (1981)
10. Pane, V., Schiffiman, R.L.: A note on sedimentation and consolidation. Geotechnique **35**(1), 69–72 (1985)
11. Yano, K.: Properties of very soft ground reclaimed by dredged marine clay and their prediction. JSCE **364**(3-4), 1–13 (1985)
12. Jeong, I.J., Yoo, K.S.: A fundamental study of dredged soils. KSCE **27**(5), 55–64 (1979)
13. Kim, S.S.: An experimental study on the settling behavior of marine fluid mud in the west seaside of Korea (Banweol area). Geotech. Eng. **3**(3), 49–61 (1987)
14. Song, J.R., Baek, S.H., Yeo, Y.H.: Consolidation properties of hydraulic fill materials of Yeocheon industrial complex. In: Proceedings of Korean Geotechnical Society Fall '92 National Conference, pp. 55–60 (1992)
15. Lee, S., Yang, T.S., Hwang, K.H.: A study on self-weight consolidation characteristics dredged and reclaimed clay. KSCE J. Civil Eng. **14**(4), 427–430 (1994)
16. Lee, K.H., Hwag, K.H., Lee, S.: A study on consolidation characteristics of dredged-reclaimed ground using seepage-induced forces. In: Proceedings of Korean Geotechnical Society Spring '95 National Conference, pp. 181–190 (1995)
17. Jee, S.H., Lee, S.W., Lee, Y.N.: A Study on sedimentation & consolidation properties of dredged soil in Yulchon industrial complex. In: Proceedings of Korean Geotechnical Society Fall '96 National Conference, pp. 111–122 (1996)
18. Yoo, N.J., Lee, J.H, Jeong, G.S., Park, B.S.: Centrifuge model experiments and numerical analysis consolidation behaviour of dredged and reclaimed ground. KSCE J. Civil Eng. **25**(4-C), 241–247 (2005)

19. Kim, H.T., Kim, S.Y., Kang, I.K., Park, J.E.: A study on the estimation method of loss ratio in dredged fills. J. Korean Geoenviron. Soc. **1**(1), 57–63 (2000)
20. Lee, S.H., Lee, J.H.: An experimental research about setting and consolidation characteristic of dredged soil in West Coast. J. Korean Geoenviron. Soc. **12**(5), 26–29 (2011)
21. Lee, C.W., Choi, H.S.: Experimental and numerical studies for sedimentation and consolidation characteristics of dredged soil in Songdo Area, Incheon. J. Korean Geoenviron. Soc. **17**(2), 13–22 (2016)
22. Jang, S.M.: A study on the determination of retention ratio of dredged soil using the result of sedimentation and self-weight consolidation test. Master's thesis, Incheon National University, pp. 35–56 (2015)
23. Mesri, G., Godlewski, P.M.: Time- and stress-compressibility interrelationship. J. Geotech. Eng. ASCE **103**(GT5), 417–430 (1977)

Chapter 5
Recommended Practice for Soft Clay Characterization with a Focus on Settlement and Stability Analysis

Vikas Thakur and Samson Abate Degago

5.1 Introduction

Safe and economical construction on soft clay deposits demands a complete understanding of parameters that govern stability and settlement issues. This is a complicated problem, as detection and characterization of soft clay can be challenging Ideally, once detected, the geotechnical properties of the soft clay deposit should be measured in the field, at in situ condition, but this is not completely feasible and therefore supplementary sampling and laboratory testing are done to gather information on soil behavior. Sampling of soft clays is an expensive affair that requires special skills and careful handling. Still, sample disturbances in soft clay samples are unavoidable. The literature suggests that sample disturbances result in lower undrained shear strength and lower pre-consolidation pressure of soft clays. Therefore, scrutiny of the laboratory test results along with assigning degree of confidence to the design parameters is crucial.

In this paper, a systematic study is presented along with the recommendations related to detection of soft clay layers, deformation and strength characteristics of soft clays and a procedure to establish representative design parameters.

V. Thakur (✉)
Norwegian University of Science and Technology, Trondheim, Norway
e-mail: Vikas.thakur@ntnu.no

S. A. Degago
Norwegian Public Roads Administration, Trondheim, Norway
e-mail: Samson.degago@vegvesen.no

M. Latha G. (ed.), *Frontiers in Geotechnical Engineering*, Developments in Geotechnical Engineering, https://doi.org/10.1007/978-981-13-5871-5_5

5.2 In Situ Methodologies to Detect Soft Clay Layers

The in situ methods applied to detect soft clays must be chosen based on the applicability of the methods for the actual ground conditions, the required soil data in the project and a cost–benefit perspective. For use in current practice, it is important to present recommendations based on the experiences and observations made with various detection methods. The most widely used methods for soil profiling are borings with standard penetration tests (SPT) that recover split spoon samples. However, in terms of cost-effectiveness, the most common method to detect soft clay layer is rotary pressure sounding. The detection of soft clay layer by conventional sounding methods may, however, be influenced by features in soil composition and layering such as laminated clays with sand- and silt lenses, soft clay layer below a layer of variable thickness and content of coarse materials, and loose, water-saturated silt and sand. In particular, thick and dominating top layers with the presence of stiff and coarse materials may influence the penetration force significantly. This may conceal deep-laying layers of soft materials. Predrilling through such layers is generally recommended.

In terms of reliability, cone penetration tests (CPT) with pore pressure measurements (CPTU) and field vane shear tests are the most accurate in situ methods for soil profiling. In Scandinavian countries, the CPT with pore pressure measurements (CPTU) is considered to be the common method for detecting soft clay layers and to characterize it. CPTU has a great potential for detection of soft clay layers through measurements of cone tip resistance, sleeve friction, and pore pressure. Figure 5.1 illustrates the results and the interpretation of four different CPTUs carried out in a central part of Norway, around Trondheim. Using the Robertson's soil behavior-type classification chart, shown also in Fig. 5.2, it is possible to interpret the ability of CPTU to provide information that can help in establishing a continuous strength and deformation parameters [20, 41, 47]. There are extension to CPTU to add more measurements such as resistivity-CPTU (R-CPTU) and seismic CPTU (SCPTU).

There exists emerging technologies such as electromagnetic resistivity tomography (ERT) and airborne electromagnetic mapping (AEM), where limited experience exists from practical use. The electric resistivity of soils is generally a function of porosity, the ion content of the pore water, salinity, clay content, and content of the charged minerals such as graphite and some sulfides. It is emphasized that local site-specific variations from the tabulated values may occur. The difference in resolution between the different geophysical methods for resistivity measurements is important to be aware of when comparing the results obtained with these methods. The measurements are representative for a soil volume ranging from some centimeters to some tens of centimeters for R-CPTU, some meters or tens of meters for ERT, and finally some tens of meters to some hundreds of meters for AEM. For example, resistivity values measured by R-CPTU for a 3 m thick clay layer over rock with high resistivity will be correct, whereas ERT measurements will be influenced by the rock, even in shallow measurements. AEM will probably not be able to detect the clay layer at all. The same will be the case for thin layers that are depicted sharply by

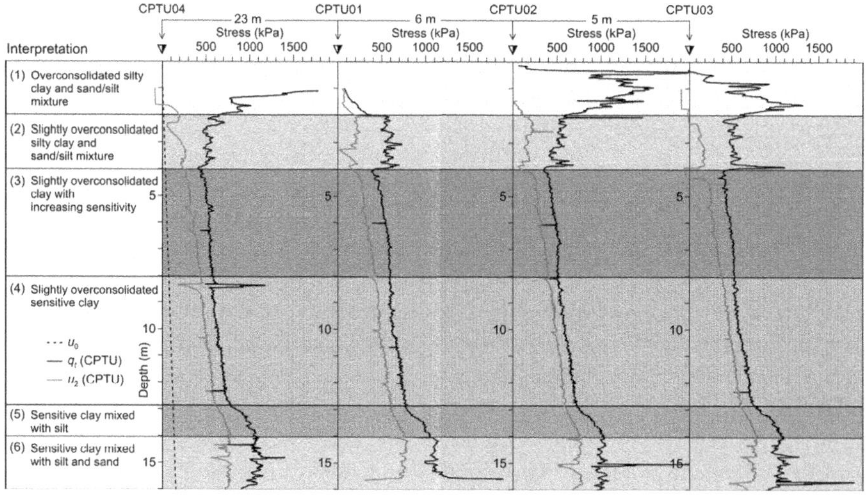

Fig. 5.1 CPTU soundings on the Tiller clay deposit. u_o is the in situ pore pressure, qt is the cone resistance corrected for pore pressure, and u_2 is the pore pressure measured just behind the cone tip [3]

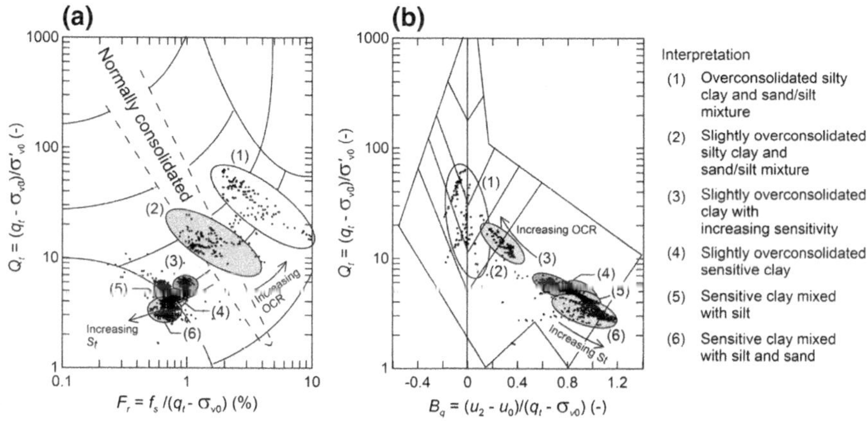

Fig. 5.2 Interpretation of the CPTU03 (Fig. 5.1) using soil identification using Robertson's diagram. Here, Q_t is the normalized net tip resistance, q_t is the corrected tip resistance, σ_{vo} σ'_{vo} is the total/effective vertical in situ stress, F_r is the normalized sleeve friction, f_s is the measured sleeve friction, B_q is the pore pressure ratio, u_2 is the measured pore pressured, and u_o is the in situ pore pressure

R-CPTU, but will show a gradual transition in presentation of ERT- and AEM-data. Except from these conceptual limitations, experience show that the measurements agree well where the soil conditions are favorable [20, 41].

5.3 Undisturbed Soft Clay Sampling and Sample Disturbance

Once the existence of soft clay deposit is detected, the next step will naturally be extracting undisturbed samples to characterize the material. The applicability of engineering parameters for geotechnical design is linked to the quality of soil sampling and testing. Over the years, significant efforts have been made to improve and refine the sampling techniques. The literature confirms that soft clays are prone to sample disturbance, especially when sampled using tube and piston samplers. Several different stages during the acquisition of a soft clay sample, e.g., drilling, sampling, transport, storage, extrusion, and handling of the sample prior to testing may have a negative effect on sample quality. Despite this conventional tube and piston samplers, they are still commonly used in soft soils. During penetration of the tube/piston sampler, shear distortions are induced by the displacement of the soil around the sampler and by friction between the tube walls and the sampled soil. As a results of this, the clay sample experiences a cycle of compression–extension strains during sampling. Excess pore pressure builds inside the soil sample leading to the reduction in effective stress.

A comparison done by Amundsen [3], shown in Fig. 5.3, confirms that soft clay samples collected using tube samples generate large amount of excess pore pressure compared to Sherbrook block samples. This could perhaps explain why block samples yields a better sample quality than samples collected using tube sampler. It is worth mentioning that the magnitude of the sampler-induced disturbance caused by the cutting-edge taper geometry of the samples depends on the sampler geometry and is directly related to the sample disturbance, meaning that thin-walled tube samplers may produce samples of higher quality than thick-walled tube samplers. Moreover, Amundsen and Thakur [2] have shown that the quality of soft clay samples deteriorates with prolonged storage time.

Varying sample quality can also give varying soil compressibility parameters with respect to stress and time actions. This is exemplified with the data from the Väsby test fill, where incremental oedometer results on samples taken from similar depth but using two different samplers are presented, Fig. 5.4. The interpreted OCR from the high-quality data (200 mm sampler) is 1.6 while an OCR of 1.1 is interpreted from the low-quality data (50 mm sampler). In addition, one can see variations in the compressibility parameters that can be interpreted from the data shown in Fig. 5.4. It is thus crucial to aim for highest possible sample quality for the best possible estimate of settlements.

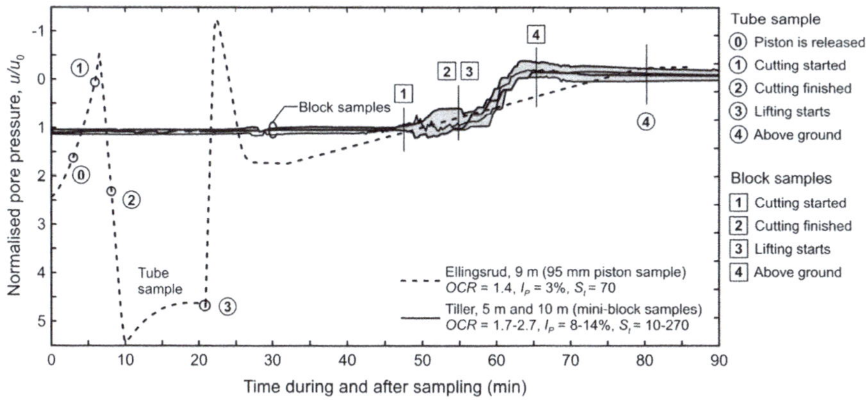

Fig. 5.3 The registered pore pressure changes in soft clay samples extracted using 95 mm piston sampler (by Schjetne [42]) and block sampler (by Amundsen [3]). u_o, and u refers to the in situ pore pressure and the recorded pore pressure inside the soil samples. +ve u/u_o refers to the excess pore pressure and −ve values refers to suction inside soil samples

Fig. 5.4 Effect of sample disturbance (after [31])

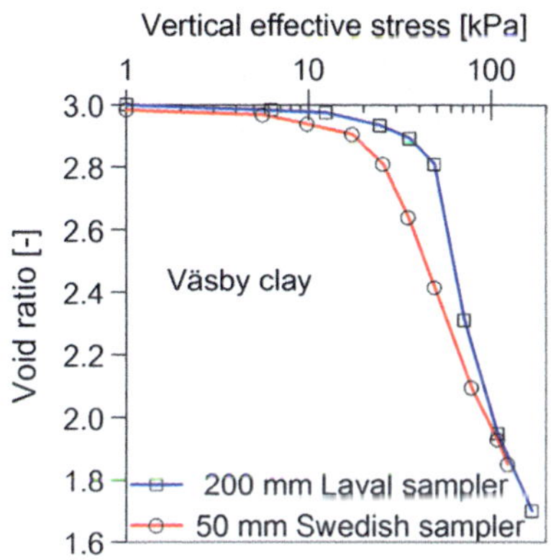

Sample disturbance is virtually unavoidable in soil samples extracted from in situ. However, the extent of sample disturbance varies from sample to sample and ideally one aims to acquire a sample with highest quality possible. Still, one could encounter situations where field settlement calculations have to be made and the available samples are of lower quality. In such cases, it is important to understand the implications of using parameters interpreted from samples of lower quality. For more details on this, the reader is refereed to Degago and Grimstad [16].

Amundsen [3] summarizes the main effects and consequences of sample disturbance in soft clay samples are as follows:

Effects

- mechanical disturbance of the soil structure (centerline and periphery strains) inflicted by the sampling equipment;

- stress relief in the sample due to extraction from the soil deposit; and
- additional disturbance after sampling caused by transport, storage duration, storage conditions, and handling.

Consequences

- inhomogeneous variation of the void ratio within the sampling tube; and
- pore pressure equalization and water migration followed by swelling of the soil structure;
- reduction of peak undrained shear strength and lower pre-consolidation pressure;
- shrinking of the bounding surface of the soil;
- chemical changes during storage.

In short, what exactly constitutes an "undisturbed sample" is unknown, as no definitive method exists to obtain a "perfect sample." Therefore, sample quality assessment becomes essential to assign a confidence level to laboratory test results regardless of the sampling methods and types. In the literature, there exists several assessment for this (see brief summary in Table 5.1). Each of these assessment criteria are intrinsically linked to one specific type of laboratory test. Triaxial and Odometer test results are commonly assessed based on change in the void ratio ($\Delta e/e_0$), a criteria proposed by Lunne et al. [35, 36]. This method is a well-adopted method for soft clays in the practice and highly recommended to use. However the method is not necessarily the best suited for the samples collected from greater depth.

5.4 Deformation in Soft Clays

5.4.1 Background and Definitions

Deformation in geomaterials can generally result due to changes in effective stress, time, chemical processes, and temperature variations. For the most commonly encountered geotechnical problems, it suffices to focus on deformation resulting due to changes in effective stress and time (creep) and this will be also the case for this article. Deformation characteristics of soft clays are an extensively studied topic in soil mechanics. In various literatures, it is noticed that some basic terminologies and concepts used to explain deformation aspects are used and interpreted differently leading to some confusion. It is thus important to start discussing some aspects of this topic by clarifying some basic terminologies and definitions.

Deformation of a saturated soil layer under loading consists of two successive phases, namely primary and secondary consolidation. During primary consolidation phase, soil deformation is accompanied by significant excess pore pressure and changes in effective stresses. Whereas, during secondary consolidation, the soil continues to deform due to time effects under approximately constant effective stresses. Deformations occurring in these phases can be divided into stress-dependent (elastic

Table 5.1 An overview over various methods to assess sample quality of soft clays [1]

Year	Method	Parameter	"Very good to excellent" quality	"Very poor" quality
Triaxial and oedometer tests				
1979–1988	Volumetric strain (ε_{v0}) at in situ effective stress (σ'_{v0})	ε_{v0}	<1%	>10% >8%
1996	Specimen Quality Designation (SQD)	ε_{v0}	<1%	>8%
1997	Change in void ratio ($\Delta e/e_0$) [36], which depends on the overconsolidation ratio (OCR)	$\Delta e/e_0$	<0.04 (OCR 1–2) <0.03 (OCR 2–4)	>0.14 OCR 1–2) >0.1Q (OCR 2–4)
2013	Oedometer stiffness ratio [25]	M_0/M_L	>2.0	<1.0
Uniaxial compression tests				
1979	Strain at failure (ε_{vf}) in an unconsolidated and undrained (UU) test on soft clay	ε_{vf} (UU)	3–5%	10%
1980	Unconsolidated and undrained shear strength, s_u(UU), measured in the laboratory	s_u(UU)	Relative assessment based on information about stress history and predicted strength using SHANSEP	
Suction and shear wave velocity measurements				
1963–2002	Residual effective stress (σ'_s) and the effective stress for a "perfect sample" (σ'_{ps})	σ'_s/σ'_{ps}	$\approx 0.25 - 0.50$ (OCR $>$ 1.5) $\approx 0.05 - 0.25$ (OCR $<$ 1.5)	
1996–2000	Soil suction (u_r)	u_r/σ'_{v0}	$\approx$ 1/5 to 1/6	
2007	Shear wave velocity (V), V_{vh} is measured in the field and V_{SCPTU} is from SCPTU	V_{vh}/V_{SCPTU}	$\geq$0.60	<0.35
2010	Combination of normalized shear wave velocity (L_{vs}) and normalized soil suction (L_u)	L_{vs} L_u	L_{vs}< 0.65 L_u < 0.4	L_{vs}> 0.8 L_u > 0.6
1985–2014	Radiography	Visual identification of sample disturbance		

and plastic deformations) and time-dependent (creep) deformations. Primary consolidation thus consists of both stress-dependent and creep deformations; whereas, secondary consolidation consists of only creep. A key aspect to notice here is that creep exists in the entire phases of primary and secondary consolidation, i.e., creep does not start with secondary consolidation.

It is worthwhile to mention that the interplay of creep and consolidation during primary consolidation has been a topic of intense discussion, especially since this was formally raised by Ladd et al. [29]. However, relatively recent and extensive studies have convincingly showed that creep during consolidation is best captured using creep hypothesis B (e.g., [10, 13, 33]). These studies at the same time have showed that hypothesis A, advocated in various works (e.g., [38, 39]), is simply wrong with various illogical implications (for details see [12]). Hypothesis B, which can be described by isotache concept [46], captures various features of soft clays convincingly and is thus used as a framework in presenting various practice recommended in this article.

5.4.2 On Framework for Modeling Deformation of Soft Clays

Soft clay deposits particularly soft marine clays are characterized by their strong tendency to exhibit high rate dependency (to undergo significant creep deformation), destructuration and anisotropy (see e.g., [5, 7, 46]). For thorough treatment of the interplay between these features, a reference is made to Grimstad et al. [17] and Grimstad and Degago [18]. However, in this work, an emphasis is given on elaborating the role of rate dependency of natural clays in a simplified way and in relation to its practical aspects.

In practice, settlement analyses of soft clay deposits in situ are commonly performed based on soil data obtained from the laboratory tests. However, the duration for completion of the primary consolidation phase is significantly different for a thin laboratory specimen as compared to a thick in situ soil layer. For a laboratory specimen, the primary consolidation duration could be in the order of minutes; whereas, it could take several decades for a thick soft clay deposits in situ [30, 33]. It is thus important to develop a clear understanding of the primary consolidation phase in order to establish deformation characteristics of soft clays. The difference in consolidation periods gives rise to different strain or deformation rates. The strain rate dependency of clays implies that the resulting effective stress–strain relationship would be different for laboratory and field condition (e.g., see Fig. 5.5). A key implication of this is that the experienced pre-consolidation stress would also be rate dependent. This concept is referred to as the isotache concept [46] and is sketched in Fig. 5.5 by considering the two cases with similar incremental loadings up to end-of-primary (EOP) state but different strain rates.

The isotache concept states that the creep strain rate is given by the current stress and strain state, i.e., there is a unique relationship between stress, strain and strain rate [5, 46]. This remains valid during the entire soil compression process (primary and secondary consolidation phases). The concept sketched in Fig. 5.5 is meant to illustrate a key isotaches principle, i.e., fast versus slow consolidation duration which could also be understood as either laboratory versus field condition; or a consolidation test conducted at a fast strain rate versus one conducted at a slower strain rate (e.g., different sample heights or drainage conditions). The implication of strain rate dependency is that the resulting pre-consolidation stress (OCR) is not unique but dependent on the strain rate (reference time) used to determine it. Thus, one should always use pre-consolidation stress or OCR along with the corresponding strain rate or reference time.

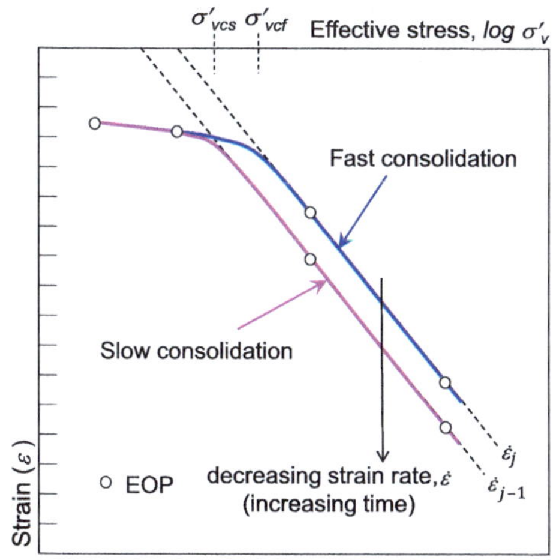

Fig. 5.5 Sketch illustrating effective stress versus end-of-primary (EOP) consolidation strain relationships for varying strain rates (consolidation periods) as implied by the isotache concept

5.4.3 *On Key Parameters for Calculation of Creep Deformation*

To accurately calculate time histories of settlement and pore pressure responses, a constitutive model with a fully coupled consolidation analysis with creep must be performed. By studying creep behavior test data, in both laboratory and field, some mathematical equations that fit the observed measurement data can be established. To begin with, the isotache concept is very appealing as it is simple and its main parameters can are easily be determined from standard laboratory tests. Demonstration of how well the isotache concept can simulate laboratory tests and field measurements can be found in, e.g., Degago et al. [10, 13].

The isotache concept [46], defines creep stain rate as a function of time as shown in Eq. 5.1.

$$\dot{\varepsilon} = \frac{\mu^*}{t} \tag{5.1}$$

where $\dot{\varepsilon}$ = the strain rate; μ^* = the modified creep parameter; and t = time. The isotache concept also uniquely relates volumetric creep strain rate ($\dot{\varepsilon}_v^{vp}$), equivalent effective stress state (p^{eq})and equivalent reference pre-consolidation stress (p_0^{eq}). This relationship as adopted in the soft soil creep (SSC) model [44] is presented in Eq. 5.2. SSC model is implemented in the commercial FE-code Plaxis.

$$\dot{\varepsilon}_v^{vp} = \frac{\mu^*}{\tau} \cdot \left(\frac{p^{eq}}{p_0^{eq}}\right)^{\frac{\lambda* - \kappa*}{\mu*}} \tag{5.2}$$

where λ^* and κ^* = the modified compression indexes for virgin compression and recompression line, respectively; τ = a reference time corresponding to the specified

OCR or (p_o^{eq}/p^{eq}). A typical value for τ is usually selected to be one day as the standard incremental oedometer tests are performed with one day increments.

5.5 Recommended Practice to Establish Deformation Characteristics of Soft Clays for Settlement Calculations

Some important aspects that are considered to contribute as recommended best practice in characterization of soft clays with a focus on settlement calculations are highlighted below by classifying them into various broad categories. A key aspect that needs to be clarified in such assessment is to evaluate the actual importance of accurate settlement calculation which could vary from project to project. The assumption made here is that there is a need for as accurate as possible settlement calculations.

5.5.1 Sample Quality

Settlement analysis of field conditions are normally performed based on soil parameters derived from laboratory tests. Hence, it is crucial that data extracted from the laboratory samples have the desired level of quality and be representative of the soil in situ to give acceptable prediction of field performances. Parameters usually needed for settlement calculations include compressibility parameters, OCR, permeability, and change of permeability and strength parameters. All these parameters can be affected by sample quality to various extents. Soft clay deposits in normally consolidated state typically exhibit an apparent over consolidation ratio owing to aging [5]. Thus, OCR is one of the most important parameters needed for accurate creep settlement calculations. However, the correct interpretation of *OCR* crucially hinges on sample quality as it can easily be affected by sample disturbance [4, 9, 28, 31, 34, 35, 36].

5.5.2 Laboratory Testing

Oedometer test is primarily used to establish key settlement parameters. However, it must be said that the standard oedometer test is not the ideal test procedure. This is due to the difficulty that arises from the inherent nature of the standard oedometer test procedure—i.e., it is not fully representative of the actual field condition due to uncertainties in the initial stress state and stress path during loading. This is particularly important around the stress state near the pre-consolidation stress (p_c')

where the effect is most pronounced. The shape of the reference surface and the selected Poisson's ratio (v) will influence the interpreted pre-consolidation stress from the simulation (thus the input OCR to the model is not the same as the interpreted OCR from the oedometer simulation). However, in the simulation of the field case, the initial stress condition and the stress path followed are more representative of the actual field condition. Hence, one should ideally aim to run oedometer tests starting from the actual initial stress state in situ. To be able to do this, one would need to perform oedo-triaxial tests, i.e., first consolidating to initial in situ stress state with a K_0-consolidated triaxial procedure then incrementally loading the sample, in oedometer condition, to the final in situ stress state. Currently, there is an ongoing research at NTNU to study this aspect. Constant rate of strain (CRS) oedometer test is usually used due to its time efficiency while incremental load (IL) oeodometer test is used to establish to creep parameter (μ^* *or* $C\alpha$). When it comes to using IL test, it is generally recommended to refine load steps around the relevant stress ranges. Common to both CRS and IL test is that the tests should preferably be performed in a temperature controlled room (temperature close to in situ) and at a room free from vibrations (especially for IL tests).

5.5.3 Interpretation of Laboratory Test

As stated earlier, oedometer test poses an inherent challenge since one cannot impose actual field stress path when exceeding the pre-consolidation stress (p_c'). However, the current state of practice still uses oedometer test and the best that can be done to get the most out of the test is to interpret the parameters in line with the numerical model to be used for field calculation. This implies modelling the laboratory test numerically with correct boundary and loading with an aim to back calculate the measured test data. In this way, one can refine and extract parameters to be used for settlement calculations. A key aspect here is that the laboratory data must be of the highest quality. CRS oedometer test is run with a chosen rate and this affects the interpreted OCR. It is thus, important to use a model that takes this into account. Different strain rates give different implied OCR [32, 33]. This should not be a problem as long as one always relates the resulting OCR with the adopted strain rate. In such cases, the numerical model to be used must be able to establish the corresponding OCR for the field strain rate. The same also applies for IL oedometer tests where the duration between load increments affects the interpreted OCR. This can also be understood as OCR without the test rate or reference time is incomplete data. It is generally preferred to interpret settlement parameters using arithmetic plots. This is especially important for OCR and creep parameter (Janbu's time resistance number (r_s) [24]). When it comes to interpretation of creep parameter (μ^* ($=1/r_s$) *or* $C\alpha$), it is crucial that one tries to avoid use of log (time) versus void ratio or strain plot as it is not objective. Regarding this and recommended alternative way of interpretation, the reader is referred to Grimstad et al. [19].

Numerical simulation of a real case naturally involves certain appropriate idealizations so that the case can readily be analyzed. In doing so, it is vital to interpret soil parameters with special emphasis on the underlying assumptions of the selected numerical model as well as the nature of the problem to be dealt with regard to the applicable effective stress range along with the associated time considerations.

5.5.4 Settlement Calculation

Numerical model to be used for settlement calculation must be based on isotache approach (hypothesis B), i.e., unique stress–strain–strain rate relationships as illustrated in Fig. 5.5. In literature and various practices, there exist several models for settlement calculation with varying degree of complexity and advancement. It is thus important to idealize the problem at hand in light of the numerical model to be used for settlement calculations. This could be important attempt to take into account features that a chosen model may not incorporate such as destructuration and anisotropy (e.g., Degago and Grimstad [14]).

In numerical idealization of settlement analysis, it is often common to assume small deformations; thus, neglecting effect of buoyancy. In many cases, this might be a reasonable assumption. However, when large settlements are expected, disregarding the effect of buoyancy will give higher value of the calculated settlements. This is because the effective stress change is not reduced in accordance with the increasing soil layer settled under the ground water table. In Degago et al. [13], sample disturbance and buoyancy effects are shown to have significant influence on the calculated settlements of a test fill with long-term measurements.

In some cases, where the main interest is to calculate the final settlement after a certain period of time, a time-independent elasto-plastic model can also be used to calculate these settlements. This can be performed by selecting a single isotache that in average meets the final expected combination of stress and strain. An isotache selected in this way would typically give a lower *OCR* and lower λ^*. If the true (undisturbed) OCR is known, then the corrected over consolidation ratio, OCR_{corr} to be used with a time-independent elasto-plastic model can be established using Eq. 5.3.

$$OCR_{corr} = \left(\frac{\tau}{t_{age}}\right)^{\frac{\mu*}{\lambda*-\kappa*}} \cdot OCR \tag{5.3}$$

As shown in Fig. 5.6, Degago et al. [13] numerically illustrated the approach shown in Eq. 5.3 for Väsby test fill using models that disregarded creep, i.e., ILLICON and a time-independent elasto-plastic model (the soft soil (SS) model in Plaxis). A detailed treatment on using a lower *OCR* (OCR_{corr}) with a time-independent elasto-plastic model to counterbalance the effect of creep can be referred to in Degago [15]. If the available sample data is of a very low quality and one is only interested in the final

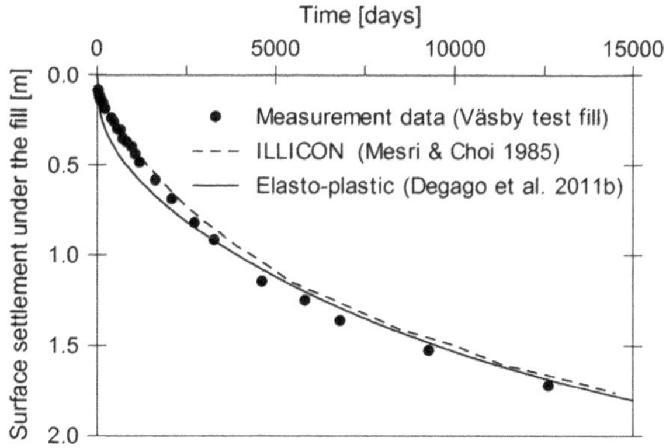

Fig. 5.6 Settlement analyses of the Väsby test fill using lower OCR and numerical models where creep and buoyancy effects are disregarded (after [13])

settlement, one can estimate settlement by simply neglecting creep and using a rate independent elasto-plastic model with an $OCR = 1.0$.

5.5.5 Evaluation of Results

An important part of practicing settlement calculations is evaluation of obtained results. This is especially important for cases when one cannot quantify the extent of sample disturbance or when there is uncertainty in quality of parameters. It is also important in cases where there is lack of experience in using a certain numerical model or if there is limitation of the numerical model. In general, it is best to be able to evaluate if the obtained results make sense or not. This can be done in various ways such as by calculating the implied initial creep strain rate (Eq. 5.2) as well as the implied creep age of the clay.

By combining Eqs. 5.1 and 5.2, it is possible to find an expression for the equivalent age of the clay, t_{age}, based on OCR (for the same K_o^{NC}-line, K_o^{NC} is the coefficient of earth pressure under virgin loading) corresponding to a reference time τ, as

$$t_{age} = \tau \cdot OCR^{\frac{\lambda* - \kappa*}{\mu*}} \tag{5.4}$$

For a given value of $(\lambda^* - \kappa^*)/\mu^*$, Eq. 5.4 can be used to estimate either the implied age of a clay when an OCR corresponding to a certain τ is known; or, to estimate an OCR corresponding to a certain reference time τ when the age of the clay, t_{age}, is known. This is based on an assumption that the ratio $(\lambda^* - \kappa^*)/\mu^*$ is constant in time (with deformation). However, in general this does not necessarily imply that λ^* or μ^* are constant with time, even though this is common assumption in some models such as SSC in Plaxis [45]. More on this can be referred to Degago and Grimstad [14].

Another important way of evaluating calculation result is to look at the settlement rate of unloaded terrain or to look at far-field settlement as well as settlement rates. Looking at the field settlements provides a benchmark for evaluating analyses results along with the parameters adopted in the analysis. This also helps to assess the contribution of the settlement resulted due to the actual fill load and settlement resulting due to pure creep. Hence, calculated far-field settlements, indirectly, indicates part of the total settlement under a fill that is actually coming due to the fill load. In practice, the far-field settlements should comparatively be insignificant. With this aspect, potential of using satellite data for evaluation of creep calculation is recently presented by Degago and Grimstad [11]. A general and a good practice that is always recommended is to make rough estimates of settlements using hand calculations. This provides a rough range of what to expect from numerical models and could be vital in evaluating obtained results.

5.6 Undrained Shear Strength of Soft Clays

The very first equation for the assessment of shear strength of soil was proposed by Coulomb in 1773:

$$\tau_f = c + \sigma \tan\varphi \tag{5.5}$$

The shear strength (τ_f) equation consisted of two parts, i.e., cohesive resistance (c) and frictional resistance (φ). However, the strength parameters c and φ was found to be difficult to determine, especially when cohesive soils were involved. Later, Hvorslev [22] proposed that cohesion (c) depends on water content; the angle of internal friction (φ) is a soil characteristic.

He replaced Eq. 5.5 with effective normal stress or the difference of total stress and pore water pressure as

$$\tau_f = c + \sigma' \tan\varphi = c + (\sigma - u)\tan\varphi \tag{5.6}$$

Here, c refers to the true cohesion; σ' refers to the effective normal stress on the failure plane; σ is the total normal stress on the failure plane; φ is the true angle of internal friction (Fig. 5.7).

Although this equation has been universally accepted to deduce the drained shear strength of geomaterials, the undrained shear strength of soft clays remains a concern. The nature of shear strength saw considerable discrepancy at the Second International Conference of Soil Mechanics in 1948. At the conference, the concept of "$\varphi = 0$ analysis" was raised by Skempton [43] because saturated cohesive soils exhibit an angle of internal resistance $\varphi = 0$ when brought to failure under undrained shearing. The corresponding strength revealed is the undrained shear strength. Specific to the $\varphi = 0$ concept, it was assumed that for a saturated clay specimen under undrained

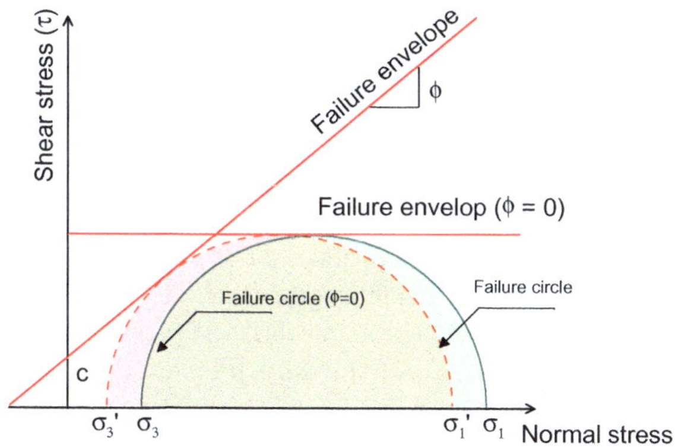

Fig. 5.7 Failure envelopes for soft clays

conditions increases in confining stress were carried by the pore water in the sample, with the effective stress in the sample remaining unchanged. This was found consistent with the Terzaghi's effective stress principle; if the effective stress in a sample does not change, the deviatoric stress required to cause failure in the sample does not change.

5.7 Recommended Practice to Establish Undrained Strength Properties of Soft Clays for Stability Calculations

Selection of the characteristic undrained shear strength (c_{uA}) under compression loading is the most important parameter in the stability analysis of various geotechnical constructions placed in or on soft clays. Thus, the choice of undrained shear strength, both too conservative and offensive, could have major positive or negative economic (and social) consequences in many projects. Therefore, the selection of c_{uA} must consider the influence of the stress history, soil fabric, strain rates, Atterberg's limits, sampling technique, and sample disturbance is great on the undrained shear strength of soft clays. Consequently, the assessment of a representative c_{uA} of soft clays has been challenging since the inception of the concept.

As a simple guidance, the undrained shear strength c_{uA} profiles with depth should be established at some selected reference boreholes where the information is collected. These c_{uA} profiles are later interpolated to the soil volume between the boreholes. Therefore, the accuracy of a c_{uA} profile at the boreholes is decisive because the estimate undrained shear strength of the soil volume between the boreholes will depend on the representativeness of the c_{uA} profiles at the reference boreholes. This paper mainly discusses the recommended procedure for assessing the c_{uA} profile at a reference borehole.

5.7.1 Laboratory Testing

The undrained shear strength of undisturbed soft clay sample can be estimated using the laboratory test such as fall cone, direct simple shear, laboratory vane shear, uniaxial, biaxial, and triaxial tests. The most reliable laboratory method to assess c_{uA} is triaxial testing provided sufficient care is taken in terms of sample preparation, selection of cell pressure, and the rate of shearing

A good practice would suggest that odometer test(s) should be performed before triaxial testing so that pre-consolidation pressure of the test material is known. There exist different methods to measure pre-consolidation pressure. The methods include Casagrande's method [8], Janbu's method [23], or Salfor's method [40].

In addition, knowledge about the in situ effective stress (pore pressure measurement) at the sample collection depths for the triaxial tests is valuable. This requires estimation of the correct K_o'. One can use the approach suggested by Brook and Ireland [6] that provides a relatively easy way to estimate K_o' based on soil plasticity (*Ip*) and the OCR. One needs to be cautious using a relatively high K_o' because this will result in a higher mean effective stress in the soil specimen and may result in a high c_{uA}.

As in the odometer test, the results of a triaxial test depend on the strain rate (rate dependence). An increase in the strain rate generally results in increased undrained strength and brittle behavior. Lunne and Andresen [37] suggest 150% increase in the undrained shear strength when the rate of shear is increased to very fast rate from a very slow rate. The experiences from Scandinavian soft clays shows that there is a little difference in undrained shear strength when the specimen are sheared at rates between 0.7 and 3.0% per hour.

The estimation of undrained shear strength from triaxial tests shall account for dilating and contracting behaviors as well. For the undrained triaxial tests exhibiting contracting behavior (positive excess pore pressure buildup during shearing), it is recommended to obtain c_{uA} at the maximum measured shear resistance. A test that first exhibits a contracting behavior but then shows dilation is an indication of sample disturbance. In such cases, it is advisable to not go beyond the c_{uA} that shows the point representing the transition between contractancy and dilatancy (see Fig. 5.8). Indeed, dilating materials (highly overconsolidated clays) often attain high strengths; however, this is normally related to high strains. In these cases, it is recommended that the c_{uA} is defined by a given strain, for example, 10% [48].

5.7.2 Field Testing

The most common field test to estimate c_{uA} is the cone penetration test (CPT) apart from vane shear testing. The cone penetrometer with measurement of pore pressure, called CPTU test (developed in the late 1970s) developed as a research tool, which gradually became commercial tool. Over the years, CPTU has tested widely and now it has become the most common in situ testing tool for determination of undrained

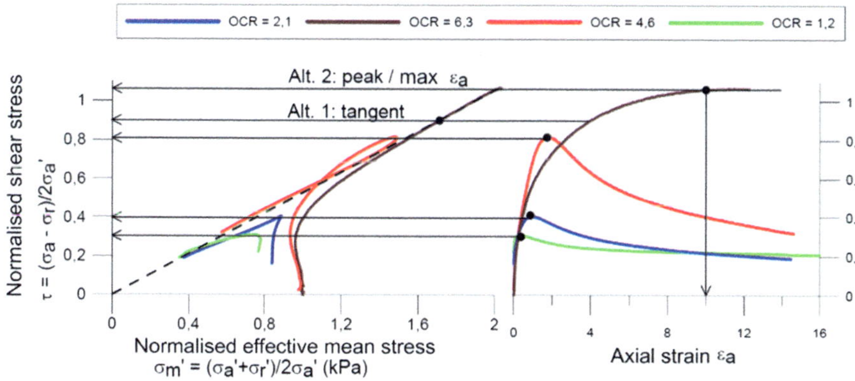

Fig. 5.8 Undrained triaxial test results for four different clays with varying OCR and recommended interpretation of c_{uA} [48]

strength of clay deposits. The major advantage of CPTU is that it helps in establishing a continuous c_{uA} profile with the depth. The value of c_{uA} is can be calculated using three different cone factors: N_{kt}, $N_{\Delta u}$, and N_{ke}.

The total tip resistance based c_{uA} is calculated as

$$c_{uA} = \frac{q_T - \sigma_{vo}}{N_{kt}} \tag{5.7}$$

The effective tip resistance based c_{uA} is calculated as

$$c_{uA} = \frac{q_T - u_2}{N_{ke}} \tag{5.8}$$

Here, q_T is tip resistance and σ_{vo} is the total vertical stress.

The pore pressure measurement based c_{uA} is calculated as

$$c_{uA} = \frac{u_2 - u_o}{N_{\Delta u}} \tag{5.9}$$

Here, u_2 is the measured pore pressure and u_o is the in situ pore pressure.

It is clear from Eqs. 5.7 to 5.9 that c_{uA} will depend on the cone factors. Several correlations exist for the N_{kt}, N_{ke}, and $N_{\Delta u}$ parameters to calculate c_{uA}. Karlsrud et al. [26, 27], who compared the CPTU with c_{uA} from block samples and suggested the cone factors N_{kt}, proposed the widely accepted method in Norway and $N_{\Delta u}$ based on the OCR and soil plasticity index (I_p). The N_{ke} parameter was correlated with B_q (pore pressure parameter).

Apart from CPTU, vane testing can be used to determine the undrained shear strength under direct simple shear conditions in clays. Both intact and remolded shear strength can be found. The vane test is the only in situ test method, which can be used to determine the remolded shear strength and the sensitivity directly. The vane test is susceptible to heterogeneities in the soil. If parts of the vane (side, top or base) are weaker or stronger, or if fragments of shells or small stones interfere with the vane, this may influence the measured values significantly [20]. One needs to be cautious in the estimation of undrained shear strength of low plasticity soft clays or silt rich clays using vane shear. Progressive failure, the shape of failure zones, and drainage of pore water pressure may affect the results.

5.7.3 Assessment of Undrained Shear Strength Profile

The interpretation of c_{uA} profiles must be done in light to laboratory and field test results to represent the soil volume it includes in the calculation. Various aspects such as topography, distance calculation profile, contour level, previous level of the terrain in the area, effective stress level, pore pressure, variation in soil conditions in the area, soil geological information, and depositional history must be accounted in the assessment.

One needs to be careful with the use of index data from routine testing (such as uniaxial tests and the fall cone) to the interpretation of the c_{uA} profile. The most likely c_{uA} profile shall follow a ranking of the measurements and the empirical data as follows:

1. Triaxial tests of good quality (very good to excellent quality)
2. CPTU (class 1)
3. Fall cone/uniaxial test/vane test
4. Experience-based c_{uA}/p_o' or SHANSEP.

The selected c_{uA} profile should normally lie between the estimated lower and upper bounds of c_{uA} profiles.

5.7.4 Strength Anisotropy

Soft clays can be anisotropic in nature. In stability analysis of an undrained situation, one needs to account for the strength anisotropy. The chosen strength anisotropy factors refer to c_{uE}/c_{uA} and c_{uD}/c_{uA} must be site-specific and deduced from high-quality soil samples. Here, c_{uA}, c_{uE}, and c_{uD} refer to the undrained shear strength obtained from active/compression (A), passive/extension (E), and direct shear (D) tests.

A recent study conducted by Thakur et al. [49] suggests that the plasticity index (I_p), the natural water content (w) is correlated with soil anisotropy. However, no clear correlation was seen between the strength anisotropy and the over consolidation ratio (OCR), the clay content (<2μ), the sensitivity (S_t), and the liquid limit (w_L). Recommended practice to select strength anisotropy for Norwegian soft clay along with a comparison with the other soft clays is presented in Fig. 5.9.

The significance of strength anisotropy in the stability analysis was studied by Thakur et al. [49] using a benchmark case of a 15 m high clay slope with an inclination

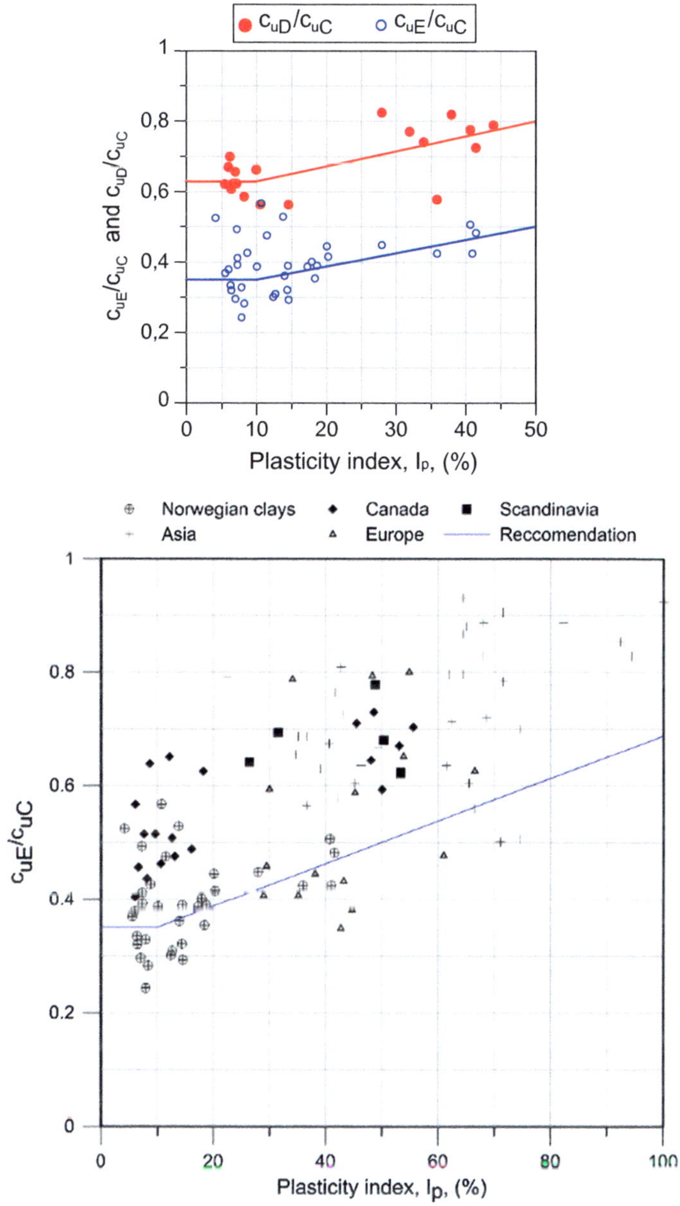

Fig. 5.9 Recommended trend lines to estimate the strength anisotropy factors for Norwegian soft clays. The red and blue trend lines are for estimate c_{uD}/c_{uC} and c_{uE}/c_{uC}, respectively. Right: A comparison between the recommendation for the passive strength anisotropy for the Norwegian soft clays and other soils reported in the literature (Based on Thakur et al. [49])

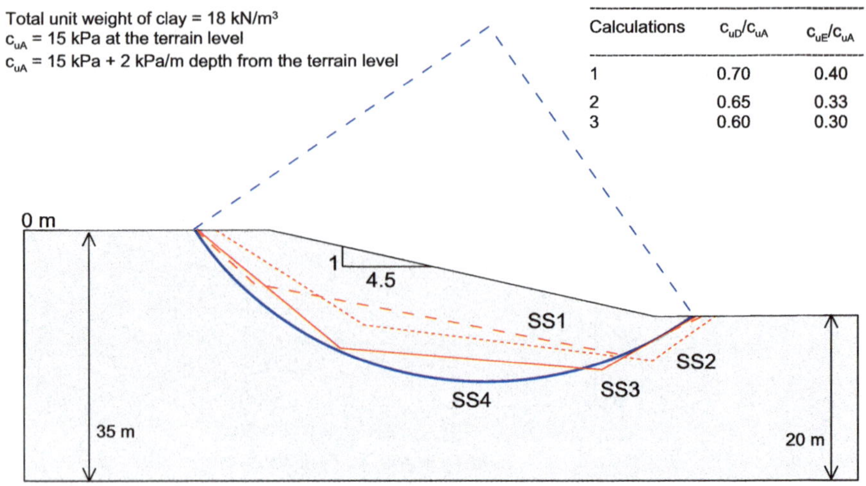

Fig. 5.10 The slope geometry and the input parameters used in the stability calculations by Thakur et al. [49]

of 1: 4.5 was considered. The slope geometry and the input parameters are shown in Fig. 5.10. Thakur et al. [49] investigated three noncircular sliding surfaces and a circular sliding surface. They are referred as SS1, SS2, SS3, and SS4, respectively. The C_{uE}/C_{uA} factor was varied between 0.3 and 0.4 and c_{uD}/c_{uA} factor was varied between 0.6 and 0.7. The study illustrated that the selected of c_{uE}/c_{uA} and c_{uD}/c_{uA} parameters within a reasonable range has relatively little effect (4–8%) on the factor of safety. This study advocates to focus on the estimation of representative c_{uA} profile.

5.7.5 Implication of Undrained Shear Strength Profile in Design Practice

Selection of undrained shear strength profile will have direct impact on the stability calculation and on the mitigation measures selected to improve stability. This can be exemplified using a real case, presented by Hermann and Jensen [21], related to the construction of a new railway track at NyKirke located at the south of city of Oslo in Norway. A 15 m high railway embankment was planned on approximately 30 m thick soft clay deposit. The design of railway fill was based on the undrained shear strength value obtained from 54-mm diameter tube sampler. The average design undrained strength profile of the soft clay deposit was established to be

$$c_{uA} = 0.25\,\sigma'_{vo} \tag{5.10}$$

As a result to this, the stability of the railway fill found to be insuffieicent. To improve the stability of the fill, concrete piles as illustrated in Fig. 5.11 were

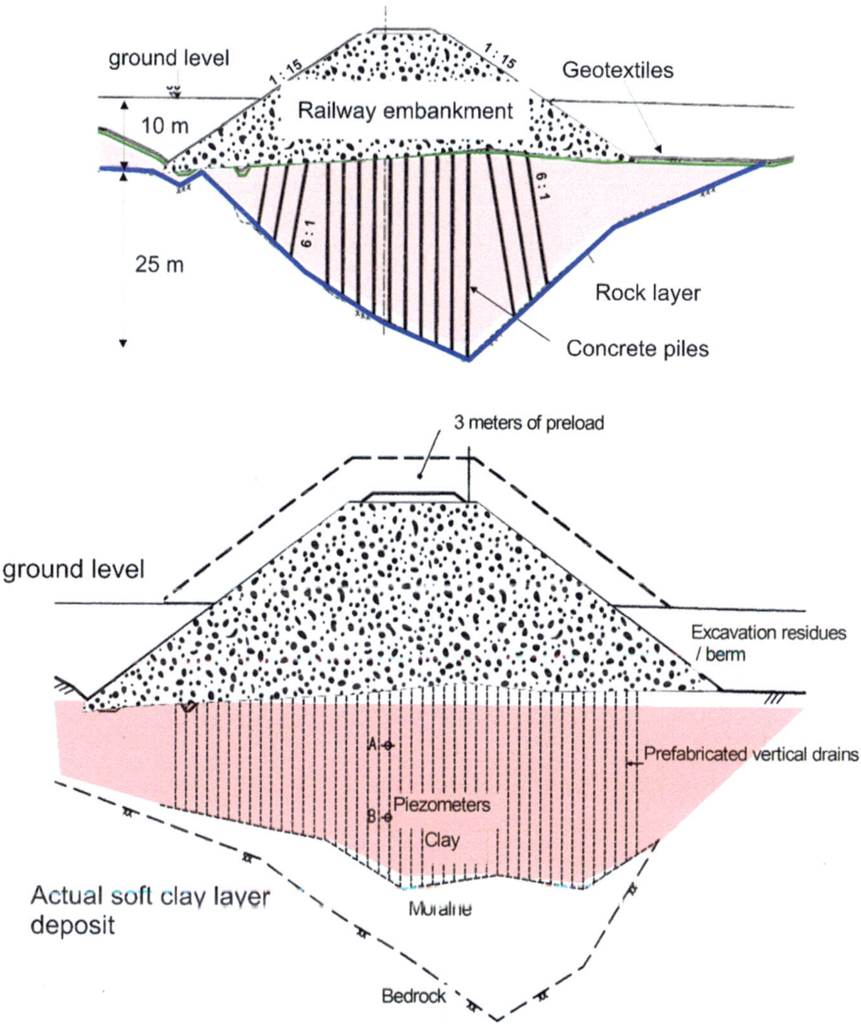

Fig. 5.11 Technical solutions to construction a railway embankment on soft clay deposit (upper) based on poor quality 54-mm diameter samples (lower) based on high-quality CPTU and block samples. Based on Hermann and Jensen [21]

suggested. This design was used in the tender document by the concerning authorities. As a part of the total contract, NGI got a task to establish technical solution to design the fill executed a series of high-quality CPTU testing and block sampling. A comparison between the traiaxial test result between the samples extracted using 54-mm diameter tube sample and block sampler is shown in Fig. 5.12. Based on the new geotechnical investigation, it was possible to upgrade the undrained shear strength profile to

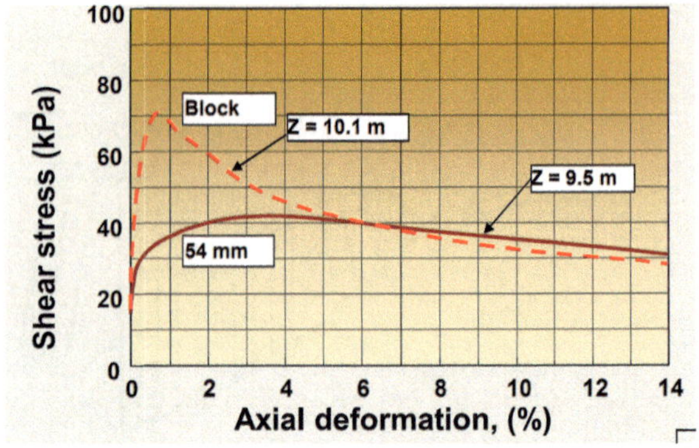

Fig. 5.12 Effect of sample disturbance on peak undrained shear strength of soft clay deposit at Nykirke [21]

$$c_{uA} = 0.6\,\sigma'_{vo} \tag{5.11}$$

From the stability viewpoint, it was not necessary to have piles to rock. Also it was suggested that settlements could be taken care of by vertical drainage combined with preloading. An updated technical solution as shown in Fig. 5.11 were suggested and the project was executed accordingely. As per Hermann and Jensen [21], the total cost savings of about USD 12 mill or 25% was made simply by optimizing the geotechnical solutions.

5.8 Summary and Conclusions

The work gives recommended practice for soft clay characterization with a focus on settlement and stability analysis. Various aspects related to selection of characteristic strength and deformation parameters are presented and discussed. It is important to realize that extracted samples constitute a key base for settlement and stability analysis. With this regard, it is vital to be aware of the role of sample disturbance and its implications. As illustrated in this work, the impact of high-quality field and laboratory investigation is positive as it helps to do cost-effective design and safer construction of the geotechnical structures.

Acknowledgements The authors of this paper acknowledge the R&D and the funding partners of Natural Hazards—Infrastructure for Floods and Slides Program—NIFS (www.naturfare.no). The OFFPhD program by the Research council of Norway (www.rcn.no) and Norwegian Public Roads Administration are gratefully acknowledged for their supports. It is worth mentioning that this paper partly presents a summary of the work that has been presented in various arenas including at 17th Nordic geotechnical meeting (NGM). Readers are encouraged to refer Thakur et al. [48] and Degago and Grimstad [14] for further details.

References

1. Amundsen, H.A., Thakur, V., Emdal, A.: Sample disturbance in block samples of low plastic soft clays. In: 17th Nordic Geotechnical Meeting, Island (2016)
2. Amundsen, H.A., Thakur, V.: Storage duration effects on soft clay samples. Accepted for publication in Geotechnical Testing Journal. https://doi.org/10.1520/GTJ20170426. ISSN 0149-6115 (2018)
3. Amundsen, H.A.: Storage duration effects on Norwegian low-plasticity sensitive clay samples. PhD Thesis, Norwegian University of Science and Technology. ISSN 1503-8181 (2018)
4. Berre, T.: Effect of sample disturbance on triaxial and oedometer behaviour of a stiff and heavily overconsolidated clay. Can. Geotech. J. **51** 896–910 (2014). https://doi.org/10.1139/cgj-2013-0077
5. Bjerrum, L.: Engineering geology of Norwegian normally consolidated marine clays as related to settlements of buildings. Géotechnique **17**(2), 81–118 (1967)
6. Brooker, E.W., Ireland, H.O.: Earth pressure at rest related to stress history. Can. Geotech. J. **2**(1), 1–15 (1965)
7. Burland, J.B.: On the compressibility and shear strength of natural clays. Géotechnique **40**(3), 329–378 (1990)
8. Casagrande, A.: The determination of the pre-consolidation load and its practical significance. In: Proceedings of the First International Conference on Soil Mechanics and Foundation Engineering Boston, vol. 3, Discussion D-34, 60 (1936)
9. DeGroot, D.J., Poirier, S.E., Landon, M.M.: Sample disturbance-Soft clays. Stud. Geotech. Mech. **27**(3–4), 91–105 (2005)
10. Degago, S.A., Grimstad, G., Jostad, H.P., Nordal, S., Olsson, M.: Use and misuse of the isotache concept with respect to creep hypotheses A and B. Géotechnique **61**, 897–908 (2011)
11. Degago, S.A., Grimstad, G.: Potential application of satellite data in evaluation of field creep calculation. In: Proceedings of the 19th International Society for Soil Mechanics and Geotechnical Engineering, Seoul, pp. 715–720 (2017)
12. Degago, S.A., Jostad, H.P., Olsson, M., Grimstad, G., Nordal, S.: Time- and stress-compressibility of clays during primary consolidation. In: Proceedings of the 7th NUMGE, Trondheim, pp. 125–130 (2010)
13. Degago, S.A., Nordal, S., Grimstad, G., Jostad, H.P.: Analyses of Väsby test fill according to creep hypothesis A and B. In: Proceedings of the 13th IACMAG, Melbourne, vol. 1, pp. 307–312 (2011)
14. Degago, S.A., Grimstad, G.: Evaluation of soil parameters for creep calculations of field cases. In: 17th Nordic Geotechnical Meeting, Island (2016)
15. Degago, S.A.: On creep during primary consolidation of clays. Ph.D. Thesis, Norwegian University of Science and Technology (NTNU), Trondheim, Norway (2011)
16. Degago, S.A., Grimstad, G.: Significance of sample quality in settlement analysis of field cases. In: Proceedings of the 8th NUMGE, Delft, The Netherlands, pp. 153–158 (2014)
17. Grimstad, G., Degago, S.A., Nordal, S., Karstunen, M.: Modeling creep and rate effects in structured anisotropic soft clays. Acta Geotech. **5**, 69–81 (2010)
18. Grimstad, G., Degago, S.A: A non-associated creep model for structured anisotropic clay (n-SAC). In: 7th European Conferences NUMGE, Trondheim, Norway, pp. 3–8 (2010)
19. Grimstad, G., Mehli, M., Degago, S.A.: Creep in clay during the first few years after construction. In: Proceedings of the 6th International Symposium on Deformation Characteristics of Geomaterials, Buenos Aires, pp. 915–922 (2015)
20. Gylland, A., Sandven, R., Emdal, A., Thakur, V.: Extended interpretation basis for the vane shear test. In: 17th NGM, Iceland (2016)
21. Hermann, S., Jensen, T.G.: CPTU combined with block sampling gave cost saving solutions for Nykirke railway crossing. In: Proceedings of Norwegian Geotechnical Day, Oslo, Paper No. 32 (In Norwegian) (2000)
22. Hvorslev, M. J.: Uber die Festigkeilesigenschaften gestorter bindiger Boden. Ingeniorvidenskabelige Skrifter, A, No. 45, Copenhagen (1937)

23. Janbu, N.: Soil compressibility as determined by oedometer and triaxial tests. In: Proceedings of the third European Conference Soil Mechanics, Wiesbadem, vol. 1, pp. 19–25 (1963)
24. Janbu, N.: The resistance concept applied to deformations of soils. In: Proceedings of 7th International Conference on Soil Mechanics and Foundation Engineering, Mexico City, vol. 1, pp. 191–196 (1969)
25. Karlsrud, K., Hernandez-Martinez, F.G.: Strength and deformation properties of Norwegian clays from laboratory tests on high-quality block samples. Can. Geotech. J. **50**(12), 1273–1293 (2013)
26. Karlsrud, K., Lunne, T., Kort, D.A., Strandvik, S.: CPTU correlations for clays. In: Proceedings of International Conference on Soil Mechanics and Foundation Engineering, 16, Osaka, vol. 2, pp. 693–702 (2005)
27. Karlsrud, K., Lunne, T., Kort, D.A., et al.: CPTU correlations for clays. In: 16th ICSMFE 16, vol. 2, pp. 693–702 (2005)
28. Lacasse, S., Berre, T., Lefebvre, G.: Block sampling of sensitive clays. In: Proceeding of the 11th International Conference on Soil Mechanics and Foundation Engineering, San Francisco, pp. 887–892 (1985)
29. Ladd, C.C., Foott, R., Ishihara, K., Schlosser, F., Poulos, H.G.: Stress-deformation and strength characteristics. State-of-the-art report. In: Proceedings of 9th ICSMFE, Tokyo, vol. 2, pp. 421–494 (1977)
30. Larsson, R., Mattsson, H.: Settlements and shear increase below embankments. SGI Rep. **63**, 88p (2003)
31. Leroueil, S., Kabbaj, M.: Discussion of 'Settlement analysis of embankments on soft clays' by Mesri & Choi. ASCE **113**(9), 1067–1070 (1987)
32. Leroueil, S.: Compressibility of clays: fundamental and practical aspects. J. Geotech. Eng. Div. ASCE **122**(7), 534–543 (1996)
33. Leroueil, S.: Šuklje Memorial Lecture: The isotache approach. Where are we 50 years after its development by Professor Šuklje? In: Proceedings of the 13th Danube European Conference on Geotechnical Engineering, Ljubljana, vol. 2, pp. 55–88 (2006)
34. Long, M., El Hadj, N., Hagberg, K.: Quality of conventional fixed piston samples of Norwegian soft clay. J. Geotech. Geoenviron. Eng. **135**(2), 185–198 (2009)
35. Lunne, T., Berre, T., Strandvik, S.: Sample disturbance effects in soft low plastic Norwegian clay. In: Proceedings of the Symposium on Recent Developments in Pavement Mechanical, Rio de Janeiro (1997)
36. Lunne, T., Robertson, P.K., Powell, J.J.M.: Cone Penetration Testing in Geotechnical Practice. Blackie Academic & Professional (1997)
37. Lunne, T. and Andersen, K.H.: Soft clay shear strength parameters for deepwater geotechnical design. In: Proceedings of the 6th OSIG, London, UK, pp. 151–176 (2007)
38. Mesri, G.: Primary and secondary compression. In: Germaine, J.T., Sheahan, T.S., Whitman, R.V. (eds.) Soil Behavior and Soft Ground Construction, vol. 119, pp. 122–166. ASCE Geotechnical Special Publication (2003)
39. Mesri, G., Feng, T.W., Shahien, M.: Compressibility parameters during primary consolidation. In: Proceedings of the International Symposium on Compression and Consolidation of Clayey Soils, Hiroshima, vol. 2, pp. 1021–1037 (1995)
40. Sallfors, G.: Preconsolidation pressure of soft high plastic clays. Ph.D. Thesis, Chalmers University of Technology, Gothenburg, Sweden (1975)
41. Sandven, R., Gylland, A., Montafia, A., Kåsin, K., Pfaffhuber, A.A.: In situ detection of sensitive clays—Part I and II: Results. In: Nordic Geotechnical Meeting, Iceland (2016)
42. Schjetne, K.: The measurement of pore pressure during sampling. In: Proceedings of the 4th Asian ISSMFE, Special Session on Quality in Soil Sampling, Bangkok, p. 12016 (1971)
43. Skempton, A.W.: The $\square = 0$ analysis of stability and its theoretical basis. In: Proceedings of the 2nd International Conference on Soil Mechanics and Foundation Engineering, vol. 1, pp. 72–78 (1948a)
44. Stolle, D.F.E., Vermeer, P.A., Bonnier, P.G.: Consolidation model for a creeping clay. Can. Geotech. J. **36**(4), 754–759 (1999)

45. Stolle, D.F.E., Vermeer, P.A., Bonnier, P.G.: Consolidation model for a creeping clay. Can. Geotech. J. **36**(4), 754–759 (1999)
46. Šuklje, L.: The analysis of the consolidation process by the Isotaches method. In: Proceedings of the 4th International Conference on Soil Mechanics and Foundation Engineering, London, vol. 1, pp. 200–206 (1957)
47. Thakur, V.: Characterization of sensitive soft clays for design purposes. Dev. Geotech. Eng. (2018). https://doi.org/10.1007/978-981-13-0505-4
48. Thakur, V.K.S., Fauskerud, O.A., Gjelsvik, V., Christensen, S.O., Oset, F., Nordal, S., Viklund, M., Strand, S.-A.: A procedure for the assessment of the undrained shear strength profile of soft clays. In: Proceedings of the 17th Nordic Geotechnical Meeting (2016)
49. Thakur, et al.: Recommended practice for use of strength anisotropy. In: 2nd IWLSC. Springer Book Series on Natural Hazards (2017)

Chapter 6
Insights into Contaminant Transport Modeling Through Compacted Bentonites

Tadikonda Venkata Bharat, Partha Das and Ankti Srivastava

6.1 Introduction

Deep geological repositories are currently gaining interest for the disposal of hazardous radioactive waste. High-level nuclear waste or spent fuel and intermediate nuclear wastes are encapsulated in metal canisters; are buried in the deeper layers of the ground—situated a few hundreds of meters beneath the ground surface [1–5]. Highly plastic clay soils such as bentonites are used in such repositories as buffer and backfill material to prevent the migration of radioactive waste to the surrounding environment and, simultaneously, to restrict the groundwater migration into the waste emplacement area [6]. Bentonites are also found its significant application in the field of landfills as a potential liner material. Compacted bentonite layer is very useful in inhibiting the migration of the generated leachate from the municipal solid waste landfills to the surrounding aquifers and ground water sources [7, 8]. The main reason for the wide application of plastic clays as buffer/backfill material for nuclear waste repositories and as liners for landfills is because of their favorable characteristics like very low saturated hydraulic conductivity, high sorption potential and very high self-sealing capacity [9–12].

The saturated hydraulic conductivity of the clay barriers is very low ($K_S \leq 10^{-9}$m/s) thus the flow though such clay layers will be governed by diffusion mechanism [8, 13–15]. Bentonite clays being impermeable to the migration of solutes under any hydraulic gradients, advection transport is deterred through such barriers. Therefore, the estimation of diffusive mass transport parameters is important for the design of radioactive nuclear waste facility in the underground under the influence of salt migration from the surrounding saturated rock mass and the design of landfills for arresting the migration of contaminants to the groundwater as the diffusion is the governing transport mechanism. Laboratory diffusion studies are undertaken on

T. V. Bharat (✉) · P. Das · A. Srivastava
IIT Guwahati, Guwahati, India
e-mail: tvb@iitg.ac.in; tvbharat@gmail.com

M. Latha G. (ed.), *Frontiers in Geotechnical Engineering*, Developments in Geotechnical Engineering, https://doi.org/10.1007/978-981-13-5871-5_6

the representative soil specimen for the estimation of mass transport parameters viz. diffusion coefficient and retardation factor for the design of the facilities.

Several laboratory techniques viz. half-cell technique [16–18], in-diffusion technique [7, 14, 19], and through-diffusion technique [8, 10, 20, 21] are extensively used to estimate the effective diffusion coefficient and retardation factors. The half-cell technique provides the spatial changes in contaminant concentration at a given time for the analysis of diffusion coefficient. In contrary, through-diffusion technique provides temporal variations in the contaminant concentrations at the boundaries of the soil specimen while in-diffusion technique provides both spatial and temporal concentration variation in the specimen and boundaries, respectively. In order to obtain the experimental concentration profile along the length of the sample in half-cell and in-diffusion techniques, the specimen is required to be extruded and sliced. However, through-diffusion test is a non-destructive technique [8, 10] thus does not require slicing of soil mass.

The estimation of mass transport parameters from the observed temporal or spatial variation in contaminant concentration variation requires numerical or analytical solution to the governing diffusion equation with appropriate initial and boundary conditions [8, 20, 21]; error minimization techniques for minimizing the error between the theoretical and experimental data [22, 23]. However, the design parameters obtained from different testing methodologies can be different because of various implications in conducting the tests.

The definition of the effective diffusion coefficient differed in various literature works. The intricacies and the associated limitations in the estimated model parameters from different laboratory techniques were discussed in this work.

6.2 Background

The diffusive flux described by Fick's first law for diffusion mass transport through the soil is given by [14, 15, 20, 24]:

$$J_d = -D_e^S \phi \frac{\partial c}{\partial x} \tag{6.1}$$

where ϕ is the accessible porosity which is the available pore space for diffusion transport, c is the concentration of the contaminating species, x is the spatial distance, and D_e^S is the effective diffusion coefficient. The superscript, s, on the diffusion coefficient indicates the effective diffusion coefficient as defined by Shackelford and Daniel [14, 15]. Effective diffusion coefficient defined by $D_e^S = \tau D_0$, where τ, is the tortuosity factor, accounting for the tortuous pathways for the solutes to diffuse through the soil; D_0 is the free solution diffusion coefficient. In contrary, few research groups lump the accessible porosity and tortuosity factor with the diffusion coefficient in defining the effective diffusion coefficient [25, 26]. The diffusion flux equation thus is written as:

$$J_d = -D_e^G \frac{\partial c}{\partial x} \tag{6.2}$$

where D_e^G is the effective diffusion coefficient which differs from the earlier definition (Eq. 6.1) by a factor ϕ.

The governing differential equation describing the one-dimensional contaminant transport through saturated soil, by invoking mass conservation principle and combining with the Fick's first law, is written as [7, 8, 10, 25, 26]:

$$\frac{\partial C}{\partial t} = \left(\frac{\phi D_e^s}{\alpha}\right)\frac{\partial^2 C}{\partial x^2} = \left(\frac{D_e^G}{\alpha}\right)\frac{\partial^2 C}{\partial x^2} \tag{6.3}$$

where c is the concentration of the contaminant in the soil pores, t is the time, x is the spatial distance along the length of the sample, and α is the capacity factor. The capacity factor or the rock capacity factor is defined as the ratio of the total mass of chemical species per unit total volume of porous medium relative to the pore-water concentration [27]. The D_e^G and D_e^S are the effective diffusion coefficients defined differently for the convenience of parameter estimation. The relationship between the two effective diffusion coefficients is expressed by:

$$\phi D_e^S = D_e^G \tag{6.4}$$

The capacity factor, or also termed as volumetric capacity of the porous medium for the chemical species, is related to the accessible porosity, distribution coefficient (K_d), and dry density (ρ_d) for the linear sorption case as

$$\alpha = \phi + K_d\rho_d = \phi\left(1 + \frac{K_d\rho_d}{\phi}\right) = \phi R_d \tag{6.5}$$

where R_d is the retardation factor which takes into account the linear sorption of chemical species on the clay surface. Often, the R_d is lumped with the D_e and a single diffusion parameter called apparent diffusion coefficient (D_a) is used for the representation of diffusion phenomenon. The governing equation thus written as

$$\frac{\partial C}{\partial t} = D_a \frac{\partial^2 C}{\partial x^2} \tag{6.6}$$

where D_a is defined as D_e/R_d.

Commonly, the porosity, n, of the clay specimen is used in place of accessible porosity in the aforementioned equations (Eqs. 6.1, 6.3–6.5) as the former term can be estimated from the compaction density of the soil. Thus applicability and usage of porosity in the diffusion analysis requires a special attention.

6.2.1 Understanding the Porosity

The accessible porosity, ϕ, in the flux equation (Eq. 6.1) is used to indicate the pore space which contributes to the mass transport [15, 27]. In general, only a fraction of the pore space (ϕ) of the total pore space (n) is available for the mass transport [27, 28]. A water saturated bentonite contains dead-end pores and occluded pockets in the pore space, the estimation and knowledge of effective porosity is vital. Further, as the anions in the chemical species cannot diffuse through the dead-end pores of the soil barriers due to anion exclusion effect [14, 15, 27] the accessible porosity is different for different species. An independent estimation of the design parameters (D_e^S, α, and ϕ or D_e^G, α) is necessary for the effective design of the liners system. The implications in estimating the effective diffusion coefficient and diffusion accessible porosity independently through different test methodologies will be brought out for the first time in this paper.

6.2.2 Laboratory Diffusion Models

6.2.2.1 Half-Cell Technique

Half-cell method enables the determination of apparent diffusion coefficient (D_a) in the laboratory. Two half-cells are filled with clay specimens and compacted to the required density. One of the half-cell is saturated with the required contaminating species to be tested and in the other half-cell the specimen is saturated with the distilled water. Once the saturation process is over, both the cells are connected together to allow the diffusion to take place from the half-cell tagged with contaminating species to the half-cell which is filled with uncontaminated soil specimen (Fig. 6.1). The concentration of the contaminant distributed to the cell containing the uncontaminated specimen with time. Once the diffusion test is over, the soil mass in each half cell is sliced and analyzed for the variation of the concentration along the complete length of soil in both cells. A detailed description of the testing method can be found elsewhere [16–18, 29]. Depending on the following two cases in this technique, proper boundary conditions should be chosen for the estimation of diffusion coefficient. The test is considered to follow infinite case if the measured chemical concentration does not reach to the end of the uncontaminated half-cell when the test is terminated and is considered to be finite case when the contaminant concentration reaches the boundary at the end of the test.

For the infinite case, the initial and boundary conditions are given by:

$$\begin{gathered} c(x \leq 0,\, t = 0) = c_0 \\ c(x > 0,\, t = 0) = 0 \\ c(x = \infty,\, t > 0) = 0 \\ c(x = -\infty,\, t > 0) = 0 \end{gathered} \tag{6.7}$$

Fig. 6.1 Illustration of laboratory half-cell

Utilizing the above initial and boundary conditions in Eq. (6.7), the solution to Eq. (6.3) is given by [30]:

$$\frac{c}{c_0} = \frac{1}{2} erfc\sqrt{\frac{x}{2D_a t}} \tag{6.8}$$

where c_0 is the initial concentration, c is the concentration at any given time at x, and D_a is the apparent diffusion coefficient.

For the finite boundary case, the concentration profile reaches to the other end of the half-cell and are described by:

$$\begin{gathered} c(x \leq 0, t = 0) = c_0 \\ c(x > 0, t = 0) = 0 \\ \partial c(x = -x_0, t > 0)\big/\partial x = 0 \\ \partial c(x = L, t > 0)\big/\partial x = 0 \end{gathered} \tag{6.9}$$

Utilizing the above initial and boundary conditions, the solution to Eq. (6.3) is given by Crank [30] as:

$$\frac{c}{c_0} = \frac{x_0}{L} + \frac{2}{\pi}\sum_{m=0}^{\infty}\frac{\exp(-D_a m^2\pi^2 t/L^2)}{m}\cos\left(\frac{m\pi x}{L}\right)\sin\left(\frac{m\pi x_0}{L}\right) \tag{6.10}$$

where m is number of iterations, x_0 is the interface distance from the contaminated soil boundary, and L is the length of the soil sample (summation of the length of two cells).

The apparent diffusion coefficient (D_a) is usually obtained by fitting the solution as described by Eqs. (6.8) and (6.10) to the measure data by method of least squares [16]. Ensuring proper connect between the two half-cells is essential for the correct analysis and thus modified half-cell method [31] is employed to ensure proper contact between the two half-cells. In the modified half-cell technique, the source half-cell is repacked and is packed directly on the uncontaminated half-cell [31]. Moreover, the test is time consuming as the saturation time for the compacted bentonite specimens is ~45 days [32].

The diffusion coefficient obtained in this technique is the apparent diffusion coefficient which is the ratio of effective diffusion coefficient to the retardation factor. Thus the apparent diffusion coefficient does not provide the knowledge of sorption potential and as such the effective diffusion coefficient value is unknown. Shackelford et al. [7] and Shackelford and Daniel [15] replaced D_a with D_e/R_d in Eqs. (6.8) and (6.10) to estimate the D_e by the substitution of R_d from the independent experiments.

6.2.2.2 In-Diffusion Technique

Laboratory in-diffusion technique is a single reservoir technique where the source reservoir is filled with the electrolyte solution and is placed over the clay specimen [33]. The soil is first placed in the diffusion cell, compacted to the required density, and then is saturated by placing water in the source reservoir as shown in Fig. 6.2 [7]. After the complete saturation of the specimen, the tracer solution is introduced into the source reservoir to initiate the diffusion transport through the clay plug. In order to assess the variation of concentration in the source reservoir, the tracer solution is monitored periodically by taking samples through the sampling port. At the end of the diffusion test, the cell is disassembled and the clay plug is sliced to analyze the concentration along the depth. For analysis the extracted soil is subjected to proper chelating agent like EDTA or DDW depending upon the type of ion under consideration [15].

The chemical concentration in the reservoir decreases with time and diffuses into the clay plug with time. As the contaminant flux can't leave through the bottom of the soil plug as shown in Fig. 6.2, the initial and boundary conditions for the mathematic analysis are given by:

$$\begin{aligned} c &= 0(0 \leq x \leq a, t = 0) \\ c &= c_0(a \leq x \leq a + L) \\ \tfrac{\partial c}{\partial x} &= 0(x = 0, t > 0) \\ R_d y + \left(\tfrac{L}{\theta}\right)\left(\tfrac{\partial x}{\partial y}\right) &= L c_0 \end{aligned} \tag{6.11}$$

where a is the length of the diffusion cell, L = effective length of the reservoir and y is the free solute per unit of soil. The solution to Eq. 6.3 with the abovementioned initial and boundary conditions is [30]:

$$\frac{c}{c_0} = \frac{\alpha}{1+\alpha} + \sum_{m=0}^{\infty} \frac{2\alpha}{1+\alpha+q^2\alpha^2} \exp\left(\frac{-D_e^S t q^2}{R_d \alpha^2}\right) \tag{6.12}$$

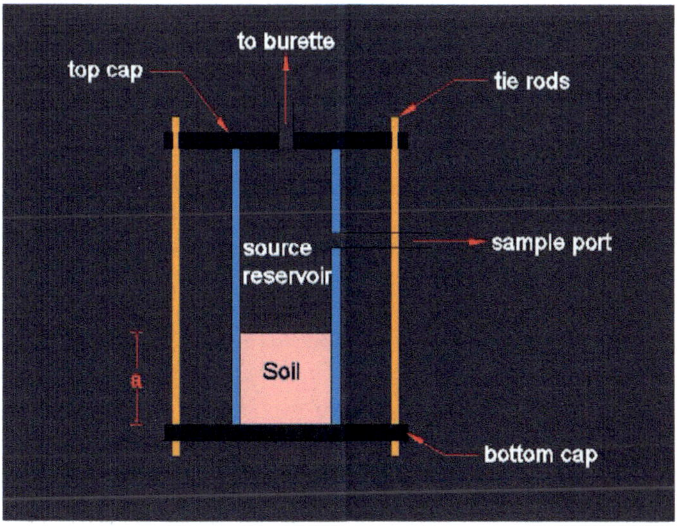

Fig. 6.2 Illustrations of in-diffusion set-up

where c is the solute concentration of the reservoir at any time t and q is the non-zero positive roots given by $\tan q = -\alpha q$ and $\alpha = \frac{L}{\phi R_d a}$.

Effective diffusion coefficient value can be calculated for each sample recovered from the source reservoir during the diffusion test. After the end of the test, the total number of samples collected will give various D_e^S values. The mean of the value is taken and theoretical curve is generated. The best fit curve of the concentration variation of the solute in the source reservoir with time from both experimental results and analytical results gives the value of effective diffusion coefficient.

Effective diffusion coefficient can also be obtained from the concentration profile along the depth of the soil. The initial and boundary conditions, when the concentrations are measured along the depth of the sample are:

$$\begin{gathered} c(x < 0, t = 0) = c_0 \\ c(x > 0, t = 0) = 0 \\ \frac{\partial c}{\partial x}(x = 0, t > 0) = \frac{-\phi R_d D_e^S}{H_f}\frac{\partial c}{\partial x} \\ c(x = \infty, t > 0) = 0 \end{gathered} \quad (6.13)$$

The solution to the in-diffusion problem using these set of initial and boundary conditions is [34]:

$$\frac{c}{c_0} = \exp\left[\frac{\phi R_d x}{H_f} + \left(\frac{\phi}{H_f}\right)^2 D_e^S R_d t\right] erfc\left(\frac{x}{2\sqrt{D_e^S t / R_d}} + \frac{\phi}{H_f}\sqrt{D_e^S R_d t}\right) \quad (6.14)$$

The mass transport parameters D_e^S and R_d are obtained by fitting the measured data of contaminant concentration variation in the source reservoir with time and variation in the soil plug with space for a given time using the analytical solutions (6.12) and (6.14). However, a major limitation of this technique lies in obtaining the contaminant concentration variation with spatial distance. Shackelford and Daniel [15] obtained the effective diffusion coefficients of kaolinite and Lufkin clay by independently obtaining the R_d using the batch experiments for different cations and anions. However, the estimated D_e^S might be smaller than the actual value due to the substitution of smaller R_d from the batch tests.

6.2.2.3 Transient Through-Diffusion (TTD) Technique

Through-diffusion technique (or double reservoir method) is similar to the aforementioned single reservoir technique, but a collector reservoir is attached to the other end of the soil plug to analyze the contaminant concentration in both source and collector reservoirs. The test is commenced by compacting the clay plug into the diffusion cell and then connecting it to two reservoirs at the two ends of the clay plug. The soil plug is saturated by filling both the reservoirs with distilled water. After the complete saturation, diffusion test is started by replacing the distilled water in the source reservoir with the electrolyte solution of known concentration. The collector reservoir is also filled with fresh distilled water to avoid any accumulation of excess salts from the clay specimen during the saturation stage. The illustration of the through-diffusion set-up is given in Fig. 6.3. After the introduction of the electrolyte solution in the source reservoir, mass transport of the chemical constituents takes place by diffusion due to concentration gradient [7]. During the diffusion process, the concentration of the source reservoir decreases but the concentration in the collector reservoir increases with time. The solutions in both the reservoirs are stirred continuously to maintain a uniform concentration at a given time. The source and collector solutions are collected at frequent intervals of time to analyze for the concentration variation with time at both the boundaries of the clay plug. The experimental concentration profile with time for both the reservoirs is used for predicting the mass transport parameters.

The initial and boundary conditions for the laboratory model as shown in Fig. 6.3 are:
Initial conditions:

$$\begin{aligned} c(x = 0, t = 0) &= c_0 \\ c(x, t = 0) &= 0 \\ c(x = L, t = 0) &= 0 \end{aligned} \tag{6.15}$$

where c_0 is the initial concentration of the solute in the source reservoir, t is the diffusion time, L is the length of the soil plug.
Boundary conditions:

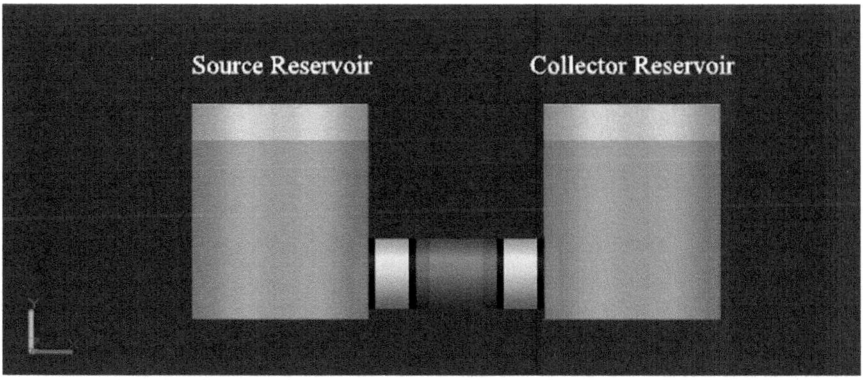

Fig. 6.3 Illustration of the through-diffusion set-up [22]

$$c(x = 0, t) = c_0 + \frac{\phi D_e^S}{H_s} \int_0^{\tau} \left(\frac{\partial c}{\partial x} \right)_{x=0} dt \tag{6.16}$$

$$c(x = L, t) = -\frac{\phi D_e^S}{H_c} \int_0^{\tau} \left(\frac{\partial c}{\partial x} \right)_{x=L} dt \tag{6.17}$$

where Hs and Hc are the equivalent heights of the source and collector reservoir respectively and τ is the integration variable. The Equations [6.3, 6.15–6.17] are solved using Laplace transformation and Cauchy's integral theorem to obtain a closed-form solution recently by Bharat [10]. The analytical solution at the source and collector boundaries is given by [10]:

$$\frac{c(x = x, t)}{c_0} = \frac{1}{2 + k} + 2 \sum_{j=1}^{\infty} \frac{\left(\alpha j^2 + K^2\right) e^{-\alpha j T}}{4K\alpha j^2(k + 1) - \left(2K - \left(\alpha j^2 - K^2\right)\right)\left(\alpha j^2 - K^2\right)} \tag{6.18}$$

$$\frac{c(x = L, t)}{c_0} = \frac{1}{2 + k} - 2K \sum_{j=1}^{\infty} \frac{\left(\alpha j^2 + K^2\right) e^{-\alpha j T}}{4K\alpha j^2(k + 1) - \left(2K - \left(\alpha j^2 - K^2\right)\right)\left(\alpha j^2 - K^2\right)} \tag{6.19}$$

where, $K = \frac{\phi R_d L}{H_r}$; $T = \frac{t D_e^S}{R_d L^2}$, H_r is the equivalent heights of the source and collector reservoir respectively and α_j are the roots of the transcendental equation:

$$\tan \alpha_j = \frac{2K\alpha_j}{\alpha_j^2 - K^2}; \alpha j \text{ are the are the eigen values.}$$

The eigen values were obtained by using the Glowworm swarm optimization technique which utilizes the simultaneous estimation of multimodal solutions [35]. The solution in Equations [17, 18] is obtained by utilizing the first 20 number of roots. The design parameters such as the effective diffusion coefficient and retardation factors are obtained by the inverse analysis by minimizing the error between the theoretical and experimental concentration data. The error minimization was performed using particle swarm optimization [10, 35, 36].

The major advantage of transient through-diffusion technique is that the test is non-destructive as the data of temporal variation in the concentration within the reservoirs are sufficient to obtain both the mass transport parameters. The convenient and non-destructive nature of the test makes the methodology popular in the current laboratory diffusion studies.

6.2.3 Numerical Analysis of Diffusion Transport

The forward solution, which gives the variation of solute concentration with space and time, is obtained by numerically solving the governing diffusion equation with appropriate boundary and initial conditions. For example, the forward solution for TTD technique gives the temporal variation of contaminant concentrations in the source and collector reservoirs by solving the Eqs. [6.3, 6.15–6.17] either analytically [17, 18] or numerically [35]. A finite difference implicit scheme is utilized to discretize the governing differential equation along with the boundary conditions. A numerical grid/mesh in space and time increments to solve the diffusion problem is shown in Fig. 6.4. The indexing in the mesh is defined as $x = x_i = i\Delta x$ and $t = t^n = n\Delta t$ in which i varies as $i = 1, 2, 3, \ldots M$ and n varies as $n = 1, 2, 3, \ldots N$. The finite difference approximation at the ith grid in time and space is computed as $c(i\Delta x, n\Delta t)$. The grid spacing is assumed to be constant in both space and time $(\Delta x = \frac{1}{M}, \Delta t = \frac{1}{N})$. At each grid points, set of equations in the form of tridiagonal matrix is obtained at every time step. The set of equations at each time steps is solved for the solute concentration in the next time level by using Thomas' algorithm [23]. A grid mesh of $M = 1000$ and $N = 100$ is used in this work. The numerical solution is in good agreement with the analytical solution [10].

6.2.4 Inverse Analysis

In the laboratory through-diffusion testing, temporal concentration variations are obtained for both source and collector reservoirs. The design parameters (D_e^S, α, and ϕ or D_e^G, α,) are obtained from these concentration profiles for the performance assessment and design purpose. Prior to the formulation of the closed-form analytical solution for TTD problem [10], semi-analytical approach is used to estimate the effective diffusion coefficient and retardation factor [20, 21]. In semi-analytical approach, the governing differential Eq. 6.3 is solved using the initial and boundary conditions by using Laplace transformation. Owing to the difficulty in converting the solution back into the time domain analytically; numerical inversion technique is used to

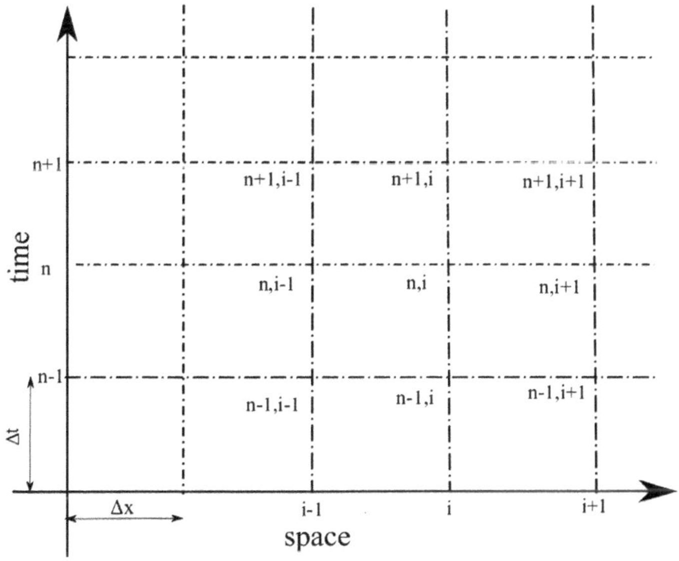

Fig. 6.4 Finite difference mesh

obtain the solution in the time domain. Commercially available software, POLLUTE, is developed based on the semi-analytical approach to compute the temporal variation of the concentration in the source and collector reservoir theoretically [21]. The mass transport parameters (D_e^S, R_d) are obtained by most of the geo-environmental studies by utilizing the commercial software, POLLUTE. The values of D_e^S and R_d are varied for number of iterations till the theoretical plot generated by POLLUTE fits the experimental concentration plots using eye judgement [19]. Bharat [37] showed on the diffusion data of volatile organic compounds through compacted bentonites that accurate estimation of design parameters by the eye-judgement techniques is not possible. Further, availability of different formulations and definitions for the effective diffusion coefficient and porosity in the literature limits the applications of such commercial software. Different optimization techniques are used to perform the inverse analysis to obtain the design parameters by minimizing the objective function. The objective function is formulated based on the difference between the oretically obtained contaminant concentration data for the assumed set of model parameters and experimental data. The smaller value of the objective function or the error implies that the assumed model parameters are close to the actual parameters. Several gradient-based and stochastic optimization algorithms are used for the estimation of model parameters [23]. Particle swarm optimization (PSO) is one of the stochastic techniques, gained significant attention for the estimation of design parameters in many engineering and science disciplines. The PSO algorithm is first introduced by Kennedy and Eberhart [36] and is a class of population-based algorithm which does not require the derivative information. The algorithm is based on the idea of boids simulation and the approach of birds towards a roosting spot [38]. In this algorithm, the flock of birds is represented by the swarm of particles/agents. The swarm flows through the objective function space for the best solution. Each particle in the solution space represents a potential solution of the target objective function. Such particles move in the search space and adaptively update the position in each iteration using the experience guided from the personal flight and from the other particles. Thus the movement of an individual particle is an aggregated acceleration towards the personal best from the own history and the best individual of whole flock. Application of probabilistic transition rules allows the algorithm to avoid local minima and helps in converging towards global minimum. Albeit many virtues, PSO is often found, on different functional spaces, to depend on the tuning parameters. A robust variant of PSO, quantum-behaved particle swarm optimization (QPSO) algorithm is used in this work for the model parameter estimation.

6.2.4.1 Qpso

In quantum behaved particle swarm optimization technique each particle is supposed to be in a bound state and have quantum behavior. Unlike the PSO technique, where the particle oscillates and converges to its local attractor, in QPSO technique, the particle is assumed to be attracted by a quantum potential well centered on its local attractor. The particle so attracted gets new position for every update of a stochastic equation. The QPSO technique samples the new position with a double exponential distribution. The update equation needs to adjust a very few parameters resulting in a very good performance of the overall algorithm. A global point known as the mean best was introduced in the algorithm for enhancement of the global search [39]. The QPSO is used to solve a wide range of optimization problems (Fig. 6.5).

6.3 Results and Discussion

6.3.1 Limitations of Optimization Techniques

Deterministic techniques such as the gradient-based optimization strategies (ex. Levenberg-Marquardt algorithm) are commonly used for obtaining the model parameters in many engineering disciplines. The fundamental principle behind these techniques is that the derivative of the objective function with respect to the model parameters (D_e, R_d) is zero at the optimum location. The unknown model parameter vector, in these methods, is searched iteratively. The application of deterministic technique for the estimation of apparent diffusion coefficient from the half-cell technique is demonstrated here using the MATLAB® call function, fmincon, for the KCl diffusion through an agricultural field soil [31]. As the deterministic techniques require an initial (guess) solution for the optimization of the model parameters, two

```
begin
    Initialize the current positions and the pbest positions of all the particles
    Set n=0; (number of iterations)
    while the termination is not met do
        Compute the mean best position (m)
        for i=1 to M do
            Evaluate the objective function value (X_i,n);
            Update pbest_i,n and pgbest_n;
            for j =1 to N do
                v_i,n^j = rand ( )*pbest_i,n + (1-rand ( ))*pgbest_n^j;
                if rand ( ) < 0.5 then
                    x_i^j,n+1 = v_i^j,n + w * (mean_i^j,n − x_i^j,n) * ln(1/ rand ( ))
                else
                    x_i^j,n+1 = x_i^j,n − w * (mean_i^j,n − x_i^j,n) * ln(1/ rand ( ))
                end
            end
        end
        Set n = n+1
    end
end
```

Fig. 6.5 QPSO algorithm (after [39])

different initial solutions were chosen as shown in Table 6.1. The predicted solutions for different initial guesses and the objective function values (i.e., RMSE) were also shown in the same table. The predicted model parameter is different by an order of magnitude when the initial guesses were different. The theoretical data from the predicted model parameters from the deterministic approach were compared against the measured data in Fig. 6.6. Theoretical data obtained for the initial guess value of $D_a = 5 \times 10^{-5}$ cm^2/s significantly deviated from the measured data. Thus the deterministic techniques are strongly influenced by the initial solution and converge to one of the local minima. The theoretical data obtained by the QPSO are also presented in Fig. 6.6. The data showed close agreement with the solution obtained by deterministic approach with an initial guess value of $D_a = 1 \times 10^{-5}$ cm^2/s. As the stochastic techniques such as QPSO do not require initial solution and gradient information, such techniques provide global minimum or solution close to the global optimum solution. Performance of the deterministic algorithms may be more detrimental for TTD and other techniques as the required number of model parameters is more than unity, which is not demonstrated here using the Levenberg-Marquardt algorithm [23].

Table 6.1 Inverse analysis by deterministic algorithm for half-cell technique

S. no.	Initial solution, D_a (cm^2/s)	Predicted solution	
		D_a (cm^2/s)	RMSE
1	5×10^{-5}	2.26×10^{-7}	0.1549
2	1×10^{-5}	2.53×10^{-6}	0.0199

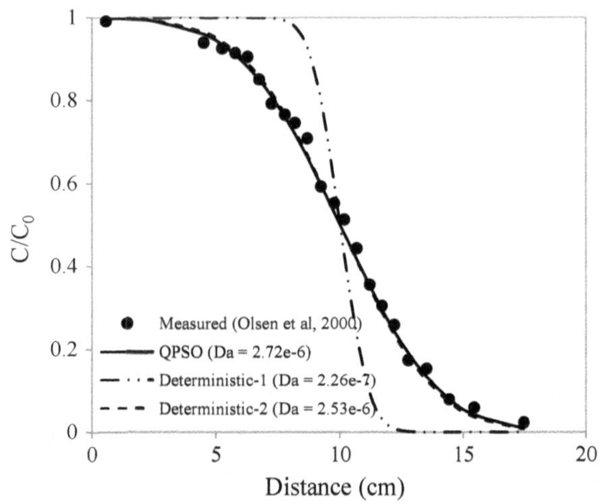

Fig. 6.6 Variation in the model parameters from the deterministic and stochastic techniques

6.3.2 Prediction of Accessible Porosity for Non-reactive Flows

As mentioned earlier, the accessible porosity is smaller than the soil porosity estimated from the dry density and specific gravity. Thus the diffusion accessible porosity is required to be estimated along with the effective diffusion coefficient and the capacity factor (or retardation factor). The soil porosity is substituted for accessible porosity in most of the literature works for the reactive transport for the estimation of D_e^S and R_d. The accessible porosity is estimated along with D_e^G for diffusion of non-reactive tracers such as HTO and Cl^- through compacted bentonites in few studies [26]. As the ϕ is independently estimated from the diffusion experiments using non-reactive tracers, the effective diffusion coefficient ($D_e^S = D_e^G/\phi$) is computed. The sensitivity of the diffusion accessible porosity on the contaminant concentration data of TTD model using different formulations Eqs. (6.1 and 6.2) was demonstrated in Figs. 6.7 and 6.8. Theoretical data were obtained by considering the clay plug thickness to be $L = 1$ cm and $H_f = H_c = 5.09$ cm for these simulations. The temporal concentration data were shown on semi-logarithmic plot to clearly distinguish different theoretical profiles. The variation in the temporal concentration profiles with

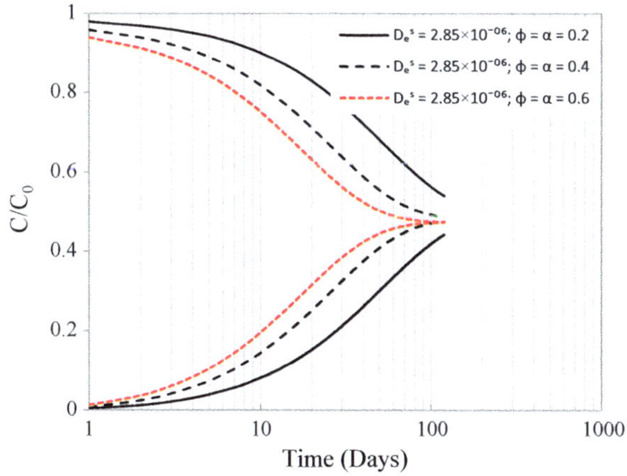

Fig. 6.7 Sensitivity of capacity factor for non-reactive transport using Eq. (6.1) in TTD tests

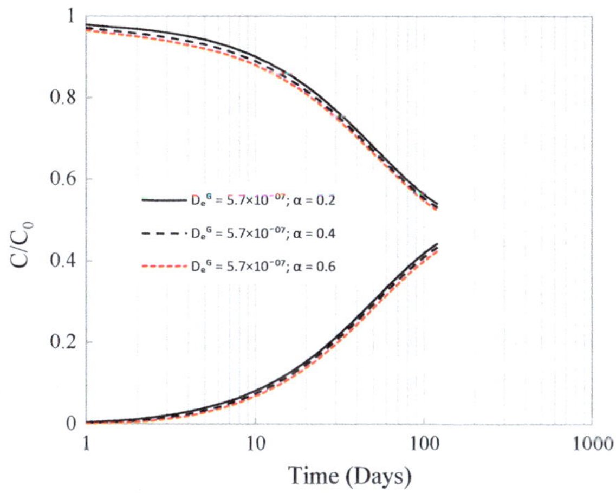

Fig. 6.8 Sensitivity of capacity factor for non-reactive transport using Eq. (6.2) in TTD tests

the accessible porosity (or capacity factor, for non-reactive diffusion flow) using the formulation based on Eq. (6.1) showed a clear influence of accessible porosity. Thus the estimation of D_e^S and ϕ is possible independently from the measured data. The variation in the temporal concentration profiles for the same diffusion coefficient ($D_e^{\mathrm{G}} = D_e^{\mathrm{S}}\phi$) as used in Fig. 6.7 with varying accessible porosity values were shown in Fig. 6.8. The variation in the temporal concentration profiles with the accessible porosity (or capacity factor, for non-reactive diffusion flow) was insignificant when the formulation using Eq. (6.2) was considered. As the concentration data appeared to be insensitive to the accessible porosity when varied over a wide range, the prediction of capacity factor may pose severe limitations.

6.3.3 Prediction of Accessible Porosity for Reactive Flows

The estimation of D_e^S for reactive tracers or salt diffusion through compacted bentonite clays requires an independent estimation of ϕ along with D_e^S and R_d from the inverse analysis of measured temporal concentration data. The possibility of estimation of D_e^S, ϕ, α (or R_d) from the synthetically generated data and measured data was demonstrated here. Synthetic data using different percentages of white noise were generated as presented in Fig. 6.9 for TTD case using $L = 1$ cm, $H_f = H_c = 5$ cm, $D_e^S = 1.52 \times 10^{-6}$ cm^2/s, $\alpha = 3.129$; $\phi = 0.694$. Inverse model based on QPSO was used to predict all three model parameters (D_e^S, ϕ, α) for synthetic data with different percentage of noise. The predicted theoretical data were in close agreement with the synthetic data for all the cases. However, two of the predicted model parameters (D_e^S, ϕ) significantly varied for different noise values albeit providing the same transient contaminant concentration data. The predicted D_e^G ($D_e^S \times \phi$), however, showed that the values converged to the inputted model parameters. Thus the TTD data were not sufficient to predict the accessible porosity for the reactive transport. Further, the three-parameter estimation exercise was conducted on the measured data of chloroform diffusion and KCl diffusion through different compacted soils.

The measured data from the diffusion experiments conducted for chloroform was taken from [19] to independently estimate the model parameters. Inverse analysis was performed to obtain the model parameters (D_e^S and α) by inputting the soil porosity. The theoretical concentration data obtained for the optimized value of the model parameters was in good agreement with the measured data. The measured data were also used to estimate the three parameters (D_e^S, ϕ, α) independently. It was observed that although the measured concentration data corroborated well with the theoretical data, but there was a significant difference in the estimated value of D_e^S (Fig. 6.8).

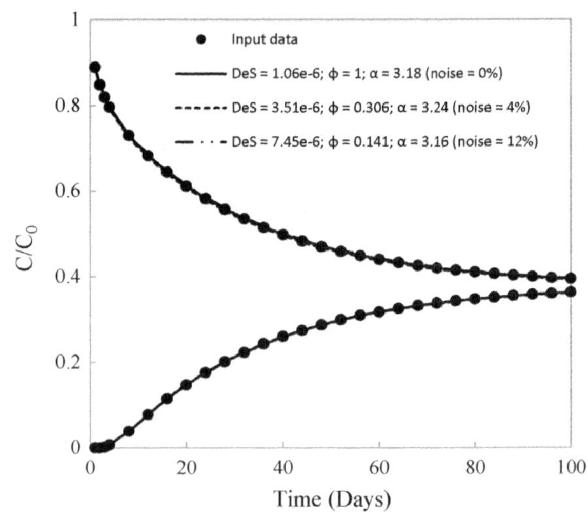

Fig. 6.9 Estimation of the model parameters from the synthetic data

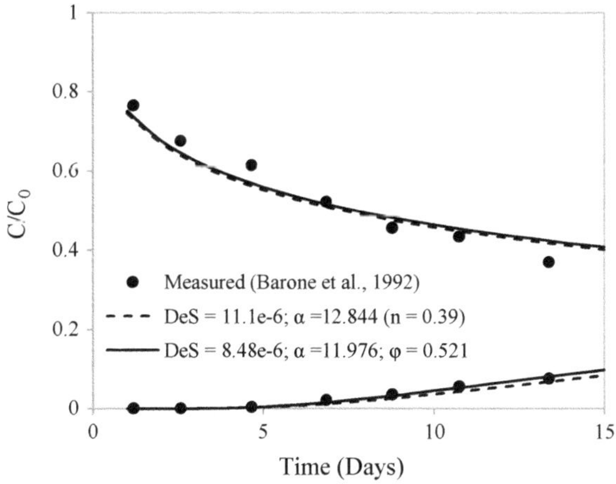

Fig. 6.10 Estimation of model parameters from the chloroform diffusion data [19]

However, the estimation of D_e^G ($D_e^S \times \phi$), would be more reliable as it leads to the estimation of same effective diffusion coefficient for both the cases (Fig. 6.8). Similar observations were found when the experimental concentration variation of the source and collector reservoirs for KCl diffusion through compacted Barmer bentonite was considered from the literature [8] for the parameter estimation. By utilizing the measured concentration data and inputting the value of soil porosity theoretical concentration profile was obtained for the optimized value of D_e^S and α (Fig. 6.9). The accurate estimation of all the three parameters was not possible and resulted in different sets of D_e^S and α. However, it was observed that the D_e^G ($D_e^S \times \phi$), was unique for both the trials. As the predicted D_e^S in the reactive case is influenced by the inputted porosity (n), an independent estimation of D_e^S similar to the non-reactive case is not possible. Therefore, the effective diffusion coefficients of reactive and non-reactive cases through the same clay specimen are not suitable for comparison (Figs. 6.10, 6.11).

6.4 Conclusions

The following conclusions are drawn based on the aforementioned diffusion analyses.

1. Different formulations for effective diffusion coefficients for the diffusion transport through compacted clays significantly influence the predicted accessible porosity (or capacity factor) for the non-reactive transport. The sensitivity of temporal concentration data to the accessible porosity is found to be different for different formulations.
2. The effective diffusion coefficients for reactive and non-reactive cases through the same clay specimen are not suitable for comparison as the effective diffusion coefficient estimation in reactive transport is influenced by the inputted porosity.

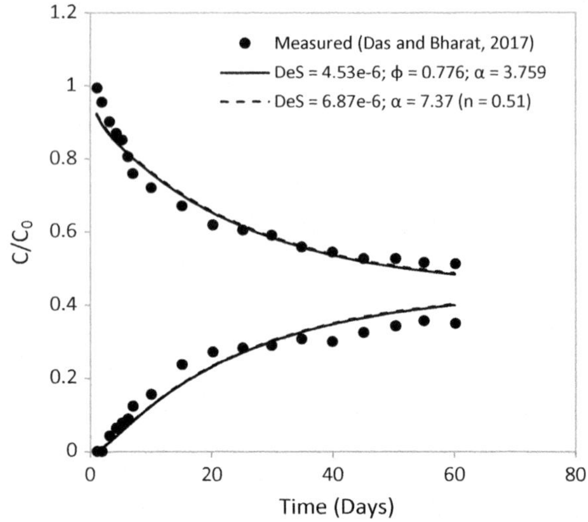

Fig. 6.11 Estimation of model parameters from the KCl diffusion data through Barmer bentonite [8]

3. Laboratory diffusion techniques should be chosen based on the required model parameters, but one technique is not substitute for the other one. For example, half-cell technique is useful for the estimation of accessible porosity, but estimation of effective diffusion coefficient is not possible. In-Diffusion and through-diffusion techniques are useful for the estimation of effective and apparent diffusion coefficients independently.

References

1. Bianchi, M., Liu, H.H., Birkholzer, J.T.: Radionuclide transport behavior a generic geological radioactive waste repository. Groundwater. National Grounwater Association (2015). https://doi.org/10.1111/gwat.12171
2. Pusch, R.: Use of bentonite for isolation of radioactive waste products. Clay Miner. **27**, 353–361 (1992)
3. Pusch, R.: The performance of clay barriers in repositories for high-level radioactive waste. Nucl. Eng. Technol. **38**, 483–488 (2006)
4. Raj, K., Prasad, K.K., Bansal, N.K.: Radioactive waste management practices in India. Nucl. Eng. Des. (2006). https://doi.org/10.1016/j.nucengdes.2005.09.036
5. Weber, W.J., Ewing, R.C., Catlow, C.R.A., Rubia, T.D., Hobbs, L.W., Kinoshita, C., Matzke, H., Motta, A.T., Nastasi, M., Salje, E.K.H., Vance, E.R., Zinkle, S.J.: Radiation effects in crystalline ceramics for the immobilization of high-level nuclear waste and plutonium. J. Mater. Res. **13**(6) (1998)
6. Wattal, P.K.: Indian programme on radioactive waste management. Sadhana 38(5) (2013). https://doi.org/10.1007/s12046-013-0170-0
7. Shackelford, C., Daniel, D.E., Liljestrand, H.M.: Diffusion of inorganic chemical species in compacted clay soil. J. Contam. Hydrol. **4**(3), 241–273 (1989)
8. Das, P., Bharat, T.V.: Effect of counter ions on the diffusion characteristics of a compacted bentonite. I. Geotech. J. (2017) https://doi.org/10.1007/s40098-017-0241-y

9. Benson, C., Zhai, H., Wang, X.: Estimating the hydraulic conductivity of compacted clay liners. J. Geotech. Eng. ASCE. **120**, 366–387 (1994)
10. Bharat, T.V.: Analytical model for 1-D contaminant diffusion through clay barriers. Environ. Geotech. **1**(4), 210–221 (2013)
11. Bohnhoff Gretchen, L., Shackelford Charles, D.: Hydraulic conductivity of chemically modified bentonites for containment barriers. In: Proceedings of the 7th International Congress on Environmental Geotechnics. iceg2014. [Barton, ACT]: Engineers Australia, pp. 440–447 (2014). ISBN 9781922107237
12. Chen, Y.G., Zhu, C.M., Ye, W.M., Cui, Y.J., Chen, B.: Effects of solution concentration and vertical stress on the swelling behavior of compacted GMZ01 bentonite. Appl. Clay Sci. **124**, 11–20 (2016)
13. Bouazza, A., Van Impe, W.F.: Liner design for waste disposal sites. Environ. Geol. **35**(1), 41–54 (1998)
14. Shackelford, C.D., Daniel, D.E.: Diffusion in saturated soil. I. Backgr. J. Geotech. Eng. **117**(3), 467–484 (1991a)
15. Shackelford, C.D., Daniel, D.E.: Diffusion in saturated soil. II. results for compacted clay. J. Geotech. Eng. **117**(3), 485–506 (1991b)
16. Robin, M.J.L., Gillham, R.W., Oscarson, D.W.: Diffusion of strontium and chloride in compacted clay-based materials. J. Soil Sci. Soc. Am. **51**, 1102–1108 (1987)
17. Oscarson., et al.: Sorption of caesium in compacted bentonite. Clays Clay Miner. **42**(6), 731–736 (1994)
18. Cho, W.J., Oscarson, D.W., Gray, M.N., Cheung, S.C.H.: Influence of diffusant concentration on diffusion coefficients in clay. Radiochim. Acta. **60**, 159–163 (1993)
19. Barone, F.S., Rowe, R.K., Quigley, R.M.: A laboratory estimation of diffusion and adsorption coefficients for several volatile organics in a natural clayey soil. J. Contam. Hydrol. **10**, 225–250 (1992)
20. Rowe, R.K., Booker, J.R.: 1-D pollutant migration in soils of finite depth. J. Geotech. Eng. **111**(4), 479–499 (1985)
21. Rowe, R.K., Caers, C.J., Barone, F.: Laboratory determination of diffusion and distribution coefficients of contaminants using undisturbed clayey soil. Can. Geotech. J. **25**(1), 108–118 (1988)
22. Bharat, T.V., Sivapullaiah, P.V., Allam, M.M.: Expert Syst. Appl. **39**(12), 10812–10820 (2012)
23. Bharat, T.V., Sivapullaiah, P.V., Allam, M.M.: Swarm intelligence-based solver for parameter estimation of laboratory through-diffusion transport of contaminants. Comput. Geotech. **36**(6), 984–992 (2009)
24. Mike, D.: Contaminant fluxes from hydraulic containment landfills-a review, environment agency science report SC0310/SR (2004)
25. Garcia-Gutieirrez, M., Missana, T., Mingarro, M., Samper, J., Dai, Z., Molinero, J.: Solute transport properties of compacted Ca-bentonite used in FEBEX project. J. Contam. Hydrol. **47**, 127–137 (2001)
26. Garcia-Gutierrez, M., Cormenzana, J.L., Missana, T., Mingarro, M.: Diffusion coefficients and accessible porosity for HTO and $36Cl^-$ in compacted FEBEX bentonite. **26**, 65–73 (2004)
27. Shackelford, C.D., Moore, S.M.: Fickian diffusion for radionuclides for engineered containment barriers: diffusion coefficients, porosities and complicating issues. Eng. Geol. **152**(2013), 133–147 (2013)
28. Skagius, K., Neretnieks, I.: Porosities and diffusivities of some non-absorbing species in crystalline rocks. Water Resour. Res. **22**(3), 389–398 (1986)
29. Gillham, R.W., Robin, M.J.L., Dytynyshyn, D.J., Johnston, H.M.: Diffusion of nonreactive and reactive solutes through fine-grained barrier materials. Can. Geotech. J. **21**(3), 541–550 (1984)
30. Crank, J.: The Mathematics of Diffusion. Clarendon Press, London (1975)
31. Olesen, T., Moldrup, P., Yamaguchi, T., Nissen, H.H., Rolston, D.E.: Modified half-cell method for measuring the solute diffusion coefficient in undisturbed, unsaturated soil. Soil Sci. **164**(11), 835–840 (2000)

32. Das, P., Manparvesh, S.R, Bharat, T.V.: Salt diffusion in compacted plastic clay: experimental and theoretical evaluation. In: 2016 Proceedings of the International Conference on Soil and Environment, ICSE, Bangalore (2016)
33. Barone, F.S., Rowe, R.K., Quigley, R.M.: Laboratory determination of chloride diffusion coefficients in an intact shale. Can. Geotech. J. **27**(2), 177–184 (1990)
34. Carslaw, H.S., Jaeger, C.: Conduction of Heat in Solids. Clarendon Press, Oxford (1959)
35. Bharat, T.V.: Agents based algorithms for design parameter estimation in contaminant transport inverse problems. In: IEEE Swarm Intelligence Symposium, pp. 1–7 (2008)
36. Kennedy, J., Eberhart, R.C.: Particle swarm optimization. In: Proceedings of the IEEE International Conference on Neural Networks, Perth, Australia, pp. 1942–1948 (1995)
37. Bharat, TV.: Metaheuristics for parameter estimation in contaminant transport modeling through soils. In: Proceedings of the 4th International Young Geotechnical Engineers (2009)
38. Clerc, M., Kennedy, J.: The particle swarm—explosion, stability, and convergence in a multidimensional complex space. IEEE Trans. Evol. Comput. **6**, 58–73 (2002)
39. Sun, J., Fang, W., Wu, X., Palade, V., Xu, W.: Quantum-behaved particle swarm optimization: analysis of individual particle behavior and parameter selection. Evol. Comput. **20**(3) (2012)

Chapter 7
Axial Load-Displacement Behavior of Energy Pipeline Systems in Sand

Sai K. Vanapalli and Mohammed Al-Khazaali

7.1 Introduction

Pipeline systems form vital infrastructure that significantly contribute to the global economy in general and Asia in particular. Recently, Asia has become the largest energy consumer in energy markets. The fast-growing economy of India and China combined demand for energy will gradually surpass half of the total global energy demands by 2030 [1]. India is the third largest fossil fuel (i.e., oil, gas, and coal) consumer in the world [2]. Oil and gas combined provide 34.4% (i.e., 27.9% from crude oil and 6.5% from natural gas) of the total consumed energy in India. To support such growing demands, new potential oil and gas suppliers in Central Asia (i.e., Kazakhstan, Turkmenistan, and Azerbaijan) in addition to Russia are expected to supplement India and China constructing new and as well as developing the existing regional energy pipeline network [3]. Table 7.1 provides a succinct summary of new and existing energy pipeline projects and size of investments for these projects in Asia and worldwide.

Onshore pipeline systems are typically buried at a shallow depth of 0.8–1.2 m from the soil surface to prevent direct expose to human daily base activities [16]. These systems are conventionally designed and analyzed following design codes, guidelines, and recommended practice procedure manuals that extend the principles of saturated soil mechanics [17–20]. In addition, BS EN 1997-1:2004 [21] recommends that for stability and serviceability requirements, the most unfavorable groundwater condition that could occur during the structure lifetime should be considered. In other

S. K. Vanapalli (✉) · M. Al-Khazaali
Department of Civil Engineering, University of Ottawa, Ottawa, Canada
e-mail: Sai.Vanapalli@uottawa.ca

M. Al-Khazaali
Building and Construction Engineering Department,
University of Technology, Baghdad, Iraq

M. Latha G. (ed.), *Frontiers in Geotechnical Engineering*, Developments in Geotechnical Engineering, https://doi.org/10.1007/978-981-13-5871-5_7

Table 7.1 Major present and prospective pipeline projects in different regions of the world: investments and benefits

Region	Investment and potential benefits	References
Africa	Africa is expected to become one of the most important natural gas production region in the near future to cater the needs of the world according to Energy Information Administration (EIA) [4]. There are several pipeline projects that connect North Africa to Europe. In addition, several projects are planned to transport natural gas within Africa, for example	EIA [4]
	A 600 km pipeline project was planned by the Nigerian government to transport gas to Benin, Togo, and Ghana. The project is expected to save $500 billion in primary energy costs over a period of 20 years [5]	Obadina [5]
	Chad-Cameroon pipeline project was constructed between 2000 and 2004 investing $3.5 billion. The pipeline project helped Chadian people to overcome poverty problems	African Development Bank and African Union [6]
	China invested $15 billion to construct 1500 km pipeline system in Sudan. This infrastructure facilitates to transport oil	
Asia	Some of the proposed and ongoing projects are listed below	Mahmood et al. [7]
	The ongoing Turkmenistan, Afghanistan, Pakistan and India (TAPI) gas pipeline project: This project will extend on 1680 km and transport 33 billion cubic meters (bcm) per year of natural gas from the Galkynysh field in Turkmenistan to Afghanistan, Pakistan and India. The project will provide 38 million cubic meters (mcm) per day of natural gas to each of India and Pakistan, and 14 mcm to Afghanistan. The construction cost of this project was estimated to be at least $7 billion [7]	MPNGGI [2]
	Jagdishpur-Haldia and Bokaro-Dhamra pipeline (JHBDPL) projects: The Indian government in 2016 granted Rs. 5,176 Core (i.e., USD $700 million) to Gas Authority of India Limited (GAIL) to develop JHBDPL pipeline projects which are known as Pradhanmantri Urja Ganga Project. The project will connect the Eastern part of the country with the natural gas grid [2]	IEA [1]
	Eastern Siberia-Pacific Ocean oil pipeline will transport 300,000 barrels per day from Eastern Siberia, Russia to China and another 300,000 barrels per day to Japan and other Asian markets	Chow et al. [3]
	A pipeline project constructed by British petroleum (BP) and partners in Azerbaijan International Oil Consortium (AIOC) to transport oil from Azerbaijan to a marine terminal at Supsa in Georgia. BP and AIOC partners invested over $500 million in construction this pipeline	
	Two pipeline projects transport the imported crude oil and natural gas from Middle East and Africa to Southwest China. The pipeline projects are 1,110 km (690 mile) crude oil pipeline starts from Kyaukphyu, Myanmar to China and 1807 km (1,123 mile) natural gas pipeline starts from the Bay of Bengal	

(continued)

Table 7.1 (continued)

Region	Investment and potential benefits	References
Europe	To supplement existing European Union energy markets, new pipeline projects have been initiated to increase the import capacity from Africa, middle east, and Asia, and ensure stable supply of petroleum products. New pipeline projects were started by Turkey and Azerbaijan (TANAP) and Trans-Adriatic Pipeline project (TAP). Natural gas from Caspian Sea Shah Deniz fields in the east would be transported to Grece via Turkey through TANAP. The project is expected to be completed in 2020 and will initially transport 16 bcm of gas per year with potential increase up to 31 bcm by 2026. The two projects, TANAP and TAP are to be connected in order to link Turkey with Italy through Greece and Albania. For EU, the new energy pipeline projects will provide safe routes and avoiding gas transit through Ukraine	Bjørnmose et al. [8] DeMicco [9]
North America	Canada has extensive pipeline infrastructure; the value of energy transported over the Canadian pipeline system within Canada and to other export markets exceeded $100 billion annually during the period 2007–2012 while the annual cost of transport through the same pipelines averaged less than $5 billion. The exported energy by Canadian pipeline system contribute to approximately one fifth of Canada's total annual merchandise export revenues [10]. According to Canadian Energy Pipeline Association (CEPA) [11], a hypothetical new natural gas pipeline investment of $1.52 billion, 1000 km located half in Alberta and half in British Columbia provinces would increase the Canadian gross domestic product (GDP) in the order of $1.2 billion. The new pipeline project would create several thousands of jobs in Alberta and British Columbia in addition to Ontario, Quebec, Saskatchewan, and Manitoba provinces. Furthermore, delaying the new pipeline construction would yield that the costs to the Canadian natural gas consumers could be as much as $57.5 billion over the period from 2006 to 2025 based on a constant 2005 dollar value	Benoit [10]
	USA is the largest natural gas producer at 65 billion cubic feet per day. In addition, USA has achieved remarkable improvement in crude oil production capacity growth, adding 1.2 million barrels per day of capacity over the 2008–2012 period [12]. The USA natural gas and oil pipeline system is the largest pipeline network with 4.18 million km (2.6 million miles) that carries 14 billion barrels of crude oil and about 25 trillion cubic feet of natural gas in 2012 [13]. New pipeline projects are needed to support the large amounts of domestically produced energy and alleviate chokepoints that prevent efficient and timely delivery of energy resources. Keystone XL pipeline project, for example, between USA and Canada is a promising project. The proposed pipeline starts from Alberta and ends at Nebraska and pass through Saskatchewan in Canada and Montana, and South Dakota in USA. The Keystone XL pipeline will serve to transport up to 830,000 barrels of oil per day from Canada and America's Bakken Shale formation to USA [13]. Keystone XL pipeline construction will create 42,000 job opportunities and provide $2 billion in workers pockets over 2 years	CEPA [11] Fullenbaum et al. [12] Gerard [13]

(continued)

Table 7.1 (continued)

Region	Investment and potential benefits	References
Middle East	Middle East region has significant oil and gas resources. According to the Organization of the Petroleum Exporting Countries (OPEC) [14] around 23.5 million barrel per day (i.e., 20% of world oil production) is pumped from this region and around 800 billion barrels are reserved in this region out of around 1,493 billion barrels world proven crude oil reserves based on 2014 statistics. In addition, around 625,800 mcm of natural gas were produced in 2014 and 80,109 bcm are reserved	OPEC [14]
	The region suffered from several conflicts and political disturbances which threatened the flowing of petroleum products to the world markets. The vulnerability of the energy transportation systems in the region led to propose several new pipeline projects, for example [15]	Mills [15]
	Ras Markaz pipeline project that is connected to export terminal and oil storage park near the port of Duqm in Oman. The project will offer an outlet to the Indian Ocean	
	Haditha-Jordan pipeline project in Iraq. This project would help to reduce the dependence of Iraqi petroleum industry on the vulnerable Basra terminals and the Gulf and Strait of Hormuz	
	Khurmala-Fishkhabour pipeline project in Iraq's Kurdistan region	
	Another pipeline project in Iran is proposed to transport oil from Neka on the Caspian Sea to Jask on the Arabian Sea	

words, soil should always be considered saturated by considering the groundwater table (GWT) is at the surface in the analyses and design assumptions.

However, assuming the presence of GWT at the natural ground surface does not necessarily assure a conservative design approach. Achten et al. [22] studies suggest that arid and semi-arid regions constitute around 40–50% of the total global land area. These regions experience significant water evaporation in comparison to infiltration which contributes to the development of a shallow unsaturated soil layer (i.e., vadose zone) that typically extends to a depth of about 2–3 m. In other words, in semi-arid and arid regions, pipeline systems are most likely to be buried in and interact with unsaturated soil in which the capillary stress or matric suction, $(u_a - u_w)$ has a direct impact on the soil–pipeline system behavior.

Limited research studies were devoted to understand the behavior of soil–pipeline system taking account of the influence of suction [23–29]. These studies highlight the mechanical behavior of a pipeline system in unsaturated soils is significantly influenced by the soil suction. Saadeldin et al. [29] investigations suggest the flexible and rigid pipes buried in an unsaturated clayey soil undergo large deformations in pipelines due to soil heave upon wetting due to a reduction in the soil suction. Robert et al. [28] numerical studies on a buried pipeline subjected to strike-slip fault movement in unsaturated sand suggest that the deformations anticipated in a pipeline buried in unsaturated sand are much higher due to contribution of $(u_a - u_w)$ towards soil stiffness and shear strength.

Al-Khazaali et al. [23] extended a modified total stress analysis approach to numerically investigate the behavior of a buried pipeline in unsaturated Indian Head till (IHT) soil nearby unsupported excavations. The study suggests that pipeline systems buried in unsaturated IHT soil may experience much higher internal forces (i.e., shear and axial forces, and bending moment) in comparison to saturated condition. The above-discussed examples provide evidence that saturated condition is not always the most unfavorable soil condition or conservative scenario that can be used in the design of pipeline systems. The higher shear strength and stiffness of soil under unsaturated conditions can contribute to extensive damages to pipeline system and likely affect its integrity.

The present study extends the principles of saturated and unsaturated soils mechanics in evaluating the response of pipeline system buried in a sand and subjected to an axial load under saturated and unsaturated conditions, respectively. The test results of the axial load-displacement tests on the pipe showed that the axial forces and stresses are significantly higher when the soil is in a state of unsaturated condition in comparison to the saturated condition. The study results highlight the need for revising the presently used design procedures which are based on conventional soil mechanics. In other words, principles of unsaturated soil mechanics should be extended towards proposing rational design procedures for pipelines buried in unsaturated soils.

7.2 Background

Axial forces on a buried pipeline arise from relative movement between the pipeline and surrounding soil. Such soil mass movement or permanent ground deformation (PGD) is induced by soil subsidence, landslides, liquefaction, frost heave or thaw settlement, or seismic loads (Fig. 7.1). According to American Lifeline Alliance (ALA) [17], longitudinal PGD causes damages to non-seismically designed steel pipelines approximately 5–10 times higher than transverse PGD. Such a statement is made based on the fact that a pipe is inherently more flexible when subjected to bending comparing to axial tension or compression.

The mobilized shear stress along the pipe body is the key parameter that is required to estimate the maximum axial load exerted on a pipeline. The mechanism of mobilizing shear stress along a pipeline body is similar to shaft resistance in pile foundation. As a result, a modified version of the classical β-method by Burland [30], Eq. (7.1) for cohesionless soils extending Terzaghi's effective stress concept and α-method by Skempton [31], Eq. (7.2) for cohesive soils extending total stress analysis approach were utilized to predict axial load per unit length, F_u exerted on a buried pipeline.

$$F_u = \pi \cdot D \cdot \left[\gamma' \cdot H \cdot \left(\frac{1 + K_0}{2}\right)\right] \cdot \tan\delta \tag{7.1}$$

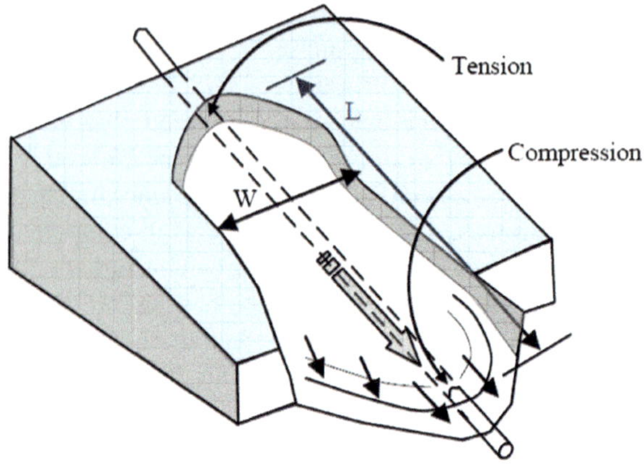

Fig. 7.1 A pipeline subjected to longitudinal PGD (modified after [19])

$$F_u = \pi \cdot D \cdot \alpha \cdot c_u \tag{7.2}$$

where D is the pipe outside diameter, K_0 is coefficient of lateral earth pressure at rest, γ' is effective unit weight, H is the soil cover thickness to the springline, δ is the friction angle for the interface, c_u is soil undrained cohesion, and α is adhesion factor. The value of δ is a function of pipeline coating materials and can be estimated as a ratio of soil effective friction angle, ϕ' using a reduction factor [20, 32].

The former equations are valid for saturated soils; due to this reason, Eqs. 7.1 and 7.2 cannot reliably estimate the axial force exerted on a pipeline body in unsaturated soils and likely result in unsafe design loads due to ignoring$(u_a - u_w)$ influence on the shear strength of soil–pipeline interface.

The shear strength for unsaturated soil is a function of two independent stress state variables, namely: net normal stress, $(\sigma - u_a)$ and matric suction, $(u_a - u_w)$ [33, 34]. Fredlund et al. [34] extended the classical Mohr–Coulomb model for interpreting the shear strength of unsaturated soils, τ_{unsat} in terms of $(\sigma - u_a)$ and $(u_a - u_w)$ (Eq. 7.3).

$$\tau_{unsat} = c' + (\sigma - u_a)\tan\phi' + (u_a - u_w)\tan\phi^b \tag{7.3}$$

where c′ is effective cohesion, σ is normal stress, u_w is pore-water pressure, ϕ^b is angle of internal friction with respect to $(u_a - u_w)$ variation under a constant $(\sigma - u_a)$, and u_a is pore air pressure.

The variation of the $\tau_{unsat}$is a function of $(u_a - u_w)$ which is directly related to moisture content. The relationship between moisture content or degree of saturation and soil suction can be described using the soil-water characteristic curve (SWCC) (Fig. 7.5). Three key zones with distinct features can be recognized from the SWCC namely: (i) In the boundary effect zone, all the pores are filled with water and the soil is essentially saturated up to the air entry value (AEV); (ii) In the transition zone, there is a gradual reduction in moisture content in response to an increase in the suction; and, (iii) In the residual zone, there is no continuity in the water phase and large changes in soil suction result in only small changes in water content [35].

The variation of the unsaturated shear strength in the zones of suction is defined by ϕ^b value. Up to the AEV, $\phi^b = \phi'$ and τ_{unsat} varies linearly with suction. As the $(u_a - u_w)$ exceeds the AEV, the contribution of ϕ^b becomes less than ϕ' due to a decrease in the wetted contact area of water within the soil particles. The variation of τ_{unsat} with respect to $(u_a - u_w)$ is nonlinear in both transition zone and residual zone suction contribution towards shear strength reduces. The variation of τ_{unsat} influences the mechanical behavior of the pipeline system buried in a soil that is in a state of unsaturated condition. Due to this reason, significantly larger loads will be exerted on the pipeline body than the expected design capacity. This background discussion suggests the conventional methods that are based on the principles of saturated soil mechanics underestimate the design capacity of pipelines.

7.3 Experimental Program

The objectives of the current study can be summarized as below: (i) to undertake an extensive experimental testing program for investigating pipeline system behavior subjected to axial load under saturated and unsaturated conditions, and (ii) evaluating the contribution of matric suction towards the axial load exerted on a pipe buried in sand under unsaturated conditions.

7.3.1 Experimental Setup

The experimental setup constituted of a relatively large soil container, a prototype pipe, several instrumentation and other features to facilitate testing under saturated and unsaturated conditions. The soil container was specially designed and fabricated using rigid steel plates with internal dimensions of 1.5 m length, 1.2 m width, and 1.1 m height.

On one of the sides of soil container, a transparent acrylic observation window that extends from the base to top of the container was installed. The window facilitates in the visual inspection of water movement during the process of saturation and desaturation (Fig. 7.2a).

Several methods are available in literature to investigate the performance of a structure under unsaturated condition, for example; spraying water on the soil surface [36] or mixing the soil before compacting it with a specific gravimetric water content that corresponds to a desired soil suction for performing the test [27]. However, such methods have limitations in ascertaining a well-defined soil-matric suction profile associated with nonuniform spatial variation of water content that are typically encountered in practice. In the current study, an efficient and straight forward test methodology is suggested to achieve a well-defined soil-matric suction profile to

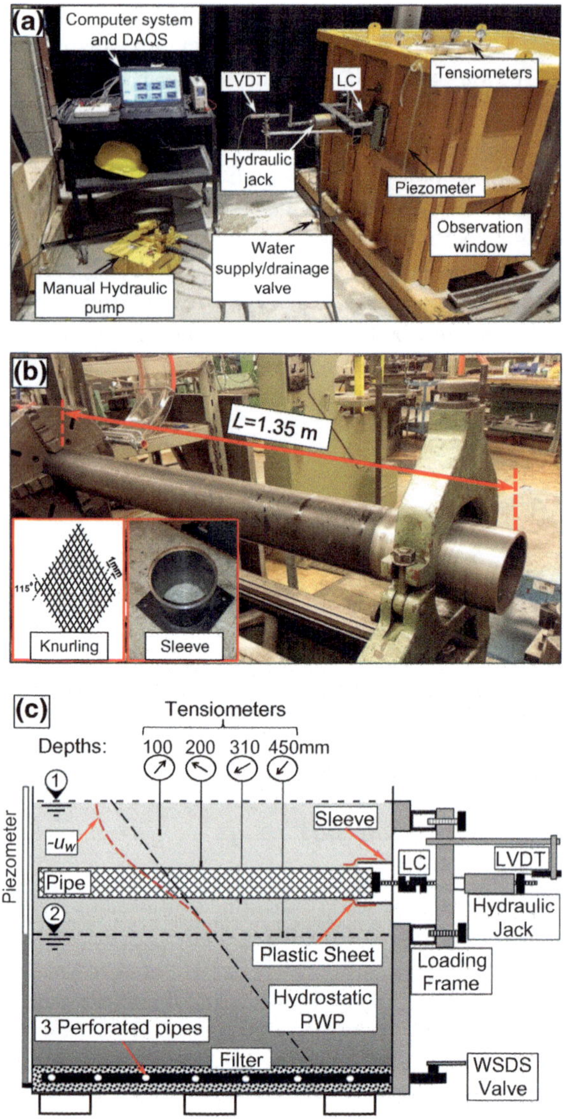

Fig. 7.2 **a** The full test setup, **b** pipe fabricating and sleeve, and **c** graphical illustration of the test setup

investigate the behavior of soil–pipe interaction under unsaturated condition. Such a profile is consistent with in situ conditions that are typically encountered in semi-arid and arid regions.

This is achieved by using a special water supply/drainage system that consists of an external valve that branches into three perforated pipes of 50.8 mm nominal diameter in the soil container. The pipes extend along the base of the container and are set apart at 400 mm center-to-center spacing. The perforated pipes are covered with a filter layer, which constitutes of a 75 mm clean gravel layer and a geofabric sheet. The filter layer serves as a barrier between the gravel and the sand. The water supply/drainage system facilitates uniform saturation and desaturation process of the sand in order to create different moisture regimes in the container.

A steel pipe of 114.3 mm outer diameter and 1.35 m length was fabricated and used to perform the axial load-displacement tests (Fig. 7.2b). The soil–pipe interface roughness has a significant impact on its behavior. Yoshimi and Kishida [37] quantified the surface roughness ($R_{max.}$) as the relative height between the highest peak and lowest valley over 2.5 mm length of a surface profile. Kishida and Uesugi [38] suggested that surface roughness can be quantified using normalized roughness ($R_{n.}$) that is defined as the ratio of $R_{max.}$ to the mean particle size of the soil D_{50}. For the current study, the pipe surface was knurled to create a rough surface using a lathe machine. Cross knurling was made along the pipe surface at 1 mm pitch with surface roughness, $R_{max.} = 0.25$ mm [37] and normalized roughness, $R_{n.} = 1$ [38].

The sand was compacted manually at optimum moisture content in several layers of 200 mm thickness using a 6.5 kg hammer. The pipe was placed and leveled horizontally at 250 mm depth from the springline (i.e., pipe center) to the soil surface during soil compaction. The front end of the pipe was accommodated in steel sleeve of 130 mm diameter with a length of 200 mm that was used to minimize the interference between the pipe front end, soil, and the box physical boundary. The sleeve served as a barrier to prevent the sand from filling the void between the pipe and the box. In addition, a plastic sheet layer of negligible strength was used to wrap the overlap pipe-sleeve gap to prevent water leakage and sand movement into the gap between the pipe and the sleeve.

The tests were performed with the aid of several instruments for gathering the required data. The instruments include: (i) A hydraulic jack of 40 mm maximum stroke that was used to apply pullout force on the prototype pipe; (ii) S-shaped Transducer Techniques load cell (LC) of 22,241 N (5000 lbs) capacity was used to measure the mobilized axial force due to soil–pipe relative movement; (iii) A linear variation displacement transducer (LVDT) of 50.8 mm maximum stroke was used to measure the corresponding displacement; (iv) Data acquisition system (DAQ) and a computer were used to collect and store the data; and (v) Four soil moisture probe 2100 F Tensiometers that were buried at depths equal to 100, 200 (i.e., top of the pipe), 310 (i.e., bottom of the pipe) and 450 mm were used to measure the soil-matric suction profiles in the soil container.

The pipe was aligned, before it was buried, horizontally with the LC, loading frame and the hydraulic jack to avoid any unbalanced forces. The axial load was applied using the hydraulic jack which was acting against an adjustable external loading frame. A19.01 mm threaded rod was used to connect the hydraulic jack, LC, and the front end of the pipe and transfer the load from the hydraulic jack to the pipe. The rod passes through a hole in the middle of the reaction beam and connect the hydraulic jack to LC from one side and through a hole in the container front side and connect the pipe to the LC from the other side.

Table 7.2 Summary of sand physical and mechanical properties

Index property	
Maximum dry unit weight (kN/m^3)	16.7
Optimum moisture content (%)	14.6
Void ratio	0.606
D_{60} (mm)	0.25
D_{30} (mm)	0.19
D_{10} (mm)	0.13
Coefficient of uniformity	1.9
Coefficient of curvature	1.1
Specific gravity	2.65
Unified soil classification system	SP
Mechanical property	
Soil peak angle of friction, ϕ' (°)	~35
Interface peak angle of friction, δ' (°)	~34

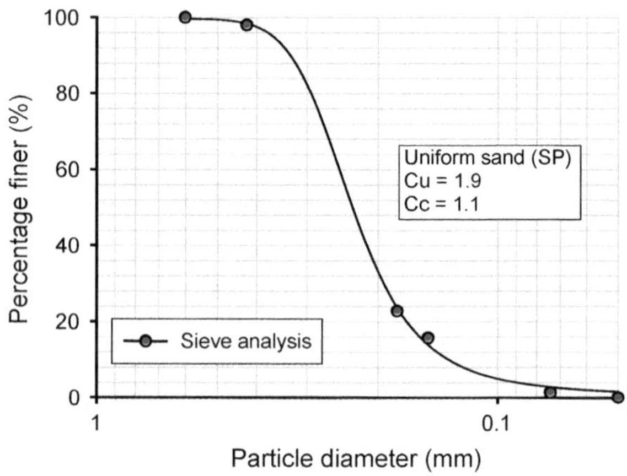

Fig. 7.3 Grain size distribution curve

7.3.2 *Soil Properties*

Poorly graded silica sand that is known commercially as Unimin 7030 was used in the current study. The key physical and mechanical properties of the silica sand are presented in Table 7.2. The grain-size distribution curve is presented in Fig. 7.3.

The SWCC was measured using hanging column method introduced by Haines [39] for coarse-grained soils. The hanging column setup was constructed following ASTM D6836-02 [40] (Fig. 7.4).

The setup consisted of height adjustable and fixed water reservoirs that were used to apply deferential pressure on the soil sample. The reservoirs were connected with a plastic tube from the bottom. The pressure difference on the sample was achieved by

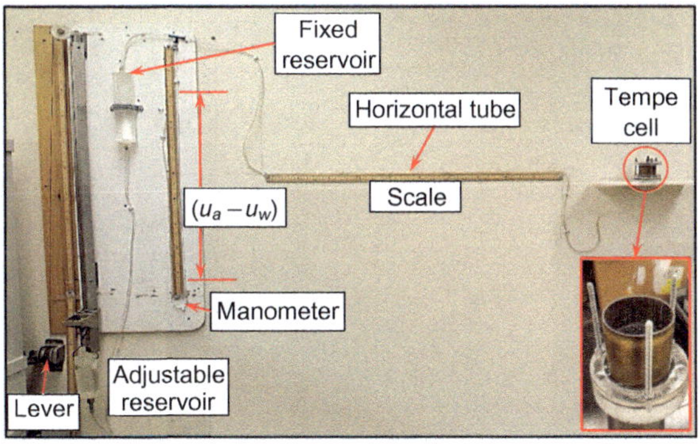

Fig. 7.4 Hanging column setup

changing the height of the adjustable water reservoir. The fixed reservoir is branched from the top into two tubes of which one is connected to a Tempe cell that hold the soil sample and the other is connected to a manometer (U-shaped tube). The pressure difference was applied on the soil sample through the first tube which was used also to collect and measure the extracted water. The manometer was used to measure the matric suction acting on the specimen by hanging column technique. The Tempe cell has a ceramic disk with an air entry value of 1 bar that holds statically compacted soil sample prepared using a triaxial shear test loading frame to achieve the desired density (i.e., $\gamma_d = 16.20–16.70$ kN/m^3).

The SWCC that was measured following the drying path is succinctly summarized below: (i) The soil sample was saturated by flooding it with deaired-distilled water; (ii) The extra moisture content was allowed to gradually drain over a period of 24 hours prior to the conducting test; (iii) Apply small differential pressure increments of 1 kPa lowering the adjustable reservoir by 100 mm; (v) The corresponding ($u_a - u_w$) was measured using the manometer; (iv) The extracted volume of water for each ($u_a - u_w$) increment from the sample was collected in the measuring tube and used to determine the water content. The measured SWCC data was fitted using Fredlund and Xing [41] model that is presented in Fig. (7.5).

7.3.3 *Test Methodology*

In the current study, water table (WT) depths was varied in the test container to achieve different soil (i.e., saturated and unsaturated) conditions. The test methodology included several steps such that different soil-matric suction profile sand stress regimes (Fig. 7.6) were achieved in the test container to study the axial load-displacement behavior of the buried pipe.

The test methodology is appropriate to study pipeline structure interaction problems in sandy soils. Repeatable soil-matric suction profile were achieved before prior to conducting each of the tests following four steps. The first step included burying

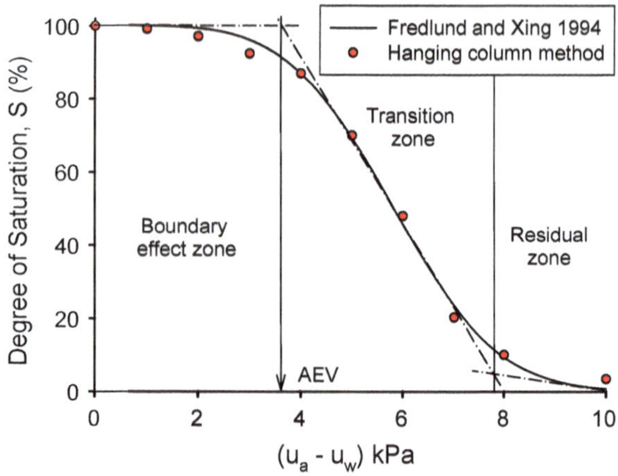

Fig. 7.5 Soil–water characteristic curve (SWCC)

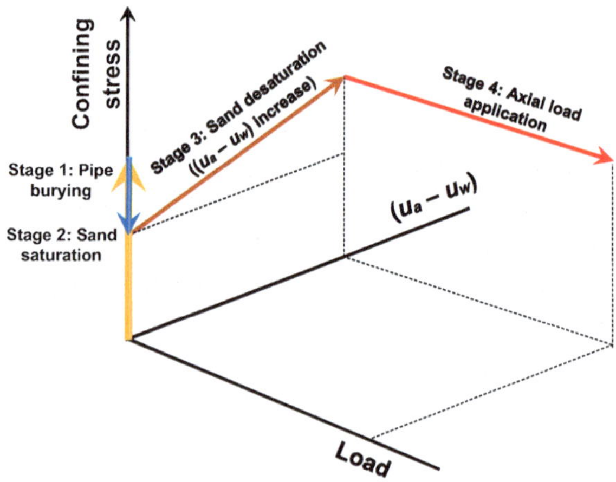

Fig. 7.6 Changes in stress regime around the pipe at each step of the test

the pipe and setting all the instrumentation. The confining stress around the pipe body increased to a certain level due to overburden pressure effect.

The soil was saturated in the test tank during the second step by raising the WT from the bottom of the soil container (i.e., to level 1 Fig. 7.2c). This is an important stage to create a reliable soil-matric suction profile for different WT levels in coarse-grained soils. If the soil is saturated by adding water from the top, it is likely to create air pockets that would contribute to a nonuniform matric suction profile. Raising the WT ensures the removal of air pockets and occluded air bubbles from the soil surface and will create uniform saturation condition. The water supply/drainage system (WSDS) was used to saturate the soil in the soil container from the bottom. The water was supplied from the inlet valve to the three perforated pipes, which provided uniform water distribution at the bottom of the box.

In the third step, unsaturated condition is created by lowering the WT to a desired depth; for example to level 2 in Fig. 7.2c. The soil above the WT is in a state of unsaturated condition while the soil under the WT is saturated. In the test, the WT was gradually lowered using the WSDS for a certain period of time which allowed to achieve uniform desaturation conditions along the cross-sectional area of the soil container. This time period was decided after several trial studies for the tested sand and was found to be approximately 18–24 h. The axial load-displacement test was initiated after achieving equilibrium condition. At this stage, the confining stress increases to a certain level due to the increase in the effective unit weight associated with desaturation.

Step four constitutes of loading the pipe which is initiated after achieving equilibrium condition of matric suction profile. The load was applied manually using the hydraulic jack. The mobilized axial resistance due to relative movement of the pipe with respect to soil and the axial displacement were measured using the LC and LVDT, respectively and recorded in a computer.

7.4 Results and Discussions

Five different soil-matric suction profiles were achieved in the testing program by varying the water table levels (i.e., WT0 (saturated), WT1, WT2, WT3, WT4, and WT5). The variation of matric suction with depth was measured using four Tensiometers. Each WT level resulted in a soil-matric suction profile and the results are presented in Fig. 7.7.

The readings of measurements from Tensiometers were found to be consistent with the matric suction values estimated indirectly. This was achieved by collecting soil samples above water table (i.e., WT2) for determining water content in six small aluminum containers. These containers were buried during preparation phase at different depths from the soil surface (i.e., at 100, 200, and 310 mm, using two containers at of each of these depths) for the third test. The depths of the containers were chosen at the top and bottom of the pipe and in the middle of the soil cover above the pipe in order to get the variation of suction within and above the pipe zone (Fig. 7.7). The soil samples were collected immediately after the test. The gravimetric water content in the collected samples was used to estimate the corresponding (u_a – u_w) from the SWCC. The estimated (u_a – u_w) from the collected samples for the third test agrees well with the Tensiometers readings as shown in Fig. 7.7.

The variation of the soil-matric suction profiles shows that as the WT depth increases, the (u_a – u_w) values increase above and below the pipe. However, the rate at which the (u_a – u_w) increases changes as the soil moves from the boundary effect zone to transition and residual zones of suction. The (u_a – u_w) increases at higher rate in the boundary effect zone which is basically a small zone above the WT. In this zone, the soil is practically still in a state of saturated condition, in spite of capillary stress. Once the soil moves to the transition zone, capillary stress will not be able to sustain free water between the soil particles at higher zones and starts to desaturate. Since (u_a – u_w) acts along the wetted area, the reduction in the wetted

Fig. 7.7 Soil-matric suction profiles for different water table levels

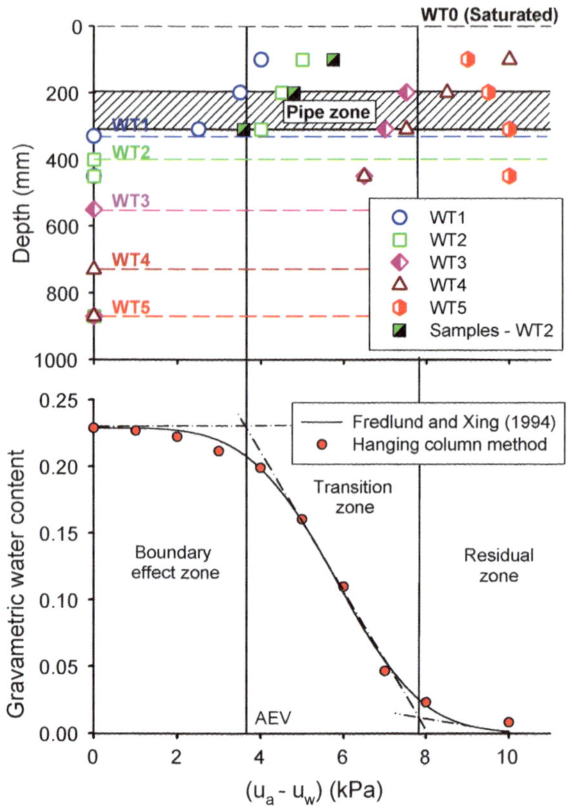

area will lead to lower increase rate in contributing to the mechanical properties of unsaturated soil in comparison to the boundary effect zone.

For WT5, the soil was allowed to drain by keeping the drainage valve open for a period of 24 hours prior to taking the Tensiometers readings. At this stage, the soil was in the residual zone of suction and approaching close to dry condition. In this zone, the wetted area is minimal and there is no continuity in the water phase. Therefore, the $(u_a - u_w)$ values were almost constant with depth.

The measured axial load-displacement behavior of the pipe for three WT levels (i.e., WT0 (saturated), WT1 and WT2) is presented in Fig. 7.8. Each WT level; WT0, WT1, and WT2 resulted in different average matric suction ($(u_a - u_w)_{av.}$) values of 0 kPa, 3 kPa, and 4.25 kPa, respectively above and below the pipe. The measured load is significantly higher for $(u_a - u_w)_{av.} = 4.25$ kPa in comparison to saturated condition (i.e., $(u_a - u_w)_{av.} = 0$ kPa). For $(u_a - u_w)_{av.} = 3.0$ and 4.25 kPa, the axial load increases significantly especially in the elastic zone due to suction-strain hardening. Han et al. [42] stated that suction is an energy potential exerting tension stress on soil skeleton bringing the soil particles and their packets together providing resistance to elastic

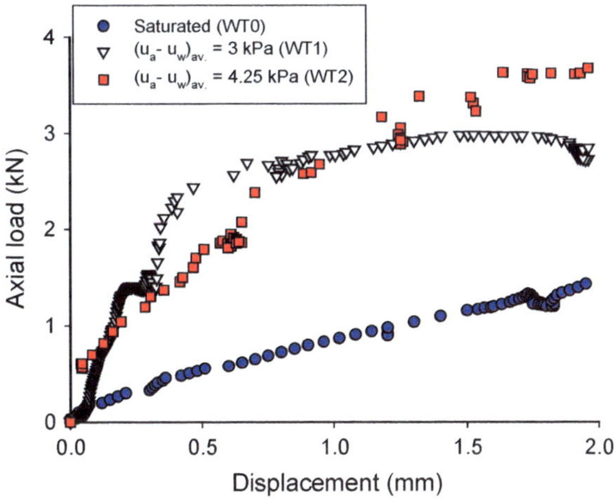

Fig. 7.8 Axial load-displacement tests results of the prototype pipeline under saturated and unsaturated conditions

and plastic deformations. Hossain and Yin [43] concluded, at low confining stress, the shear strength of unsaturated interface exhibit strain hardening in the pre peak zone due to matric suction contribution.

7.5 Conclusion

In the present study, an extensive experimental investigation was undertaken to study the behavior of a prototype steel pipeline of 114.3 mm outer diameter and 1.35 m length in a sand that was subjected to a parallel soil mass movement in the longitudinal direction under saturated and unsaturated conditions. This was achieved following systematic testing procedure in a specially designed soil container system with water/supply drainage provisions. Reproducible soil-matric suction profiles for different WT levels were achieved using different WT levels to create different stress regime (i.e., saturated and unsaturated) conditions.

The axial load test results of the prototype pipeline in unsaturated sand were found to be 2–2.5 times greater than saturated condition. Results from the present study demonstrate that pipeline systems in unsaturated soil experience higher external axial forces. The design of pipeline systems based on conventional approaches underestimates the external, and internal forces. Due to this reason, design procedures extending the principles of saturated soil mechanics for pipelines buried in unsaturated soils is not conservative; it is unsafe.

This study highlights the need for revisiting design procedures of pipeline systems and other soil-structure interaction problems that are in direct contact with unsaturated soils that are based on conventional soil mechanics. The study summarized in this paper suggests that the rational interpretation of the pipeline systems is possible by extending the principles of unsaturated soil mechanics.

Acknowledgements The authors gratefully acknowledge their appreciation to the Iraqi Ministry of Higher Education and Scientific Research, which funded the second author for his Ph.D. research program. Funding received from NSERC, Canada by the first author supported the experimental studies of the research summarized in this article.

References

1. International Energy Agency (IEA): World energy outlook. International Energy Agency and Organisation for Economic Co-operation and Development (OECD) (2008)
2. Ministry of Petroleum and Natural Gas: Government of India (MPNGGI). Annual report 2016–2017 (2017)
3. Chow, E., Hendrix, L.E., Herberg, M.E.: Pipeline politics in Asia: the intersection of demand, energy markets, and supply routes. The National Bureau of Asian Research (NBR), Report 23 (2010)
4. Energy Information Administration (EIA): Annual energy outlook 2007 with projections to 2030. Energy Information Administration, Office of Integrated Analysis and Forecasting, U.S. Department of Energy Washington, DC (2007)
5. Obadina, T.: Harnessing abundant gas reserves: Nigeria alone could provide the power needs of all West Africa. Africa Recovery **13**(1), 16 (1999)
6. African Development Bank and African Union: Oil and Gas in Africa. Joint study by the African Development Bank and African Union. Oxford University Press Inc., New York, USA (2009)
7. Mahmood, A., Javaid, N., Zafar, A., Ali Riaz, R., Ahmed, S., Razzaq, S.: Pakistan's overall energy potential assessment, comparison of LNG, TAPI and IPI gas projects. Renew. Sustain. Energy Rev. Elsevier **31**(2014), 182–193 (2014). https://doi.org/10.1016/j.rser.2013.11.047
8. Bjørnmose, J., Roca, F., Turgot, T., Hansen, D.S.: An Assessment of the Gas and Oil Pipelines in Europe. European Parliament, Brussels (2009)
9. DeMicco, P.: Changing Pipelines, Shifting Strategies: Gas in South-Eastern Europe, and the Implications for Ukraine. European Parliament, Brussels (2015)
10. Benoit, L.: Current and Future State of Oil and Gas Pipeline and Refining Capacity in Canada. Repot of the Standing Committee on the Natural Resources, Parliament of Canada, Ottawa, Ontario, Canada (2012)
11. Canadian Energy Pipeline Association (CEPA): The Importance of Timely Construction of New Pipeline Infrastructure to Canada and Canadians (2005)
12. Fullenbaum, R., Fallon, J., and Flanagan, B.: Oil & natural gas transportation & storage infrastructure: status, trends & economic benefits. IHS Global Inc. and American Petroleum Institute (API), Report (2013)
13. Gerard, J.N.: The State of American Energy: America's Energy. America's Choice, American Petroleum Institute (API), Report (2014)
14. Organization of the Petroleum Exporting Countries (OPEC): OPEC Annual Statistical bulletin, 50th edn. Austria, Vienna (2015)
15. Mills, R.: Risky Routes: Energy Transit in the Middle East. Brookings Institution, Washington DC, USA (2016)
16. BSI: BS EN 14161:2003. Petroleum and natural gas industries-pipeline transportation systems. BSI, London, UK (2003)
17. ALA: Seismic guidelines for water pipelines. American Lifelines Alliance (ALA) in partnership with the Federal Emergency Management Agency (FEMA) (2005)
18. ASCE: Guidelines for the seismic design of oil and gas pipeline systems. American Society for Civil Engineering (ASCE), New York, Committee on Gas and Liquid Fuel Lifelines (1984)
19. IITK-GSDMA: Guidelines for seismic design of buried pipelines. Indian Institution of Technology Kanpur—Gujarat State Disaster Management Authority (IITK-GSDMA), National Information Center of Earthquake Engineering, Kanpur, India (2007)

20. Pipeline Research Council International (PRCI): Guidelines for constructing natural gas and liquid hydrocarbon pipelines through areas prone to landslide and subsidence hazards. In: Report Prepared for the Design, Material, and Construction Committee of Pipeline Research Council International, Inc (2009)
21. BSI: BS EN 1997-1:2004. Geotechnical design—part 1: general rules. BSI, London, UK (2004)
22. Achten, W.M.J., Trabucco, A., Maes, W.H., Verchot, L.V., Aerts, R., Mathijs, E., Vantommc, P., Singh, V.P., Muys, B.: Global greenhouse gas implications of land conversion to biofuel crop cultivation in arid and semi—arid lands—lessons learned from Jatropha. J. Arid Environ. **89**, 135–145 (2013). https://doi.org/10.1016/j.jaridenv.2012.06.015
23. Al-Khazaali, M., Vanapalli, S.K., Oh, W.T.: Estimation of deformations in soil-pipeline systems nearby unsupported trenches excavated in unsaturated soils. In: Proceeding of the 69th Canadian. Geotechnical. Conference, Vancouver, BC (2016)
24. Jung J.K., O'Rourke T.D., Olson, N.A.: Lateral soil-pipe interaction in dry and partially saturated sand. J. Geotech. Geoenviron. Eng., ASCE **139**(12): 2028–2036 (2013). https://doi.org/10.1061/(asce)gt.1943-5606.0000960
25. Olson, N.A.: Soil performance for large-scale soil-pipeline tests. Ph.D. thesis, Cornell University (2009), NY, USA
26. O'Rourke, T.D.: Geohazards and large, geographically distributed systems. Géotechnique **60**(7), 505–543 (2010). https://doi.org/10.1680/geot.2010.60.7.505
27. O'Rourke, T.D., Jezerski, J.M., Olson, N.A., Bonneau, A.L., Palmer, M.C., Stewart, H.E., O'Rourke, M.J., Abdoun, T.: Geotechnics of pipeline system response to earthquakes. In: Proceedings of Geotechnical Earthquake Engineering and Soil Dynamics IV (GEESD), ASCE, Sacramento, CA, pp. 1–38 (2008)
28. Robert, D.J., Soga, K., O'Rourke, T.D.: Pipelines subjected to fault movement in dry and unsaturated soil. Int. J. Geomech. (2016). https://doi.org/10.1061/(asce)gm.1943-5622.0000548
29. Saadeldin, R., Hu, Y., Henni, A.: Numerical analysis of buried pipes under field geo-environmental conditions. Int. J. Geo-Eng. **6**(1) (2015). https://doi.org/10.1186/s40703-015-0005-4
30. Burland, J.B.: Shaft friction of piles in clay-a simple fundamental approach. Ground Eng. **6–3**, 30–42 (1973)
31. Skempton, A.W.: Cast-in-situ bored piles in London clay. Géotechnique **9**, 153–173 (1959)
32. ALA: Guidelines for the design of buried steel pipe. American Lifelines Alliance (ALA) in partnership with the Federal Emergency Management Agency (FEMA) and American Society for Civil Engineers (ASCE) (2001)
33. Bishop, A.W.: The principle of effective stress. Technisk Ukeblad **106**(39), 859–863 (1959)
34. Fredlund, D.G., Morgenstern, N.R., Widger, R.A.: The Shear Strength of unsaturated Soils. Can. Geotech. J. **15**(3), 313–321 (1978). https://doi.org/10.1139/t78-029
35. Vanapalli, S.K., Fredlund, D.G., Pufahl, D.E., Clifton, A.W.: Model for the prediction of shear strength with respect to soil suction. Can. Geotech. J. **33**, 379–392 (1996). https://doi.org/10.1139/t96-060
36. Ravichandran, N., Krishnapillai, S.H., Machmer, B.: A novel procedure for physical modeling of unsaturated soil-pile system using geotechnical centrifuge. J. Earth Sci. Geotech. Eng. **3**(1), 119–134 (2013)
37. Yoshimi, Y., Kishida, T.: A ring torsion apparatus for evaluating friction between soil and metal surfaces. Geotech. Test. J. ASTM. **4**(4), 145–152 (1981). https://doi.org/10.1520/GTJ10783J
38. Kishida, H., Uesugi, M.: Tests of the interface between sand and steel in the simple shear apparatus. Géotechnique **37**(1), 45–52 (1987). https://doi.org/10.1680/geot.1987.37.1.45
39. Haines, W.B.: The hysteresis effect in capillary properties and the modes of moisture distribution associated there with. J. Agric. Sci. **20**, 96–105 (1930)
40. ASTM: D 6836-02; Standard Test Methods for Determination of the Soil Water Characteristic Curve for Desorption Using Hanging Column, Pressure Extractor, Chilled Mirror Hygrometer, or Centrifuge. ASTM International, West Conshohocken, PA, USA (2008)
41. Fredlund, D.G., Xing, A.: Equations for the soil-water characteristic curve. Can. Geotech. J. **31**(4), 521–532 (1994). https://doi.org/10.1139/t94-061

42. Han, Z., Vanapalli, S.K., Kutlu, Z.N.: Modelling the behavior of a friction pile in compacted glacial till. J. Geomech. ASCE, Int (2016). https://doi.org/10.1061/(ASCE)GM.1943-5622.0000659
43. Hossain, M., Yin, J.: Dilatancy and strength of an unsaturated soil-cement interface in direct shear tests. Int. J. Geomech. ASCE (2014). https://doi.org/10.1061/(asce)gm.1943-5622.0000428,04014081

Chapter 8
Reflection Crack Assessment Using Digital Image Analysis

Sireesh Saride and V. Vinay Kumar

8.1 Introduction and Background

Reflection crack is one of the major modes of distress witnessed in the hot mix asphalt (HMA) overlays over the decades. Cleveland et al. [1] defined the reflection cracking phenomenon as a process of propagation of prevailing cracks and distress from the existing old damaged pavement layers into and through the new asphalt overlays. The major driving factors are the traffic induced wheel loads and the seasonal/temperature variations in the overlays and the old pavement layers [2, 3]. Reflection cracking is a complex process consisting of two major stages: crack initiation and crack propagation [4]. Besides, there is no unique solution to arrest the crack growth completely and safeguard the HMA overlays. However, there are numerous solutions to restrict the crack growth for a substantial duration, thereby, enhancing the overlay service life effectively [5–7]. The most widely employed solution is to provide a stress-absorbing membrane interlayer (SAMI) [7, 8], paving fabric [6, 9, 10] geogrid [11–13], road mesh [11], and geocomposites [14–16] at the interface of old and new pavement layers.

Further, to understand the influence of these inclusions on the reflection crack mechanism, a large number of laboratory and field studies have been performed. A few important test techniques include wheel tracking test [17], the École Nationale des Travaux Publics de l'État (ENTPE) test [18], the Belgian road research center (BRRC) test performed at Belgium road laboratory [19], reflective wheel cracking test [20], UGR-FACT test [21], flexural beam fatigue tests [11–13] and direct tensile strength tests [10]. However, these techniques failed to understand the reflection cracking phenomenon in the HMA mixtures. Besides, additional instrumentation in

S. Saride (✉) · V. V. Kumar
Department of Civil Engineering, IIT Hyderabad, Kandi, Telangana, India
e-mail: sireesh@iith.ac.in

V. V. Kumar
e-mail: christite.vinay@gmail.com

M. Latha G. (ed.), *Frontiers in Geotechnical Engineering*, Developments in Geotechnical Engineering, https://doi.org/10.1007/978-981-13-5871-5_8

the form of sensors such as strain gauges and linear variable displacement transducers (LVDT) have been incorporated to predict the crack initiation and propagation stages. These solutions failed to make an impact due to the inaccurate results, as it is impossible to trace the measurement location prior to the test. To overcome these issues, Saride and Kumar [6] and Kumar and Saride [9] have adopted the digital image analysis program to study the full-field displacements and strains mobilized in the specimen.

Digital image correlation (DIC) is a two-dimensional, non-interferometric, optical measurement technique developed in the early 1980s with an aim to estimate and analyze the heterogeneities in the composites, and concrete infrastructure [22, 23]. Choi and Shah [24] were the first in the construction field to adopt the DIC technique to study the deformation characteristics of concrete infrastructure. Further, Kim and Wen [25] made an attempt to study the characteristics of asphalt concrete specimens, using digital image analysis. In addition, Safavizadeh et al. [26], Wargo et al. [27], Saride and Kumar [6], and Kumar and Saride [9, 10] have successfully adopted the digital image technique to analyze the cracking characteristics of the asphalt mixtures. These findings suggest that the digital image analysis could be effectively adopted to understand and assess the reflection crack mechanism in the two-layered asphalt system. In this regard, the current study aims to assess the reflection crack phenomenon in a two-layered asphalt system, consisting of an old distressed bottom layer and an HMA overlay treated with a tack coat and geosynthetics, using beam fatigue tests and digital image analysis program.

8.2 Materials

8.2.1 Binder and Mix

The binder adopted in the asphalt mix and as a tack coat in the current research is of a penetration grade (PG) 60/70. The properties of the binder have been determined under various laboratory tests and the results are presented in Table 8.1. The asphalt mix consists of a nominal aggregate size of 13 mm, as per Ministry of Road Transport and Highways (MORTH) specifications [28]. The properties of the mix were determined under the Marshall stability tests performed as per ASTM D6927 [29]. It was found that the optimum binder content (OBC) of the mix was 5.5% by the weight of aggregates. In addition, the strength and the flow values of the mix with OBC were determined to be 14.25 kN and 2.5 mm, respectively.

Table 8.1 Binder properties

Properties	Values
Penetration value (1/10th mm)	66
Specific gravity	1.01
Ductility (cm)	100+
Softening point (°C)	52
Viscosity at 60 °C (cP)	400
Flash point (°C)	340
Fire point (°C)	360

8.2.2 *Grid and Composite*

The current study aims to understand the reflection crack mechanism in asphalt overlays with and without geosynthetic-reinforcements. In this regard, a biaxial polyester grid (PET) and a glass-grid composite (GGC) were adopted as geosynthetic-reinforcements. It was reported that these geosynthetic-interlayers would perform better in terms of antireflection cracking systems and would provide high resistance to fatigue and rutting of asphalt layers [6, 26]. To understand the working tensile and mechanical properties of the grid and composites, a wide-width tensile strength test was performed as per ASTM D4595 [30] along the machine (MD) and cross-machine directions (CMD). These tests were conducted initially at room temperature (25 °C) and repeated after a temperature test. In the temperature test, to measure the deviation in the tensile strength of the interlayers due to high placement temperatures of HMA and tack coat during the installations, the geosynthetic-interlayers were first placed on the old pavement layer and an HMA concrete layer was placed at a temperature of about 150 °C and then the geosynthetic-interlayers were carefully extruded after a duration of 24 h. The latter exercise is to verify the influence of temperature and other physical damages on the working mechanical properties of the geosynthetics which are typical during the construction process.

The polyester grids are manufactured by knitting together high molecular weight and high tenacity polyester yarns to form a grid with a square aperture of 18 mm. The grids are coated with a polymer-modified binder to enhance their bonding characteristics. Based on the assessment of tensile and mechanical characteristics of the grid, it was found that the PET has an ultimate tensile strength of 48 kN/m (MD) and 52 kN/m (CMD) at a failure strain of 18–20% before the temperature test. Similarly, the tensile strength tests performed after temperature tests resulted in an ultimate tensile strength of 48 kN/m (MD) and 52 kN/m (CMD) at a failure strain of 28–30%. No reduction in the tensile strength was observed, however, the strain corresponding to the failure has increased from 18 to 28%. This observation infers that the PET grids would not show failure until 28% strain, or in other words, PET grids would not mobilize the peak tensile strength under normal working strain conditions of 2–5%.

On the other hand, the glass-grid composite is manufactured by mechanically bonding together the fiberglass grids and nonwoven geotextiles. The assessment suggests that the GGC has an ultimate tensile strength of 28 kN/m at a failure strain of 2% and 25 kN/m at a failure strain of 1.75% along the MD and CMD, respectively. It is interesting to note that the influence of temperature was observed to be negligible on the working tensile and mechanical properties of the composites.

8.2.3 Specimen Preparation

The straight and prismatic beam specimens adopted in the current study consists of a two-layered system with a binder tack coat and a geosynthetic-interlayer at the interface. The bottom layer in the two-layered specimen is an old pavement layer extruded during a highway rehabilitation program and the top layer comprise of an asphalt concrete layer compacted using static weight compactor. A detailed specimen preparation procedure for the fatigue analysis and corresponding digital image analysis are presented by Saride and Kumar [6] and Kumar and Saride [9, 10, 16]. Besides, to assess the reflection crack phenomenon in the two-layered system, the current study considers different conditions such as a crack/notch of 40, and 25 mm depth and a notch-less condition, in the bottom layer of the two-layered system. Figure 8.1 presents the typical flexural test setup consisting of the two-layered specimen and the test apparatus with the dimensions and loading conditions for all the three conditions considered in the study. It can be visualized that the final specimens have a span length of 360 mm, width of 50 mm and a thickness of 90 mm (45 mm old layer ad 45 mm overlay). In addition, the notch-less and a notch depth of 25, and 40 mm are presented in Fig. 8.1a, b and c, respectively.

8.3 Test Methods

8.3.1 Beam Fatigue Test (BFT)

Beam fatigue tests were performed under a load-controlled mode on the two-layered asphalt specimens discussed under Sect. 8.2.3. The tests were performed on specimens with and without reinforcements under the three different conditions to understand their flexural fatigue behavior. In addition, the influence of notch depths and the type of reinforcement on the performance of overlays against the cracking phenomenon were studied. Figure 8.1 presents the typical test setup adopted in the study. The load is applied on the beam specimens placed in a four-point test rig. The load corresponding to a moving traffic condition replicating a 550 kPa single-axle contact pressure was considered in the study. A detailed procedure of the flexural fatigue tests performed on the beam specimens under stress (load) controlled mode is avail-

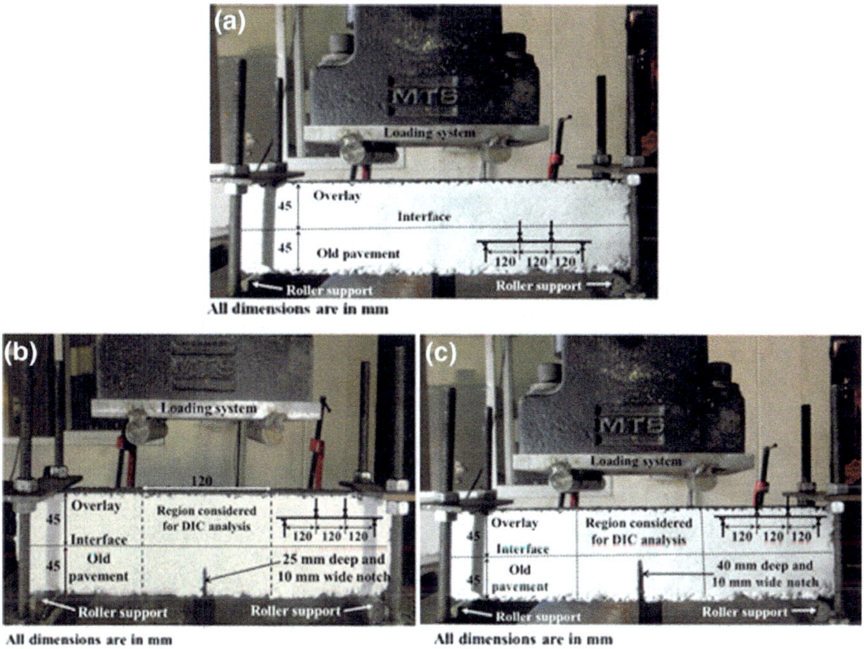

Fig. 8.1 Typical beam fatigue test setup adopted in the study **a** Notch-less; **b** 25 mm notch depth; and **c** 40 mm notch depth

able in Saride and Kumar [6, 16], and Kumar and Saride [9]. In addition, to assess the influence of notch depth and geosynthetic-reinforcements on the reflection crack phenomenon, a two-dimensional full-field crack monitoring technique was adopted.

8.3.2 Digital Image Correlation (DIC)

The digital image correlation and tracking is an optical method that employs tracking and image registration techniques for accurate 2D and 3D measurements of changes in the images. DIC is a data analysis procedure that uses the mathematical correlation method to analyze digital images of a specimen undergoing deformation. This technique offers the advantage of obtaining a spatially continuous measurement of displacements, deformations, strain, optical flow and crack initiation and propagation.

Figure 8.2 presents a schematic of a typical DIC setup adopted in the current study to record the images continuously during the beam fatigue test program. The images of the specimens under various number of load cycles are recorded with the help of a high-definition digital camera connected to the constant power supply and a recording system. These images, recorded at various load cycles, are compared with

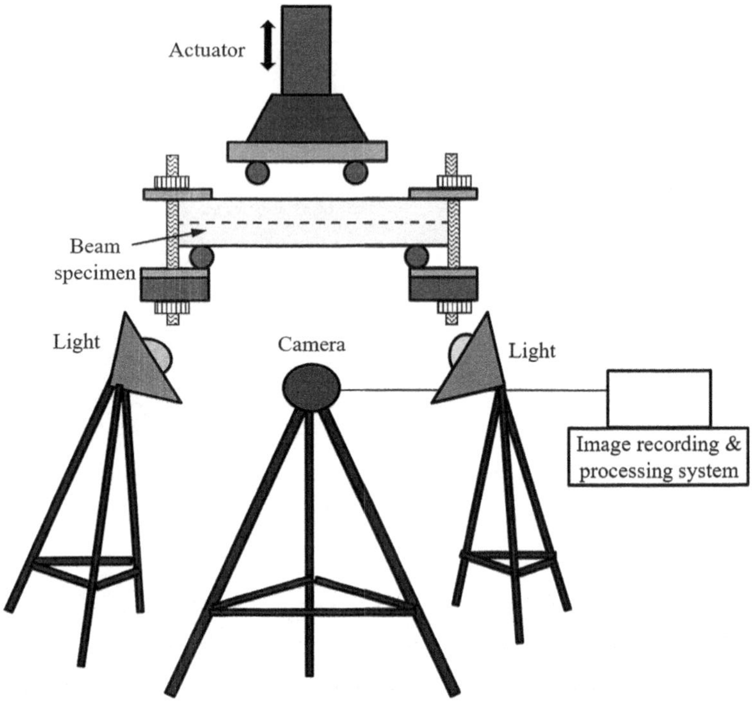

Fig. 8.2 Schematic of a typical DIC setup adopted in the study

that of an undeformed image of the specimen to track the changes in their speckle patterns. The changes in the speckle pattern of deformed images are carefully tracked with the help of VIC-2D program, which adopts an iterative mathematical correlation technique to track the change in the images. The changes are presented in the form of deformations and displacements along the "x" and "y" directions. A detailed analysis of these deformation fields would result in the calculation of strains along the "x" and "y" directions, respectively.

8.4 Results and Discussion

8.4.1 General

The beam fatigue test results of two-layered asphalt specimens are presented in Fig. 8.3, which depicts the variation of vertical deformation with number of load cycles (N), also referred to as fatigue life. In all the test configurations, the control specimen (CS) could not resist a large number of load repetitions before failure, i.e., cracks propagated into the overlays at about 330 number of load cycles

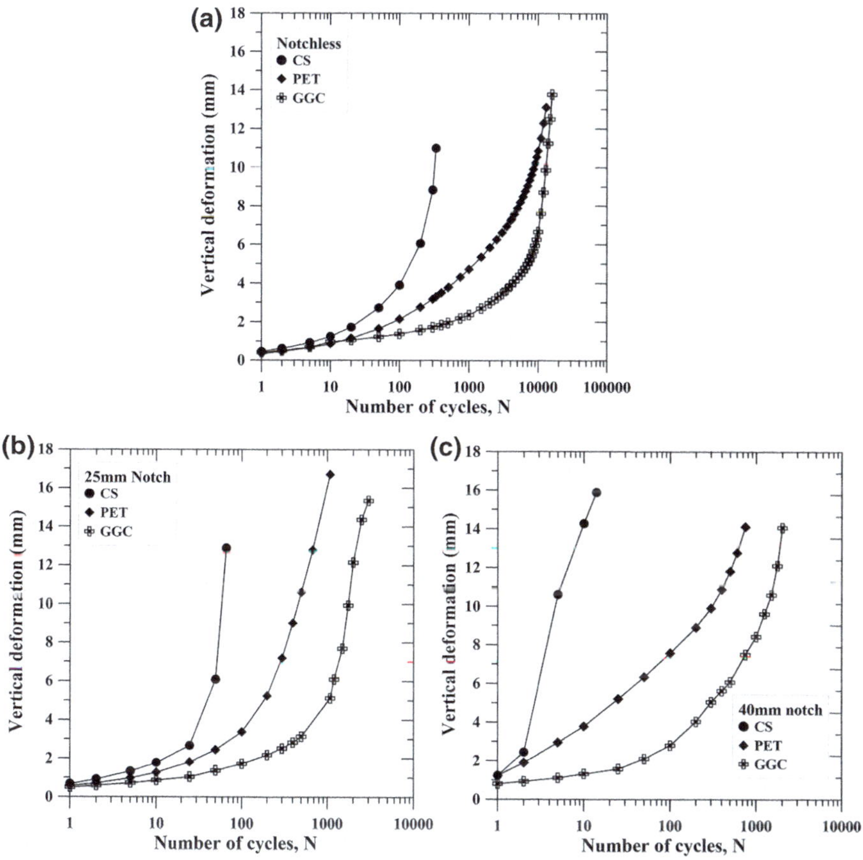

Fig. 8.3 BFT results: variation of vertical deformation with number of cycles **a** Notch-less; **b** 25 mm notched; and **c** 40 mm notched specimens

(N ~ 330). However, as expected, the specimens without notch (Fig. 8.3a) have sustained a higher number of load repetitions, followed by the specimens with 25 mm notch depth (Fig. 8.3b) and finally the specimens with 40 mm notch depth (Fig. 8.3c). The fatigue life of beam specimens for different notch cases and interlayer configurations are presented in Table 8.2. The fatigue life of beam specimens surely depends on the notch depth (existing crack) and the type of interlayer present at the interface. The beam specimens without notch (intact old bottom layer) have shown higher fatigue life than with preexisting notch. The specimens with 40 mm notch have shown low fatigue life than the rest, indicating that the propagation of existing cracks is much faster which reflects on the top surface. The variation in fatigue life of beam specimens with different notch depths may be attributed to their flexural stiffness, which is higher for notch-less specimens followed by specimens with 25 mm notch depth and 40 mm notch depths, respectively.

Table 8.2 Fatigue life (N) of beam specimens (VD ~ 14mm)

Interlayer type	Notch-less	25 (m)	40 (m)
CS	335	66	14
PET	13116	1077	750
GGC	16,000	3000	2000

Besides, the specimens with geosynthetic-reinforcement have shown higher fatigue life, irrespective of the notch depth. This enhanced performance may be attributed to the tensile properties of the geosynthetic-reinforcement, which restricts the vertical deformations and in-turn, crack propagation into the overlays, effectively. Among the specimens with geosynthetics, the fatigue life of composite (GGC) specimens is superior to that of grid (PET) specimens, irrespective of the notch depth. The superior performance in GGC specimens is due to the ability of GGC interlayer to mobilize a tensile strength of about 28 kN/m at a very low strain of about 2%.

The digital images captured during the BFT program are analyzed and the full-field deformations are presented in the form of vertical deformation bands in Fig. 8.4. The vertical deformation at any point on the specimen surface can be determined from the displacement fields obtained at different load cycles during the BFT. Note that in each color band, the vertical deformations are uniform and the downward deformation is considered as negative in the analysis. As the curvature of the beam at failure is maximum, the deformation band at the mid-span of the beams show a maximum value. The deformation bands in the CS are continuous throughout the depth of the beam, indicating that the deformations are uniform throughout the depth, for all the notch conditions (Fig. 8.4). Whereas, the deformation bands are observed to be discontinuous in all the specimens with geosynthetics, which are very prominent in notched specimens (Figs. 8.4b, c). The discontinuities in the deformation bands are due to the presence of geosynthetics at the interface zone, which restrains the crack propagation while absorbing the strain energy into the interlayer and controls the vertical deformation. Hence, maximum deformations are observed in the bottom layer, below the interlayer zone.

The corresponding number of load repetitions (fatigue cycles, N) at failure are also presented in each case in Fig. 8.4. The fatigue life of specimens with geosynthetics are observed to be higher than that of CS, irrespective of the notch depth. Among the specimens with geosynthetics, the performance of composite-reinforced specimens is observed to be significant to that of grid-reinforced specimens. It can be seen that the notch-less, 25 mm notched, and 40 mm notched composite-reinforced specimens have a fatigue life of 16,000, 3000, and 2000 cycles, respectively. At same deformation level, the grid-reinforced specimens have sustained a load repetition of 13,116, 1077, and 750 cycles in notch-less, 25 mm notched and 40 mm notched conditions, respectively.

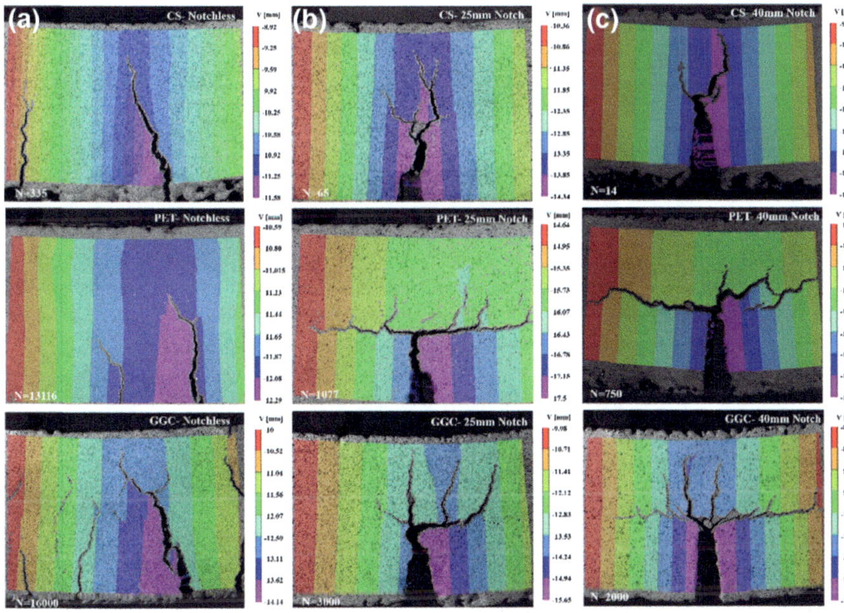

Fig. 8.4 Vertical deformation bands at failure **a** Notch-less; **b** 25 mm notched; and **c** 40 mm notched specimens

Before analyzing the DIC data further, the vertical deformations obtained at the mid-span of the specimen from an external linear variable deformation transducer (LVDT) was compared with the deformations obtained from DIC analysis. Figure 8.5 presents the variation of the vertical deformations obtained at mid-span from both the methods for all the notch-less beam specimens, as a function of number of load cycles (N). It can be observed that the deformation values obtained from DIC analysis and that recorded from LVDTs are very similar and the differences can be termed as negligible. These observations suggest that the adopted digital image analysis program is suitable to understand and assess the reflection cracking mechanism in unreinforced and geosynthetic-reinforced asphalt overlays, under repeated load conditions. In this regard, Safavizadeh et al. [26], Saride and Kumar [6], and Wargo et al. [27] have successfully employed DIC analysis to evaluate the fatigue behavior of geosynthetic-reinforced notched asphalt beams under repeated load tests.

From Figs. 8.3, 8.4 and 8.5, it can be observed that the geosynthetic-reinforced asphalt specimens have consistently performed better than the control specimens, irrespective of the notch depths. In order, to evaluate the enhanced performance in geosynthetic reinforced beam specimens with respect to the control specimens, a non-dimensional performance factor known as fatigue life improvement ratio (I_{NF}) is introduced. The fatigue life improvement ratio can be defined as a ratio of number of load repetitions on a reinforced beam to the number of load repetitions on control

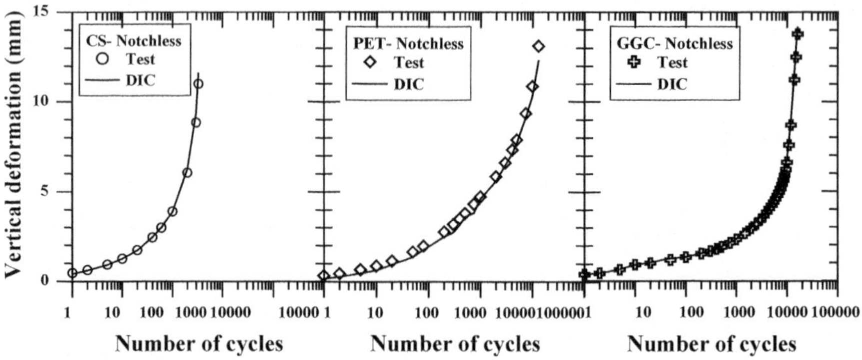

Fig. 8.5 Comparison of vertical deformations obtained from DIC analysis and LVDT

beam calculated at same vertical deformation values and mathematically expressed as

$$I_{NF} = \frac{N_R}{N_U}, \tag{1}$$

where, N_R and N_U are the fatigue lives of geosynthetic-reinforced and unreinforced specimens, respectively.

Figure 8.6 presents the variation of fatigue life improvement ratio (I_{NF}) with vertical deformation values for all the test configurations. It can be observed that the I_{NF} has increased with an increase in the vertical deformation, irrespective of the notch depth and geosynthetic type. The improvement in PET specimens is less up to a 4 mm vertical deformation, thereafter there is an invariable increase in the I_{NF} of PET specimen. Whereas, in the GGC specimens, the improvement in fatigue life can be witnessed as early as at a vertical deformation of 2 mm. For instance, at a VD of 2 mm, an I_{NF} of 19 is observed in notch-less GGC specimens, against an I_{NF} of 3 in the notch-less PET specimens (Fig. 8.6a).

Similar observations can also be witnessed in the notched specimens as well (Figs. 8.6b, c). Among the notched specimens, the I_{NF} achieved is higher in the specimens with 40 mm notch depth than the specimens with 25 mm notch depth. As the stiffness of the specimens with 25 mm notch is generally higher than that of 40 mm notch depth, the former beams have sustained a higher number of load repetitions than the latter ones (Fig. 8.3b, c), thus the decrease in the improvement ratio. It can be clearly seen from Fig. 8.6b, c, that for the GGC specimen with 40 mm notch depth, an I_{NF} of about 120 is achieved against an I_{NF} of 25 for specimens with 25 mm notch depth, respectively at 6 mm vertical deformation. The superior performance of GGC specimens may be attributed to the reinforcing effects of the interlayer, which can be witnessed as early as a vertical deformation of 2 mm is reached and continued till the failure.

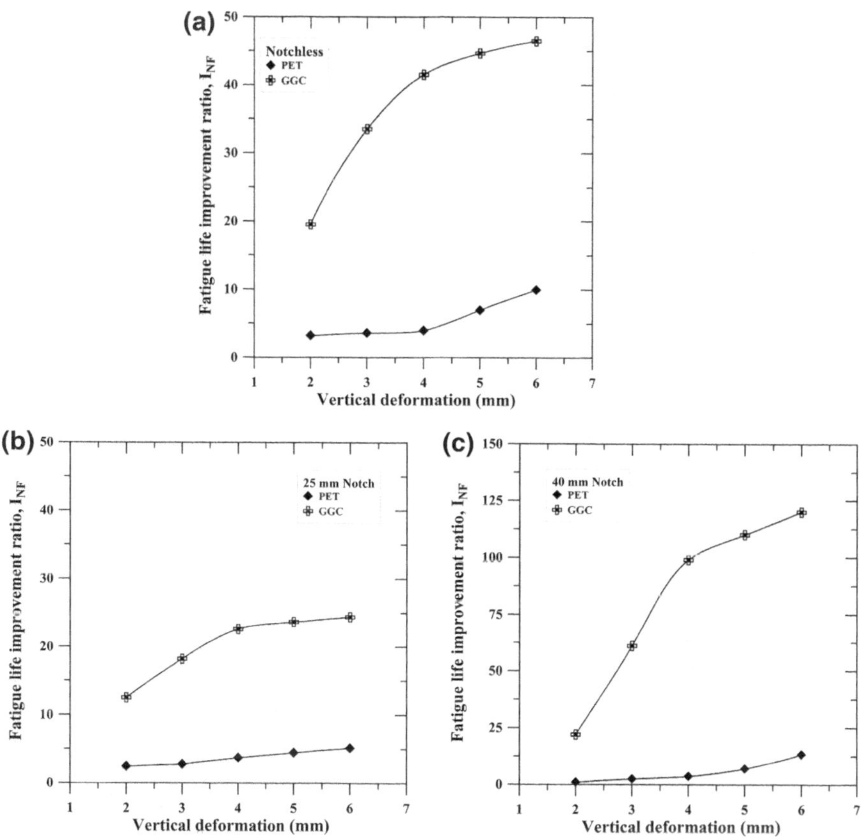

Fig. 8.6 Variation of fatigue life improvement ratio with number of cycles **a** Notch-less; **b** 25 mm notched; and **c** 40 mm notched specimens

Besides, the deformation fields presented in the form of vertical deformation bands are analyzed further to calculate the corresponding maximum tensile (horizontal) strains at the crack tip. Based on the BFT results, it is understood that the presence of geosynthetics help to minimize the crack propagation in the vertical direction by controlling the maximum tensile strain at the crack tip. The tensile strain contours for the control, grid-reinforced and composite-reinforced specimens with notch-less, 25 mm notch depth, and 40 mm notch depth conditions are presented in Fig. 8.7. It can be observed that the tensile strains in CS are consistently higher than the specimens with geosynthetics, irrespective of notch depth. The tensile strains as high as 74%, 30%, and 4.75% were witnessed in a CS with 40 mm notch depth, 25 mm notch depth and notch-less conditions, respectively. Whereas, the tensile strains in specimens with geosynthetics are in the range of 9–19% (40 mm notch depth), 5–11% (25 mm notch depth), and 1.5–3.4% (notch-less). A reduction in the tensile strain of about 35–70% can be witnessed in specimens with geosynthetics

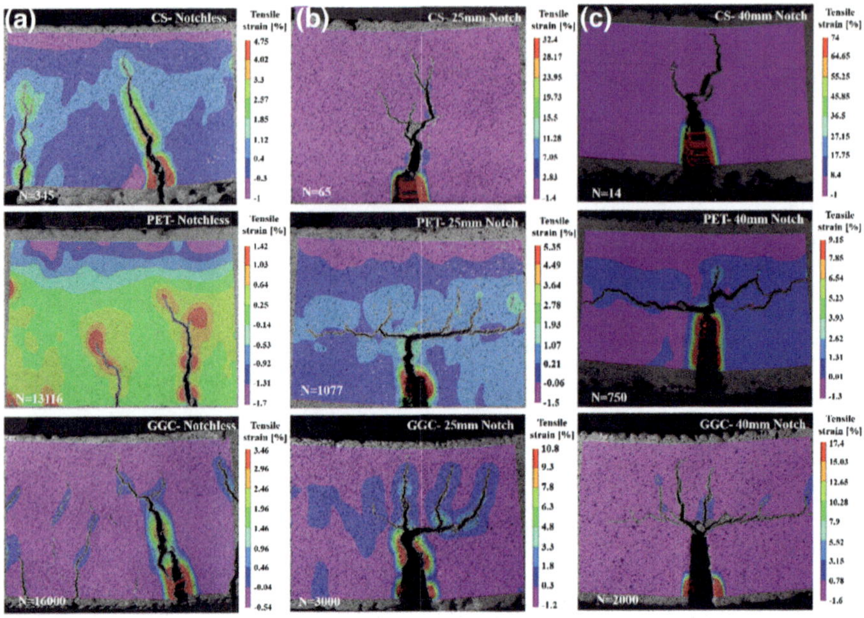

Fig. 8.7 Tensile strain contours at failure **a** Notch-less; **b** 25 mm notched; and **c** 40 mm notched specimens

and this may be attributed to the ability of the geosynthetics to absorb the strain energy responsible for crack propagation into the overlays effectively. It can also be observed that the tensile strains in the 40 and 25 mm notch depth specimens are consistently higher than the notch-less specimens, irrespective of the geosynthetics type. This may be due to the presence of an initial crack in the specimens with 25 and 40 mm notch depth, which propagates into the overlay with a less number of load repetitions. Whereas, in the case of notch-less specimens, the crack has to be initiated first to propagate into the overlays.

From Fig. 8.7a, it can be visualized that in notch-less control specimens, the crack has traversed through the interface zone and the overlay and lead to failure at 335 load cycles. Whereas, the grid-reinforced and composite-reinforced, notch-less specimens have resisted the crack propagation effectively till 13,116 and 16,000 load cycles, before failure. It is also interesting to note that the cracks have mostly reached the interface zone and propagated laterally along the in-plane direction of geosynthetics at failure. The cracks in the horizontal direction are prominent in composite-reinforced specimens (Figs. 8.4 and 8.7), as there are no apertures in composites to enhance the bonding mechanism between the old and new pavement layers. On the other hand, the polyester grids have an 18 mm square aperture and are also coated with a polymer modified binder on their surface to enhance the interface bond strength of the grids. With the incorporation of geosynthetics, the fatigue performance of two-layered asphalt beam specimens (notch-less) is improved by about

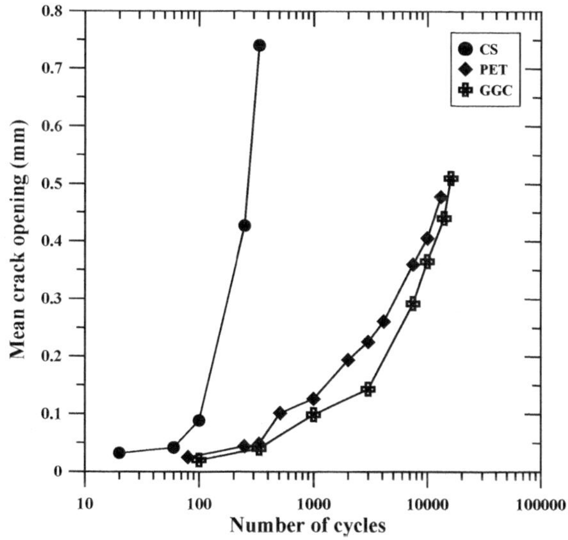

Fig. 8.8 Variation of mean crack opening with number of cycles-Notch-less specimens

47-fold and 38-fold for composite and grid-reinforced specimens, respectively. These observations suggest that the improvement in fatigue performance depends on the working tensile and physical properties of the geosynthetic-reinforcements.

Further, the crack openings and the maximum vertical crack growth in the specimens at any instant could be determined with the help of detailed image analysis. The crack opening can be estimated with the help of a crack opening displacement (COD) tool. In the current study, the COD tool was employed to estimate the mean crack opening in the specimens, effectively. The variation of mean crack openings, as a function of number of cycles, for notch-less condition, is presented in Fig. 8.8. An increase in the mean crack opening with an increase in the number of load cycles can be noticed. It can also be visualized that a crack opening as high as 0.74 mm is witnessed in notch-less control specimen at failure (N = 335). Whereas, the grid-reinforced and composite-reinforced notch-less specimens have a crack opening of 0.47 mm and 0.52 mm at 13,116 and 16,000 cycles, respectively. This reduction in the crack opening suggests that the presence of geosynthetics have restricted the crack growth by absorbing the tensile stress responsible for crack propagation.

The maximum vertical crack growth was estimated based on the color bands obtained postprocessing of DIC images. The vertical crack growth for all the specimens at different load cycles was determined and the variation of maximum crack growth, as a function of number of load cycles, for notch-less condition, is plotted (Fig. 8.9). It can be observed that the presence of geosynthetics have restricted the crack propagation in the vertical direction, until the geosynthetics were effective. The vertical crack growth as high as 71 mm was witnessed in the notch-less control specimens, at failure (N = 335). Whereas, the grid-reinforced and composite-reinforced

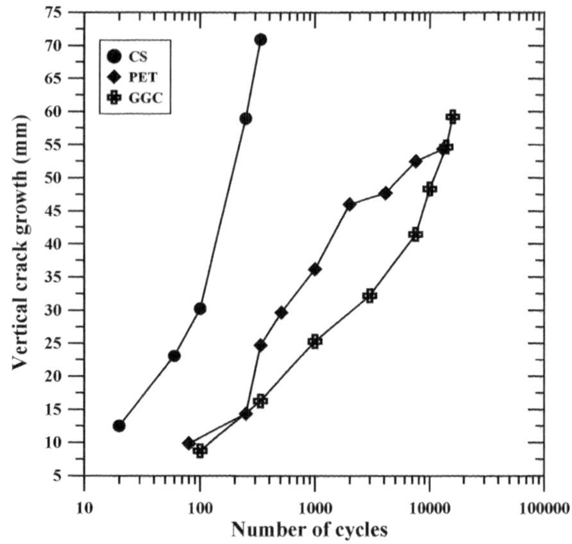

Fig. 8.9 Variation of vertical crack growth with number of cycles-Notch-less specimens

notch-less specimens showed a vertical crack growth of 55 mm and 60 mm at 13,116, and 16,000 cycles, respectively, before failure.

Table 8.3 presents the summary of crack growth in all the specimens obtained from DIC analysis. The mean crack opening (CO) and maximum vertical crack growth (CG) are highest for control specimens, irrespective of notch depths. This observation suggests that the geosynthetics incorporated at the interface of old and new pavement layers absorb the strain energy from the crack tip and restrict the crack growth into the overlays effectively. It is to be noted that the mean crack openings and maximum vertical crack growths reported for the 25 and 40 mm notch specimens are the total crack openings and depths, inclusive of their initial notch widths and depths. The crack widths and crack heights are presented with respect to vertical deformations in specimens. At a given vertical deformation value, the load cycles sustained by composite-reinforced specimens are higher, followed by grid-reinforced specimens, and finally control specimens, irrespective of the notch depths.

8.4.2 Discussion of Service Life of Asphalt Overlays

Researchers, Caltabiano [31] and Khodaii et al. [32], have initially attempted to predict the service life of asphalt overlays corresponding to the crack growth information obtained from the beam fatigue test results. The models developed were complex and required a number of input parameters to estimate the service life of the overlays successfully. In addition, these models failed to predict the service life of asphalt overlays based on the results obtained from independent test data. Recently, Saride

Table 8.3 Summary of crack growth- DIC analysis

Specimens		VD = 4 mm		VD = 8 mm		Failure	
		CO (mm)	CG (mm)	CO (mm)	CG (mm)	CO (mm)	CG (mm)
Notch-less	CS	0.09	30.2	0.42	59	0.75	70.9
	PET	0.10	23	0.28	46	0.48	54.5
	GGC	0.10	24.1	0.39	50	0.52	59.2
25 mm notch	CS	13	60	16	73	18	80
	PET	10.8	50.4	13.4	60.2	15	65
	GGC	10.3	43	13	62.8	14.8	70
40 mm notch	CS	15	68	19	78	24.5	85
	PET	11.4	52	16.1	70	19	75
	GGC	10.5	45	15.3	67	18.6	78

and Kumar [19] have developed a simple correlation model to estimate the service life of geosynthetic-reinforced asphalt overlays based on the beam fatigue test and the digital image analysis program. The model developed suggest to conduct a simple beam fatigue test on asphalt specimens to obtain the normalized complex modulus × cycles (NCM) and the corresponding number of cycles to predict their service life. In addition, the model has been successful in estimating the service life of overlays from independent test data and hence, validating the proposed model.

From Table 8.3, it can be observed that all the control specimens, irrespective of notch depth have been completely cracked. Whereas, in the case of geosynthetic-reinforced specimens, the full-depth cracks have not been witnessed and the service life would be estimated corresponding to the maximum crack growth. In this regard, the service life of control specimens for a full-depth cracking condition and the geosynthetic-reinforced specimens for a half-depth cracking condition could be predicted with the help of beam fatigue test results and the corresponding digital image analysis. Saride and Kumar [19] have estimated a service life of 0.26 million standard axles (msa) for control specimens with 40 mm notch depth at full-depth crack conditions. In addition, the predicted service life of PET and GGC specimens with 40 mm notch depth and half-depth crack conditions as 1.56 msa and 2.18 msa, respectively, which is 6 and 8.5 fold higher service life than the control sections.

Overall, it can be summarized that, with the introduction of geosynthetics between the old and new pavement layers, the performance of asphalt overlays has been improved invariably, by restricting the crack propagation into the overlays. However, the rate of performance improvement depends upon the working mechanical and tensile properties of the adopted geosynthetics type. The presence of geosynthetics not only improve the performance life of overlays but also accelerates the possibility of delamination between the pavement layers.

8.5 Conclusions

A field scenario of HMA overlays with and without geosynthetics placed over an old pavement layer was considered, so that the reflection crack phenomenon could be studied by using the BFT and digital image analysis program. Based on the BFT results and the corresponding digital image analyses, the following conclusions can be drawn from the study:

The grid and composite reinforcements have improved the fatigue lives of two-layered asphalt beam specimens, against the reflection cracks, irrespective of the notch depth. An improvement in fatigue life of about 47-fold and 38-fold is achieved for notch-less composite and grid-reinforced specimens, respectively with reference to the control specimens. Similarly, an improvement of 45-fold (GGC) and 15-fold (PET) were achieved for specimens with 25 mm notch depth, at failure. In addition, the improvement in fatigue life of 40 mm notched specimens are about 120-fold (GGC) and 54-fold (PET), with respect to control specimens, at failure.

The DIC technique has been effective in detecting the crack evolution and its patterns; this information helps in understanding and assessing the reflection crack phenomenon in two-layer asphalt specimens. The digital image analysis suggested that the tensile strains at the crack tip were very high in CS compared to the grid and composite-reinforced specimens, irrespective of the notch depths. An average reduction in tensile strain of about 35–70% was noticed in grid and composite-reinforced specimens against the control specimens.

Geosynthetic-reinforcements absorb the mobilized tensile stresses and restrict the crack growth in terms of both width and height. A mean crack opening as high as 0.74 mm was observed in CS, at failure (N = 335), whereas, a mean crack opening of 0.52 mm and 0.47 mm were observed in GGC (N = 16,000) and PET (N = 13,116) specimens, respectively, at failure. Similarly, a maximum vertical crack growth as high as 70.89 mm was observed in the CS against a crack growth of 60 mm, and 55 mm in composite and grid-reinforced specimens, respectively, at failure.

The beam fatigue test results (NCM and N) and the corresponding crack growth information from digital image analysis could be adopted together to predict the service life of asphalt overlays, corresponding to full-depth or half-depth cracking conditions.

Overall, the improvement in fatigue life of geosynthetic-reinforced HMA overlays, against reflection cracks, depend upon the type of geosynthetics, their working mechanical and tensile properties, which can be quantified with DIC technique.

References

1. Cleveland, G.S., Button, J.W., Lytton, R.L.: Geosynthetic in Flexible and Rigid Pavement Overlay. Texas Transportation Institute, Texas A&M University System. Report, pp. 1777–1781 (2002)
2. De Bondt, A.: Effect of reinforcement properties. In: Proceedings of 4th International RILEM Conference on Reflective Cracking in Pavements, pp. 13–22
3. Kim, J., Buttlar, W.G.: Analysis of reflective crack control system involving reinforcing grid over base-isolating interlayer mixture. J. Trans. Eng. **128**(4), 375–385 (2002)
4. Bandaru, R.: Cost-Effective Prevention of Reflective Cracking in Composite Pavements. MS thesis Submitted to Louisiana State University, Baton Rouge (2010)
5. Button, J.W., Lytton, R.L.: Guidelines for using geosynthetics with HMA overlays to reduce reflective cracking. Report 1777–P2. Texas Department of Transport Austin, Texas (2003)
6. Saride, S., Kumar, V.V.: Influence of geosynthetic-interlayers on the performance of asphalt overlays on pre-cracked pavements. Geotext. Geomemb. **45**(3), 184–196 (2017)
7. Zamora-Barraza, D., Calzada-Perez, M.A., Castro-Fresno, D., Vega-Zamanillo, A.: Evaluation of anti-reflective cracking systems using geosynthetics in the interlayer zone. Geotext. Geomemb. **29**(2), 130–136 (2011)
8. Ogundipe, O.M., Thom, N., Collop, A.: Investigation of crack resistance potential of stress absorbing membrane interlayers (SAMIs) under traffic loading. Constr. Build Mater. **38**, 658–666 (2012)
9. Kumar, V.V., Saride, S.: Use of digital image correlation for the evaluation of flexural fatigue behavior of Asphalt beams with geosynthetic interlayers. Trans. Res. Rec. J. Transp. Res. Board 2631, 55–64 (2017)
10. Kumar, V.V., Saride, S.: Evaluation of cracking resistance potential of geosynthetic-reinforced asphalt overlays using direct tensile strength tests. Constr. Build. Mater. **162**, 37–47 (2018)
11. Sanders, P.J.: Reinforced Asphalt Overlays for Pavements. Ph.D. Thesis Ssubmitted to University of Nottingham (2001)
12. Virgili, A., Canestrari, F., Grilli, A., Santagata, F.A.: Repeated load test on bituminous systems reinforced by geosynthetics. Geotext. Geomemb. **27**, 187–195 (2009)
13. Ferrotti, G., Canestrari, F., Pasquini, E., Virgili, A.: Experimental evaluation of the influence of surface coating on fiberglass geogrid performance in asphalt pavements. Geotext. Geomemb. **34**, 11–18 (2012)
14. Pasquini, E., Bocci, M., Canestrari, F.: Laboratory characterisation of optimised geocomposites for asphalt pavement reinforcement. Geosynth. Intl. **21**(1), 24–36 (2014)
15. Pasquini, E., Pasetto, M., Canestrari, F.: Geocomposites against reflective cracking in asphalt pavements: laboratory simulation and field application. Road Mater. Pavement Des. **16**(4), 815–835 (2015)
16. Saride, S., Kumar, V.V.: Estimation of service life of geosynthetic-reinforced asphalt overlays from beam and large-scale fatigue tests. J. Test. Eval. **47**(4), 20170605 (2019)
17. Livneh, M., Ishai, I., Kief, O.: Bituminous pre-coated geotextile felts for retarding reflection cracks. In: Proceedings of 2nd International RILEM Conference on Pavements, pp. 343–350
18. De Benedetto, H., Nejl, J., Antonie, J., Pasquier, M.: Apparatus for laboratory study of cracking resistance. In: Proceedings of 2nd International RILEM Conference on Pavements, pp. 179–186 (1993)
19. Francken, L., Vanelstraete, A.: On the thermorheological properties of interface systems. In: Proceedings of 2nd International RILEM Conference on Pavements, pp. 206–219 (1993)
20. Prieto, J.N., Gallego, J., Perez, I.: Application of the wheel reflective cracking test for assessing geosynthetics in anti-reflection pavement cracking systems. Geosynth. Int. **14**(5), 287–297 (2007)
21. Moreno-Navarro, F., Rubio-Gamez, M.C.: UGR-FACT test for the study of fatigue cracking in bituminous mixes. Constr. Build. Mater. **53**, 182–189 (2014)
22. Grediac, M.: The use of full-field measurement methods in composite material characterization: interest and limitations. Compos. A Appl. Sci. Manuf. **35**, 751–761 (2004)

23. Romeo, E.: Two-dimensional digital image correlation for asphalt mixture characterization: interest and limitations. Road Mater. Pavement Des. **14**(4), 747–763 (2013)
24. Choi, S., Shah, S.P.: Measurement of deformations on concrete subjected to compression using image correlation. Exp. Mech. **37**, 307–313 (1997)
25. Kim, Y.R., Wen, H.: Fracture energy from indirect tension test. J. Ass. Asph. Pav. Technol. **71**, 779–793 (2002)
26. Safavizadeh, S.A., Wargo, A., Guddati, M., Kim, Y.R.: Investigating reflective cracking mechanisms in grid-reinforced asphalt specimens. Transp. Res. Rec. J. Transp. Res. Board 2507, 29–38
27. Wargo, A, Safavizadeh, S.A., Kim, Y.R.: Comparing the performance of fiberglass grid with composite interlayer systems in asphalt concrete. Transp. Res. Rec. J. Transp. Res. Board 2631, 123–132
28. MORTH. Specifications for Road and Bridge works. Ministry of Road Transport and Highways, Government of India
29. ASTM D6927. Standard Test Method for Marshall Stability and Flow of Asphalt Mixtures. Annual Book of ASTM Standards, ASTM International, West Conshohocken, PA
30. ASTM D4595. Standard Test Method for Determining Tensile Properties of Geotextiles by the Wide-Width Strip Method. Annual Book of ASTM Standards, ASTM International, West Conshohocken, PA
31. Caltabiano, M.A.: Reflection Cracking in Asphalt Overlays. Master of Philosophy Thesis submitted to the University of Nottingham (1990)
32. Khodaii, A., Fallah, S., Nejad, F.M.: Effects of geosynthetics on reduction of reflection cracking in asphalt overlays. Geotext. Geomemb. **27**, 1–8 (2009)

Chapter 9
Centrifuge Model Studies on the Performance of Geosynthetic-Reinforced Soil Structures

B. V. S. Viswanadham

9.1 Introduction

The method of soil reinforcement dates back to antiquity, as many ancient historic structures were observed to incorporate layers of natural tensile elements for reinforcing the soil mass. Presently, with the advent of geosynthetic materials composed of strong and durable polymers, reinforcement of soil has become a viable and cost-effective technology. Under the impact of external loading, strains in the soil mass induce strains within the geosynthetic layers and generate tensile loads, which restrict soil movements and control possible deformations. As a result, the reinforced soil system possesses significantly greater shear strength and stability than the soil mass alone.

9.1.1 Geosynthetic Types

Geosynthetics are capable of performing multiple functions like filtration, drainage, reinforcement, separation and containment, and can be broadly subdivided into seven categories, viz. geotextiles, geogrids, geonets, geomembranes, geosynthetic clay liners, geopipes and geofoams. Geotextiles perform one of the following discrete functions among filtration, separation, reinforcement and drainage, whereas geogrids perform exclusively as reinforcing materials. Geonets are employed to convey liquids of all types, while geomembranes are used primarily as linings and covers for liquid and solid storage facilities. Geosynthetic clay liners, on the other hand , are a

B. V. S. Viswanadham (✉)
Department of Civil Engineering, Indian Institute of Technology Bombay, Powai, Mumbai 400076, India
e-mail: viswam@civil.iitb.ac.in

M. Latha G. (ed.), *Frontiers in Geotechnical Engineering*, Developments in Geotechnical Engineering, https://doi.org/10.1007/978-981-13-5871-5_9

combination of woven/nonwoven geotextiles and natural soil materials, and function primarily as containment systems. Geopipes focus entirely on providing drainage function, whereas, geofoam is effective in minimizing settlement on underground utilities, reducing earth pressure behind retaining walls and for slope stabilization purposes.

A special variety of geosynthetic is hybrid geosynthetics or geocomposites, which are capable of serving two or more functions simultaneously. They are manufactured or assembled materials having at least one geosynthetic product among the components, and are fabricated by combining the best features of different materials in such a way that specific applications are addressed optimally and at minimum cost [10].

9.1.2 Geosynthetic-Reinforced Soil Structures

Construction of geosynthetic-reinforced soil structures are now well established in conventional applications, and their uses in non-conventional projects are fast expanding. A widely employed application of geosynthetics and allied products involve construction of retaining walls, as they bring about significant savings of time and cost. There have been several surveys regarding the cost of various types of soil walls over the years, the most recent one being conducted by the Geosynthetic Institute [9]. Based on area of wall facing, it was seen that MSE walls with geosynthetic reinforcement are the least expensive of all the wall types, whereas, the classical gravity wall is more than twice as expensive as any other wall type. This may be attributed to the fact that, use of geosynthetics within the soil walls facilitates the use of locally available fill material, and low material and installation costs, thereby saving typically up to almost 50%, as compared to conventional cast-in situ concrete retaining walls. They also allow a variety of facing options like treated timbers, wraparound and vegetated facings, gabions, precast concrete panels, wire baskets, natural stones and so on, to provide superior aesthetics compared to conventional concrete walls.

In addition to the above, a few typical applications of geosynthetic-reinforced soil structures include highway embankments, airport runway embankments, containment dykes, flood control levees, earth dams, berms, mine reclamations, municipal solid waste landfills, segmental retaining walls, bridge abutments, coastal protection walls, unpaved temporary haul roads, paved roads over soft soils, foundation support, industrial access roads, control of subsidence over pipes and subsurface structures, secondary reinforcement for soil retaining structures and so on [10, 17]. Selected applications from the above are outlined briefly in Fig. 9.1.

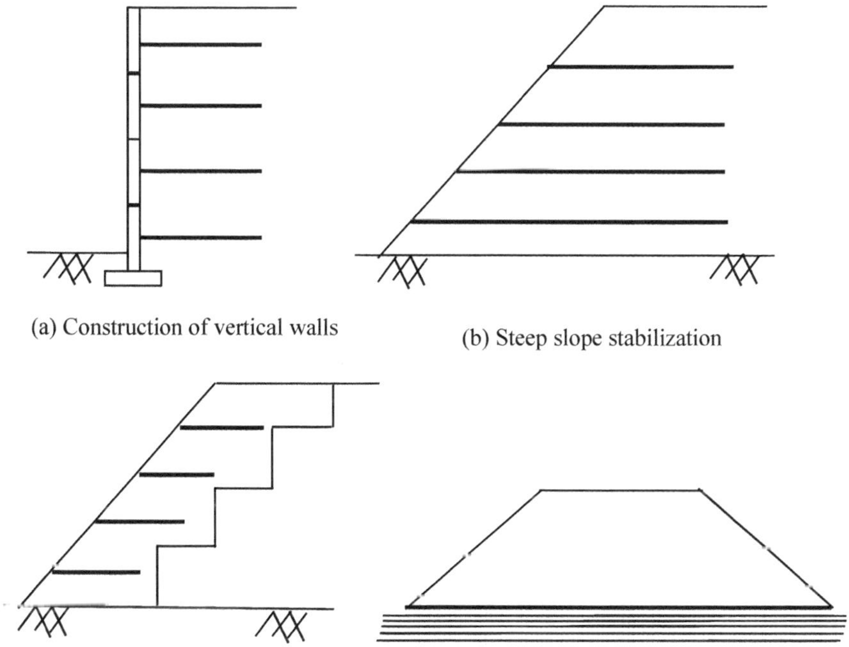

(a) Construction of vertical walls

(b) Steep slope stabilization

(c) Slip prevention and remedial measures

(d) Embankment construction on soft soil

Fig. 9.1 Applications of geosynthetic-reinforced soil structures (modified after Jewell [8])

9.2 Modelling Considerations of Geosynthetics in a Geocentrifuge

Physical modelling of geotechnical engineering problems can be executed at full-scale level or as a reduced scale model to replicate various field situations. However, a limitation of small-scale laboratory models under normal gravity is that the stress levels are much smaller than in prototype structures. Only full-scale physical models can include all these complexities, but they are expensive and time-consuming, and difficult to replicate with natural hazards like flooding, drawdown, seepage and rainfall. In such situations, geotechnical centrifuge modelling can be used as an effective tool to replicate identical stress–strain response in the model as that of the prototype, and to study the behaviour of engineered earth structures in a controlled environment by deriving standard scaling laws linking the model behaviour to that of the corresponding prototype. However, a challenging aspect of modelling geosynthetics in a geocentrifuge is that the similitude condition does not allow identical geosynthetic materials to be incorporated in the prototype and model tested. The idealized scaling considerations involved in this regard from hydraulic and mechanical point of view are discussed in subsequent sections.

9.2.1 Hydraulic Scaling Considerations: Geotextiles

The hydraulic scaling considerations of nonwoven geotextile is governed by the transmissivity (θ), defined as the product of geotextile thickness (t_g) and the in-plane hydraulic conductivity of geotextile (k_g), as represented in Eq. (9.1) as

$$\theta = k_g t_g, \tag{9.1}$$

where θ is the transmissivity (m^2/s), k_g is the coefficient of permeability in the plane of geosynthetic layer (m/s). As per the findings of Raisinghani and Viswanadham [11], for achieving similarity of in-plane flow characteristics in the model and prototype geosynthetics, the transmissivity should be identical in either case, as depicted by Eq. (9.2) as:

$$\theta_m = \theta_p, \tag{9.2}$$

where the subscripts m and p have been used to denote the centrifuge model and prototype respectively. Similarly, the scale factor for in-plane permeability coefficient of model geotextile can be evaluated as represented in Eq. (9.3):

$$\frac{(k_g)m}{(k_g)p} = \frac{\left[\frac{q}{Wt_g i}\right]_m}{\left[\frac{q}{Wt_g i}\right]_p} = \frac{q_m}{q_p}\frac{W_p}{W_m}\frac{(t_g)p}{(t_g)m}\frac{i_p}{i_m} = \frac{1}{N}NN(1) = N \tag{9.3}$$

Based on Eq. (9.3), it can be inferred that the in-plane permeability coefficient of model hybrid geosynthetic subjected to centrifuge testing is N times higher than the in-plane permeability for corresponding prototype geosynthetic. The discharge (q) *is* defined as $q = V/t$, where V is the volume in (m^3) and t the time in (s), thereby giving the scale factor for time as presented in Eq. (9.4):

$$\frac{t_m}{t_p} = \left[\frac{q_p}{q_m}\right]\left[\frac{v_m}{v_p}\right] = N\frac{1}{N^3} = \frac{1}{N^2} \tag{9.4}$$

Further, cross-plane flow characteristics of geotextiles are evaluated based on permittivity (ψ), defined as the ratio of cross-plane permeability of geosynthetic (k_n) to its thickness(t_g) as presented in Eq. (9.5):

$$\psi = \frac{k_n}{t_g} \tag{9.5}$$

As per the findings of Rajabian et al. [12], for similarity of cross-plane flow characteristics in model and prototype geosynthetics, the geotextile permittivity should be identical for both cases, as depicted by Eq. (9.6) as

$$\psi_m = \psi_p \tag{9.6}$$

9.2.2 Mechanical Scaling Considerations: Geogrids

Modelling of reinforcing geogrid component in a geocentrifuge is done based on scaling of the tensile load–strain behaviour and frictional bond behaviour, by adopting similar approach as Viswanadham and König [21], Viswanadham and Jessberger [20], Viswanadham and Rajesh [23] and Balakrishnan and Viswanadham [1]. Idealized characteristics of the parent geogrid material to be scaled down by a factor N are shown in Fig. 9.2, where the notation a denotes the grid opening size, b indicates the width of the ribs and suffixes l and t denote the longitudinal and transverse direction respectively. In order to replicate geometric similitude condition, the parameters a and b are scaled in transverse and longitudinal directions as presented in Eqs. (9.7a)–(9.7b)

$$\frac{[a_t]_m}{[a_t]_p} = \frac{[b_l]_m}{[b_l]p} = \frac{[t_l]_m}{[t_l]_p} = \frac{1}{N} \tag{9.7a}$$

$$\frac{[a_l]_m}{[a_l]p} = \frac{[b_t]_m}{[b_t]_p} = \frac{[t_t]_m}{[t_t]_p} = \frac{1}{N}, \tag{9.7b}$$

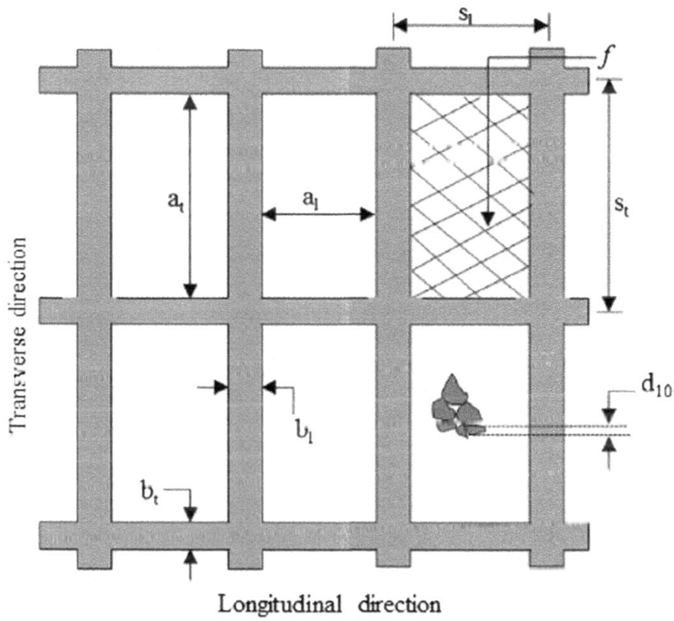

Fig. 9.2 Geometric characteristics of geogrid (after Balakrishnan and Viswanadham [1])

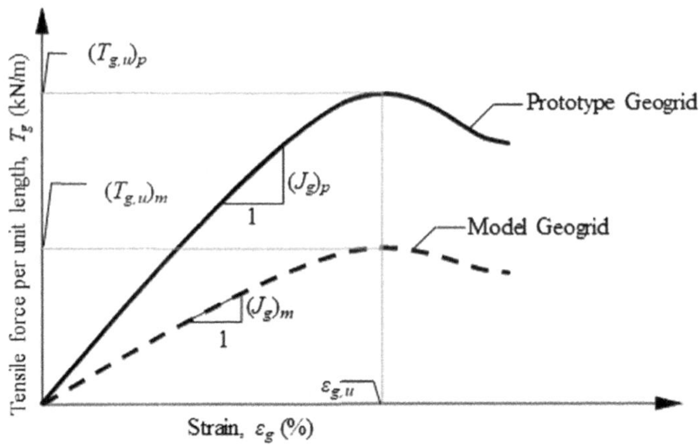

Fig. 9.3 Idealized tensile strength–strain behaviour of geogrid reinforcement

where t_l and t_p denote the average thickness. In addition, to prevent the possibility of losing contact between soil-geogrid interface, identical percentage open areas (f) should be ensured for the model and prototype geogrids, as presented in Eqs. (9.8a)–(9.8b) respectively

$$f_m = f_p \tag{9.8a}$$

$$f = \frac{a_l a_t}{(a_l + b_l)(a_t + b_t)} \tag{9.8b}$$

Figure 9.3 depicts the idealized tensile strength–strain behaviour of model reinforcement layers replicating the corresponding behaviour of prototype reinforcement layers. As can observed from the Fig. 9.3, for similar material composition prevailing in the model and prototype, the tensile strength–strain behaviour of geogrid reinforcement is modelled using Eqs. (9.9) and (9.10) as follows:

$$\left(T_{gu}\right)_m = \left(T_{gu}\right)_p / N \tag{9.9}$$

$$(\varepsilon_{gu})_m = (\varepsilon_{gu})_p, \tag{9.10}$$

wherein $T_{g,u}$ is the ultimate tensile strength (defined as load per unit width) of reinforcement and $\varepsilon_{g,u}$ denotes the ultimate strain in the reinforcement layer. In a similar manner, by using the validity of $T_g = J_g\ \varepsilon_g$, the secant stiffness J_g of the model geogrid is scaled by $1/N$ times as presented in Eq. (9.11)

$$\left(J_g\right)_m = \left(J_g\right)_p / N \tag{9.11}$$

The second similarity requirement is the frictional bond simulation between soil and geogrid, which is accomplished by using Eq. (9.12), where δ denotes the soil-reinforcement interface friction angle.

$$\delta_m = \delta_p \tag{9.12}$$

This implies that the tensile load and secant stiffness of selected model geogrid will be scaled by 1/N times compared to of that of commercially available prototypes. Such reduced tensile load and stiffness requirements can only be achieved by considerably reducing rib dimensions and thickness at rib junctions. In addition, the percentage open area f shall have to be identical in model and prototype geogrids for ensuring perfect soil-geogrid interaction (Fig. 9.2) and maintenance of identical frictional bond area. In this context, the percentage open area, f is defined as the ratio of area formed by grid opening sizes to area formed by grid opening sizes measured up to centre of width of ribs.

$$f_m = f_p \text{ where } f = \left(\frac{a_l a_t}{(a_l + b_l)(a_t + b_t)}\right) \tag{9.13}$$

The tensile load T_g and secant stiffness J_g of the model reinforcement material were scaled down to 1/N times that of the prototype material for a geogrid-reinforced wall model, maintaining the $(\phi_{sg})_m = (\phi_{sg})_p$ and $f_m = f_p$, where N is the scale factor or g-level, ϕ_{sg} the soil-geogrid interface friction angle and the subscripts m and p represent model and prototype, respectively. Further, the ratio of soil particle size of backfill soil to geogrid opening size affects the soil-geogrid interaction. Springman et al. [18] suggested that $s_t/d \gg 5$, where s_t is the spacing between transverse ribs (shown in Fig. 9.1) and d is the particle size. Izawa and Kuwano [7] suggested $s_t/d_{50} \gg 10$ where d_{50} is the average particle size. Izawa and Kuwano [7] had reported that it is not necessary to achieve similitude of geometric characteristics, if appropriate mechanical properties (i.e., tensile stiffness and soil-reinforcement interaction) are achieved. Springman et al. [18], Viswanadham and Jessberger [20], Izawa and Kuwano [7] and Rajesh and Viswanadham [15] had adopted polymeric mesh type materials representing bandwidth of characteristics of commercially available geogrids. Izawa and Kuwano [7] used muster geogrids made out of pre-punched Poly Vinyl Chloride and Polycarbonate sheets.

9.2.3 Hydro-mechanical Scaling Considerations: Hybrid Geosynthetics

Hybrid geosynthetics or geocomposites are assembled materials developed by integrating the reinforcement action of geogrids with the drainage function of nonwoven geotextiles to avert possible slope instability underwater infiltration [3, 11, 19]. The model hybrid geosynthetic material to be incorporated in a geocentrifuge should thus be developed to serve the dual functions of reinforcement and drainage, and needs to be scaled appropriately to satisfy the reinforcement function of parent geogrid, as well as drainage aspects of parent geotextile discussed in previous sections. Detailed scaling considerations of hybrid geosynthetics in a geotechnical centrifuge are outlined in Bhattacherjee and Viswanadham [4].

9.2.4 Mechanical Scaling Considerations: Geomembranes

The important aspects of scaling down commercially available geomembranes for centrifuge study include simulation of tensile load–strain behaviour, soil-geomembrane interface characteristics and thickness of model geomembrane material.

If t_g is the thickness and l_g is the length of geomembrane, then the scaling laws for length and thickness of geomembrane are represented as in Eq. (9.14), where suffix "*g*" stands for prototype geomembrane and "*mg*" for model geomembrane.

$$(t_g)/(t_{mg}) = (l_g)/(l_{mg}) = N \tag{9.14}$$

Further, if E_g is the modulus of elasticity of the geomembrane material and σ_g is the tensile stress, the geomembrane strain can be obtained using Eq. (9.15):

$$\varepsilon_g = \sigma_g/E_g \tag{9.15}$$

Assuming that tensile strains in the geomembrane with identical material characteristics in model and prototype are identical, i.e. ($\varepsilon_g = \varepsilon_{\mathrm{mg}}$ and $E_g = E_{mg}$), the stresses in model and prototype geomembranes are also equal, implying $\sigma_g = \sigma_{mg}$ [Eq. (9.14)]. In addition, if the tensile stiffness of the geomembrane is represented by $J = Et$, and T_{mg} is the tensile load in a model geomembrane in geocentrifuge, then as reported by Divya et al. [6], the relevant scaling laws are represented by Eqs. (9.16a)–(9.16b) respectively

$$(J_g)/(J_{mg}) = (E_g t_g)/(E_{mg} t_{mg}) = N \tag{9.16a}$$

$$(T_g)/(T_{mg}) = (\varepsilon_g E_g t_g)/(\varepsilon_{mg} E_{mg} t_{mg}) = N \tag{9.16b}$$

Further, a geomembrane overlying a compacted clay barrier needs to achieve identical magnitude of subsidence as that of the clay barrier, and the associated tension developed is resisted by the interface frictional strength between the soil and geomembrane for preventing pull-out. If c_g is the adhesion and δ_g is the interface friction angle at the soil-geomembrane interface, then as per identical stress

Table 9.1 Scaling laws for modelling geosynthetics in geocentrifuge

Parameter	Unit	Prototype	Ng model
Tensile strength (T_g)	kN/m	1	1/N
Strain (ε)	%	1	1
Secant modulus (J_g)	kN/m	1	1/N
Bond-stress (τ_b)	kN/m^2	1	1
Percentage open area (f)	%	1	1
Soil-geosynthetic interface angle (δ)	°	1	1
Soil-geosynthetic interface stiffness (J_i)	kN/m^2	1	N
Transmissivity (θ)	m^2/s	1	1
Time for in-plane flow (t)	s	1	1/N^2
In-plane permeability (k_g)	m/s	1	N

N Gravity level or scale factor; For example, γ_m/γ_p = N; m = model; p = prototype

considerations in model and prototype, the scaling laws can be represented by Eqs. (9.17a)–(9.17b) respectively

$$\delta_g = \delta_{mg} \tag{9.17a}$$

$$c_g = c_{mg} \tag{9.17b}$$

The basic scale factors involved in modelling the hydraulic and tensile characteristics of geosynthetics, along with consideration of soil-geosynthetic interaction effects in a geocentrifuge are summarized in Table 9.1.

9.3 Centrifuge Modelling of Geogrid-Reinforced Clay Liners

Design of a proper containment system for landfills is essential to encapsulate municipal solid wastes, toxic wastes and low-level radioactive wastes. However, differential settlement of liner results in tension cracking of the cover soil/clay barrier and shear failure of barrier materials, accompanied by the formation of sinkhole-type depressions. In order to mitigate the same, researchers have resorted to different methods for averting possible differential settlements, including mixing of discrete fibres with clay barrier [24], or placing geogrids within the clay barrier [13, 15, 16, 20]. Since the deformation and cracking characteristics of clay barriers is highly influenced by prototype stress conditions including thickness, material properties and overburden pressure, centrifuge modelling technique can be effectively employed to study the response of unreinforced and geogrid-reinforced clay barriers subjected to various distortion levels.

9.3.1 Model Test Package and Simulator Assembly

A Motor-based Differential Settlement Simulator (MDSS) was designed and developed by Rajesh and Viswanadham [14] to induce continuous differential settlements on a clay barrier layer under high-gravity environment (Fig. 9.4). Centrifuge tests were performed using a 4.5 m radius large beam geotechnical centrifuge facility available at Indian Institute of Technology Bombay. The tests were conducted at 1 mm/min using the MDSS system at 40*g* on clay barriers of two different thicknesses of 0.6 and 1.2 m, under a nominal overburden pressure of 12.5 kN/m^2 and at a higher pressure of 25 kN/m^2, close to the closure system of low-level radioactive waste disposal. The model unreinforced clay barrier layer of required thickness (d) was prepared by mixing kaolin and sand in the ratio of 4:1, moist-compacted at a moulding water content of OMC +5% at a dry unit weight of 14.2 kN/m^3. In the case of geogrid-reinforced clay barrier, model geogrid was placed at a distance of 0.25d from the top surface of clay barrier. Upon completion of model, a seating pressure of 5 kN/m^2 (static load of 1.4 kN) was applied uniformly for about 5 h on clay barrier in order to achieve proper contact between soil and the geogrid layer. Identical seating pressure was also applied to unreinforced clay barriers. Further details related to material properties, model preparation and testing procedure are discussed elaborately in Rajesh and Viswanadham [13].

9.3.2 Instrumentation

Five numbers of Pore Pressure Transducer (PPTs) of type PDCR81 were placed above the clay barrier surface to monitor water levels and to determine water breakthrough characteristics as an indicative of water infiltration through the clay barrier system. In addition, six numbers of Linear Variable Differential Transformers (LVDTs) were placed to monitor the deformation of the clay barrier with and without geogrid layer in terms of induced central settlement (a) and the settlement rate.

9.3.3 Results and Discussion

The influence of geogrid reinforcement on the deformation behaviour of clay barriers subjected to differential settlements was evaluated by Rajesh and Viswanadham [13] in terms of water breakthrough characteristics, strain distribution and crack width

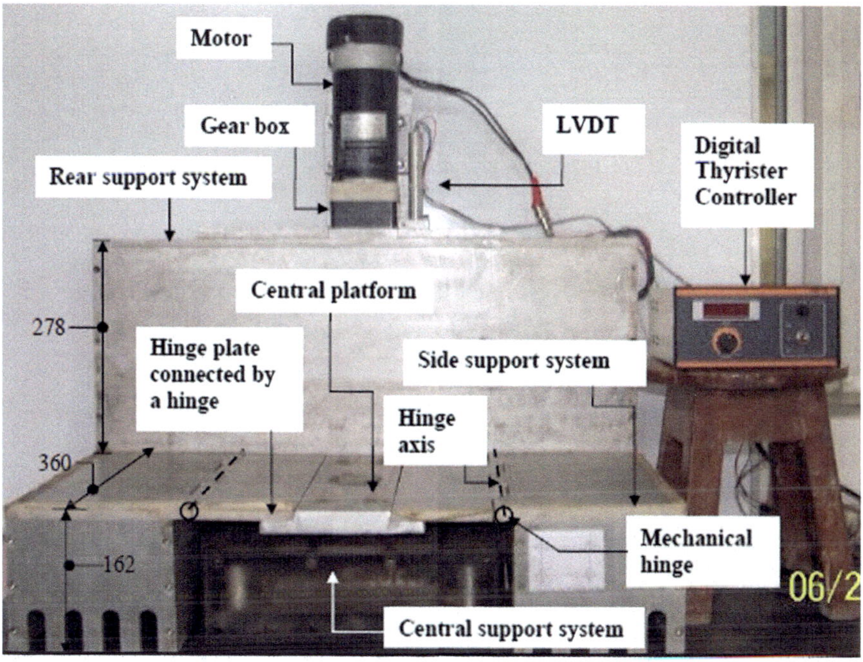

Fig. 9.4 Components of MDSS system (after Rajesh and Viswanadham [14])

measurements. The deformation profiles of the top surface of the clay barrier as well as geogrid layer for a typical centrifuge model test is shown in Fig. 9.5 corresponding to various stages of central settlements, as measured from LVDT data and by image analysis of discrete markers. It can be observed from Fig. 9.5 that both geogrid layer and the clay barrier experienced identical deformation profiles at various stages of central settlements, indicating good compatibility between soil and geogrid at the onset of differential settlements. The deformation profiles were subsequently used to determine strain along the top-most surface of the clay barrier and to observe the phenomena of crack formation, by following the procedure outlined in Viswanadham and Rajesh [23].

It was observed that the 1.2 m thick unreinforced clay barrier at a nominal overburden of 12.5 kN/m^2 experienced multiple cracks extending up to the full depth, as indicated in Fig. 9.6a, accompanied by a catastrophic water breakthrough at a settlement ratio of 0.42 and distortion of 0.052. In contrary, the 1.2 m thick geogrid-reinforced clay barrier subjected to 12.5 kN/m^2 overburden was reported to experience narrow cracks extending only up to the geogrid level, even at a distortion level of 0.125 as indicated in Fig. 9.6b. Similar findings indicating superior performance of geogrid layer in terms of restraining cracks and water breakthrough characteristics were reported for a 0.6 m thick clay barrier, and upon increasing the overburden

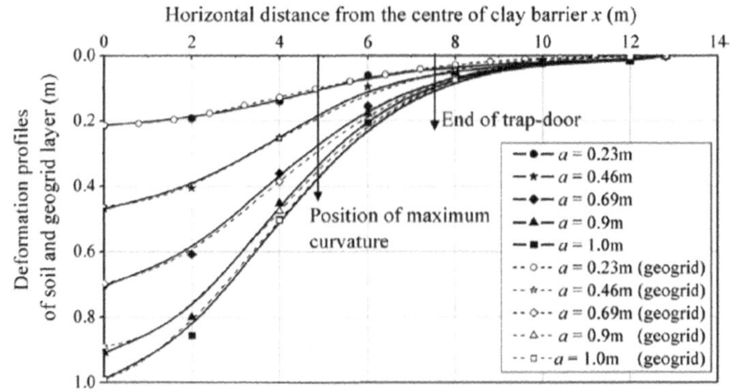

Fig. 9.5 Variation of measured deformation profiles with horizontal distance from the centre of clay barrier (after Rajesh and Viswanadham [13])

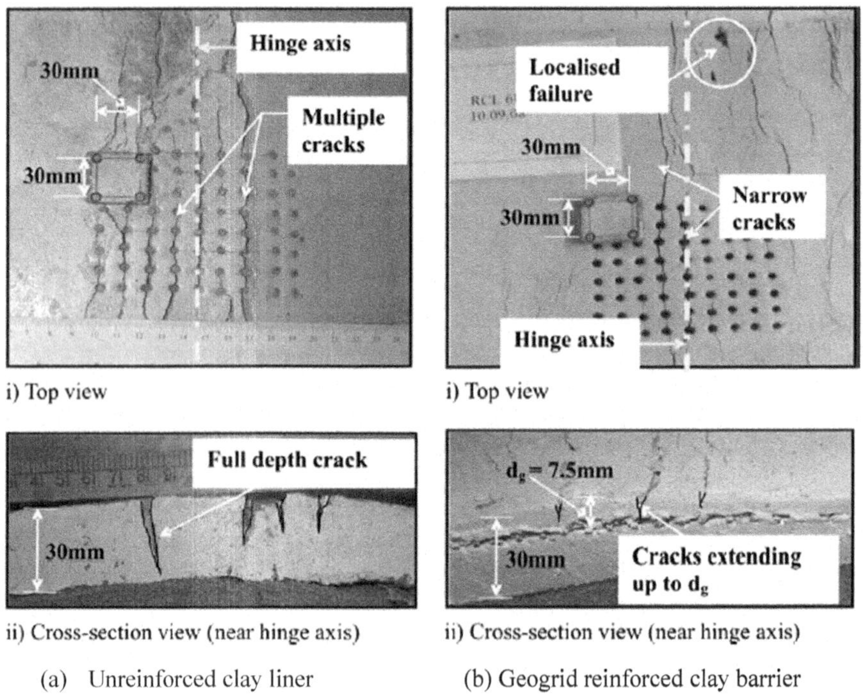

Fig. 9.6 Status of 1.2 m thick unreinforced and geogrid reinforced clay barriers with nominal overburden at the zone of maximum curvature (after Rajesh and Viswanadham [13])

pressure to 25 kN/m^2. The above results highlight the significance of geogrid layers in maintaining the integrity of clay barrier of cover systems at the onset of differential settlements.

Further, as reported by Rajesh and Viswanadham [15], increase in the thickness of soil barrier resulted in the enhanced ability withstand higher distortion levels for both unreinforced and geogrid-reinforced soil barriers. Substantially improved characteristics of the soil barrier was also observed upon increasing the strength and stiffness of the geogrid, and it was mentioned that the thickness of soil barrier can be reduced from 1.2 to 0.6 m by adopting suitable tensile load–strain characteristics of geogrid reinforcement.

9.4 Centrifuge Modelling of Geogrid-Reinforced Retaining Walls with Marginal Backfill

Mechanically stabilized earth (MSE) walls reinforced by geogrids and geotextiles have seen a tremendous growth over the past 30 years. However, along with this growth numerous cases of failures have been reported in the field dealing with excessive deformation and, in some cases, actual collapse of retaining walls. The primary cause of failure in reinforced soil walls (especially those constructed with on-site marginal backfill) may be attributed to the absence of properly designed drainage system at the back and bottom of the reinforced soil zone. Drawing parallels to the above, centrifuge-based physical model tests were conducted by Viswanadham et al. [25] to ascertain the deformation characteristics of geogrid-reinforced soil retaining walls with precast panel facing and marginal backfill subjected to seepage conditions arising out of increasing groundwater conditions. Further, the effect of chimney sand drain in improving the performance of geogrid-reinforced soil walls was also investigated through centrifuge model tests.

9.4.1 *Model Test Package and Simulator Assembly*

Appropriate choice of facing element is one of the main criteria of any geogrid-reinforced soil wall system, which can lead to significant reduction in wall deformation and loads transferred to reinforcement layers [2]. In this case, precast concrete panel was selected as the facing type in view of the fact that full-height concrete facing and precast concrete panels are frequently adopted in the field, and ensure more rigidity in comparison with other facing types. In addition to scaling consideration for flexural rigidity, lightweight, low water absorption as well as high stiffness should be considered for modelling the facing panel in a geotechnical centrifuge. Accordingly, three layers of 3 mm thick marine plywood sheets were joined together using a special wood glue to make the facing panel shown in Fig. 9.7, as marine plywood was found to be the best-suited for these requirements. In addition, a special geogrid-facing anchorage system was custom-fabricated and adopted in the wall

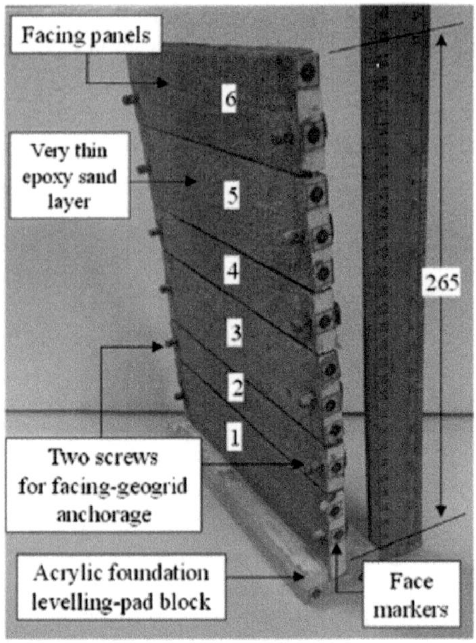

Fig. 9.7 Facing panel details curvature (after Viswanadham et al. [25])

models to ensure proper load transfer between panel and geogrid and for preventing any rupture of the geogrid part within the connection.

The centrifuge models were constructed with a blended low-permeable soil formulated by combining fine sand and commercially available kaolin in the proportion of 4:1 by dry weight to represent RE walls with marginal backfill. The soil walls were reinforced with two geogrid types of varying tensile stiffness and were 250 mm high (H) with a 50 mm thick base layer, tested at 40 gravity loading in a 4.5 m radius large beam geotechnical centrifuge facility at Indian Institute of Technology Bombay. Six layers of geogrid reinforcement of length 200 mm (0.8H) were placed after compaction of each soil layer. In order to avert damage of geogrid-facing connections, adequate care was taken while imparting compactive effort near the facing. In addition, plastic markers made from thin transparent sheets were stuck along the geogrid layers and facing panels (Fig. 9.8) to obtain strain along the geogrid layers by tracing their coordinates in the images captured during experimentation. Further, a seepage tank was placed towards left-hand side of the model container along with a water reservoir and a solenoid valve for simulating rising groundwater condition during centrifuge test, as illustrated in Fig. 9.8.

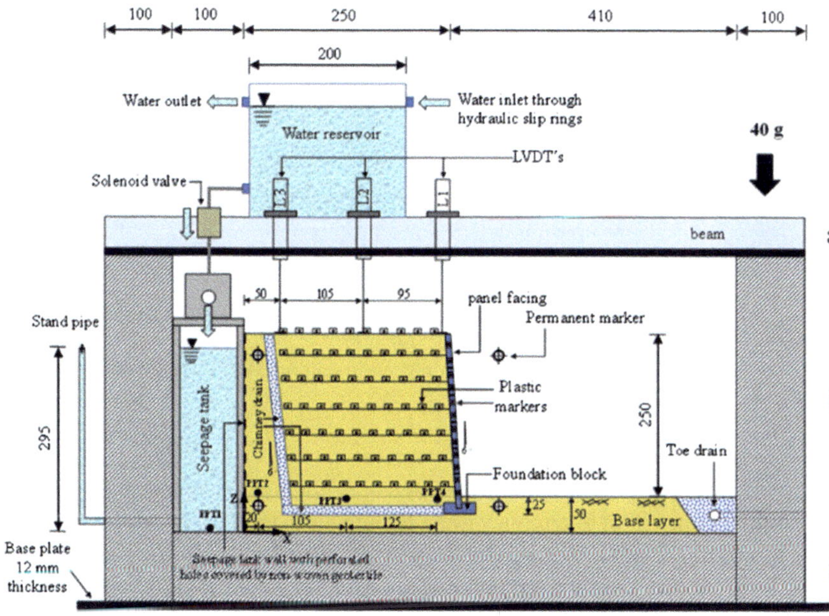

Fig. 9.8 Schematic view of model adopted for centrifuge test; All dimensions are in mm (after Viswanadham et al. [25])

9.4.2 Instrumentation

The geogrid-reinforced soil wall models were instrumented with three Linear Variable Differential Transformers (LVDTs) and four miniature Pore Pressure Transducers (PPTs). LVDTs were placed at the top of wall at a distance of 0 mm (L1), 95 mm (L2) and 200 mm (L3) from the crest of the wall, while one PPT (PPT1) was placed within the seepage tank, and three PPTs were placed within the wall model along the base layer at a distance of 20 mm (PPT2), 125 mm (PPT3) and 250 mm (PPT4) from the seepage tank, as indicated in Fig. 9.8. PPT 3 was thus kept at the middle portion of the wall and PPT4 was placed near the toe region of the wall. The data obtained from these two PPTs were normalized and used for comparison purposes between models.

9.4.3 Results and Discussion

The geogrid-reinforced soil wall models were monitored for surface settlements, pore water pressures and wall face movements at the onset of seepage at 40 g on the basis of instrumentation data and image analysis results. Figure 9.9a and b depict the phreatic surfaces developed within the wall models at an interval of 1 day after

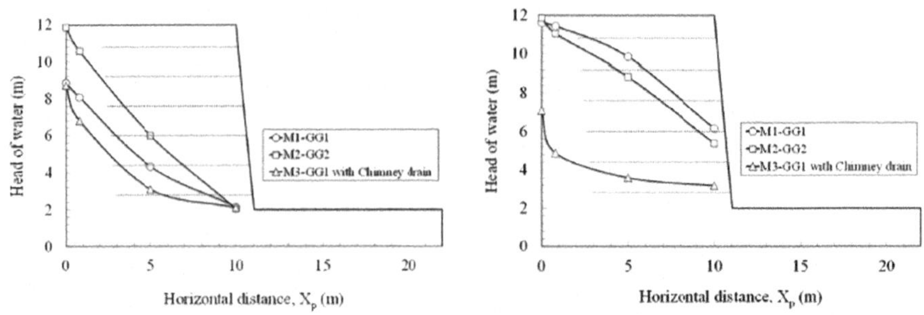

Fig. 9.9 Development of phreatic surfaces within geogrid reinforced soil walls with and without chimney drain (after Viswanadham et al. [25])

inducing seepage and during penultimate stage of centrifuge test respectively. As can be observed, the inclusion of chimney drain having high drainage capability resulted in considerable dissipation of pore water pressures through the reinforced soil zone at the wall toe, and at a section located mid-distance from the wall toe.

Figure 9.10 presents the variation of wall face movements with normalized height (z/H) for models M1 and M2 constructed with geogrids of varying tensile stiffness. The vertical axes indicate the location of markers at the facing along the wall height with origin at the toe. Model M1 (with weaker geogrid) was observed to undergo a global stability failure of a bulging type with increasing face movements from 1.03 m to 2.84 m at the bottom portion of the wall between 11.61 days and 12.68 days of seepage time. It was further observed that the value of z_{max}/H (location of maximum facing displacement) changed from 0.233 to 0.134 at the time of failure. In contrast, the maximum face movement was restricted to 0.06 m for Model M2 (with stiffer geogrid), thereby indicating that marginal soil can be used as a backfill in reinforced soil walls, provided it has geogrid layers of adequate stiffness and/or associated chimney drains.

9.5 Centrifuge Modelling of Geotextile-Reinforced Slopes

In recent years, increasing infrastructural developments have necessitated construction of geosynthetic-reinforced soil structures, especially slopes and embankments reinforced with geosynthetic inclusions. It is therefore imperative to understand the behaviour of geosynthetic-reinforced slopes, both at pre-failure state and failure. The current design methods for reinforced soil vertical walls assume a triangular distribution of peak tensile strength in the reinforcement layers for resisting active earth pressures in the reinforced soil zone. The same rationale is usually extended to geosynthetic-reinforced slopes, where a linear distribution of reinforcement peak tension with height is assumed, with zero tension at the crest and maximum peak tension

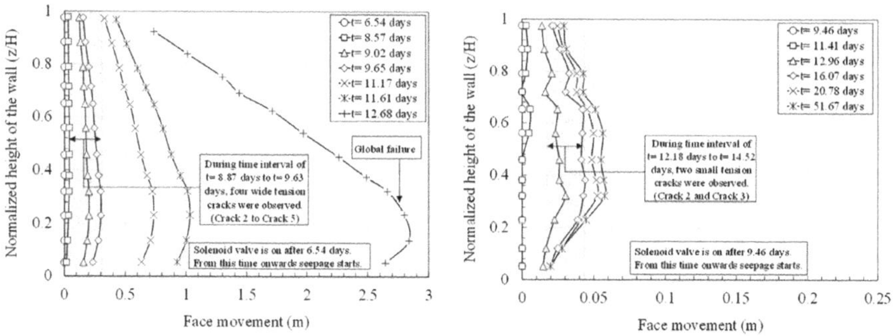

(a) Model M1
[Geogrid GG1; T_g= 1.014 kN/m]

(b) Model M2
[Geogrid GG2; T_g= 2.08 kN/m]

Fig. 9.10 Variation of face movements with normalized wall height (after Viswanadham et al. [25])

at the toe of the structure. Although this assumption may be appropriate for the design of vertical geosynthetic-reinforced walls, little evidence was observed for the design of geosynthetic-reinforced slopes [26]. The following section highlights the effect of slope inclination, vertical spacing of reinforcement layers and type of reinforcement layer on the stability and deformation behaviour of geosynthetic-reinforced slopes through a series of centrifuge model tests conducted by Viswanadham and Mahajan [22] in a 4.5 m radius large beam geotechnical centrifuge facility available at Indian Institute of Technology Bombay.

9.5.1 Model Test Package and Simulator Assembly

The test set-up is illustrated schematically in Fig. 9.11, which depicts a geotextile-reinforced model slope of 270 mm height resting over a 30 mm thick firm foundation layer. The model slopes were constructed using clean, uniformly graded Goa sand (in dry state) at a consistent relative density of 85%. In the reinforced zone and backfill zone, the sand was pluviated in layers to achieve a relative density of 55% and void ratio of 0.733. To some extent, this also represents poor densification and marginal material characteristics of a backfill material in the field. In all the centrifuge tests conducted, the ratio of reinforcement length to the height of the slope (L_R/H) was adopted equivalent to 0.85, as the study focused on the evaluation of internal stability and rupture aspects of geotextile reinforcement layers. Further, for each geotextile layer, an anchorage length equivalent to 0.22H was used. All the geotextile layers were placed layer by layer and trimmed along their weak direction to form a flexible facing using the wraparound technique. Lateral flaps were incorporated at the slope face for averting sloughing of the sand during testing, and the top layer had an additional wraparound length equivalent to 0.37H to prevent local failures. In this case, each test involved loading a model by gradually increasing its self-weight at

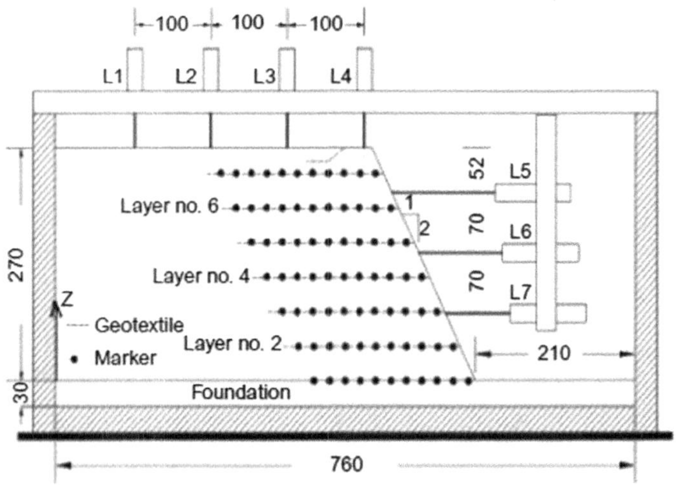

Fig. 9.11 Cross-sectional view of geotextile-reinforced slope; All dimensions are in mm (after Viswanadham and Mahajan [22])

5 g intervals, from an initial value of 10 g up to a maximum of 75 g unless failure occurred first.

9.5.2 Instrumentation

The model was instrumented with seven LVDTs, wherein four (L1, L2, L3 and L4) were installed on the top surface of the reinforced slope on 100 mm centres from the crest of the slope, and three (L5, L6 and L7) were oriented horizontally and placed against the slope face to measure lateral displacements of the slope face (Fig. 9.11). Figure 9.12 shows a perspective view of a model geotextile-reinforced slope before performing centrifuge test.

9.5.3 Results and Discussion

Based on the tracked coordinates of marker positions, displacement vectors and strain distributions were calculated along the reinforcement layers for each g-level. In order to compute the strain between any two adjacent markers at Ng, relative positions were always compared with respect to position of the marker at 1 g. Figure 9.13a–d present the strain distribution obtained for reinforcement layers 3, 5, 7 and 9 (top-most layer) of a typical model MRS7 having a slope inclination of 2 V:1H, and constructed with nine layers of model geotextile fabrics. The curves have been plotted at 10 g, 15 g,

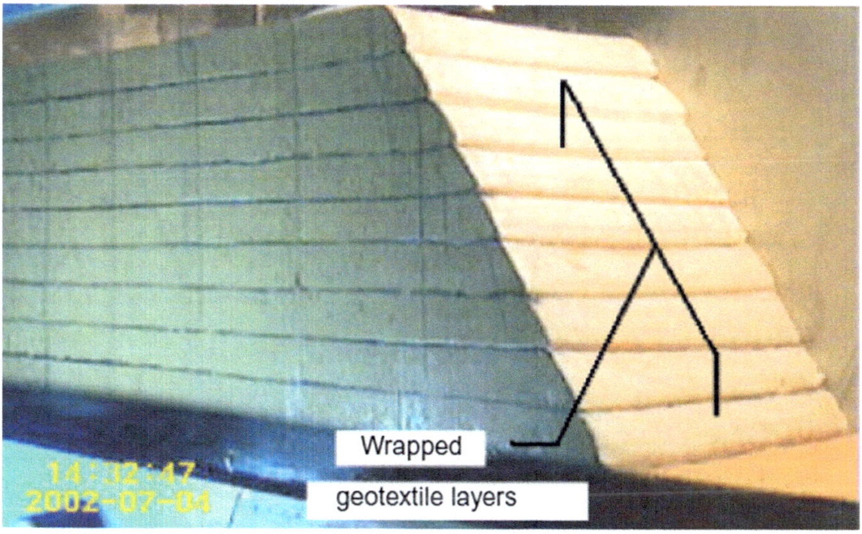

Fig. 9.12 Perspective view of geotextile-reinforced slope (after Viswanadham and Mahajan [22])

20 g, 25 g, 30 g, 35 g and at failure of the centrifuge test, respectively. As can be observed, reinforcement strain increases with increasing g-level, and from toe to crest of the slope, a shift in the location of the formation of peak strain can be observed. For reinforcement layer 3, a peak was observed to occur at 35 g along the reinforcement layer at x/H = 0.185 (where x is the distance from slope face), whereas, for reinforcement layers 5, 7 and 9, it was observed at 0.259, 0.26 and 0.27, respectively. Further, at 35 g, reinforcement layers 5, 7 and 9 were observed to experience strains greater than the ultimate strain of the model fabric, indicating rupture of layers.

On the basis of series of centrifuge tests conducted by Viswanadham and Mahajan [22], it was concluded that for a geotextile-reinforced slope tested in centrifuge, maximum peak strain in the reinforcement layers would occur at 0.44–0.55 times the slope height as measured vertically from the toe. For the same reinforcement type and layout, increase in slope inclination resulted in higher magnitude of peak strains at failure. The location of maximum peak strain was found to be within the bottom half of the slope, whereas for a slope with 2V:1H slope inclination, the maximum peak strain was observed to occur within the top half of the slope. This implies that distribution of reinforcement forces with depth is different from the conventional triangular distribution of reinforcement tension with a maximum tension at the toe of the slope. It was also observed that reinforcement tension distribution was a function of slope inclination in a reinforced slope. In addition, the formation of a kink in reinforcement layers at about mid-height for all the models tested indicated that

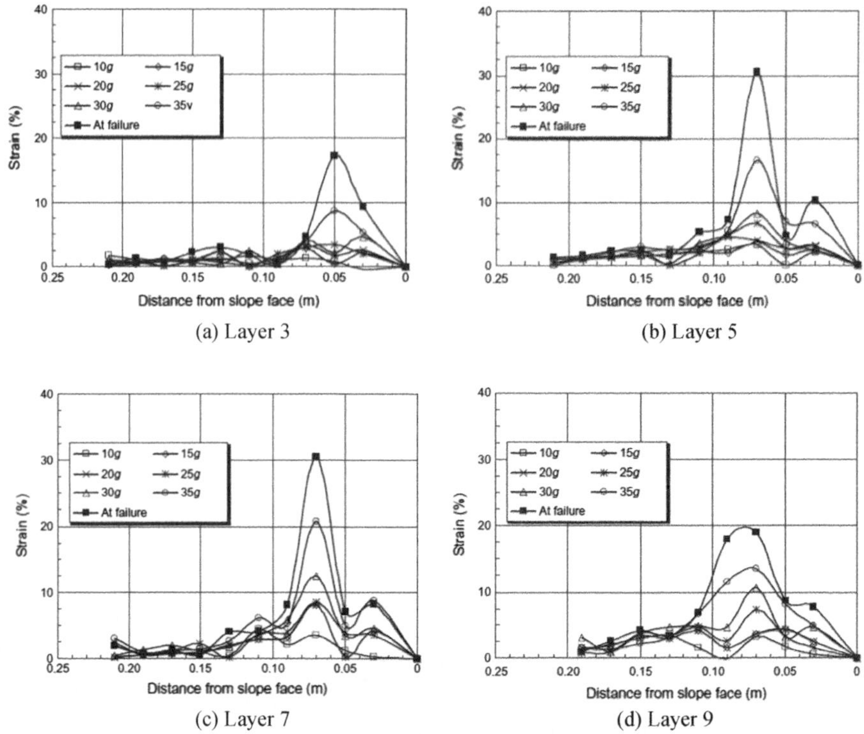

Fig. 9.13 Variation of strain with distance from slope face in a typical geotextile-reinforced slope (after Viswanadham and Mahajan [22])

failure was initiated at mid-height of the slopes. It was further deduced that the location of the failure surface and that of maximum peak strain were unaffected by the reinforcement type or slope inclination.

9.6 Centrifuge Modelling of Hybrid Geosynthetic-Reinforced Slopes Subjected to Rainfall

Recently, innumerable cases of landslides and slope failures induced by heavy rainfall were reported globally, coincident with global warming and associated climatic changes. The instability arises due to the inefficiency of the soil to dissipate the excess pore water pressure generated within the soil structures during the rainfall event. In this regard, structural mitigation measures through the incorporation of geosynthetics are widely adopted in the field to prevent rain-triggered landslides and damages to human life and property. The following segment highlights the possible inclusion of a special type of geosynthetic material, referred herein as hybrid geosynthetic (which possess both reinforcement and drainage capabilities) within a typical low-permeable soil slope, to enhance its performance and stability aspects under various rainfall intensities.

9.6.1 Model Test Package and Simulator Assembly

Fine sand and commercially available kaolin in the ratio of 4:1 by dry weight were blended for formulating the model soil, which represented the geotechnical properties of locally available low-permeable soils found in major portions of India. Centrifuge model tests were performed at 30 g on unreinforced and hybrid geosynthetic-reinforced slopes of 240 mm height and 2V: 1H inclination subjected to a heavy rainfall intensity of 14.4 m/day (20 mm/h) triggered in a 4.5 m radius large beam geotechnical centrifuge facility available at Indian Institute of Technology, Bombay, India. The model hybrid geosynthetic material was developed by integrating drainage function of permeable nonwoven geotextile (N1) with reinforcement of woven geogrid (G1), so as to depict the characteristics of commercially available prototype hybrid geosynthetic materials. A robust rainfall simulating system was designed and developed for inducing rainfall at high gravities, by considering the relevant scaling laws related to modelling of rainfall in centrifuge, as outlined in Bhattacherjee and Viswanadham [5].

The simulator (shown in Fig. 9.14) replicates rainfall in the form of fine mist at high gravities using specially designed air-atomizing nozzles. The lower and upper limits of rainfall intensities can be varied from 2 to 80 mm/h in this case, which are the maximum and minimum recorded intensity till date globally. Further, special measures were incorporated in the design to nullify the effects produced by Coriolis force on droplet trajectory at high gravities, details of which are outlined in Bhattacherjee and Viswanadham [5]. A seepage tank with perforations was placed on the right-hand end of the model container (Fig. 9.14) to induce initial groundwater table within the soil slopes up to the base prior to rainwater infiltration, and a run-off collector was placed on the swing basket of centrifuge to collect the excess water during tests.

9.6.2 Instrumentation

Four pore pressure transducers (PPTs) were placed above the base layer at distances of 20 mm (PPT4), 125 mm (PPT3), 250 mm (PPT2) and 350 mm (PPT1) from the perforated face of the seepage tank. Thus, PPT2 was kept at a point vertically below the crest of the slope, PPT3 was at a point vertically below the mid-length of the slope top surface, and PPT1 was placed almost at the toe of the slope. The model was further instrumented with three Linearly Variable Differential Transformers (LVDTs) placed at the top of the slope at a distance of 50 mm (L1), 155 mm (L2) and 240 mm

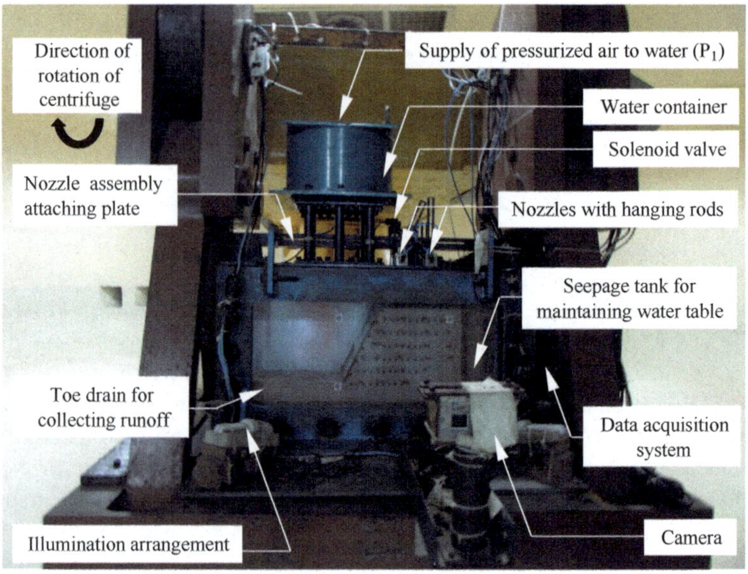

Fig. 9.14 View of model test package with rainfall simulator mounted on the swing basket of centrifuge (after Bhattacherjee and Viswanadham [5])

(L3) from the edge of the seepage tank for measuring surface settlements induced by rainfall.

9.6.3 Results and Discussion

Figure 9.15 presents the variation of normalized pore pressure ($u_{toe}/\gamma h$) with rainfall as measured by PPT 1 placed near the slope toe. As evident from Fig. 9.15, the unreinforced slopes registered high values of $u/\gamma h$ due to rapid loss of matric suction and subsequent build-up of excess positive pore water pressures induced by rainfall, thereby necessitating suitable drainage measures within the slope. Further, the $u/\gamma h$ values reduced by 47% on an average with the inclusion of dual-function geocomposites within the slope, thereby highlighting the importance of drainage function provided by nonwoven geotextile component of hybrid geosynthetics in maintaining slope stability under rainfall.

Figure 9.16a, b present the variation of surface settlements with horizontal distance from the slope crest with the progress of rainfall, till failure/penultimate stage of test. The above was computed by tracking the top row of plastic markers in conjunction with LVDT data captured during centrifuge tests. As evident from Fig. 9.16a, the surface settlements increased with rainfall in the unreinforced slope due to the advancement of the wetting front into the slope body, until failure

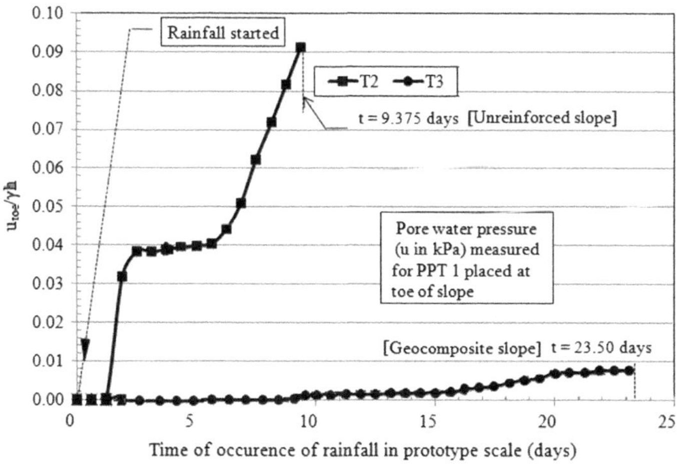

Fig. 9.15 Variation of pore water pressure with rainfall in unreinforced and hybrid geosynthetic-reinforced slopes (after Bhattacherjee and Viswanadham [4])

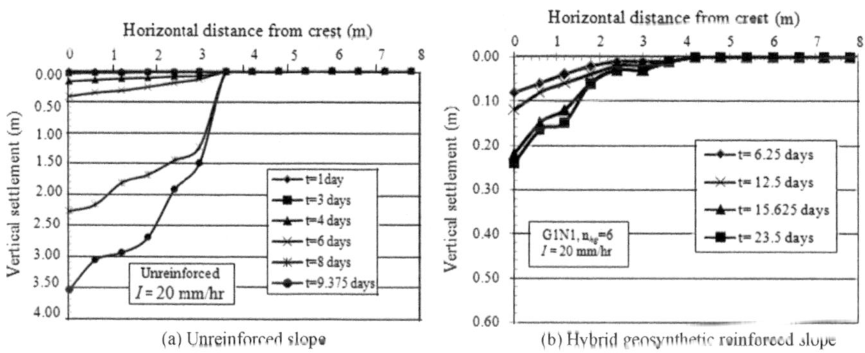

Fig. 9.16 Variation of surface settlements measured from crest of the slope with rainfall (in prototype dimensions) (after Bhattacherjee and Viswanadham [4])

occurred at t = 15 min (9.375 days). On the contrary, the hybrid geosynthetic (G1N1) reinforced slope exhibited negligible deformation under the influence of rainfall (Fig. 9.16b) with a maximum magnitude of 7.33 mm (0.22 m) at t = 37.60 min (23.5 days). The above observations hereby reinstate the effectiveness of geogrid component in providing reinforcement effects under rainfall, and mitigating settlements in low-permeability soil slopes.

The gradual movement of slope face with the progress of rainfall was obtained by tracking the coordinates of inclined markers in the unreinforced slope, and that of plastic markers stuck to geocomposite layers towards the slope face for the reinforced slope. Figure 9.17a, b predicts the face displacements thus obtained through Digital Image Analysis (DIA) technique, plotted considering the slope face to be vertical,

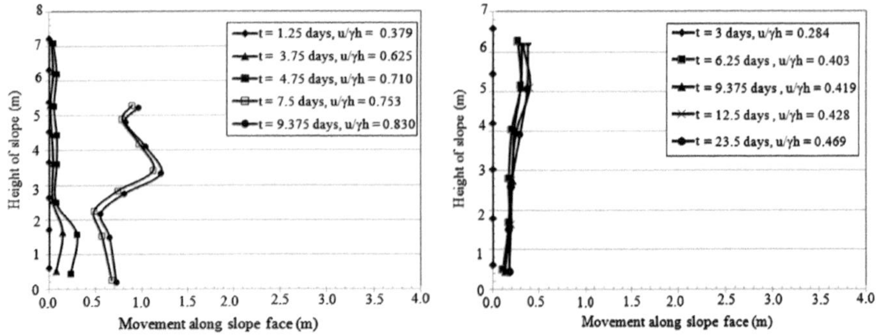

(a) Unreinforced slope (b) Hybrid geosynthetic reinforced slope

Fig. 9.17 Displacements at the slope face during rainfall (in prototype dimensions) (after Bhattacherjee and Viswanadham [4])

and coinciding with the vertical axis and origin at the toe. The increasing trend of u/γh values [for PPT 4 placed at 350 mm distance (in model dimensions) from the slope toe] in Fig. 9.17a, b are further indicative of the slope movements experienced with the progress of rainfall. As evident from Fig. 9.17a, the unreinforced slope model recorded a sudden displacement at the toe from an initial value of 0.15 m at t = 3.75 days to 1.20 m in prototype dimensions at the time of failure (t = 9.375 days). This may be attributed to pore water pressure build-up within the slope during rainfall, generating positive seepage forces. On the contrary, the hybrid geosynthetic-reinforced slope depicted negligible lateral movements along the slope face during rainfall, with a maximum value of 0.36 m in prototype scale (Fig. 9.17b). The above study thus highlights the importance of coupling reinforcement and drainage functions simultaneously within low-permeability slopes under rainfall condition. In addition, use of hybrid geosynthetics facilitates the use of locally available on-site soils in reinforced earth construction, thereby economizing the project and catering to the problem of scarcity of good quality permeable granular materials in recent times.

9.7 Summary and Conclusions

In this paper, scaling considerations required to be adopted for modelling geosynthetic-reinforced soil structures are presented. These scaling considerations help in selecting muster materials for simulating corresponding commercially available geosynthetics in the field. Further, this paper presents various centrifuge model studies on geosynthetic-reinforced soil structures undertaken at IIT Bombay using a 4.5 m radius large beam centrifuge facility. The above studies could bring out the importance of centrifuge modelling along with digital image analyses technique the performance of geosynthetic-reinforced soil structures before and at failure. The analysis and interpretation of centrifuge test results have yielded valuable information for understanding the failure mechanism of geosynthetic-reinforced soil structures. Particularly, the effect of geogrid reinforcement layer in restraining clay-based cracking, effect of chimney drain on the performance of geogrid-reinforced soil walls subjected to seepage, failure mechanism of geotextiles reinforced slopes at the onset of failure and the performance of geosynthetic-reinforced slopes subjected to rainfall using in-flight rainfall simulator could be postulated distinctly. This elucidates that the centrifuge-based physical modelling is one of the viable options for understanding the performance of geosynthetic-reinforced soil structures.

Acknowledgements The author would like to acknowledge doctoral research works of **Dr. Ratnakar Mahajan, Dr. S. Rajesh, Dr. D. V. Raisinghani, Dr. P. V. Divya, Dr. A. Rajabian, Dr. Sreeja Balakrishnan, Dr. Dipankana Bhattacherjee, Mr. Jaber Mamaghanian** and Master students **Mr. Subhro Senguptha, Mr. Bibek Kumar Jha** and **Mr. Ch. Manikumar**. Special thanks are due to **Dr. Dipankana Bhattacherjee** for providing excellent help in editing this manuscript. The authors would like to thank all the funding agencies (like DST, MHRD, IFCPAR) for sponsoring studies and the centrifuge staff at the large beam centrifuge facility of the Indian Institute of Technology Bombay, Mumbai, India for their active involvement and untiring support during the course of this endeavour.

References

1. Balakrishnan, S., Viswanadham, B.V.S.: Performance evaluation of geogrid reinforced soil walls with marginal backfills through centrifuge model tests. Geotext. Geomembr. **44**(1), 95–108 (2016)
2. Bathurst, R.J., Vlachopoulos, N., Walters, D.L., Burgess, P.G., Allen, T.M.: The influence of facing stiffness on the performance of two geosynthetic reinforced soil retaining walls. Can. Geotech. J. **43**(12), 1225–1237 (2006)
3. Bhattacherjee, D., Viswanadham, B.V.S.: Numerical studies on the performance of hybrid-geosynthetic-reinforced soil slopes subjected to rainfall. Geosynth. Int. **22**(6), 411–427 (2015)
4. Bhattacherjee, D., Viswanadham, B.V.S.: Effect of geocomposite layers on slope stability under rainfall condition. Indian Geotech. J. **48**(2), 316–326 (2018)
5. Bhattacherjee, D., Viswanadham, B.V.S.: Design and performance of an in-flight rainfall simulator in a geotechnical centrifuge. Geotech. Test. J. (ASTM) **41**(1), 72–91 (2018)
6. Divya, P.V., Viswanadham, B.V.S., Gourc, J.P.: Influence of geomembrane on the deformation behaviour of clay-based landfill covers. Geotext. Geomembr. **34**, 158–171 (2012)
7. Izawa, J., Kuwano, J.: Centrifuge modeling of geogrid reinforced soil walls subjected to pseudo-static loading. Int. J. Phys. Model. Geotech. **10**(6), 1–18 (2010)
8. Jewell, R.A.: Soil Reinforcement with Geotextiles. Constrution Indistry Research and Information Association, 332 pp. Special Publication 123, Thomas Telford (1996)
9. Koerner, J., Soong, T.Y., Koerner, R.M.: Earth retaining wall costs in the USA. GRI Report#20, Geosynthetic Institute, 38 p. Folsom, PA (1998)
10. Koerner, R.M.: Designing with Geosynthetics, 6th edn. Xlibris Publishing (2012)
11. Raisinghani, D.V., Viswanadham, B.V.S.: Centrifuge model study on low permeable slope reinforced by hybrid geosynthetics. Geotext. Geomembr. **29**(6), 567–580 (2011)
12. Rajabian, A., Viswanadham, B.V.S., Ghiassian, H., Salehzadeh, H.: Centrifuge model studies on anchored geosynthetic slopes for coastal shore protection. Geotext. Geomembr. **34**, 144–157 (2012)

13. Rajesh, S., Viswanadham, B.V.S.: Evaluation of geogrid as a reinforcement layer in clay based engineered barriers. Appl. Clay Sci. **46**(2), 153–165 (2009)
14. Rajesh, S., Viswanadham, B.V.S.: Development of a motor-based differential settlement simulator setup for a geotechnical centrifuge. Geotech. Test. J. (ASTM) **33**(6), 507–514 (2010)
15. Rajesh, S., Viswanadham, B.V.S.: Hydro-mechanical behavior of geogrid reinforced soil barriers of landfill cover systems. Geotext. Geomembr. **29**(1), 51–64 (2011)
16. Rajesh, S., Viswanadham, B.V.S.: Centrifuge modeling and instrumentation of geogrid reinforced soil barriers of landfill covers. J. Geotech. Geoenviron. Eng. (ASCE) **138**(1), 26–37 (2012)
17. Sarsby, R.W.: Geosynthetics in Civil Engineering, 1st edn. The Textile Institute, CRC Press, Woodhead Publishing Limited, Cambridge, England (2007)
18. Springman, S., Bolton, M., Sharma, J., Balachandran, S.: Modelling and instrumentation of a geotextile in the geotechnical centrifuge. In: Ochiai, H., Hayashi, S., Otani, J. (eds.) Proceedings of International Symposium on Earth Reinforcement Practice, pp. 167–172. A. A. Balkema, Rotterdam (Pubs)
19. Viswanadham, B.V.S., Bhattacherjee, D.: Studies on the performance of geocomposite reinforced low-permeable slopes subjected to rainfall. Jpn. Geotech. Soc. Spec. Publ. **2**(69), 2362–2367 (2016)
20. Viswanadham, B.V.S., Jessberger, H.L.: Centrifuge modeling of geosynthetic reinforced clay liner of landfills. J. Geotech. Geoenviron. Eng. (ASCE) **131**(5), 564–574 (2005)
21. Viswanadham, B.V.S., König, D.: Studies on scaling and instrumentation of a geogrid. Geotext. Geomembr. **22**(5), 307–328 (2004)
22. Viswanadham, B.V.S., Mahajan, R.R.: Centrifuge model tests on geotextile reinforced slopes. Geosynth. Int. **14**(6), 365–379 (2007)
23. Viswanadham, B.V.S., Rajesh, S.: Centrifuge model test on clay based engineered barriers subjected to differential settlements. Appl. Clay Sci. **42**(3–4), 460–472 (2009)
24. Viswanadham, B.V.S., Jha, B.K., Sengupta, S.S.: Centrifuge testing of fiber-reinforced soil liners for waste contained systems. Pract. Period. Hazard. Toxic Radioact. Waste Manag. (ASCE) **13**(1):45–58 (2009)
25. Viswanadham, B.V.S., Razeghi, H.R., Mamaghanian, J., Manikumar, C.H.S.G.: Centrifuge model study on geogrid reinforced soil walls with marginal backfills with and without chimney sand drain. Geotext. Geomembr. **45**(5), 430–446 (2017)
26. Zornberg, J.G., Arriaga, F.: Strain distribution within geosynthetic-reinforced slopes. J. Geotech. Geoenviron. Eng. (ASCE) **129**(1), 32–45 (2003)

Chapter 10
Soft Ground Improvement—Theoretical, Experimental, Numerical and Field Studies

Buddhima Indraratna, Pankaj Baral, Cholachat Rujikiatkamjorn and Thanh Trung Nguyen

10.1 Introduction

The increasing population and associated development in metropolitan and coastal areas necessitate ground improvement and land reclamation for future infrastructure, especially the very soft soils which are less permeable but highly compressible [1, 2]. The low bearing capacity associated with highly compressible soil directly affects the long-term stability of building superstructures and transport infrastructure [3]. Controlling the excessive differential settlements of these soft soils before starting construction is therefore imperative and fortunately, there are many ground improvement techniques currently available; stone columns, gravel piles, vibro-compaction, soil–cement stabilisation, and Prefabricated Vertical Drains (PVDs) with preloading, etc. Among these techniques, PVDs combined with surcharge and vacuum preloading is regarded as the most cost-effective and efficient method of ground improvement, especially for highly compressible soil with an undrained shear strength of less than 5 kPa.

Another reason for using PVDs with preloading is their relatively low cost and easy installation; this is a great advantage because PVDs can reduce the preloading period by almost 10 times by accelerating radial consolidation due to reducing the drainage path. Furthermore, PVDs can be installed rapidly with minimal environmental implications and quarrying requirements compared to semi-rigid inclusions.

Several researchers have studied the behaviour of vertical drains and vacuum preloading worldwide in terms of their theoretical, experimental, and numerical development. Developing the fundamental aspects of radial consolidation theory started with Carillo [4] and followed by Barron [5], Akagi [6], Hansbo [7–9], Onoue [10], Zeng and Xie [11], Indraratna and Redana [12, 13] and Mohamedelhassan and Shang [14]. Apart from these, many other researches on this topic also considered

---

B. Indraratna (✉) · P. Baral · C. Rujikiatkamjorn · T. T. Nguyen
Centre for Geomechanics and Railway Engineering (GRE), University of Wollongong, Wollongong, NSW, Australia
e-mail: indra@uow.edu.au

M. Latha G. (ed.), *Frontiers in Geotechnical Engineering*, Developments in Geotechnical Engineering, https://doi.org/10.1007/978-981-13-5871-5_10

the types of vertical drains and effects of installation, the smear zone, design procedures with practical guidelines, natural PVDs; they also considered the effects of biodegradation, soil de-structuration, PVDs with cyclic loading, large strain analysis and elastic visco-plastic behaviour. These developments have been verified worldwide using instrumented and monitored case studies.

This paper deals with those aspects of research carried out at University of Wollongong over the last 20 years in the area of soft ground improvement combined with preloading (surcharge or vacuum). The theoretical, experimental and numerical development work carried out at UOW through higher education training is explained and validated with various case studies.

10.2 Theoretical Developments of PVD and Vacuum Consolidation

10.2.1 Vacuum Consolidation with Smear Effect

Based on the law of superposition, Mohamedelhassan and Shang [14] developed a combined vacuum and surcharge load system using conventional 1-D consolidation theory by Terzaghi. The mechanism shown in Fig. 10.1 indicates that the average degree of consolidation (DOC) for combined surcharge and vacuum preloading is given by

$$U_{vc} = 1 - \sum_{m=0}^{\infty} \frac{2}{M} e^{-M^2 T_{vc}} \tag{10.1}$$

$$T_{vc} = c_{vc} t / H^2, \tag{10.2}$$

where

T_{vc} time factor corresponding to combined preloading (vacuum plus surcharge), and

c_{vc} coefficient of consolidation for combined preloading (vacuum plus surcharge).

The distribution of vacuum pressure in the field as discussed previously by Indraratna et al. [15], indicates there is a reduction in the suction head along the length of PVDs that reduces their efficiency, and while this distribution is absent in the laboratory due to shorter length PVDs, it is very prominent in field studies. Some studies also suggest a vacuum can quickly propagate an almost 30 m long drain, if properly instrumented [16, 17], however, as well as drain length, the loss of vacuum also depends on the type of PVD (filter and core properties). These researches also

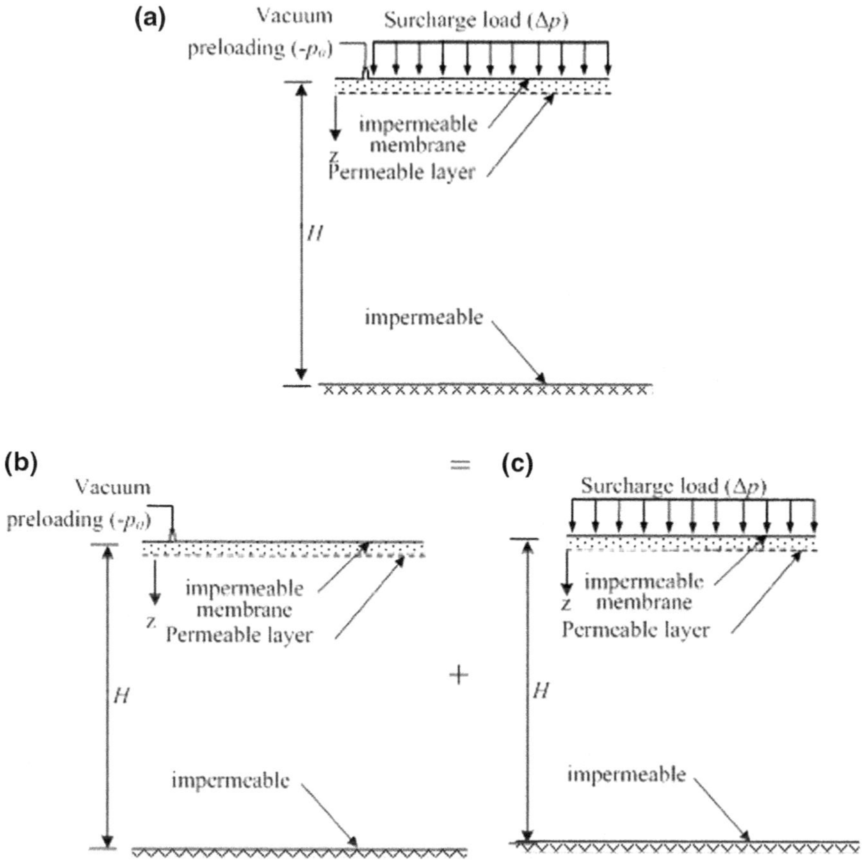

Fig. 10.1 Principal of superposition for vacuum preloading system; **a** combined system; **b** vacuum only; and **c** surcharge only (After Mohamedelhassan and Shang [14])

propose a trapezoidal distribution of vacuum pressure along the length of PVD to quantify the loss of vacuum, so based on this distribution, the average excess pore pressure ratio ($R_u = \Delta p/\overline{u_0}$) of a soil cylinder for radial consolidation incorporating vacuum preloading can be formulated as

$$R_u = \left(1 + \frac{p_0}{\overline{u}_o}\frac{(1+k_1)}{2}\right)\exp\left(-\frac{8T_h}{\mu}\right) - \frac{p_0}{\overline{u}_0}\frac{(1+k_1)}{2} \tag{10.3}$$

and

$$\mu = \ln\left(\frac{n}{s}\right) + \left(\frac{k_h}{k_s}\right)\ln(s) - 0.75 + \pi z(2l - z)\frac{k_h}{q_w}\left\{1 - \frac{k_h/k_s \quad 1}{(k_h/k_s)(n/s)^2}\right\}, \tag{10.4}$$

where

u_0	initial excess pore water pressure,
k_h	permeability in horizontal direction for undisturbed zone,
k_s	permeability in horizontal direction for smeared zone
T_h	time factor,
d_e, $2r_e$	diameter of the equivalent soil cylinder,
d_w, $2r_w$	diameter of the drain,
n	d_e/d_w
d_s	$2r_s$; is the diameter of the smear zone with constant permeability,
s	d_s/d_w
l	equivalent length of drain,
z	depth,
q_w	drain discharge capacity,
p_0	vacuum applied at the top of the drain, and
k_1	vacuum distribution coefficient.

10.2.2 Overlapping Smear Zone

Installing a vertical drain with a mandrel disturbs the surrounding subsoil, but while the permeability of this disturbed soil decreases further, compressibility increases due to remoulding. This concept of reduced permeability by arbitrarily lowering the coefficient of consolidation first proposed by Barron in 1948 has been improved with a smear zone model with reduced permeability around the drain proposed by Hansbo [8]. The very common expression for smear zone based on permeability and geometric parameter μ is normally used in design, although laboratory evidence demonstrates that this reduction of permeability towards the drain is gradual. Since obtaining in situ smear parameters is very difficult, preference is given to the simple expression for the permeability distribution obtained from many analytical solutions (many of them are from the first author). The method of using such simple expressions produces numerical value that also considers the geometric and permeability effect to a single parameter of μ.

In case of overlapping, there is little chance of overlapping when the constant permeability smear zone is compared with gradual changes in the permeability smear zone. The possibility of overlapping smear zones has been investigated by Walker and Indraratna [18] by considering the linear permeability distributions. Walker and Indraratna [18] also proposed an expression for μ for a single non-overlapping smear zone (see Fig. 10.2),

For $s \neq \kappa$,

$$\mu_L = \ln\left(\frac{n}{s}\right) - \frac{3}{4} + \frac{\kappa(s-1)}{s-\kappa}\ln\left(\frac{s}{\kappa}\right) \tag{10.5}$$

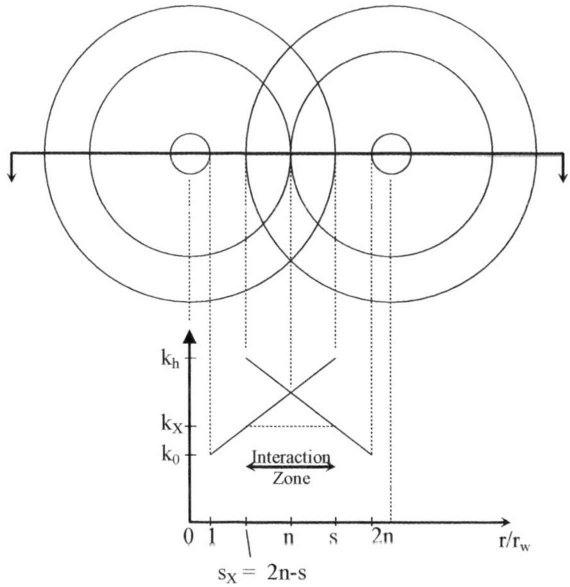

Fig. 10.2 Overlapping smear zone (Walker and Indraratna, [18], with kind permission from Thomas Telford)

For $s = \kappa$,

$$\mu_L = \ln\left(\frac{n}{s}\right) - \frac{3}{4} + s - 1 \tag{10.6}$$

10.2.3 Large Strain Effect

A small strain model cannot accurately predict the settlement of very soft estuarine deposits under extensive surcharge and vacuum preloading; it can only be predicted with a large strain model. Indraratna et al. [19] proposed a model for large strain consolidation by considering non-Darcian flow with variations of permeability and compressibility for radial consolidation facilitated by PVDs with preloading. Darcy's law may not be valid for soils with low porosity under very low hydraulic gradients and non-Darcian flow is the best way to describe the flow of pore water in this situation. An exponential relationship between the seepage velocity and hydraulic gradient describes the non-Darcian behaviour as postulated by Hansbo [20].

A governing equation for large-scale consolidation has been established based on large strain coordinate system, it is given by

$$\frac{1}{1+e_i} \cdot \frac{\partial e}{\partial t} \cdot \mathrm{d}a + v_r(r) \cdot \frac{2r}{r_e^2 - r^2} \cdot \frac{\partial \xi}{\partial a}\mathrm{d}a = 0 \tag{10.7}$$

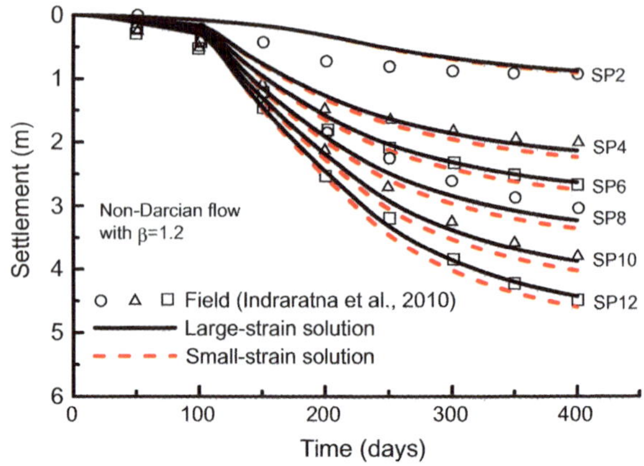

Fig. 10.3 The performance of small and large strain solutions with field data at Ballina Bypass (Indraratna et al. [19], with permission from ASCE)

The following relationship between the Lagrangian coordinate a and the convective coordinate ξ exists in a large strain coordinate system:

$$\frac{\partial \xi}{\partial a} = \frac{1+e}{1+e_0}, \tag{10.8}$$

where e is the void ratio, e_0 is the initial void ratio, r is the radius, r_e is the drain influence radius, t is the time and $v_r(r)$ is the inward seepage velocity at radius r.

To solve Eq. 10.7, the settlement predicted from large strain and the small strain analyses can be compared with the field data obtained from Ballina Bypass trial embankment (Fig. 10.3). Note here that the results from the large strain analysis provide more acceptable prediction at each settlement plate location compared to the small strain solution, it is, therefore, better to use large strain under extensive surcharge preloading.

10.2.4 Elastic Visco-Plastic Behaviour

The debate regarding whether the soil element follows Hypothesis A or B when dealing with the elastic visco-plastic behaviour of soft soil is ongoing, however, Hypothesis A suggests that creep occurs only after primary consolidation and during secondary consolidation, whereas Hypothesis B suggests that creep occurs during the whole consolidation process. Moreover, according to Hypothesis A, there is no relationship between the thickness of the sample and effective stress. Several researchers did try to investigate the viscoelastic behaviour of soft soil with theoretical and laboratory evidence but the exact phenomenon is still unknown. Some researchers agree with the Hypothesis A, some with Hypothesis B, while some agree with A and B.

Yin and Graham [21] did propose an elastic visco-plastic model (EVP) based on the equivalent timeline concept, but this has since been successfully calibrated by others using laboratory and field data. This model has been further modified by Indraratna et al. [22] using non-Darcian flow for radial consolidation and solved using a Finite difference method called the Peaceman–Rachford ADI-FDM scheme, where the matrix form for the predictor and corrector are written in MATLAB to facilitate the computation of settlement and dissipation of excess pore water pressure. The governing equation can be expressed as

$$m_v \frac{\partial u}{\partial t} - g(u, \varepsilon_z) = \frac{\alpha_c \beta}{\gamma_w^{\beta}} \left[\left| \frac{\partial u}{\partial r} \right|^{\beta-1} \left\{ \frac{\partial^2 u}{\partial r^2} + \frac{1}{r} \frac{\partial u}{\partial r} \right\} + || \frac{\partial u}{\partial z} |^{\beta-1} \frac{\partial^2 u}{\partial z^2} \right] \tag{10.9}$$

&

$$g(u, \varepsilon_z) = \frac{\psi}{t_0 V} exp\left[-(\varepsilon_z)\frac{V}{\psi} \right] \left(\frac{(\sigma_x - u)}{\sigma'_{z0}} \right)^{\frac{\lambda}{\psi}} \tag{10.10}$$

After further simplification using the Peaceman–Rachford (P-R) scheme, the above equations can be estimated using the following steps:

The matrix form of the predictor is

$$u_{i,j}^{t+\frac{\Delta t}{2}} = C^I + C^{II} u_{i+1,j}^{t+\frac{\Delta t}{2}} + C^{III} u_{i-1,j}^{t+\frac{\Delta t}{2}} \tag{10.11}$$

For each j,

$$\begin{bmatrix} 1 & -C^{II}_{1,j} & 0 & & & \\ -C^{III}_{2,j} & 1 & -C^{II}_{2,j} & & & \\ 0 & -C^{III}_{3,j} & 1 & & & \\ \vdots & & & \ddots & & \\ & & & -C^{III}_{N-1,j}, & 1 & -C^{II}_{N-1,j} \\ & & & & \left(-C^{III}_{N,j} - C^{II}_{N,j}\right) & 1 \end{bmatrix}^t \begin{bmatrix} u_{1,j} \\ u_{2,j} \\ u_{3,j} \\ \vdots \\ u_{N-1,j} \\ u_{N,j} \end{bmatrix}^{t+\frac{\Delta t}{2}} = \begin{bmatrix} C^I_{1,j} + C^{III}_{1,j} u_{0,j,t+\frac{\Delta t}{2}} \\ C^I_{2,j} \\ C^I_{3,j} \\ \vdots \\ C^I_{N-1,j} \\ C^I_{N,j} \end{bmatrix}^t \tag{10.12}$$

where

$$C^I = \frac{(1 - 2\Delta T_v)}{\left(1 + 2\Delta T_r + \Delta T_r.\frac{\Delta r}{r}\right)} u^t_{i,j} + \frac{\Delta T_v}{\left(1 + 2\Delta T_r + \Delta T_r.\frac{\Delta r}{r}\right)} \left(u^t_{i,j+1} + u^t_{i,j-1}\right)$$

$$
\begin{aligned}
&+\frac{\Delta t}{\left(1+2\Delta T_r+\Delta T_r.\frac{\Delta r}{r}\right)m_v}g(u,\epsilon_z)_{i,j}\\
C^{II} &= \frac{\Delta T_r.\left(1+\frac{\Delta r}{r}\right)}{\left(1+2\Delta T_r+\Delta T_r.\frac{\Delta r}{r}\right)}\\
C^{III} &= \frac{\Delta T_r}{\left(1+2\Delta T_r+\Delta T_r.\frac{\Delta r}{r}\right)}
\end{aligned}
$$

The matrix form of the corrector is

$$
u_{i,j}^{t+\Delta t} = C^{I} + C^{II}u_{i,j+1}^{t+\Delta t} + C^{III}u_{i,j-1}^{t+\Delta t} \tag{10.13}
$$

For each i,

$$
\begin{bmatrix}
1 & -C_{i,1}^{II} & 0 & & & \\
-C_{i,2}^{III} & 1 & -C_{i,2}^{II} & & & \\
0 & -C_{i,3}^{III} & 1 & & & \\
\vdots & & & \ddots & & \\
 & & & -C_{i,M-1}^{III} & 1 & -C_{i,M-1}^{II} \\
 & & & & \left(-C_{i,M}^{III}-C_{i,M}^{II}\right) & 1
\end{bmatrix}^{t+\frac{\Delta t}{2}}
\begin{bmatrix} u_{i,1}\\ u_{i,2}\\ u_{i,3}\\ \vdots\\ u_{i,M-1}\\ u_{i,M}\end{bmatrix}^{t+\Delta t}
=
\begin{bmatrix} C_{i,1}^{I}+C_{i,1}^{III}u_{i,0,t+\Delta t}\\ C_{i,2}^{I}\\ C_{i,3}^{I}\\ \vdots\\ C_{i,M-1}^{I}\\ C_{i,M}^{I}\end{bmatrix}^{t+\Delta t/2} \tag{10.14}
$$

where

$$
\begin{aligned}
C^{I} &= \frac{\left(1-2\Delta T_r-\Delta T_r.\frac{\Delta r}{r}\right)}{(1+2\Delta T_v)}u_{i,j}^{t+\frac{\Delta t}{2}} + \frac{\Delta T_r.\left(1+\frac{\Delta r}{r}\right)}{(1+2\Delta T_v)}u_{i+1,j}^{t+\frac{\Delta t}{2}} + \frac{\Delta T_r}{(1+2\Delta T_v)}u_{i-1,j}^{t+\frac{\Delta t}{2}}\\
&+\frac{\Delta t}{m_v(1+2\Delta T_v)}g(u,\epsilon_z)_{i,j}\\
C^{II} &= \frac{\Delta T_v}{(1+2\Delta T_v)}\\
C^{III} &= \frac{\Delta T_v}{(1+2\Delta T_v)}
\end{aligned}
$$

This elastic visco-plastic model has been validated using a *Class C* [23] prediction exercise from the Ballina Embankment. A *Class A* prediction of the Ballina prediction embankment [22] took place before construction and the *Class C* prediction took place after the field data became available. The *Class A* Prediction is based on industry standard data and a large-scale consolidometer (350 mm large specimen) which matches the initial rate of settlement based on the staged construction (Fig. 10.4a); ultimate settlement is predicted quite accurately (see Fig. 10.4b), although a large error arose due to the absence of a creep model in our *Class A* prediction method. The inclusion of visco-plastic behaviour (i.e. creep) made the prediction more accurate so it is defined as a *Class C* prediction. Similarly, if the excess pore water pressure dissipates, these *Class A* predictions [19, 24, 25] cannot predict the dissipation (Fig. 10.4c) behaviour because these models do not have viscosity parameters in their analytical form, whereas the prediction using Case D (*Class C*) can predict the dissipation of excess pore water pressure due to the inclusion of visco-plastic behaviour. Although excess pore water pressure cannot be predicted accurately after one year (due to the retarded excess pore water pressure), the Case D (*Class C*) approach has advantages in predicting excess PWP dissipation compared to *Class A* [19, 24, 25] predictions.

10.2.5 Natural Prefabricated Drains and Clogging

Biodegradable PVDs (BPVDs) imply that drains made from naturally occurring materials such as jute and coconut fibres have better engineering characteristics and are also environmentally friendly due to their biodegradability. Although these novel drains have been in use for almost 30 years [26], their practical applications are limited due to uncertainty with regards to their performance if subjected to adverse conditions such as soils with cellulose-degrading bacteria. The question of how a severe biodegradation of natural fibres can affect soil consolidation has not really been addressed properly, which is why this issue needs investigation to widen the application of these types of drains.

Previous studies [27, 28] indicate that the discharge capacity of BPVDs is enough to facilitate soil consolidation through laboratory and field investigations. However, being exposed to severe degradation which can destroy the porous structure and reduce the discharge capacity of a drain [29] and have a significant effect on the consolidation of soil (Fig. 10.5). A theoretical revision of conventional radial consolidation induced by vertical drains usually assumes that drains require a constant discharge capacity over time.

To evaluate the effect that the biodegradation of BPVDs can have on consolidation, Indraratna et al. [30] assume a reduced discharge capacity of drains while parameters such as the diameter and length remain unchanged. Since a reduction in discharge capacity of natural fibre drains has not been clarified very well, a general time-dependent discharge capacity $q_w(t)$ is proposed, whereby the average excess pore pressure of a unit cell using a vertical drain as written by Indraratna et al. [17] is as follows:

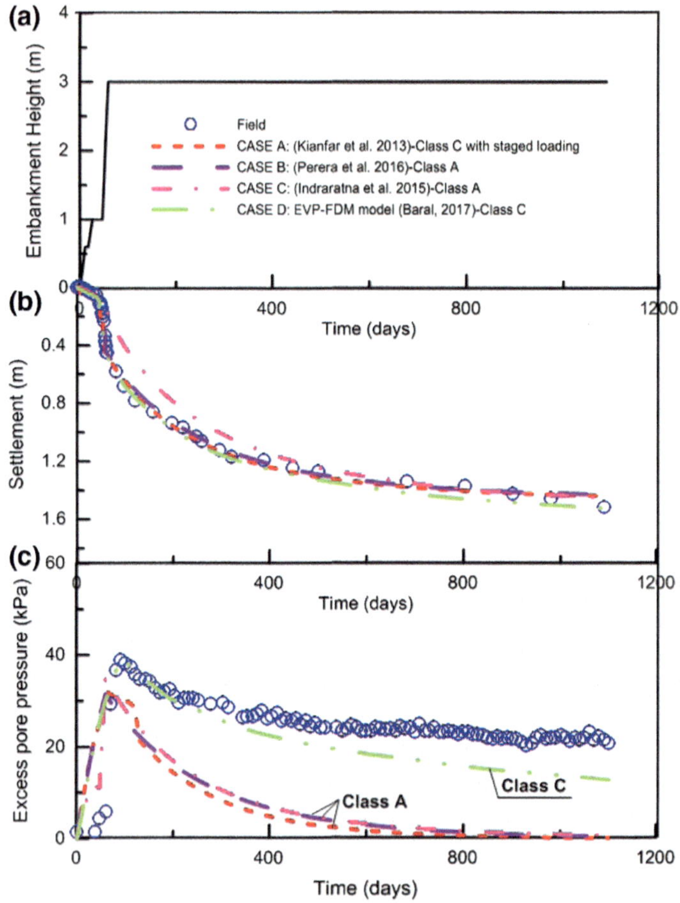

Fig. 10.4 Predictions (Class A and C) using integrated cell approach for Ballina Embankment at NFTF (Indraratna et al. [22], with kind permission from Elsevier)

$$u = \frac{\int_0^l \int_{r_w}^{r_s} u_s(2\pi r)drdz + \int_0^l \int_{r_s}^{r_e} u_u(2\pi r)drdz}{V} \tag{10.15}$$

Integrating the right side with $q_w(t)$ results in:

$$u = \frac{\gamma_w}{2k_{h,a}} \frac{d_e^2}{4} (\mu_{n,s,a} + \mu_{q,a}) \frac{d\varepsilon}{dt}, \tag{10.16}$$

where $\mu_{q,a}$ is a function of $q_w(t)$; other parameters can be found in Indraratna et al. [30]. The general solution for Eq. (10.16) can be written as follows:

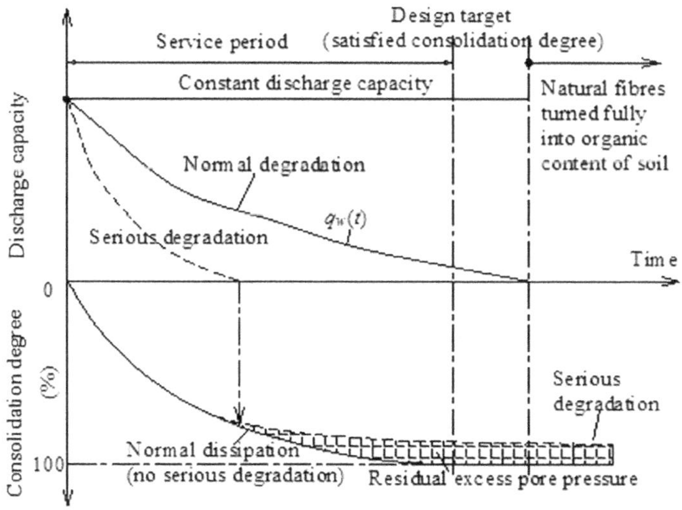

Fig. 10.5 Influence that the serious degradation of drains has on soil consolidation

$$u_a(t) = u_o exp\left(-\int_0^t \frac{1}{f(t)} dt\right), \tag{10.17}$$

where $f(t)$ is a function of $q_w(t)$ which can incorporate various forms of drain degradation. For example, with an exponential degradation $q_w(t) = q_{wo} \times e^{-\omega t}$ where ω is the degradation rate, Eq. (10.17) becomes

$$\frac{u(t)}{u_o} = exp\left\{\frac{-8T_{h,p}}{\mu_{n,s,p}} + \frac{1}{\chi_p \mu_{n,s,p}\omega}\left[ln\left(\frac{\mu_{n,s,p}}{\mu_{qo,p}} + e^{\omega t}\right) - ln\left(\frac{\mu_{n,s,p}}{\mu_{qo,p}} + 1\right)\right]\right\} \tag{10.18}$$

Details of this solution are presented in Indraratna et al. [30] and further developments for multi-forms of drain degradation (e.g. polynomial degradations) are discussed in Nguyen et al. [29].

10.3 Experimental Studies

10.3.1 Laboratory and Theoretical Based Smear Zone

There are several literatures explaining the smear zone that are based on assumptions, and testing of small-scale samples, as well as large-scale testing of soil specimens on different clay. The diameter and method of a smear zone based on several researchers are tabulated in Table 10.1.

Table 10.1 Extent of smear zone

Extent of smear zone (r_s)	Permeability aspects	Remarks	Source
$1.6\,r_m$	$\frac{k_h}{k_s} = 3$	Assumption based	[5]
$1.5–3\,r_m$	open	Literature based	[7]
$1.5\,r_m$	$\frac{k_h}{k_s} = 3$	Assumption based	[8]
$2\,r_m$	$\frac{k_h}{k_v} = 1$	Back analysis for Bangkok clay and experimental investigation	[31]
$1.6\,r_m$	$\frac{k_h}{k_s} = 3$	Based on test interpretation	[32]
$1.5–2\,r_m$	$\frac{k_h}{k_s} = 3–6$	Based on authors experience	[33]
$4–5r_m$	$\frac{k_h}{k_v} = 1.15$	Based on laboratory investigation of Sydney clay	[12]
$2 - 3\,r_m$	$\frac{k_h}{k_s} = C_f\left(\frac{k_h}{k_s}\right)$	Based on lab and field values variation	[34]
$1.6\,r_m$	$\frac{k_h}{k_s} = 3$	Design recommendation	[35]
$4\,r_m$	$\frac{k_h}{k_s} = 1.3$	Based on laboratory investigation for Kaolin clay	[36]

10.3.2 Laboratory and Theoretically Based Smear Zone

A smear zone based on field testing was first evaluated using a full-scale test embankment made from soft Bangkok soil [31]. In total, 168 100 mm by 6 mm band drains has been installed in an area covering 15 by 19 m using two different mandrels (i.e. 150 by 45 mm and 150 by 50 mm). The depth of installation is 8 m and the spacing between the drains is 1.2 m, in a square pattern. The dissipation of pore water pressure and extensometer data are used to analyse the effect of smear for two different sets of mandrels; this analysis indicates that the dissipation rate for the larger mandrel is slower than the smaller mandrel; the extensometer reading also shows a higher settlement rate with the smaller mandrel. This shows that the size of a mandrel is a function of the smear zone, and it plays a large role in the effectiveness of ground improvement techniques.

To further understand the smear zone from field testing, Indraratna et al. [37] obtained tube samples taken at different radii from the PVD by the National Field Testing Facility (NFTF), Ballina, for single and multi-drain cases; these samples are then evaluated based on permeability and a normalised water content. The locations where these samples had been taken in the field are shown in Fig. 10.6.

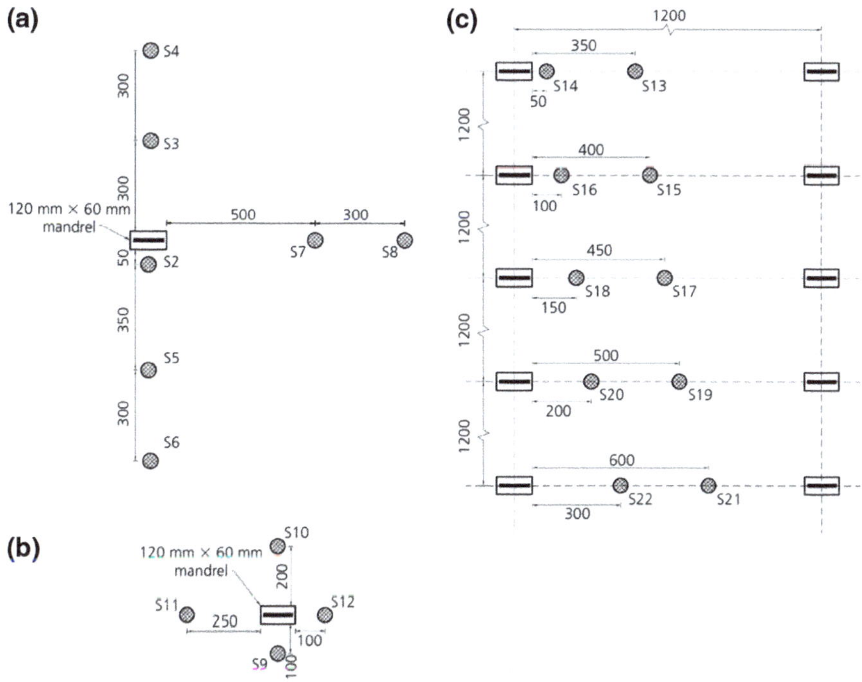

Fig. 10.6 Sampling (**a**), **b** Single-drain case, and **c** multi-drain case (Indraratna et al. [37], with kind permission from ICE)

Baral et al. [38] took a large-scale undisturbed specimen from the same location and carried out a large-scale consolidation test with a scaled down drain and mandrel to mimic the field conditions, and the normalised water content approach to investigate the smear zone of large-scale specimens with natural parting and seashells. They also compared the result with single-drain and multi-drain field studies [37] and laboratory studies using remoulded samples; the results from the large-scale consolidometer are consistent with the single and multi-drain field studies ([37]; see Fig. 10.7). Furthermore, the extent of the smear obtained from the remoulded sample using Sathananthan and Indraratna [39] confirms that the r/rm (where rm is the ratio of mandrel) ratio is half of the size of the large-scale undisturbed specimens; in fact they estimate the diameter of the smear zone is 2.5 times larger than the diameter of the mandrel.

10.3.3 Large-scale Consolidation with Vacuum Preloading

The corer cum consolidometer (two-in-one) shown in Fig. 10.8 was built at UOW following Indraratna and Redana [12] and then used to investigate the consolidation

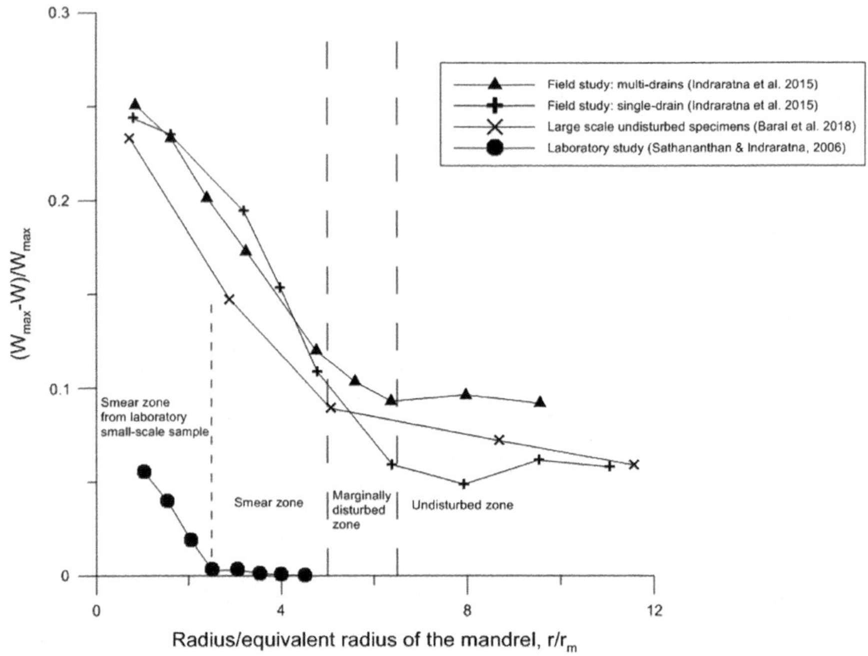

Fig. 10.7 Determining the smear zone using the normalised water content approach (Baral et al. [38], with kind permission from Elsevier)

of large-scale specimen under surcharge and vacuum preloading. It has three primary parts which are described below

(a) Cylindrical corer: This 350 mm diameter by 700 mm long cylinder is manufactured by rolling 5 mm thick steel plate and then cutting it longitudinally into two halves. A custom made 345 mm diameter piston with two 5 mm diameter O-ring grooves can easily transfer the loads from a pneumatic air pressure chamber to the top of the sample. The O-rings are designed to prevent air or water leakage during experiments. The corer also has two pore water pressure transducers at the bottom base plate; one in the centre and another 96.25 mm from the centre.
(b) The loading rig and platform: The upper part of the corer together with the piston and air pressure chamber is attached to the loading rig, while the lower part of the corer is attached to the base of the platform.
(c) The pneumatic air pressure chamber: an air pressure chamber is used to apply a load to the sample during testing, while the load cell measures the load applied to the system.

A traditional band-shaped drain (100 mm × 3 mm), based on the equivalent diameter and spacing in the field is scaled down for a laboratory simulation, so a model prefabricated vertical drain (PVD) 25 mm × 3 mm (i.e. equivalent wick drain radius, $r_w = 7$ mm) is used. A hollow rectangular mandrel (30 mm × 5 mm) is used

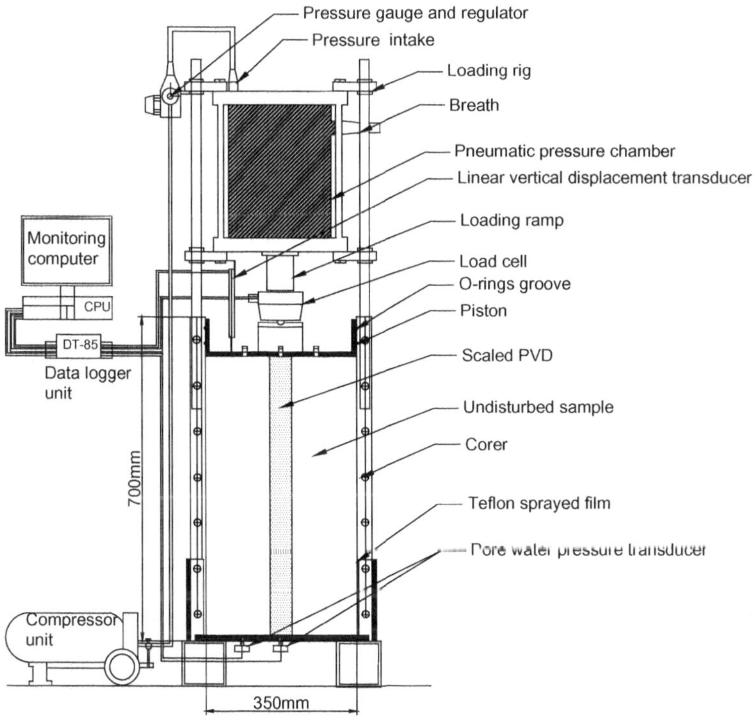

Fig. 10.8 Schematic diagram of large-scale consolidometer (Baral et al. [38], with kind permission from Elsevier)

to insert the model PVD into the laboratory sample. This hollow rectangular mandrel has an equivalent mandrel diameter (d_m) of 13.8 mm. The model PVD is installed by pushing the mandrel at a constant speed along a vertical guide to the required depth inside the corer that contains an undisturbed sample; the mandrel is withdrawn as soon as it reaches the base of the sample. A linear variable differential transformer (LVDT) is connected on top of the piston to measure the vertical displacement of the sample, and there are two pore water pressure transducers connected to the bottom base at different locations. To investigate the effect of a vacuum in the large-scale consolidometer, there is a 1.5 mm diameter porous plastic sheet around the inner wall of the corer and a 1.5 mm thick rubber membrane between the sample and the porous plastic sheet. There is a rim drain around the piston that is connected to a pressure–volume controller. The rubber membrane and the wall of the corer can be separated as the soil retracts due to vacuum consolidation. Based on the volume of water flow in the gap between the wall and the rubber membrane over time, the lateral deformation of the sample can be calculated. A surcharge load of 20 kPa and a vacuum pressure of 60 kPa is applied to the consolidometer to maintain a vacuum surcharge ratio (VSR) of 0.75.

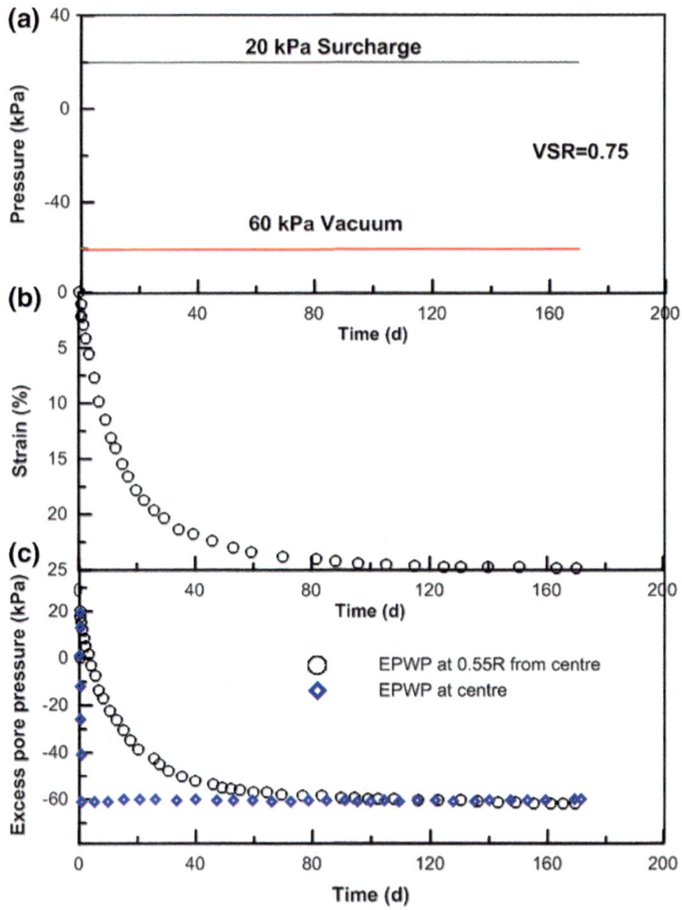

Fig. 10.9 Large-scale consolidometer with vacuum preloading at VSR = 0.75 **a** staged construction **b** Settlement, and **c** Dissipation of excess pore water pressure (Baral et al. [38], with kind permission from Elsevier)

The pressure applied along with settlement and the dissipation of excess pore water pressure over time at two different locations is shown in Fig. 10.9. The volumetric strain at the end of primary consolidation for large-scale specimen under VSR of 0.75 is 25.1%, resulting in a volume compressibility of 3.138×10^{-03} m^2/kN. The coefficient of secondary compression for this large-scale specimen under a vacuum decreased to 0.0005, unlike the specimen without a vacuum pressure. The coefficient of horizontal consolidation (c_h) is based on the reading taken by different pore water pressure transducers (11c) is 4.82×10^{-03} m^2/day, which is 36% more than the surcharge only case. This shows that using a vacuum accelerates the rate of radial consolidation and reduces the long-term secondary consolidation coefficient.

10.3.4 Large-scale Testing Subjected to Cyclic Loading

Under static and repeated loading clays with high plasticity can exhibit high excess pore water pressure but this can be prevented by installing PVDs, as shown by Indraratna et al. [40]. The practical implications of this can be found in transport geotechnics, especially when dealing with high-speed trains travelling on very soft clay along coastal areas, however, a phenomenon known as mud pumping has recently emerged as a potential problem that is mainly due to cyclic loading on soft soils.

Large-scale triaxial test (as shown in Fig. 10.10) which can accommodate a sample 300 mm diameter and 600 mm high has been used at UOW laboratory to investigate the effect of cyclic load on radial consolidation facilitated with PVDs. The pore water pressure is monitored at critical locations via miniature pore pressure transducers in a sample of remoulded estuarine clay compacted to maintain the unit weight of 17 kN/m^3 and k_o of 0.6–0.7; this resembles the in situ conditions of many Australian coastal regions. The tests are carried out at frequencies of 5–10 Hz and a cyclic loading of 25 kPa. This amplitude simulates a train speed that varies from 60 to 100 km/h with an axle load of 35–30 tonnes. Under these conditions, the dissipation rate of excess pore water pressure near the PVD (i.e. T2) is much faster than the T1 (cell boundary), as shown in Fig. 10.11.

The excess pore water pressure ratio (R_u) versus the number of cycles (N) plotted in Fig. 10.12 shows that an undrained type of failure occurs with a rapid increase of excess pore pressure ($R_u = 0.9$), when soil is not stabilised with PVD. The axial strain reported in Fig. 10.13 shows that without PVD, it took 200 and 100 cycles to develop cyclic axial strains large enough to cause the cyclic CK_0U test and UC test to fail. Furthermore, there is also a concave (downward) curve of $\varepsilon_a - \log N$ (as shown in Fig. 10.13b) when the soil is not stabilised with PVD. However, when the soil is stabilised with PVD, the axial strain gradually increases and there is no failure, even after 3000 cycles (Fig. 10.13 a, b).

It can, therefore, be concluded that the build-up of excess pore water pressure can be reduced when the soil is stabilised with PVDs under cyclic loading; it also helps to accelerate dissipation during a rest period and prepares the track for the next loading stage. The rise of cyclically induced pore pressure will be at a maximum when the shear strain exceeds 1.5–2% and soft clays, if facilitated with PVDs, can withstand a higher level of cyclic stress than the critical cyclic stress ratio via radial drainage without causing potential undrained failure.

10.4 Numerical Simulation

10.4.1 Plane Strain Conversion

Indraratna and Redana [13] propose a method to convert the axisymmetric unit cell to plane strain model, and while this method mainly describes the axisymmetric unit

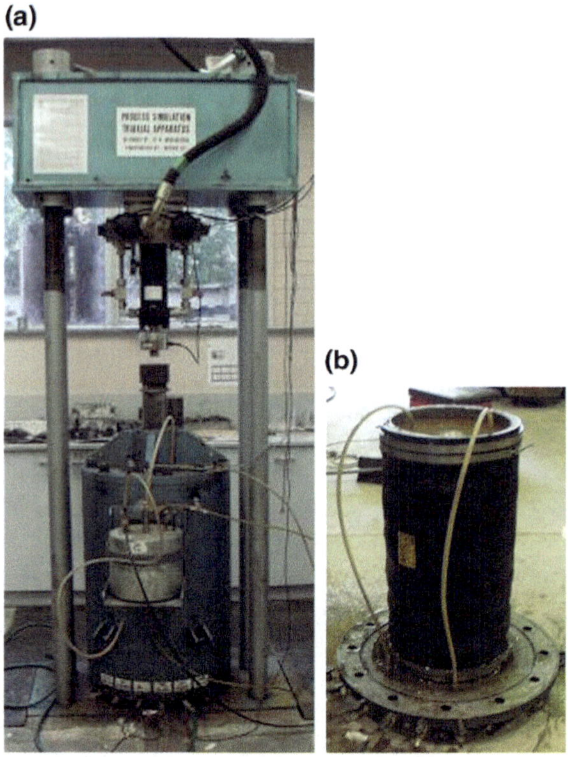

Fig. 10.10 **a** Triaxial rig; **b** large-scale soil specimen (Indraratna et al. [40], with kind permission from ASCE)

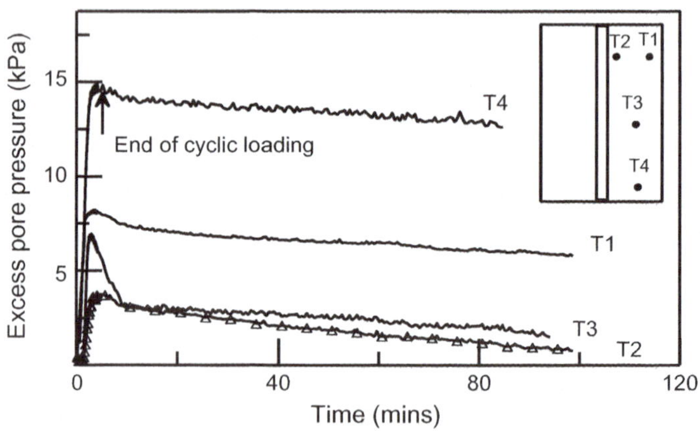

Fig. 10.11 Dissipation of excess pore water pressure at various locations from the PVD (Indraratna et al. [40], with kind permission from ASCE)

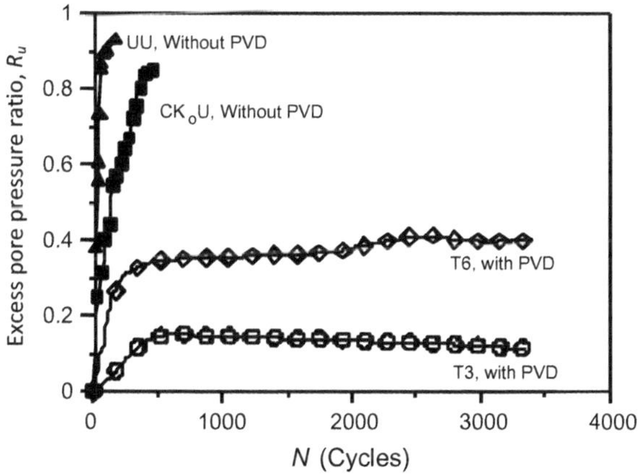

Fig. 10.12 Excess pore pressure generated under cyclic loading with and without PVD (Indraratna et al. [40], with kind permission from ASCE)

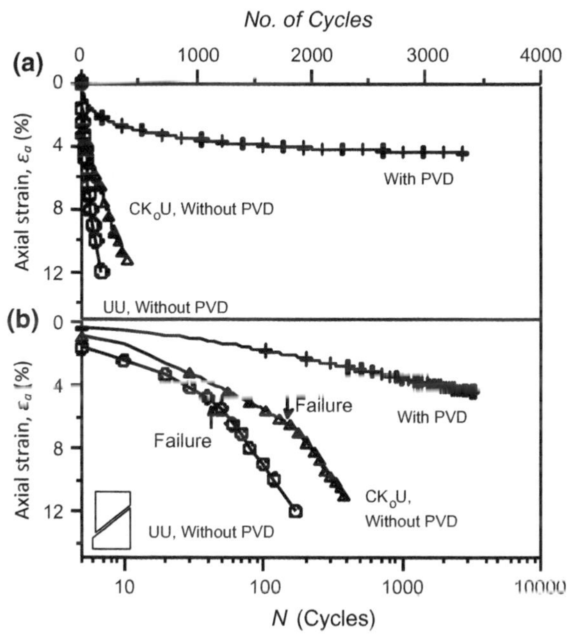

Fig. 10.13 Cyclically loaded axial strains with and without PVD versus number of loading cycles (N): **a** arithmetic; **b** semi logarithmic scales (Indraratna et al. [40], with kind permission from ASCE)

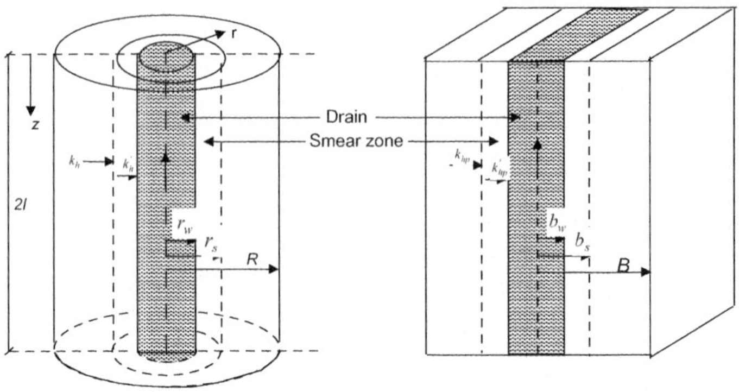

Fig. 10.14 Conversion of an axisymmetric unit cell into plane strain condition (Indraratna et al. [17], with kind permission from ASCE)

cell considered before, plane strain finite element analysis can easily be adopted for most field situations [13, 41] to represent realistic behaviour by converting the axisymmetric unit cell using permeability and geometrical matching. The plane strain model can also be used to model the vacuum pressure described by Gabr and Szabo [42]. With these modifications to the original theory by Indraratna and Redana [12], Indraratna et al. [17] proposed an equivalent plane strain approach to simulate vacuum pressure in a system facilitated with PVDs. The equivalent plane strain conditions can be determined: (a) based on permeability and keeping drain spacing constant, (b) based on geometry (i.e. by changing the drain spacing) and keeping soil permeability constant, and (c) based on permeability and the geometric approach (permeability is determined using convenient drain spacing).

Indraratna et al. [17] propose an average degree of consolidation for plane strain by assuming that the plane strain cell (width of $2B$), the half width of the drain b_w and the half width of the smear zone b_s may be kept the same as their axisymmetric radii r_w and r_s, respectively. This implies that $b_w = r_w$ and $b_s = r_s$ (Fig. 10.14). The excess pore pressure can be determined from

$$\frac{\overline{u}}{\overline{u}_0} = \left(1 + \frac{p_{0p}}{\overline{u}_o}\frac{(1+k_1)}{2}\right)\exp\left(-\frac{8T_{hp}}{\mu_p}\right) - \frac{p_{0p}}{\overline{u}_0}\frac{(1+k_1)}{2} \tag{10.19}$$

$$\mu_p = \left[\alpha + (\beta)\frac{k_{hp}}{k'_{hp}}\right], \tag{10.20}$$

where $\bar{u}_0$ = the initial excess pore pressure, $\bar{u}$ = the pore pressure at time t (average values) and T_{hp} = the time factor in plane strain, and k_{hp} and k'_{hp} are the equivalent undisturbed horizontal and corresponding smear zone permeability, respectively. The geometric parameters α and β are given by

$$\alpha = \frac{2}{3} - \frac{2b_s}{B}\left(1 - \frac{b_s}{B} + \frac{b_s^2}{3B^2}\right) \tag{10.21}$$

$$\beta = \frac{1}{B^2}(b_s - b_w)^2 + \frac{b_s}{3B^3}\left(3b_w^2 - b_s^2\right) \tag{10.22}$$

At a given level of effective stress, and at each time step, the average degree of consolidation for the axisymmetric ($\overline{U}_p$) and equivalent plane strain ($\overline{U}_{p,pl}$) conditions are made equal.

By making the magnitudes of R and B the same, Indraratna et al. [17] presented a relationship between k_{hp} and k'_{hp}. The smear effect can be captured by the ratio between the smear zone permeability and undisturbed permeability, hence

$$\frac{k'_{hp}}{k_{hp}} = \frac{\beta}{\frac{k_{hp}}{k_h}\left[\ln\left(\frac{n}{s}\right) + \left(\frac{k_h}{k'_h}\right)\ln(s) - 0.75\right] - \alpha} \tag{10.23}$$

Ignoring the effects of smear and well resistance in the above expression would lead to the simplified solution proposed earlier by Hird et al. [43]

$$\frac{k_{hp}}{k_h} = \frac{0.67}{[\ln(n) - 0.75]} \tag{10.24}$$

Indraratna et al. [17] compared two different distributions of vacuum along a single drain for the equivalent plane strain (2D) and axisymmetric conditions (3D). Varying the vacuum pressure in PVDs installed in soft clay would be more realistic for long drains, but a constant vacuum with depth is justified for relatively short drains.

10.4.2 Numerical Modelling of Biodegradable Drains

The finite element method (FEM) is used widely to model conventional drains with a constant discharge capacity, but to capture a reduction in the discharge capacity of degradable drains; a sub-routine [30] has been created. In this approach the permeability of drains is assumed to decrease over time, i.e. $k_w(t)$ while their cross sectional area and length remain the same, so the discharge capacity of drains is estimated with respect to the permeability and equivalent diameter of drain elements. A function describing k_w with time is incorporated into the governing equations of soil and excess pore water dissipation in FEM (i.e. Abaqus).

To illustrate how the proposed solutions can predict how biodegradation influences the consolidation of soil, their parameters are assumed in this investigation; in this instance the unit cell parameters are: $d_e = 0.6$ m, $l = 20$ m, $s = 3.4$, $k_h/k_s = 3$, $k_h = 1 \times 10^{-9}$ m/s, $c_h = 0.63$ m^2/year. A natural fibre drain made from jute and coconut fibres with an initial discharge capacity of 0.43 m^3/day is considered in this study.

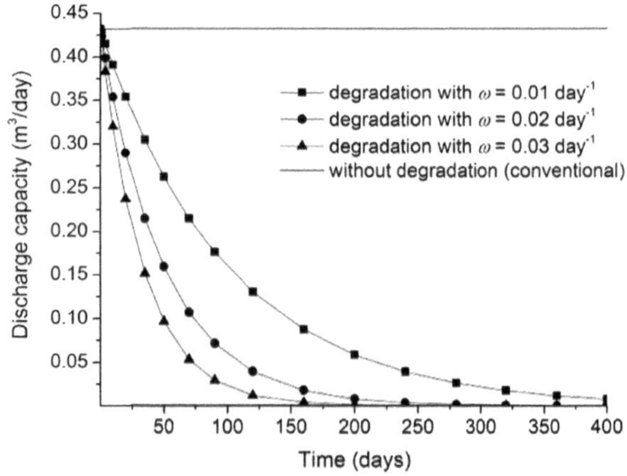

Fig. 10.15 Exponential reduction in discharge capacity with different rates of degradation

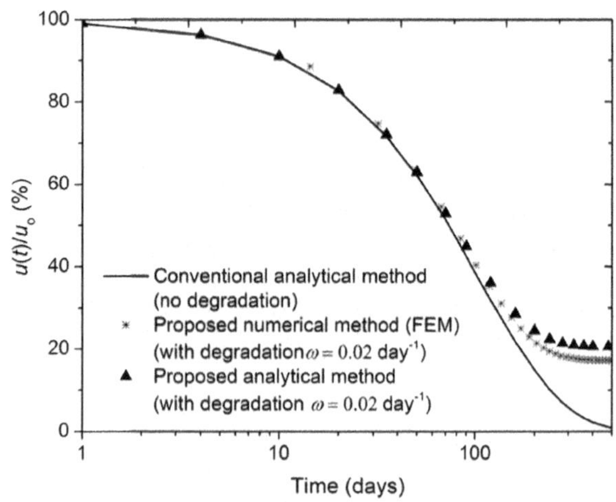

Fig. 10.16 Retarded dissipation of excess pore pressure induced (Indraratna et al. [30], with kind permission from Elsevier)

Details of this type of BPVDs with laboratory testing can be found in Nguyen et al. [29]. Figure 10.15 shows the exponential degradation in the discharge capacity of BPVDs with different rates, i.e. $\omega = 0.01$, 0.02 and 0.03 day^{-1} over 400 days.

Figure 10.16 shows how an exponential degradation in the discharge capacity of BPVDs with $\omega = 0.02$ day^{-1} can affect consolidation. While the dissipation curves of excess pore pressure induced by conventional and biodegradable drains are almost identical in the first 90 days they do begin to deviate from each other after 100 days. After 400 days where the discharge capacity of BPVDs drops to 0.003 m^3/day, the residual excess pore pressure by BPVDs is about 20%, but with conventional drains it has almost completely dissipated; this indicates that as fibre drains degrade they actively retard consolidation. Figure 10.16 also shows how well numerical and analytical methods can predict the consolidation of soil and the reduction in discharge capacity (less than 5% difference).

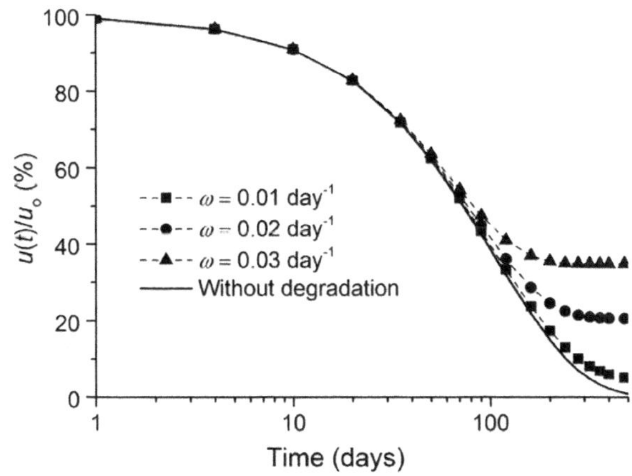

Fig. 10.17 Influence of different degradation rates on soil consolidation (Indraratna et al. [30], with kind permission from Elsevier)

Figure 10.17 shows how the different rates of degradation affect the dissipation of excess pore pressure ($u(t)/u_o$); fundamentally, the earlier and faster the reduction in discharge capacity leads to more retarded dissipation. For example, using $\omega = 0.03$ day^{-1} results in more than 30% of excess pore pressure being blocked, as predicted by the proposed analytical method. The smallest degradation rate $\omega = 0.01$ day^{-1} does not make much difference compared to the conventional method, in fact there is less than 5% deviation after 400 days.

Note that only exponential degradation is addressed in this paper, the actual reduction in the discharge capacity of BPVDs in the field might have more complex characteristics which are discussed in a recent study by Nguyen et al. [29]. Apart from biodegradation which is stemmed from biological activities, BPVDs made by combining individual natural fibres in a certain arrangement can experience considerable physical degradation (i.e. soil clogging), so a more detailed and comprehensive investigation into the degradation of BPVDs is essential.

10.5 Field Studies

10.5.1 Port of Brisbane (POB), Queensland

Works at the Port of Brisbane involves expanding the port by reclaiming 235 ha of land using materials dredged from the Brisbane River and Moreton Bay Channels. This site contains highly compressible problematic soft clays up to 30 m deep, so in an effort to reduce the consolidation period, several techniques are in use by three different contractors. These techniques include consolidation using PVDs and surcharge, PVDs with surcharge and vacuum (membrane and membrane less), and

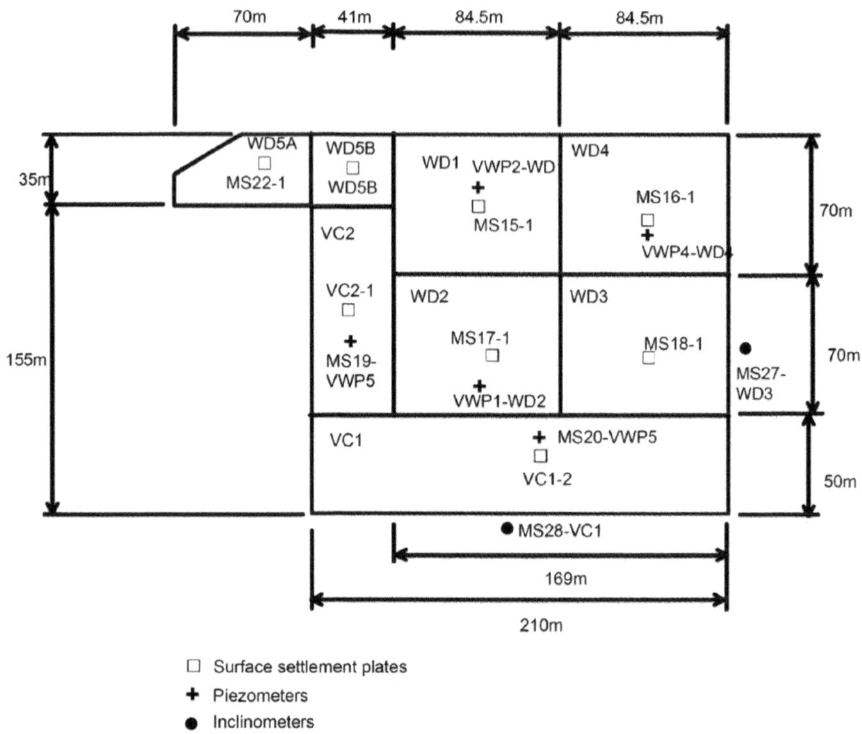

Fig. 10.18 Layout and design specification of trial area (Indraratna [44], with kind permission from AGS)

Table 10.2 Soil properties at POB, Queensland (Indraratna [44], with kind permission from AGS)

Soil layer	Soil type	γ_t (kN/m^3)	$c_c/(1+e_0)$	c_v(m^2/year)	c_h(m^2/year)	k_h/k_s	$s = d_s/d_w$	$C_\alpha/(1+e_0)$
1	Dredged Mud	14	0.3	1	1	1	1	0.005
2	Upper Holocene Sand	19	0.01	5	5	1	1	0.001
3	Upper Holocene Clay	16	0.18	1	2	2	3	0.008
4	Lower Holocene Clay	16	0.235	0.8	1.9	2	3	0.0076

PVDs with different types, diameters and spacing. The layout of the trial area is shown in Fig. 10.18 and the relevant soil properties are tabulated in Table 10.2.

For a typical settlement plate (TSP3), the behaviour of an embankment in terms of its settlement and excess pore water pressure is shown in Fig. 10.19. In terms of

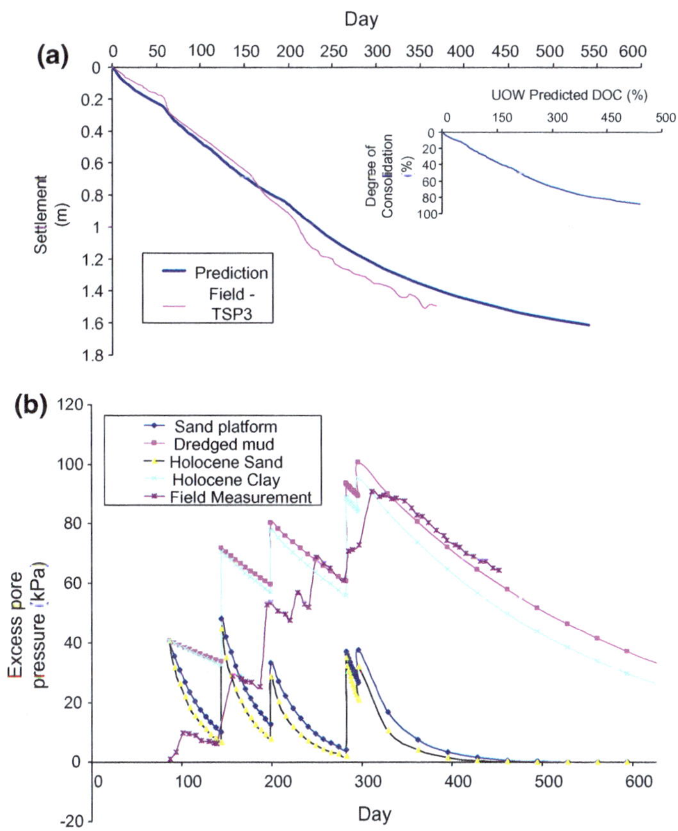

Fig. 10.19 **a** Settlement and **b** EPWP predictions and field data (Indraratna [44], with kind permission from AGS)

settlement, the predicted settlement curve agrees with the field data, although the excess pore water pressure dissipation curve has some deviation in terms of the rate of dissipation (slower than the actual one); this slower rate of dissipation is mainly due to the highly viscous nature of soft soil in every section monitored.

In terms of lateral deformation, the lateral displacements normalised with applied effective stress at two different positions (MS 24 and MS34) of inclinometers are shown in Fig. 10.20. Note here that lateral displacement is insignificant at shallow depths (below 10 m) but it is very significant in the upper Holocene shallow clay depths. Moreover, the membrane-less vacuum system (MS34) did control lateral displacement very well compared to the surcharge only section (MS24), thus confirming how effectively vacuum pressure improves soft ground. Figure 10.21 describes the behaviour of embankments in terms of stability, and when the incremental rate of change of the lateral deformation/settlement ratio μ becomes small, it reflects towards higher reliability with stability, whereas a higher value of μ reflects

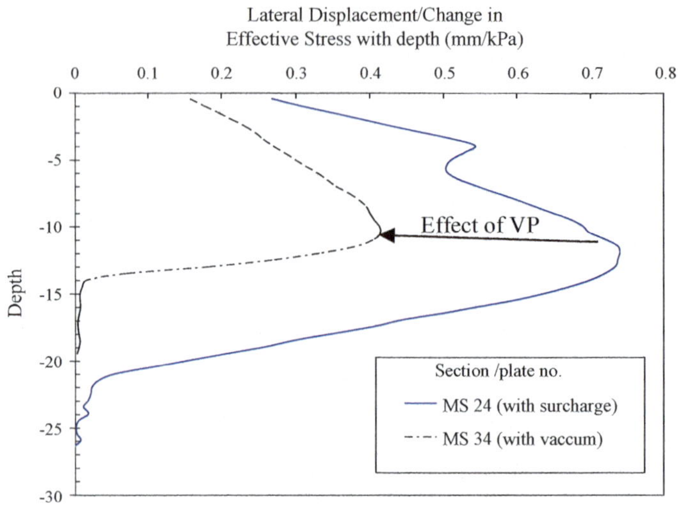

Fig. 10.20 Comparison between the lateral deformation in vacuum and non-vacuum areas (Indraratna [44], with kind permission from AGS)

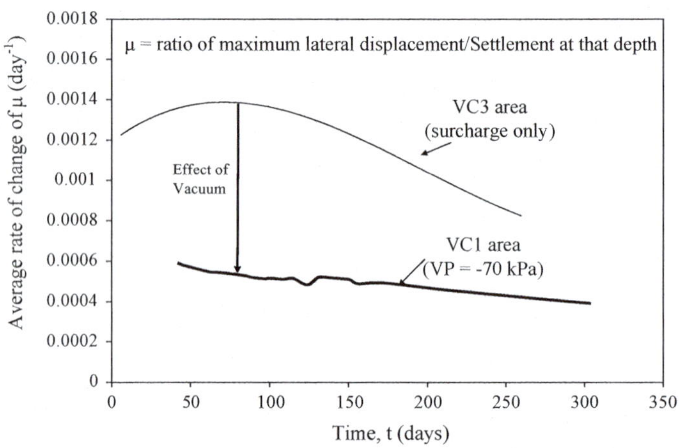

Fig. 10.21 Rate of change of lateral displacement to settlement (Indraratna [44], with kind permission from AGS)

potential lateral instability. In Fig. 10.21, the gradient in the non-vacuum area VC3 increases initially possibly due to the final surcharge loading being placed quickly while the clay was still at early stages of consolidation. However, as the PVDs become fully active and settlement increases at a healthy rate, the gradient of μ decreases, as expected. Figure 10.21 shows that the vacuum pressure generally provides a relatively unchanging gradient of μ with time.

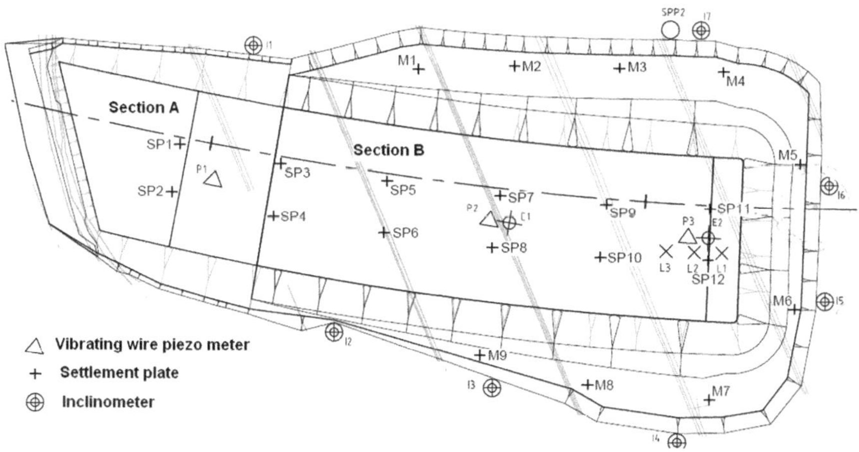

Fig. 10.22 Instrumentation layout for Ballina Bypass (Indraratna et al. [45], with kind permission from ICE)

10.5.2 Ballina Bypass, NSW

The Australian east coast which links Sydney and Brisbane bypasses the town of Ballina, but this bypass route crosses a floodplain of soft, highly compressible marine clay up to 40 m thick. A combination of vacuum with surcharge preloading will be used to improve the ground at one of the bridge abutment, but to determine its effectiveness, there is a trial embankment at Ballina where 34 mm diameter PVDs are placed at 1 m spacing in a square pattern, and membrane-type vacuum consolidation with horizontal transmission pipes, peripheral trenches, and heavy-duty vacuum pumps are used. A range of instruments including settlement plates, inclinometers and piezometers are installed in the trail embankment; their locations are shown in Fig. 10.22.

Two sections (A & B) are shown within the embankment area, with Section A being allowed to improve without vacuum pressure and section B being allowed to improve with combined vacuum and surcharge pressure. Since the layers of soft clay vary from 7 m to 25 m thick, the embankment is from 4.3 to 9 m thick to limit post consolidation settlement, and a vacuum pressure of 70 kPa has been applied to section B at the drain interface and then removed after 400 days. The geotechnical parameters of these three layers of subsoil obtained from standard odometer tests are given in Table 10.3. A profile of this subsoil and the relevant geotechnical properties are shown in Fig. 10.23. There is a 10 m thick layer of soft silty clay underlain by moderately stiff silty clay 10–30 m deep, followed by firm clay. The groundwater table is assumed to be at the surface. The water content is 80–120%, which is higher than the liquid limit, ensuring that the soils are fully saturated. The behaviour of this embankment in terms of settlement and the dissipation of excess pore water pressure is shown in Fig. 10.24, along with the construction schedule for the embankment. The

Table 10.3 Soil properties at Ballina Bypass (Indraratna et al. [45], with kind permission from ICE)

Depth (m)	Soil type	λ	κ	γ (kN/m^3)	e_0	$k_{h,ax}$ (10^{-10} m/s)	OCR
0.0–0.5	Clayey Silt	0.57	0.06	14.5	2.9	10	2
0.5–15.0	Silty Clay	0.57	0.06	14.5	2.9	10	1.7
15.0–24.0	Stiff Silty Clay	0.48	0.048	15.0	2.6	3.3	1.1

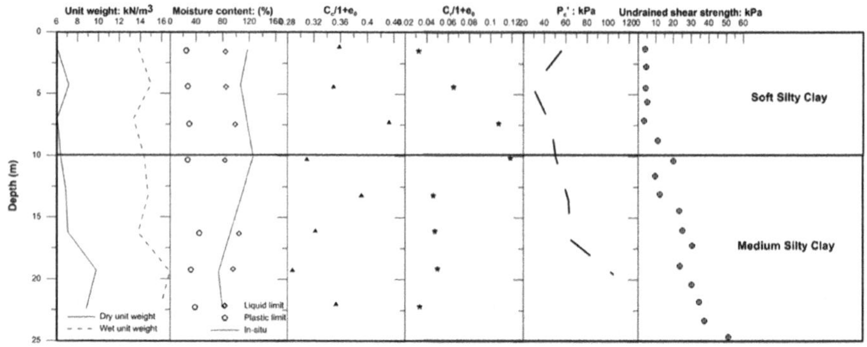

Fig. 10.23 Sub-soil profile with the properties at Ballina Bypass

vacuum pressure varies from −70 to −80 kPa, as measured by miniature piezometers embedded inside the drain. The settlement noted at differential settlement plates varies from 1 to 5 m and the dissipation of excess pore water pressure is retarded in most piezometer reading due to the highly viscous nature of soft soil as well as potential biological and chemical clogging of the piezometer tips mainly in the region prone to have acid sulphate soil.

The rate of surcharge loading rate is faster in the vacuum areas (when raising the embankment) than the non-vacuum areas, so the lateral displacement of the embankment must be examined to check its stability. The profile of lateral displacement with depth is shown in Fig. 10.25a, and the normalised lateral displacement plot (by embankment height) is shown in Fig. 10.25b. Inclinometer I1 is located in the non-vacuum areas while all the other inclinometers are located at the edge of the vacuum areas. Figure 10.25a, b shows that lateral displacement under vacuum pressure is smaller despite the high embankments, and the use of vacuum pressure undoubtedly reduced lateral displacement.

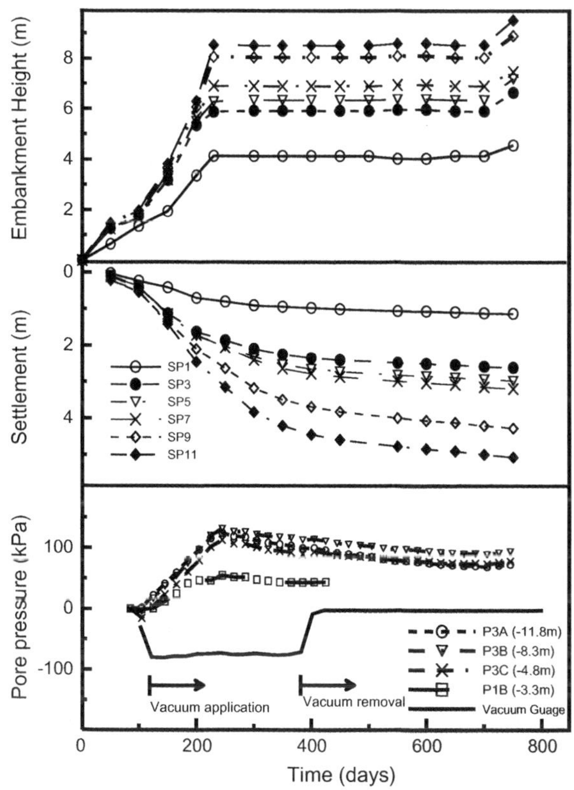

Fig. 10.24 Staged construction of the embankment along with associated settlement and dissipation of pore water pressure at different instrument location (Indraratna et al. [45], with kind permission from ICE)

10.6 Conclusions

Ground improvement methods using PVDs and surcharge and vacuum preloading is now common practice and is regarded as the best and most cost-effective and friendly technique for improving soft clay. Analytical and numerical modelling related to vacuum preloading are ongoing to solve complexities such as: PVD bending with settlement in shallow depths, the biodegradation of natural fibre drains, and the soil–drain interface. Many predictions have been made to describe the behaviour of embankments stabilised with PVDs and vacuum pressure but there is always a discrepancy between field data and predictions in terms of settlement and the dissipation of excess pore water pressure; these discrepancies are mainly due to the uncertainty of soil properties, soil de-structuration, the smear effect, inaccurate distribution of vacuum pressure through the drain depths, large strains, inaccurate

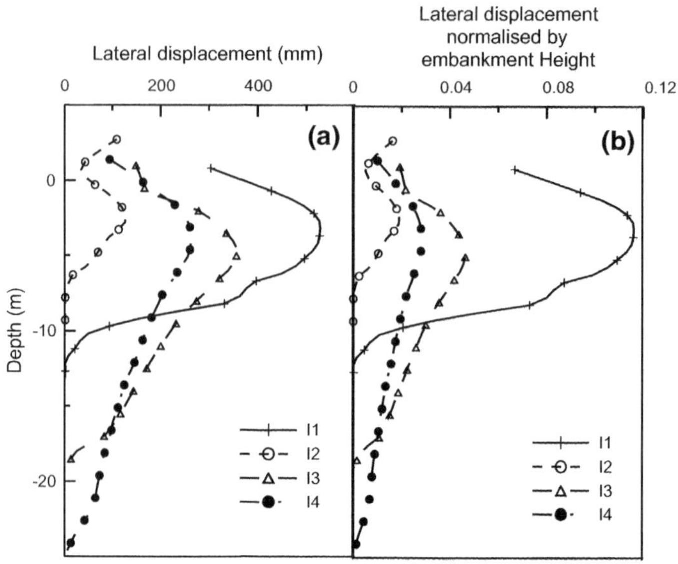

Fig. 10.25 General and normalised lateral displacement profile shown by inclinometers (Indraratna et al. [45], with kind permission from ICE)

conversion of axisymmetric to plane strain conditions, and the elastic visco-plastic behaviour of soft clay.

When there are very soft soils in reclamation areas, vacuum consolidation for very large projects has been used successfully because the amount of surcharge fill can be reduced to achieve the same ultimate settlement and lateral yield, but only with help from PVDs and vacuum pressure. The efficiency of vacuum-assisted consolidation depends on the air-tightness, how well the membrane is anchored, the groundwater level, and soil conditions. A detailed evaluation is needed to determine how effective the type of membrane and membrane-less vacuum technology will be. Without a proper quantitative and comprehensive analysis, the study of proper methods for vacuum-assisted consolidation becomes imperative, numerically, experimentally, and more significantly in the field through trial sites.

A large-scale consolidometer has been used to study how well vacuum-assisted consolidation will deliver more realistic parameters (soil–drain interface, extent of smear zone, and distribution of vacuum pressure through drain depth). The main reason for using a large-scale consolidometer is because conventional small-scale equipment such as a Rowe cell cannot capture the above-mentioned properties. The studies carried out at University of Wollongong indicate that the equivalent radius of smear zone will be 2–3 times larger than the radius of the mandrel, and in the same way, soil permeability in the smear zone is reduced due to the remoulding phenomenon and is generally smaller than the undisturbed zone by 1.5–2 times. In the absence of boundary effects (which is predominant in laboratory conditions) but are absent in the field, the smear zone can be larger, which means soil sampling and analysing field piezometer data to predict exact behaviour in terms of the extent of the smear zone and soil disturbances. Suitable drain spacing patterns also need to be determined to minimise the effects of overlapping smear zones.

Analytical modelling for PVDs combined with vacuum preloading under plane strain as well as axisymmetric conditions that simulate unit cell consolidation has been developed, as has the distribution of vacuum pressure along the length of drains, the occurrence of non-Darcian flow and the nonlinear properties of soil to represent more realistic conditions. Large strain analysis based on convective coordinate systems reveals that large strain analysis has a notable difference predicting settlement when preloading is higher than 120 kPa, but for most traditional road embankments where preloading is not very high, a small strain solution is still valid and acceptable.

2D plane strain analysis is usually good enough to model embankments for large construction sites that will have many PVDs. The method of conversion proposed by Indraratna to convert axisymmetric to plane strain conditions agrees very well with the measured data for many case histories, including the Port of Brisbane (POB) in Queensland, Australia. Another advantage of the plane strain method is its readiness to be incorporated into numerical analysis (FEM) due to its simplicity and efficiency. When the permeability and vacuum pressure has been converted correctly from the 3D model to the 2D model, the time-settlement curves are similar to the more complex 3D analysis except when a plane strain condition cannot be assumed due to specific boundaries and there is sudden change in the properties and terrain (i.e. curved sections of road and rail embankments); in this instance a 3D analysis is only one option. Field case studies and model predictions indicate that the efficiency of PVDs is a function of the drain discharge capacity, the magnitude and distribution of the vacuum, and the extent of smear caused by a vertically penetrating mandrel. The drain discharge capacity is very relevant for natural PVDs which can potentially decay over time, so if NPVDs can consolidate enough prior to decay, they can be regarded as the best environmentally friendly ground improvement technique due to their self decaying nature, unlike conventional synthetic drains.

Vacuum preloading successfully reduces surcharge fill heights while giving the same degree of consolidation as surcharge only embankments, provided there are no air leaks. Vacuum consolidation is also very important where lateral displacement must be controlled due to critical boundary conditions such as at the Port of Brisbane where the marine boundary compels contractors to limit lateral displacement; in fact, if a vacuum is not used the boundaries may pollute the water by increasing the turbidity level. Moreover, PVDs also help to stabilise rail embankments in coastal area by dissipating excess cyclic pore pressure and curtailing lateral displacement.

The dynamic triaxial testing with CSR at frequencies representing fast heavy haul trains that have taken place at the University of Wollongong over the past decade indicates how important the rapid dissipation of cyclic pore water pressure is to stabilise saturated subgrade. It is also important to study PVD under cyclic loading to discover how to avoid mud pumping (formation of subgrade slurry). Current findings from laboratory experiments prove that high-speed trains cannot operate on tracks built over soft subgrade unless it is consolidated beforehand. While a lot of this research is ongoing, the laboratories at UOW tend to focus on transport geotechnics.

In Australia, a lot of attention is now being given to ground improvement techniques because the government has approved a 2000+ km long M-B Inland rail project.

Acknowledgements The author would like to thank more than a dozen PhD students and research fellows who contributed to the field of soft ground improvement directly and indirectly during their time at UOW. The assistance from laboratory technicians Alan Grant, Cameron Neilson and Ritchie McLean during the laboratory and field work is also appreciated. The authors acknowledge the Australian Research Council (ARC) and Industry partners for providing support through the ARC Industrial Transformation Training Centre for Advanced Technologies in Rail Track Infrastructure (ITTC-Rail). The author would also like to acknowledge Australian Research Council (ARC) for providing funding to the research via several linkage projects in the field of soft ground improvement over the last two decades. Most of the contents produced in this paper are reproduced with kind permission from the Australian Geomechanics Journal (e.g. Author's 2009 EH Davis Lecture), Journal of Geotechnical & Geoenvironmental Engineering ASCE, International Journal of Geomechanics, ASCE, Proceedings of the Institution of Civil Engineers – Ground Improvement, ASTM Geotechnical Testing Journal, Géotechnique and Canadian Geotechnical Journal among others.

References

1. Indraratna, B., Balasubramaniam, A.S., Balachandran, S.: Performance of test embankment constructed to failure on soft marine clay. J. Geotech. Eng. ASCE **118**(1), 12–33 (1992)
2. Indraratna, B., Rujikiatkamjorn, C., Adams, M., Ewers, B.: Class A prediction of the behaviour of soft estuarine soil foundation stabilised by short vertical drains beneath a rail track. J. Geotech. Geoenviron. Eng. ASCE **136**(5), 686–696 (2010)
3. Johnson, S.J.: Precompression for improving foundation soils. J. Soil Mech. Foundat. Div. ASCE **1**, 111–114 (1970)
4. Carrillo, N.: Simple two and three dimensional case in the theory of consolidation of soils. Stud. Appl. Math. **21**(1–4), 1–5 (1942)
5. Barron, R.A.: Consolidation of fine-grained soils by drain wells. Trans. ASCE **113**, 718–754 (1948)
6. Akagi, T.: Consolidation caused by mandrel-driven sand drains. In: Proceedings of the 6th Asian Regional Conference on Soil Mechanics and Foundation Engineering, Singapore, Southeast Asian Geotechnical Society, Bangkok, vol. 1, pp. 125–128 (1979)
7. Hansbo, S.: Consolidation of clay by band-shaped pre-fabricated drains. Ground Eng. **12**(5), 16–25 (1979)
8. Hansbo, S.: Consolidation of fine-grained soils by prefabricated drains and lime column installation. In: Proceedings of 10th International Conference on Soil Mechanics and Foundation Engineering, Stockholm, A.A. Balkema, Rotterdam, The Netherlands, vol. 3, pp. 677–682 (1981)
9. Hansbo, S.: Aspects of vertical drain design: Darcian or non-Darcian flow. Géotechnique **47**(5), 983–992 (1997)
10. Onoue, A.: Consolidation by vertical drains taking well resistance and smear into consideration. Soils Found. **28**(4), 165–174 (1988)
11. Zeng, G.X., Xie, K.H.: New development of the vertical drain theories. In: Proceedings of 12th ICSMFE. Rio de Janeiro, pp. 1435–1438 (1989)
12. Indraratna, B., Redana, I.W.: Laboratory determination of smear zone due to vertical drain installation. J. Geotech. Eng. ASCE **125**(1), 96–99 (1998)
13. Indraratna, B., Redana, I.W.: Numerical modeling of vertical drains with smear and well resistance installed in soft clay. Can. Geotech. J. **37**, 132–145 (2000)

14. Mohamedelhassan, E., Shang, J.Q.: Vacuum and surcharge combined one-dimensional consolidation of clay soils. Can. Geotech. J. **39**, 1126–1138 (2002)
15. Indraratna, B., Bamunawita, C., Khabbaz, H.: Numerical modeling of vacuum preloading and field applications. Can. Geotech. J. **41**, 1098–1110 (2004)
16. Bo, M.W., Chu, J., Low, B.K., Cho, V.: Soil improvement; prefabricated vertical drain techniques. Thomson Learning, Singapore (2003)
17. Indraratna, B., Rujikiatkamjorn, C., Sathananthan, I.: Analytical and numerical solutions for a single vertical drain including the effects of vacuum preloading. Can. Geotech. J. **42**, 994–1014 (2005)
18. Walker, R., Indraratna, B.: Vertical drain consolidation with overlapping smear zones. Geotechnique **57**(5), 463–467 (2007)
19. Indraratna, B., Zhong, R., Fox, P., Rujikiatkamjorn, C.: Large-strain vacuum-assisted consolidation with non-darcian radial flow incorporating varying permeability and compressibility. J. Geotech. Geoenviron. Eng. 04016088. https://doi.org/10.1061/(asce)gt.1943-5606.0001599 (2016)
20. Hansbo, S.: Consolidation equation valid for both darcian and non-darcian flow. Geotechnique **51**, 51–54 (2001)
21. Yin, J.H., Graham, J.: Viscous-elastic-plastic modelling of one-dimensional time-dependent behaviour. Can. Geotech. J. **26**(2), 199–209 (1989)
22. Indraratna, B., Baral, P., Rujikiatkamjorn, C., Perera, D.: Class A and C predictions for Ballina trial embankment with vertical drains using standard test data from industry and large diameter specimens. Comput. Geotech. **93**, 232–246 (2018)
23. Lambe, T.W.: Predictions in soil engineering. Geotechnique **23**, 149–202 (1973)
24. Kianfar, K., Indraratna, B., Rujikiatkamjorn, C.: Radial consolidation model incorporating the effects of vacuum preloading and non-darcian flow **63**, 1060–1073 (2013)
25. Perera, D., Indraratna, B., Leroueil, S., Rujikiatkamjorn, C., Kelly, R.: An analytical model for vacuum consolidation incorporating soil disturbance caused by mandrel-driven drains. Can. Geotech. J. (2016)
26. Lee, S.L., Ramaswamy, S.D., Aziz, M.A., Das Gupta, N.C., Karunaratne, G.P.: Fibredrain for consolidation of soft soils. In: Proceedings of the post- vienna conference on geotextiles, Singapore, vol. 2, pp. 238–258 (1987)
27. Jang, Y.S., Kim, Y.W., Park, J.Y.: Consolidation efficiency of natural and plastic geosynthetic band drains. Geosynt. Int. **8**(4), 283 (2001)
28. Asha, B.S., Mandal, J.N.: Absorption and discharge capacity tests on natural prefabricated vertical drains. Geosynth. Int. **19**, 263–271 (2012)
29. Nguyen, T.T., Indraratna, B., Carter, J.: Laboratory investigation into biodegradation of jute drains with implications for field behaviour. J. Geotech. Geoenviron. Eng. **144**(6), 0401802-1:15
30. Indraratna, B., Nguyen, T.T., Carter, J., Rujikiatkamjorn, C.: Influence of biodegradable natural fibre drains on the radial consolidation of soft soil. Comput. Geotech. **78**, 171–180 (2016)
31. Bergado, D.T., Asakami, H., Alfaro, M.C., Balasubramaniam, A.S.: Smear effects of vertical drains on soft bangkok clay. J. Geotech. Eng. ASCE **117**, 1509–1530 (1991)
32. Onoue, A., Ting, N.H., Germaine, J.T., Whitman, R.V.: Permeability of disturbed zone around vertical drains. Proc, pp. 879–890. ASCE Geotech. Enggr. Congress, Colorado (1991)
33. Almeida, M.S.S., Ferreira, C.A.M.: Field in situ and laboratory consolidation parameters of a very soft clay, predictive soil mechanics. In: Proceedings of the Worth Memorial Symposium, London. Thomas Telford, pp. 73–93 (1993)
34. Chai, J.C., Miura, N.: Investigation of factors affecting vertical drain behavior. J. Geotech. Geoenviron. Eng. **125**, 216–226 (1999)
35. Hird, C.C., Moseley, V.J.: Model study of seepage in smear zones around vertical drains in layered soil. Geotechnique **50**, 89–97 (2000)
36. Sharma, J.S., Xiao, D.: Characteristics of a smear zone around vertical drains by large-scale laboratory tests. Can. Geotech. J. **37**, 1265–1271 (2000)

37. Indraratna, B., Perera, D., Rujikiatkamjorn, C., Kelly, R.: Soil disturbance analysis due to vertical drain installation. Proc. Institut. Civil Eng. Geotech. Eng. **168**, 236–246 (2015)
38. Baral, P., Rujikiatkamjorn, C., Indraratna, B., Kelly, R.: Radial consolidation characteristics of soft undisturbed clay based on large specimens. J. Rock Mech. Geotech. Eng. https://doi.org/10.1016/j.jrmge.2018.06.002 (2018)
39. Sathananthan, I., Indraratna, B.: Laboratory evaluation of smear zone and correlation between permeability and moisture content. J. Geotech. Geoenviron. Eng. **132**, 942–945 (2006)
40. Indraratna, B., Attya, A., Rujikiatkamjorn, C.: Experimental investigation on effectiveness of a vertical drain under cyclic loads. J. Geotech. Geoenviron. Eng. ASCE **135**(6), 835–839 (2009)
41. Hird, C.C., Pyrah, I.C., Russell, D., Cinicioglu, F.: Modelling the effect of vertical drains in two-dimensional finite element analyses of embankments on soft ground. Can. Geotech. J. **32**, 795–807 (1995)
42. Gabr, M.A., Szabo, D.J.: Prefabricated vertical drains zone of influence under vacuum in clayey soil. In Conference on In Situ Remediation of the Geoenvironment, ASCE, pp 449–460 (1997)
43. Hird, C.C., Pyrah, I.C., Russel, D.: Finite element modeling of vertical drains beneath embankments on soft ground. Geotechnique **42**(3), 499–511 (1992)
44. Indraratna, B.: 2009 EH Davis Memorial Lecture: Recent advances in the application of vertical drains and vacuum preloading in soft soil stabilization. Australian Geomech. J. AGS **45**(2), 1–43 (2010)
45. Indraratna, B., Rujikiatkamjorn, C., Kelly, R., Buys, H.: Soft soil foundation improved by vacuum and surcharge loading. Proc. ICE Ground Improv **165**, 87–96 (2012)

Chapter 11
Issues on Seismic Site Characterization

S. K. Prasad, K. V. Vijayendra and Sitaram Nayak

11.1 Introduction

Earthquakes have been the most devastating of all the natural disasters in terms of both human loss and damage to built environment. Time and again, it has been demonstrated that the damaging effects of earthquake are more pronounced in places that are ill-equipped in managing earthquake disasters compared to those which have good practice of earthquake-resistant construction. For a structure to be earthquake resistant during strong earthquakes, each of its components should perform well and most importantly the foundation and the foundation soil should perform adequately. The foundation soil can be from strong and intact rock to a very soft soil having sufficient overburden thickness.

It has been well established that the overburden soil influences the ground motion considerably. There have been many instances during past earthquakes in which the performance of a particular type of building was influenced by the overburden soil on which the structure rested. Michiokan earthquake of Mexico (1985), Lomaprieta earthquake of USA (1989), Hyogo Ken Nambu earthquake of Japan (1995) and Bhuj earthquake of India (2001) are a few examples of overburden soil influencing the damage to many structures. There have been instances of near resonance situation in many cases leading to damaging effects to particular types of buildings. Hence, studies on determination of ground motion at the surface due to travel of seismic waves

S. K. Prasad (✉)
Sri Jayachamarajendra College of Engineering, Mysuru 570006, India
e-mail: prasad_s_k@hotmail.com

K. V. Vijayendra
Bangalore Institute of Technology, Bangalore 560004, India
e-mail: vijayendra.kv@gmail.com

S. Nayak
National Institute of Technology, Surathkal 575025, Karnataka, India
e-mail: snayak65@yahoo.co.in

M. Latha G. (ed.), *Frontiers in Geotechnical Engineering*, Developments in Geotechnical Engineering, https://doi.org/10.1007/978-981-13-5871-5_11

in layered soil are extremely important. Site characterization, site effect, ground amplification are some terminologies used for such study.

The seismic waves generated at the hypocenter travel upward through layered soil experiencing reflection and refraction at layer interfaces resulting in considerable changes in wave motion, both in terms of amplification and frequency content. The energy dissipating characteristics of soil will also influence the ground motion. Site characterization and seismic ground response involve in the determination of expected ground surface motions at a site from a specified control motion at some other point where an observed or estimated motion is available considering the factors such as magnitude of shaking, epicentral distance, focal depth, thickness of overburden, number of layers, type of soil, presence of groundwater table, shear wave velocity of ground, topography of ground, stiffness , damping characteristics of ground, etc. The procedure [24] involves,

- Determination of the characteristics of motions likely to develop in the rock formation underlying the site and selection of an accelerogram with these characteristics for use in the analysis.
- Determination of the dynamic properties of the soil deposit.
- Computation of the response of the soil deposit to the base rock motions. A one-dimensional method of analysis can be used if the soil structure is essentially horizontal.

Figure 11.1 explains the procedure involved in estimating motion at one place from known motion anywhere else. In general, the results of seismicity evaluations are presented as the peak acceleration expected for a given return period. The acceleration usually corresponds to the shaking at a rock outcrop, not at the surface of a soil profile. Using the estimated acceleration time history of the rock outcrop as input motion at the bedrock and propagating through soil media, the site effects must then be evaluated as a function of soil parameters such as type of soil deposit, deposit thickness, stiffness, and damping properties of deposit. There are many empirical, simple, and complex procedures available to compute site-specific dynamic soil response. The site-specific seismic ground response analyses include characterizing the modification in the frequency and amplitude of the seismic waves. Figures 11.1 and 11.2 show the procedure involved in seismic site characterization where layered soil profile is considered.

It is interesting to note that the ground motion at the surface during an earthquake can be very different from that at some depth below in hard rock depending on geotechnical factors of ground. Depending on the situation, there can be changes in amplitude and frequency content of motion at the surface. In many situations, a considerable amplification in ground motion is a possibility at the surface. The amplified ground motion will be responsible for damaging effects during earthquakes.

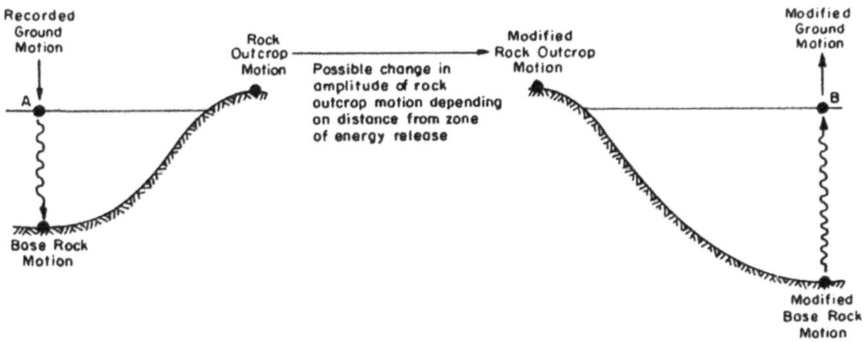

Fig. 11.1 Energy transfer from rock to soil for computing local soil effect on ground motion (extracted from Schnabel et al. [43])

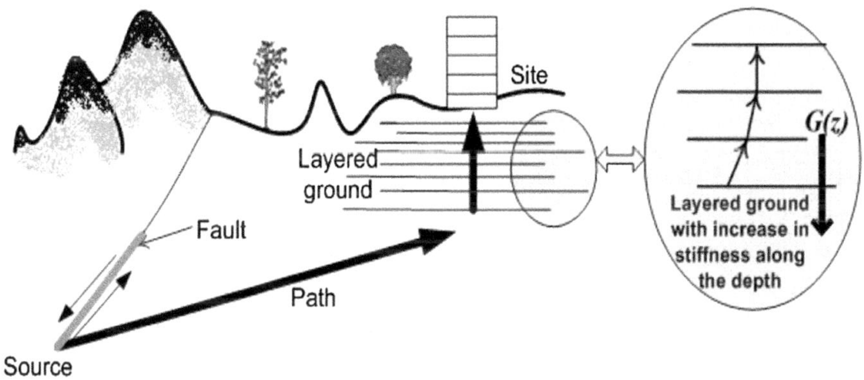

Fig. 11.2 Propagation of shear wave through layered soil (modified from Kramer [24])

11.2 Lessons Learnt from Past Earthquakes

During the past major earthquakes across the globe, it has been learnt that ground motion can amplify and frequency content of ground motion can be altered depending on several factors discussed resulting in huge damage to a particular type of structure due to near resonance situation. A few case studies in this regard are briefly discussed.

11.2.1 Michiokan Earthquake, Mexico, 1985

The earthquake of magnitude 8.3 struck on the September 19, 1985 in the Pacific Ocean, off the coast of Mexican state of Michoacan, at a distance of more than 350 km from Mexico City. Mexico City is vulnerable to earthquakes, though not near any fault line like San Francisco or Los Angeles. The main reason for this is the surface geology of the area, especially the downtown area, mostly resting on the silt and volcanic clay sediments which are between seven and thirty-seven meters deep and have a high water content. The city grew from an island in the middle of lake Texcoco outward, as this lake was eventually drained. On the bed of the lake, the prevailing silt and volcanic clay sediments amplify seismic shaking [1, 6]. Further, it was observed that many taller structures were either destroyed or severely damaged. Very low natural frequency of the system (tall structure resting on soft ground of sufficient thickness) matched with the low-frequency components of earthquake motion to cause near resonance effect in addition to amplification in ground motion causing severe damage to tall structures. Many of the older colonial buildings have survived hundreds of years on the lakebed simply because they are not tall enough to be affected by the resonance effect.

11.2.2 Lomaprieta Earthquake, USA, 1989

On October 17, 1989, the devastating Lomaprieta earthquake, measuring 7.1 on the Richter scale, rocked the city of Oakland, causing billions of dollars worth of damage and a death toll of 64. Over half the victims were on the Cypress Street Viaduct that was severely damaged during the quake. Hanks and Brady [15] studied the acceleration, velocity, and displacement waveforms of this earthquake at two rock sites in San Francisco, a rock site on Yerba Buena Island, an artificial-fill site on Treasure Island, and three sites in Oakland underlain by thick sections of poorly consolidated Pleistocene sediments. The waveforms at the three rock sites display a strong coherence, as do the three sedimentary sites in Oakland. The duration of strong motion at the rock sites is very brief, while the records in Oakland show strong amplification effects due to site geology. The *S*-wave group at Treasure Island is phase coherent with the Oakland records, but at somewhat diminished amplitudes, signaling the onset of liquefaction.

Cypress Street Viaduct, was a 2 km two-tier highway with five lanes per deck, and traffic flowing at ground level. A certain portion of Cypress Street Viaduct was constructed on soft mud where the bedrock was over 150 m below the surface [57]. The bridge was constructed using cast-in-place concrete with multi-celled reinforced box girders that typically spanned about 25 m [32].

During the Lomaprieta earthquake, a 1.4 km section of the Cypress Viaduct collapsed as shown in Fig. 11.3. Dashed red line in the figure indicates collapsed section of Cypress structure in 1989. During the Lomaprieta earthquake, the entire Viaduct structure began to vibrate tremendously [35]. Well-graded soils helped to dampen vibrations [57], the soft "bay mud" upon which most of the structure was constructed actually amplified the ground motion by up to five times in comparison to that of the rest of the freeway which was built on rock. In addition, it was later determined that the frequency of the seismic waves almost matched a natural frequency of the individual horizontal sections of the structure [14]. It is felt that these sediment resonances could have played a significant role in contributing to the freeway collapse,

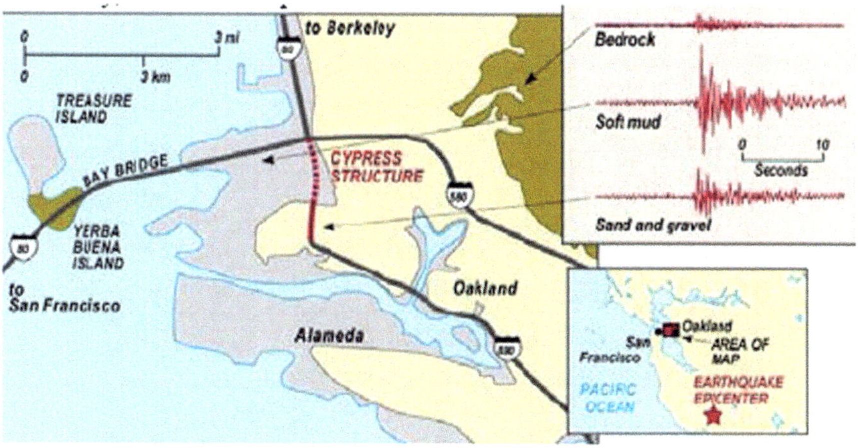

Fig. 11.3 Plan of Oakland area indicating the type of ground, amplification of ground motion at different sites and the location of cypress viaduct that was severely damaged (https://pubs.usgs.gov/fs/1999/fs151-99/)

since the resulting forces were not anticipated in the original design. The problem of matched resonant frequencies was not known at that time and may not have been considered in the design.

11.2.3 Hyogo ken Nanbu Earthquake, Kobe, Japan, 1995

Hyogo ken Nanbu earthquake also known as The Great Hanshin earthquake or Kobe earthquake occurred on the January 17, 1995 at 05:47 am 20 km away from Kobe city on the northern end of Awaji Island in the southern part of Hyogo Prefecture, Japan. It measured 6.9 on the moment magnitude scale at a focal depth of 17 km. About 6,400 people lost their lives, with more than 4,000 of them from Kobe. This was Japan's worst earthquake in the twentieth century after the Great Kanto earthquake in 1923. Figures 11.4 and 11.5 clearly illustrate relation between site effect and intensities of damage during this event (personal communication, [17]). Figure 11.4 indicates the aerial satellite view of Kobe city along with manmade islands Port Island and Rokko Island. The yellow strip in the figure represents the severely damaged location where most buildings had collapsed. Also, in the figure red line indicates the location whose geological cross section is provided in Fig. 11.5. It can be seen that a length of about 3 km comprises of hilly terrain in the north gradually sloping downwards toward seashore in the south by about 75 m in elevation. From the borehole investigation of geological formation, it was evident that the hilly terrain possessed dense gravelly

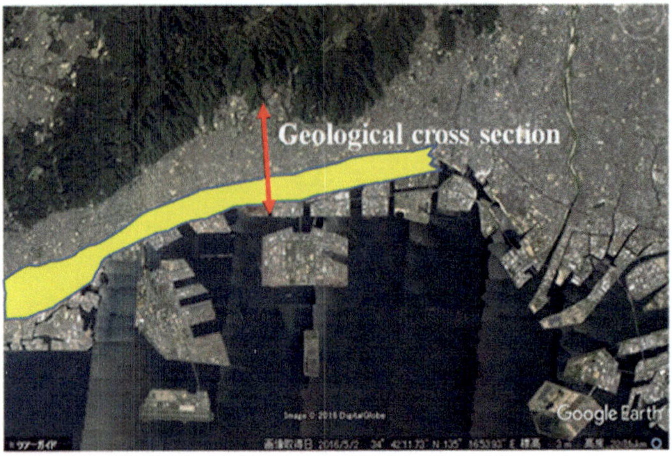

Fig. 11.4 Satellite picture of Kobe city with man-made islands [47]

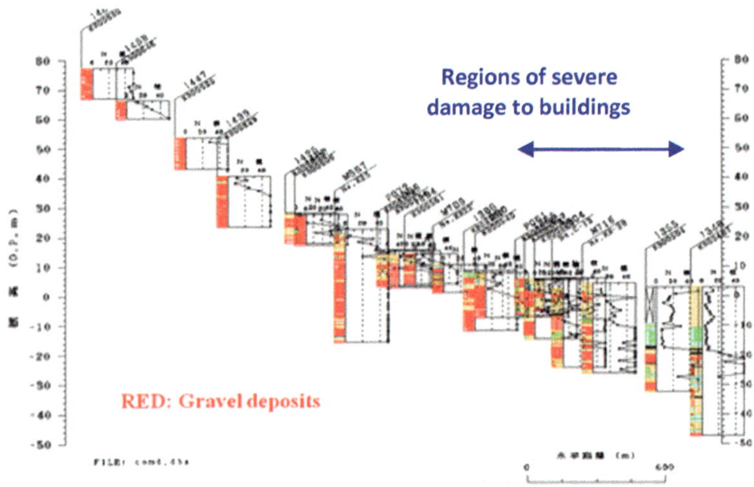

Fig. 11.5 Cross section of ground from hilly region to the shore in Kobe area [47]

deposit and the ground near the sea was low lying and possessed soft marine clay for a sufficient thickness. Most of the damage to buildings occurred at a place having thick soft overburden suggesting the amplification in ground motion.

11.2.4 Failure of Residential Flats in Ahmedabad During Gujarat Earthquake of 2001

The **2001 Gujarat earthquake** struck western part of India on January 26, 2001, India's 51st Republic Day, at 08:46 am with a moment magnitude of 7.6 that lasted for almost two minutes. The epicenter was about 50 km west of Bhuj in Kutch District of Gujarat, India. The earthquake had a maximum felt intensity of X on the Mercalli intensity scale and killed around 20,000 people, injured another 167,000, and destroyed nearly 400,000 homes. This was an intraplate earthquake, one that occurred at a distance from an active plate boundary. The shock waves spread 700 km. 21 districts were affected and 600,000 people were left homeless.

This earthquake caused widespread damage to civil engineering structures in Ahmadabad [37]. It is natural that the damage will be severe close to the epicenter and the effects reduce with the increase in distance from epicenter. Bhuj, Gandhidham, and Bhachau were close to the epicenter and hence suffered severe damage. However, the most surprising aspect of this event was the damage suffered by buildings in Ahmadabad city situated over 250 km east of epicenter. During the field investigation carried out in Ahmadabad after the earthquake, it was found that the distribution of damaged buildings in Ahmadabad was situated in a particular band and such buildings were close to Sabarmati river. Many multi-storied buildings (generally four storey and above), that were recently built, were reduced to rubble and the damage was restricted to a small portion of the city (less than 10% of total city area). Most other cities situated at similar range of distance from epicenter, including Gandhinagar had little impact of the earthquake. Thus, in order to find the reasons for failure of multi-storied structures on soft alluvial soil, following analyses were carried out.

- The available accelerogram recorded at passport office in Ahmadabad was processed and the predominant frequencies were evaluated.
- The natural frequency of framed structure mainly depends on the type of the material used; the cross-sectional details of different members, number of storeys, inertia characteristics, etc. The natural frequencies of bare multi-storied R.C. frames with varying number of storeys and number of bays were found to be between 1.32 to 2.11 Hz for multistoreyed structures with 5 to 7 floors.
- Using the soil properties of the area close to the damaged buildings in Ahmadabad, an equivalent linear analysis was performed using SHAKE [43] from the above acceleration time history and the natural frequencies of the ground were estimated.

The corrected ground acceleration and derived velocity and displacement time histories of a typical ground motion are shown in Fig. 11.6. The corresponding predominant frequencies are presented in Fig. 11.7. Sabarmati river runs from north to south across the city separating the old city from the recent formation. Around the river, soil is generally of soft alluvial type consisting of loose fine silty sand or clayey deposits for over 20 m depth. A typical soil profile (the borehole was located near the right bank of Sabarmati River between Gandhi Bridge and Nehru Bridge) along with N value (corrected) for a depth of around 30 m was available and the same

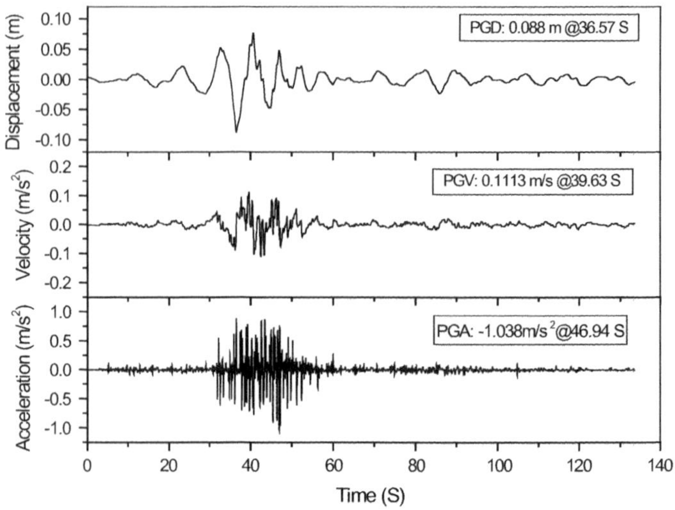

Fig. 11.6 Corrected acceleration and derived velocity and displacement time histories recorded at Ahmedabad during Bhuj earthquake of 2001 (component: N 78°E)

is presented in Fig. 11.8. Below this depth, very stiff weathered rock (N > 50) was encountered. Generally, the hard stratum is situated at a depth of more than 30 m. The groundwater table was inferred to be at around 2 m below ground level. It can be seen that the N values as low as 2–20 were recorded showing low strength and poor deformation characteristics of soil. Empirical equations of Japan Road Association (JRA) were used to compute the shear wave velocity [50]. Equivalent Linear Analysis [43] was performed and the results are presented in Fig. 11.9 and Table 11.1. The amplification between ground motion and motion at the base in frequency domain is shown in Fig. 11.9. It can be seen that the amplification as high as 13.3 was observed at 1.49 Hz. This suggests that the average natural frequency of 32 m deep soft soil was around 1.49 Hz. Table 11.1 compares the natural frequencies of soil and structure with the predominant frequency of earthquake. It is interesting that all the frequencies fall within a close range (1.20–2.11 Hz). Hence, it was inferred that site amplification and near resonance conditions were the two major causes for failure of some structures in the city of Ahmadabad in addition to many other causes already reported. This phenomenon, in addition to soft soil behavior seemed to be responsible for the vulnerable response of the entire system causing collapse of many multi storeyed buildings. Other factors such as bad workmanship, bad design, poor material used, illegal construction, etc. may have contributed to the disaster.

The above case histories help us to make inferences that thickness and stiffness of overburden, type of structure, number of layers of overburden, etc. play significant role in amplifying the ground motion at the surface. If the frequency of ground motion nearly coincides with the natural frequency of ground at a site, near resonance effect can be felt at sites, especially when shaking is for long duration causing widespread

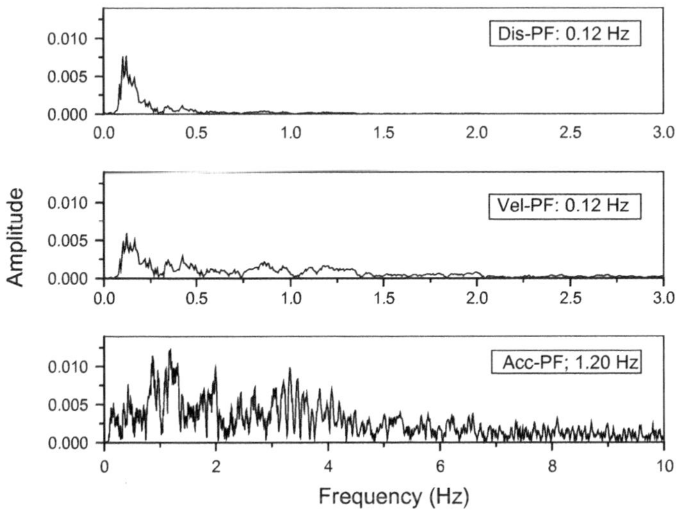

Fig. 11.7 Predominant frequencies of acceleration, velocity, and displacement time histories (component: N 78°E)

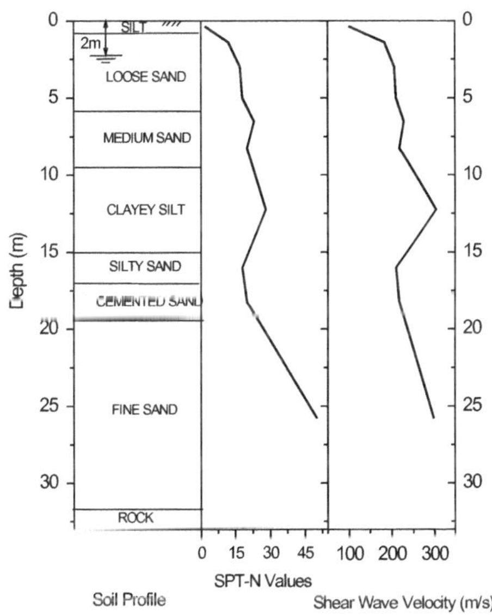

Fig. 11.8 Soil profile and variation of SPT-N and shear wave velocity with depth

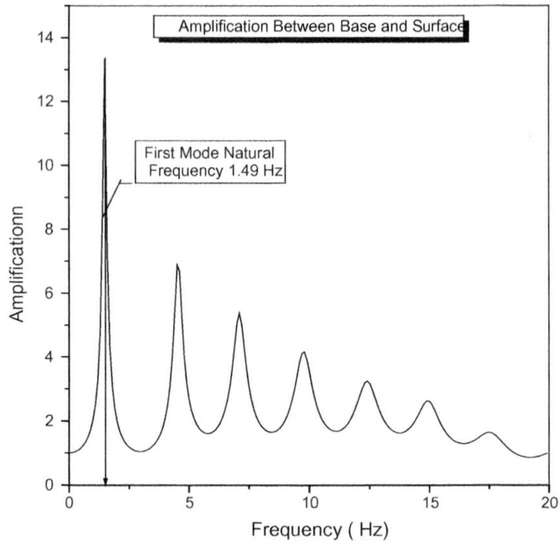

Fig. 11.9 Amplification between surface and base motion in frequency domain

Table 11.1 Natural frequencies of ground, structures, and predominant frequencies of earthquake

Natural frequency of ground estimated from SHAKE		1.49 Hz
Natural frequency of five to seven storeyed R. C. Structures		1.32–2.11 Hz
Predominant frequency of 2001 Gujarat earthquake felt at Ahmadabad	N 78°E component	1.20 Hz
	N 12°W component	1.40 Hz

damage to the structures founded on them. These effects of amplification in ground motion and resonance-like phenomenon can be termed as site effect [38].

Further, there is a possibility that one or more layers of soil may liquefy leading to reduction in strength and stiffness of soil. In the worst case, the strength of soil in that layer may become zero. Increase in excess pore water pressure, increase in strain level, effect of vibration, inertial effect are responsible for the ground to soften leading to bearing capacity failure, settlement, changes in lateral earth pressure among many.

There are a number of mathematical formulations, mostly based on wave propagation theory to assess the amplification in ground motion at the surface based on the input motion at some depth below in hard rock. The analysis can be for level ground or ground with topographical changes (basin or mound), for homogeneous or horizontally layered profiles, for one-, two- or three-dimensional idealizations, etc.

In spite of many sophistic formulations, there is considerable scatter in the results from different analyses and huge gap exists between actual behavior at site and predicted data suggesting that there is scope for further studies in this area. The following are some of the issues related to one-dimensional seismic response analysis using equivalent linear analysis.

11.3 Consideration of Continuous Variation in Soil Properties Instead of Abrupt Change in Properties as in Layered Soil

In has been well established that soil profile can be idealized as a number of horizontal layers, with properties in each layer being homogeneous. The geological formation of soil is also gradual and fairly uniform permitting such an idealization. However, it is difficult to believe that soil properties change abruptly between layers in natural soil. It should be noted that the properties change gradually between layers especially when the formation is due to geological process and gradual. Hence, it may be appropriate to idealize the non-homogeneity of the surface deposit to continuous variation of stiffness and density rather than distinctly layered formation on many occasions. Recognizing this fact, many investigators have attempted to treat the prevailing condition of non-homogeneity and computed the dynamic response of the deposit subjected to harmonic base excitations. Among these, important contributions are from Gazetas [11], Towhata [46], Roviths et al. [42], Vrettos [53] etc. Hence, there is scope and need for improving equivalent linear analysis for predicting seismic ground response. The analytical and numerical studies, reported in literature, have pointed out that main drawback of one-dimensional modeling of the ground arises out of the assumption of perfectly horizontal layers of varying depths. This assumption leads to unrealistic and contrasting impedance ratio between adjacent layers. In turn, this may result in poor simulation of ground response. Hence, there is a need for overcoming this deficiency by modeling the ground profile with continuous variation of soil properties, particularly shear wave velocity, which introduces a smooth transition zone at layer interfaces instead of abrupt variation.

It has been well established that soft soil is more hazardous during earthquakes than firm ground [33]. The maximum acceleration experienced by the ground is affected by the surface topography and frequency of incident wave in addition to the soil type. Past experience revealed that soft deposits undergo greater maximum acceleration than the hard ground. Towhata [46] illustrated the distribution of maximum acceleration recorded during the 1989 Lomapreita earthquake in California and showed that the recorded acceleration was greater on the soft coastal deposit and smaller in the hilly areas made of stiffer soils. He also calculated maximum acceleration at many sites in central Tokyo using an equivalent linear technique in the frequency domain considering layered deposit and showed that maximum acceleration was large on the firm deposits, smaller in the alluvial deposits and the smallest in the reclaimed areas. He inferred that the maximum acceleration in central Tokyo area analyzed from equivalent linear technique considering layered ground contradicted the observed behavior in some cases.

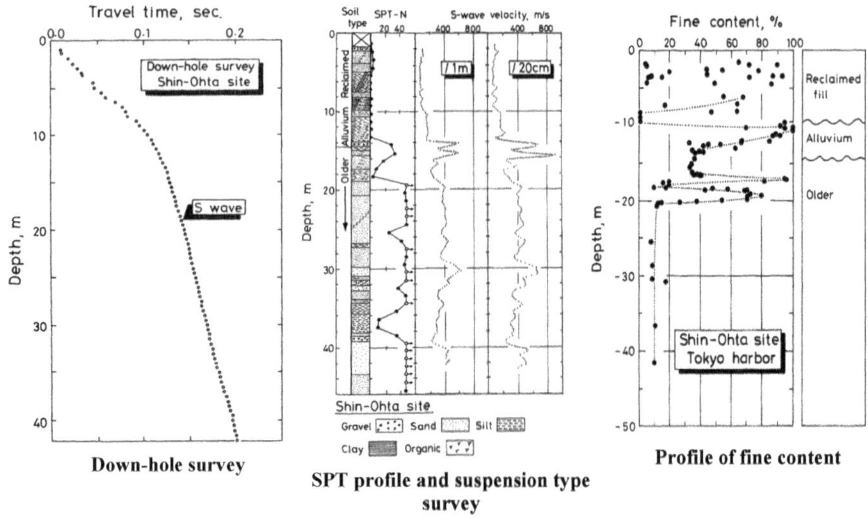

Fig. 11.10 Geotechnical investigation at Shin-Ohta site in Tokyo Harbor. *Source* Towhata [46]

Natural soil deposit shows continuous variation of deformation characteristics with depth, though not uniform. As the deposition of natural ground is by sedimentation, type of materials deposited change gradually except in case of occasional flood deposits. Hence, the change of soil properties with depth is generally continuous. Figure 11.10 shows the results of downhole survey at Shin-Ohta site in a reclaimed island in Tokyo harbor [46]. In the analysis considering layered deposits, it seems a common practice to divide the top 40 m into two layers—top layer from 0 to 10 m at Vs = 94 m/s and the remaining layer from 10 to 40 m at a Vs of around 388 m/s. This results in an impedance ratio of more than 4 between the layers assuming constant unit weight. The conventional analysis demonstrates that more than 30% of incident energy is reflected back. However, the variation in modulus is more continuous around 10 m depth and the use of layered ground concept may result in a high impedance ratio between layers, thereby reducing the energy transferred to the surface. The figure also indicates the SPT profile and suspension type survey profile at Shin-Ohta site. Though SPT results show sudden variation between two consecutive points (example around 13 and 18 m depths), Vs measurements show more continuity. This fact was better established from Vs measurements at 20 cm center to center spacing. Further, it can be observed from the figure that the fine content in the reclaimed land is randomly distributed while it follows a continuous and definite distribution in the deeper natural ground. Therefore, assumption of continuity in soil properties with depth in natural deposits is more rational than dividing the ground into layers.

In layered ground idealization, properties of ground in a layer are assumed to be constant. If the soil modulus is evaluated from SPT results, general tendency is to consider two different layers whenever there is a sharp variation in the measured

N values. This results in large impedance ratio between the layers. Hence, analysis using layered soil assumption shows the reflection of larger magnitude of wave energy at the interface, allowing a fraction of energy to be incident to the surface. Extent of reflection increases with the increase in the discontinuity of impedance. Therefore, during the analysis, if large discontinuity is considered between the layers, it may result in underestimating the displacement at the surface. Hence in dynamics, considering continuous variation of soil properties compared to layered ground idealization may be appropriate.

11.4 Computation of Effective Strain to Update Dynamic Properties of Soil

Equivalent linear analysis utilizes effective strain to update the soil properties after every iteration and these updated shear moduli and damping values are employed in the next iteration. Throughout a particular iteration, these values remain constant. There is no technically rational procedure available to convert the resulting maximum strain to effective strain. In their computer program SHAKE91, Idriss, and Sun [19] computed effective strain as equal to the product of R and maximum strain where R is an empirical parameter which depends on magnitude of the earthquake (M) as $R = (M - 1)/10$, while Schnabel et al. [43] recommended a value of 0.55 to 0.65 for R in their program SHAKE. It is observed that single value of parameter R is not capable of reproducing the entire response history at complete range of frequencies. Several other methods for computing R have been proposed. However, they are aimed at altering R depending on excitation frequency instead of acceleration or strain amplitude [22, 45, 58]. Hence, evolving a rational procedure to compute effective strain based on acceleration or strain amplitudes of the corresponding iteration is imperative.

As mentioned earlier it is customary to set the value of R based on magnitude (M) of earthquake. For the earthquake magnitude ranging between 4 and 9, R value would vary from 0.3 to 0.8. To illustrate the effect of R value used in equivalent linear analysis on the overall response of the soil deposit, a soil deposit of 30 m thickness with continuous variation of shear modulus (G) given by $G(z) = A(z_0 + z)^n$ with $G(z = 0) = 18,000$ kPa, $G(z = 30\,\text{m}) = 162,000$ kPa and $n = 4$ is considered. The nonlinear analysis is carried out till convergence of strains is achieved. Figure 11.11 shows the amplification transfer function between surface and base of the deposit considered above. Figure 11.12 shows variation in maximum acceleration along the depth for various values of R. In both of these figures, effect of R value chosen for the analysis on computed response quantities is evident.

A new method was proposed [51], which is based on the level of shaking a particular layer experiences in each of the iteration. In the past, several popular relationships relating Peak Ground Acceleration (PGA) and Modified Mercalli Intensity (MMI) have been developed. Some of these are site-specific while others can be used

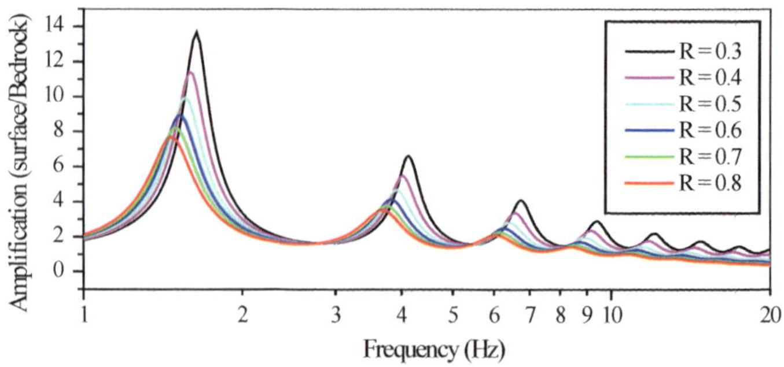

Fig. 11.11 Effect of R value on amplification between surface and bedrock (after eight iterations)

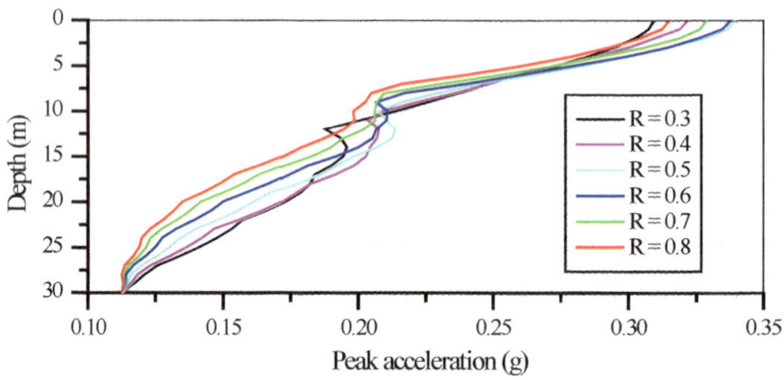

Fig. 11.12 Effect of R value on peak acceleration across the thickness of ground (after eight iterations)

universally. Compilation of such relationships between PGA and MMI are given by Linkimer [26]. These prediction equations are plotted in Fig. 11.13. It can be observed in Fig. 11.13b that mean trend of these prediction equations almost exactly match with the most popular equation proposed by Trifunac and Bardy [48]. Knowing peak acceleration at a layer during any iteration, it is possible to estimate MMI, which in turn can be used to arrive at the value of R, as $R = (MMI - 1)/10$.

For the same soil deposit considered previously, the analysis was carried out using both the methods, i.e., analysis with constant value of $R = 0.50$ and continuous profile analysis by computing R value using this alternative method. In this method, the value of R is updated at the end of each iteration for every layer unlike traditional method wherein R value remains constant during all iterations for all the layers. Figure 11.14 depicts the variation in R computed in each of the iterations based on intensity of shaking experienced by that layer. Thus, for every iteration soil properties in each of the layer are updated more realistically based on prevailing level of shaking in the previous iteration.

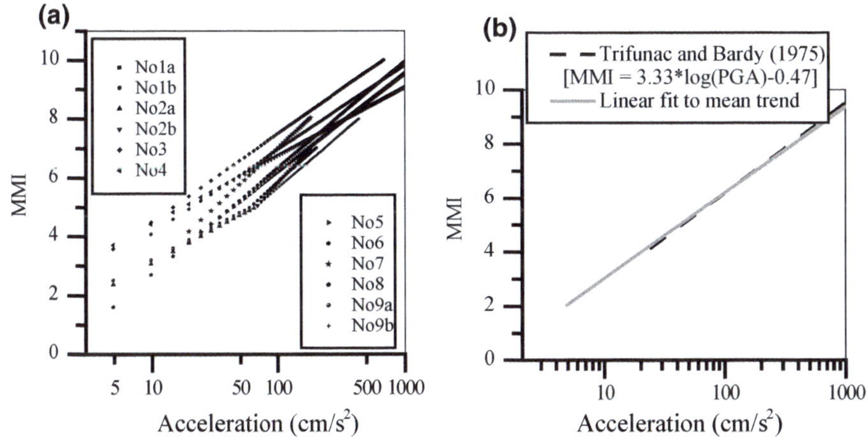

Fig. 11.13 **a** Peak ground acceleration and modified Mercalli intensity correlations. **b** Comparison of mean trend of empirical correlations and Trifunac and Bardy [48] correlation

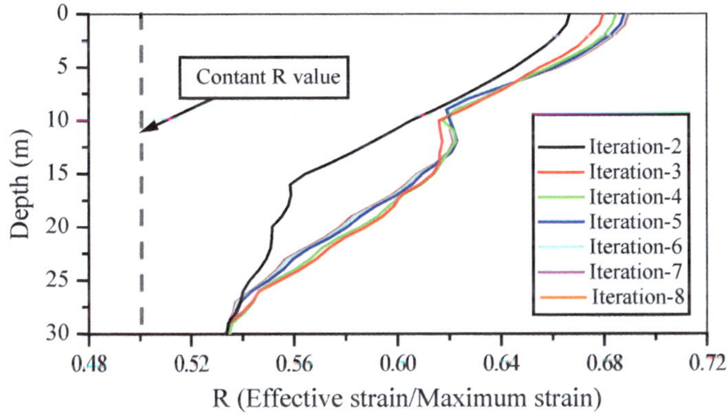

Fig. 11.14 Variation R value over each of the eight iterations and thick dashed line represents the constant R value used in routine equivalent linear method

11.5 Frequency-Dependent Soil Properties

It has been observed that equivalent linear analysis fails to simulate the measured ground motions and in general the analysis does not yield consistent results under weak and strong ground shakings. Generally, the response obtained from Equivalent Linear (EQL) analysis is overestimated in the vicinity of fundamental resonant frequencies and attenuated at higher frequencies. Equivalent linear analysis is iteratively employed with the help of strain-dependent modulus degradation and damping curves. These curves for particular type of soil are essentially obtained from laboratory tests. Earlier methods of laboratory procedures had several limitations which lead to disputable understanding of frequency dependence of dynamic soil properties. In recent times, some researchers have succeeded in recognizing the effect of loading frequency on dynamic soil properties. However, many issues are yet to be addressed in this regard; particularly, range of frequency and its quantitative effect on energy dissipation characteristics of soil.

The routine laboratory methods of determining dynamic soil properties have indicated that dynamic deformation characteristics were observed to be almost insensitive to loading frequency within the frequency range that is vital in seismic response analysis. Determination of dynamic soil properties in the laboratory are usually conducted using sinusoidal excitation of soil specimen. Hence, very often the investigator has limited scope to simulate the actual field conditions, particularly, the seismic event. In case of earthquake loading, soil is excited by multiple frequencies which are very much difficult to reproduce in routine laboratory setups. Among others, resonant column and cyclic triaxial tests are the most popular means of determining dynamic soil properties in the laboratory. These tests are performed at contrastingly different strain amplitudes and frequencies [56]. Thus, the comparison between the results from resonant column and cyclic triaxial tests should be done at similar frequencies. In resonant column tests, the resonant frequency of the specimen is used to evaluate its dynamic properties. Also, conventionally, in case of resonant column test the analysis of test results is carried out by assuming the test specimen is an undamped elastic system to evaluate shear modulus. Then, independently using shear modulus, damping ratio is computed from free vibration or half-power bandwidth method or transfer function method. Hence, it is impossible to evaluate frequency dependent dynamic properties using conventional resonance tests.

Owing to reasons cited above with regard to routine laboratory testing procedures, results presented by many researchers indicated that loading frequency in the range of 0.1–250 Hz had no significant effect on the shear modulus of both clayey and sandy soils [16]. On the other hand, some have (e.g., [40]) concluded that strain rate or loading frequency significantly affects the dynamic properties of clays but its effect is potentially insignificant in case of sandy soils. An important contribution concerning frequency dependent dynamic soil properties published by Isenhower and Stokoe [22] questioned the internal consistency of resonant column tests results under variable strain rates used to measure the stiffness of the soil at prescribed strain level. Their results obtained using cyclic torsional shear tests clearly indicated that, for medium plasticity silty clay, measured shear modulus increases with increase in strain rates.

Several researchers, through experimental evidences, discuss the effect of frequency/strain rate on the dynamic properties of the soil [3, 8, 25, 28, 44, 49, 54, 55, 60 and others]. According to Shibuya et al. [44], the damping ratio decreases in the low-frequency range, increases in the high-frequency range and it is almost constant in the medium range of frequency as conceptually represented in Fig. 11.15. Zovoral and Campanella [60], using the results of limited number of tests with resonant column and torsional cyclic tests, indicated that the increase in stiffness with strain rate/frequency is significant in case of plastic soils, while in sands it is much smaller.

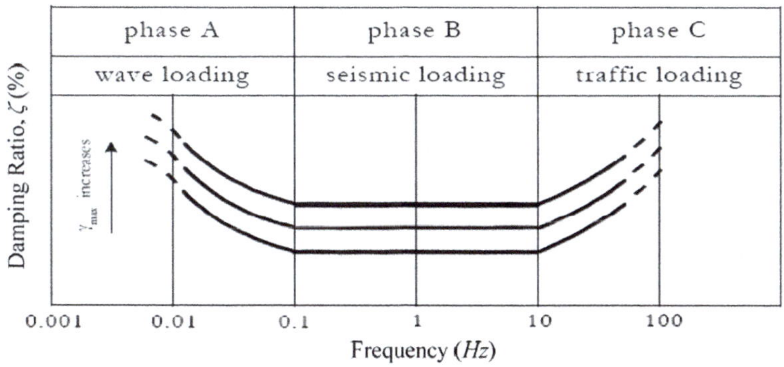

Fig. 11.15 Frequency dependence of energy dissipated within a soil mass [44]

They concluded that frequency dependence of damping property is negligible for both cohesive and cohesionless soils.

Laboratory investigation regarding effect of rate of strain on dynamic soil properties by Matesic and Vucetic [30] corroborated the results of Zovoral and Campanella [60] through more detailed and systematic study. They reported that stiffness at very small strains does not seem to be affected by the strain rate in low-plasticity soils. However, in case of medium to high plastic soils an increase of the stiffness at very small strain with increasing strain rate is observed. Also their results clearly indicated that, at higher strain levels, the stiffness reduction curves are affected by the strain rate, soil stiffness generally increases with increasing strain rate. It is important to note that all these tests were conducted at relatively small strain levels in the order of less than 0.1%/s. Their main objective was to evaluate shear strain modulus factor, α_G defined as ratio of increase in shear modulus and increase in logarithm of rate of strain ($\dot{\gamma}$) that is, $\alpha_G = (\Delta G)/(\Delta \log \dot{\gamma})$ as discussed in Isenhower and Stokoe [21].

Effect of loading frequency or strain rate on damping ratio is yet to be ascertained clearly. However, as mentioned earlier, Shibuya et al. [44] have confirmed that effect of loading frequency on damping ratio is evident. But, the researchers are still debating about general trend that can model dependency of damping property with respect to loading frequency. Rix [41], through experimental study and followed by analytical verification, reaffirmed the general trend of effect of frequency on damping suggested by Shibuya et al. [44]. Results presented by Rix [41] with respect to shear modulus agree well with that of other published results, that is shear modulus increases with frequency of loading at all strain levels. According to Bray et al. [3] and Gookin et al. [13] the damping ratio systematically decreases with increasing frequency. On the other hand, a series of cyclic torsional shear tests on Ottawa sand conducted by Lin and Huang [25] revealed that the internal damping of the soil at various strain levels increased linearly between 2.5 and 8% in the frequency range of 0.1–20 Hz. Khan et al. [23] reported that there is increase in damping in low frequency range (<60 Hz), while it decreases in frequency range of 60–150 Hz and it becomes independent for frequencies greater than 150 Hz. Hence, in the studies reported in literature, the trend of dependency of damping on frequency is put up with contradictory results. Finally, to substantiate frequency effect on dynamic properties of soil, Carvajal et al. [5] described a procedure of estimating the error between measured and computed ground response using system identification procedure. The error thus estimated is attributed to frequency dependence of shear modulus and damping. Accordingly, these properties are adjusted as frequency dependent properties so as to match the measured response. The trend is shear modulus increases with increase in frequency and no such particular trend is observed for damping but it is observed that damping ratio generally has a decreasing trend with increase in frequency.

In order to show that both nonlinear and frequency effects are significant in ground motion prediction, Idriss [17] considered the difference in site amplification of soft soil profile subjected to ground motions of different magnitude ($M = 5.5$ and $M = 7$) earthquakes scaled to same peak ground accelerations. The difference in ground response for the input motions of same PGA but derived from earthquakes of different magnitude, is mainly attributed to frequency content of the ground motions. Cameron and Green [4] studied the effect of near field and far field earthquakes at the same location on site amplification. Using the data available they separated the sites into groups namely, near field and far field based on hypocentral distances of less than and greater than 50 km, respectively. The near field motions are considerably more erratic and richer in high frequencies than far field motions. From their study Cameron and Green [4] observed that, in case of near field sites site amplification is considerably less than site amplification in case of far field sites. Hence they concluded that frequency effect on site amplification is significant. In equivalent linear method of analysis, damping is kept independent of frequency of excitation. This has serious consequence on the computed results, even though computed response closely matches with that of observed data in case of convolution process (case of input motion being prescribed at base of the soil deposit), the high-frequency response at the free surface results in unrealistically low spectral values. In case of de-convolution process (i.e., case of prescribing input motion at the surface) the response solutions of equivalent linear model diverge when the input motion is of high amplitude or rich in high frequencies [22].

Prasad [36] has shown that the modulus is small and damping is large during shaking at low confining pressures in laminar box tests on shaking table at 1 G environment. It was also observed that the damping increases with the increase in frequency of loading. Even the area of stress–strain loop, which represents the energy dissipation in a cycle, was observed to increase with an increase in frequency. Prasad opined that, perhaps the frequency effects are not visible at small strain levels (as observed in the field tests and resonant column tests or cyclic shear tests at low frequency). At larger strain levels, this effect is more. One possible reason for this is that at large strain levels, magnitude of acceleration will be very high. Increase in frequency increases the magnitude of acceleration further. Higher acceleration on the particles of soil reduces the modulus due to inertia effect. Based on the observations, he concluded that both inertia effect and frequency effect were visible in the tests carried out. However, it was difficult to separate the two effects clearly.

Effects of inertia are to reduce the modulus and increase the damping, whereas the effects of frequency are to increase both modulus and damping. Though the absolute effect of inertia was to reduce the modulus considerably, additional reduction in modulus due to increase in frequency (which increases the acceleration) was probably compensated for by the frequency effect. Thus, we can conclude that though effect loading frequency on soil properties is debatable, in order to improve the accuracy of response prediction of EQL method, particularly in the high-frequency range, it may be appropriate to incorporate frequency effects into the analysis procedure.

The reasons for unrealistic frequency response of EQL analysis is mainly attributed to employing constant values of shear modulus and damping throughout a particular iteration for all frequency ranges. Since the amplitudes of strain associated with low and high-frequency responses are distinctly different, apparently far from actual values of shear modulus (G) and damping ratio (ζ) are assigned when they are obtained for single value of effective strain. Following reasons elucidate the cause for these discrepancies:

- The maximum acceleration response is the contribution of high-frequency response, hence the displacement and thus strain at these frequencies is smaller.
- The effective strain computed using constant R value is lower than the strains resulting from low frequency response and higher than the strain level corresponding to high-frequency oscillations. Hence higher value of G and small value of ζ are assigned to compute response at low frequency range.
- In the high-frequency range the strains are smaller than effective strain used to obtain updated soil properties from their reference curves, obviously G is underestimated and ζ is overestimated.
- Although, high value of G is expected to result in stiff response of low magnitudes of strain and vice versa is anticipated in high-frequency range, the effect of damping in frequency response is significant to give enhanced and attenuated responses at low- and high-frequency ranges, respectively.

Therefore, in order to obtain more realistic responses at different ranges of frequencies, the damping values must be altered pragmatically. Incorporating frequency dependent damping and shear modulus may not resolve the problem at hand because the studies reviewed in this aspect have clearly indicated that the damping increases in the frequency region of concern. Though studies have indicated that there is significant effect of loading frequency on these properties, the database is too limited to accommodate it in routine EQL analysis.

11.6 Natural Period of Layered Ground

The current state of research in the field of earthquake engineering mainly focuses on modeling the seismic site effects in an appropriate and simplest possible manner. Hence in the recent past, evolving reliable method to assess site effects is the primary task of earthquake geotechnical engineering research throughout the world. One of the simplest analytical methods of assessing the site effects, in terms of amplification corresponding to fundamental frequency of the soil deposit overlying bedrock, is to consider the heterogeneous layered soil deposit overlying the bedrock as one-dimensional homogeneous layer having equivalent constant shear wave velocity.

Often in practice, one-dimensional seismic response analysis of inhomogeneous soil deposit is carried out for the case of vertically incident shear waves propagating through horizontally layered soil deposit. Through several studies on free vibration characteristics of two dimensional soil deposits, it has been recognized that the fundamental period computed from one-dimensional analysis fairly represents the fundamental period of wide and shallow sediment filled valleys of low shape factor (ratio of depth of the valley to its half width is less than 0.3) [34]. Moreover, it has been established by Bard and Bouchon [2] that the fundamental period of two-dimensional soil deposits may be approximately related to one-dimensional fundamental period of the deposit at the valley center and its shape factor.

Most of the international codes of practices prescribe average shear wave velocity of top 30 m of soil deposit (v_{s30}) as parameter for site classification [9, 10]. On the other hand, some of the seismic design codes such as building laws of Japan adopt natural period of soil deposit as the criteria for seismic site classification [29]. Essentially, based on these site classification data, codes will provide soil amplification or attenuation factors (scaling factors) to arrive at scaled elastic design response spectra which are compatible with local soil conditions. In this regard, some related studies have indicated that fundamental natural period of the surface layer above the bedrock is more appropriate parameter to predict amplification ratios than v_{s30} criteria particularly for long period deposits [31, 59]. It may be pointed out that such a rational method of computing design forces by appropriately accounting for local soil condition is absent in Indian code of practice IS-1893 [20]. Currently, Indian code classifies foundation soil into four categories depending on range of SPT-N values and type of soil. There is immediate need for updating Indian seismic design codes commensurate with other national codes by incorporating local site effects in computation of design forces. For this purpose, fundamental natural period of the soil deposit may be used as key parameter because it accounts for both stiffness and depth of the surface soil deposit above the bedrock. Hence reliable assessment of fundamental period of the soil deposit is an important primary step in seismic site characterization particularly in the process of microzonation of urban area.

Perceiving its importance in seismic site characterization tasks, many methods have been formulated and proposed to estimate reliably the fundamental period of one-dimensional horizontally layered soil deposit. While implementing these methods, it is essential to idealize the soil deposit as a layered profile as far as possible closely representing the actual shear wave velocity distribution. Uncertainty is evident in idealizing the actual profile into an equivalent layered system because of complex variability of soil properties along the depth. As pointed out earlier, the soil deposit may exhibit distinct continuous variation of soil properties depending upon its genesis, stress history and other related process. However, questionably, even such deposits displaying distinct continuous variation in soil properties are approximated as layered system. In such situations, employing the methods derived for layered profiles may be inappropriate for estimation of fundamental period.

Methods based on weighted average of shear wave velocities of the layered soil profile are widely employed in practice to compute fundamental natural period. There are methods which are accurate and more reliable than weighted average methods, but they are iterative in procedure and hence tedious. Consequently, they are unpopular for quick estimation of fundamental period of soil deposits.

Since shear wave velocity and fundamental period are directly interrelated, many empirical relationships have been proposed to compute average shear wave velocity as an equivalent substitute to complex shear wave velocity profile of inhomogeneous layered deposit. Summary and relative comparison of all empirical methods available till then are given in the paper by Dobry et al. [7]. Among these methods weighted average shear wave velocity method is most commonly employed in practice. For a homogeneous soil deposit of total thickness H and constant shear velocity v_s, the fundamental period is given by,

$$T = 4H/v_s.$$

In case of layered soil deposit, average shear wave velocity, v_s is obtained to compute fundamental period. Usually, for a soil deposit of N layers with shear wave velocities v_{si}, and corresponding layer thicknesses H_i, $\bar{v}_s$ is calculated as weighted average of the layer shear wave velocities from,

$$\bar{v}_s = \frac{1}{H}\sum_{i=1}^{N} v_{si} H_i$$

Then, fundamental period is computed using $\bar{v}_s$ instead of v_s. Thus,

$$T = \frac{4H^2}{\sum_{i=1}^{N} v_{si} H_i}$$

Madera [27] proposed an alternative method in which fundamental period obtained for two-layered system overlying bedrock is used repeatedly to analyze multilayered soil deposit. This procedure based on closed form analytical solution is comparatively reliable but it is tedious compared to simple weighted average shear wave velocity method. Dobry et al. [7] have proposed a non-iterative alternative approach basically derived from exact Rayleigh's procedure. For the profiles considered in their study, they noted that Rayleigh's iterative procedure usually converges, within acceptable error range (within ±3%), to fundamental period in the first iteration itself. Based on this observation, a set of closed-form equations was formulated for computation of the fundamental period as obtained in the first iteration of Rayleigh's equation.

An alternative method for computing natural fundamental period of layered ground was proposed by Vijayendra et al. [52]. The proposed method primarily involves approximating the layered shear wave velocity profile with an equivalent linearly varying shear wave velocity profile. The closed form exact analytical solution is used to compute the fundamental period of the deposit of linearly varying shear wave velocity profile. For any layered profile with distinct shear wave velocity for each layer, curve fitting process using regression analysis can be employed to get an equivalent linear fit. Thus, obtained curve fitting constants, consisting of nonzero shear wave velocity at the surface and rate of change of shear wave velocity along the depth, are used to derive an approximate relationship for the estimation of fundamental period of the layered profile as

$$T = \frac{2\pi H}{v_{s0}(0.324 + 1.254\mu^{0.853})}$$

where, v_{s0} and v_{sH} are the shear wave velocities at the surface (ie., $z = 0$) and at the bottom (ie., $z = H$) of the soil deposit obtained as curve fitting parameters for linearly varying shear wave velocity profile representing the layered profile. $\mu = v_{sH}/v_{s0}$ is the ratio of base to surface shear wave velocities. It has been shown that maximum error in computation of fundamental period of deposit having linearly varying shear wave velocity profile using the above equation is about 0.25%.

In order to ascertain the efficiency of the method proposed by Vijayenda et al. [52] over other methods, fundamental period computed from proposed method is compared with that obtained by other methods such as weighted average method, simplified Rayleigh's method and with more rigorous Madera's approach [7]. Also, for the purpose of comparing these estimated results with realistic and exact values, fundamental periods were computed using amplification function of transformation matrix approach (SHAKE analysis).

In order to validate this method, data pertaining to four instrumented geotechnical downhole arrays was employed. They are La-Obregon Park, La-Cienega, Eureka Somoa, and El Centro-Meloland Geotechnical Arrays. These geotechnical downhole arrays were established as part of a project on site amplification studies by the California Strong Motion Instrumentation Program (CSMIP) in California, USA. Geotechnical data available at these stations include suspension logging shear wave velocity profiles. The data related to shear wave velocity profile and earthquakes recorded by these instrumented sites are available in CSMIP website. More details about these arrays are given in Graizer et al. [12] and references cited therein.

The fundamental period computed from representative earthquake data recorded at each of these four arrays is reaffirmed by comparing these results with periods corresponding to first peak of the amplification transfer function obtained for the corresponding site using SHAKE. The natural period computed from theory of multiple reflections of waves employed in SHAKE is almost same as the exact solution [7]. Amplification transfer function between surface and depth H is obtained for low damping value of 2% for all four sites and is as shown in Fig. 11.16. Also shown are the periods corresponding to peaks representing fundamental period of the particular soil deposit. The amplification shown in Fig. 11.16 is normalized using respective maximum values such that amplification ratio is one for all the cases.

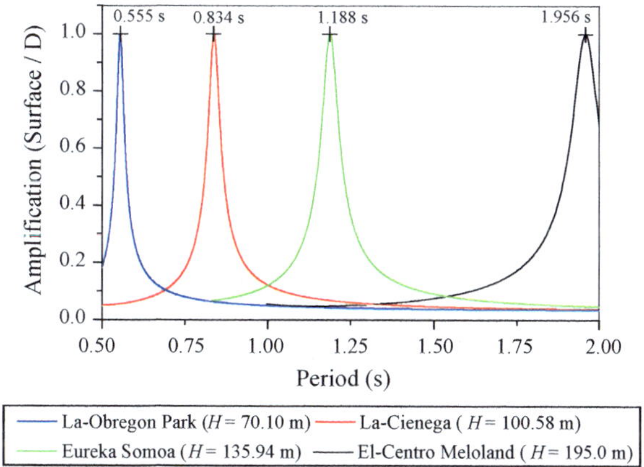

Fig. 11.16 Fundamental period computed from amplification transfer function of SHAKE program

Table 11.2 The fundamental periods computed from different methods

Soil deposits–geotechnical arrays	$T^{(1)}$ (s)	$T^{(2)}$ (s)	$T^{(3)}$ (s)	$T^{(4)}$ (s)	$T^{(5)}$ (s)
La-Obregon Park	0.568	0.559	0.548	0.531	0.555
La-Cienega	0.930	0.850	0.822	0.853	0.834
Eureka Somoa	1.341	1.248	1.136	1.213	1.188
Elcento Meloland	2.184	2.011	2.597	2.020	1.956

$T^{(1)}$—Weighted average method
$T^{(2)}$—Simplified Rayleigh's Method [7]
$T^{(3)}$—Madera [27] method
$T^{(4)}$—Method proposed by Vijayendra et al. [52]
$T^{(5)}$—SHAKE analysis

Fundamental period computed from different methods including proposed method along with result of SHAKE analysis are tabulated in Table 11.2. As noted earlier, the values obtained from SHAKE analysis closely represent exact results.

Figure 11.17 compares the results obtained from different methods including new method proposed by Vijayendra et al. [52] with respect to exact value of fundamental period computed from SHAKE analysis. Percentage error in the fundamental period computed using different methods, including newly proposed method of this study, with respect to results obtained from SHAKE analysis $\left(T^{(5)}\right)$ is presented using bar graph in Fig. 11.18. The newly proposed alternative method presented here, has been able to estimate the fundamental period $\left(T^{(4)}\right)$ which closely agrees (less than 5% error) with the exact values in case of all the four deposits considered in this study. The error associated with $T^{(4)}$ compared to exact value $T^{(5)}$ are 4.32%, 2.28%, 2.10%, and 3.27%, respectively for La-Obregon Park, La-Cienega, Eureka Somoa, and El Centro-Meloland soil deposits.

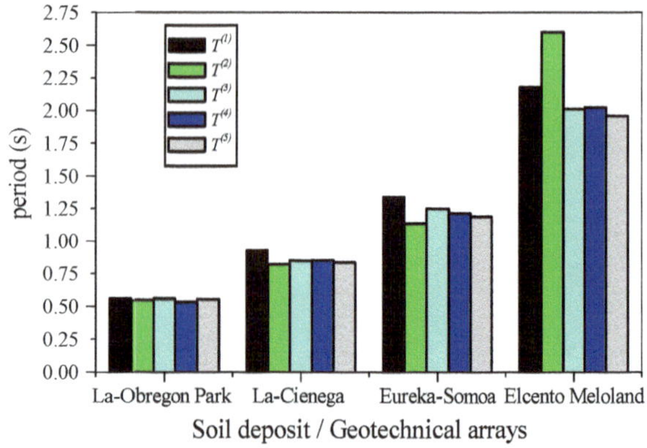

Fig. 11.17 Comparison of fundamental periods computed from different methods

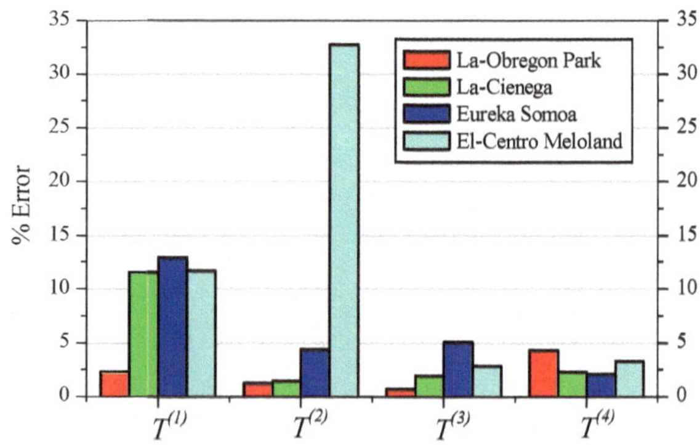

Fig. 11.18 Percentage error in fundamental periods computed from different methods

11.7 Shake Table Test on Performance of Model Structure in Soft and Dense Ground

Saturated ground always poses problems during earthquake. One extreme situation is when either the entire overburden or a part of it liquefies. One method of understanding the ground behavior for the purpose of site characterization is by model testing to simulate the actual condition in the laboratory.

Model testing in earthquake geotechnical engineering is very expensive. It involves (i) A large sized container that houses the model ground with structure, (ii) A shaking table that simulates earthquake by creating vibration on which the container is placed, and (iii) Transducers that record the response. A manual one-dimensional shaking table fabricated at Sri Jayachamarajendra College of Engineering, Mysore was used for the present work. The best features of this shaking table included absolute economy, shaking of sufficient magnitude (0.5 g level of acceleration, 50 mm double amplitude displacement at 2 Hz frequency) with minimum effort (10 kgf of manual shaking), and pay load carrying capacity of 7.5 kN. The transparent container made of acrylic plates was 1.67 m long, 0.5 m wide, and 0.6 m deep. Figure 11.19

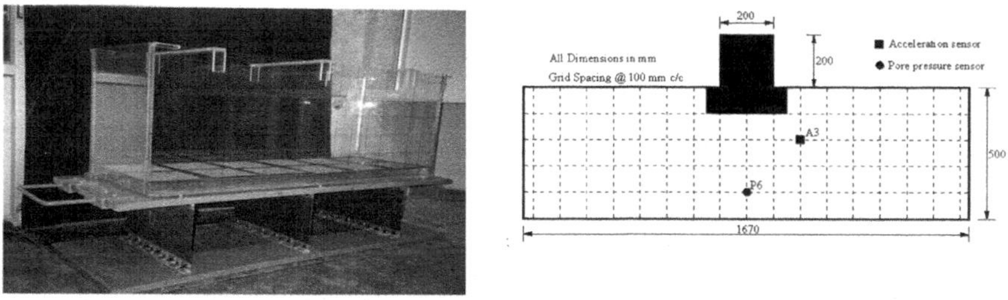

Assembly of Manual Shaking Table and Transparent Container

Layout of model ground with positions of structure and sensors

Fig. 11.19 Manual shaking table with schematic of transducer locations

shows the assembly of shaking table and acrylic container. Transducers consisted of acceleration sensors and pore pressure sensors. The sensors were light in weight (30 g), small in size (30 mm × 30 mm × 10 mm), and water tight. Signal conditioners and data loggers could record the response at 500 Hz. All transducers were placed inside the model ground so that they recorded the ground behavior better. Sand used in the present study to prepare the model ground was fine and uniform. Colored sand prepared using a black dye was utilized to form a grid of 100 mm square on the front face to present the deformed shape of ground during shaking. Model structure was made of 10 mm thick plywood. It was 300 mm × 300 mm square at the base and possessed a height of 300 mm. The inside of it was filled with sand to achieve a total weight of the structure of 210 N. The ground was prepared at different densities to understand the effects of relative density on the seismic behavior. It was saturated to consider the effect of excess pore water pressure generation and worst condition. The baffle walls, filters, and inlet assembly at the base of the container helped gradual saturation with minimum disturbance to model. Shaking was performed for 12 s. Sinusoidal one directional shaking at 2 Hz was achieved.

It should be noted that there are many changes happening in ground under the dynamic force of earthquake. Alternate horizontal force leads to stress axis transformation. Increased amplitude of motion during strong shaking will result in increased strain level. As a result, there may be rise in excess pore water pressure if the soil is fully saturated. The rigorous shaking can result in large inertial force. All these factors are responsible for causing reduction in strength and stiffness of soil. This has to be properly evaluated for accurate prediction of seismic performance. Table 11.3 and Fig. 11.20 indicate the result from shaking table test on model structure resting on foundation soil of different relative densities. Dynamic bearing capacity was always less than the static bearing capacity. The margin of difference was more at looser densities and the difference reduced when the density increased. The details are presented in Prasad et al. [39].

It may be essential to consider the effects of strength and stiffness degradation with time in one or more layers of ground depending on the situation. When some layers in level ground liquefy, there may be deamplification in ground acceleration at

Table 11.3 Static and seismic bearing capacities (q_f) of model ground at different densities

Sl. no.	I_d (%)	ϕ_{static}	$\phi_{dynamic}$	Bearing pressure (kPa)	$q_{fstatic}$ (kPa)	$q_{fdynamic}$ (kPa)
1	25.41	33.26	8.70	2.322	37.92	1.68
2	26.37	33.51	9.77	2.400	39.62	1.91
3	40.22	34.00	12.38	2.333	44.07	2.71
4	62.52	34.40	23.52	2.322	50.44	11.55
5	72.24	34.70	32.66	2.277	54.57	40.59

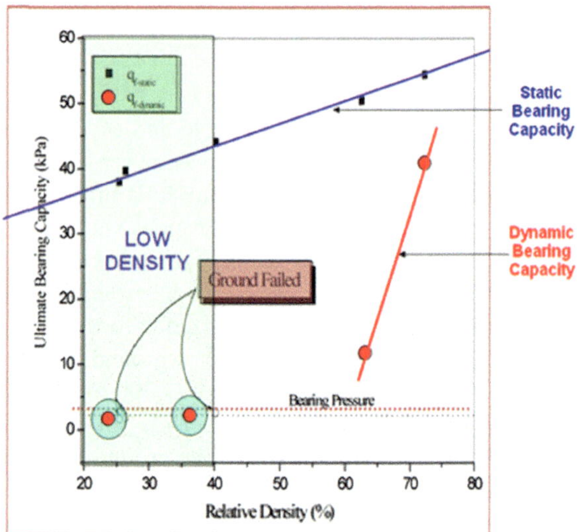

Fig. 11.20 Degradation in strength of foundation soil under dynamic load

the surface as shear waves may not be transferred through liquefied ground in level ground. However, the situation can be completely different in a sloping ground (even if it has gentle slope) when the lateral movement of soil can be catastrophic. Most available analysis procedures fail to consider this effect.

11.8 Comments on IS 1893-2016

Indian Standard code of practice for criteria for earthquake resistant design of structures has seen six revisions so far. The major changes between the previous version and the recent one with regard to geotechnical aspects are presented in Table 11.4. The main addition in the recent code is the introduction of simplified procedure for Liquefaction Analysis and classification of soil into four types instead of three. Further, no revision is made with regard to site effect. It is however accepted that there are exceptions where resonance-like conditions have been seen to occur between long-

Table 11.4 Comparison of geotechnical aspects in the recent two editions of IS 1893

Description	IS 1893 [19]	IS 1893 [20]	Remarks
Code revision	First published in 1962. It is the fifth revision	First published in 1962. It is the sixth revision	
Type of structure	Covers various structures such as buildings, elevated structures. There is no clarity on important or special structures such as refineries, nuclear power plants, large dams etc.	Provision is applicable to all structures including critical and special structures such as refineries, nuclear power plants, large dams	For critical and special structures site-specific hazard assessment shall be made. It is the minimum condition specified
Temporary structures and excavations	Need not be designed for earthquake forces	Minor or temporary structures such as parking structures, security cabins, ancillary structures, scaffolding, excavations need to be designed for earthquake forces	No clear specification is given. It may escalate the cost
Assumptions	Earthquake causes impulsive ground motions, which are complex and irregular in character, changing in period and amplitude each lasting for a small duration. Therefore, resonance of the type as visualized under steady-state sinusoidal excitations, will not occur as it would need time to build up such amplitudes	Earthquake ground motions are complex and irregular, consisting of several frequencies and of varying amplitudes each lasting for a small duration. Therefore, usually, resonance of the type as visualized under steady-state sinusoidal excitations will not occur, as it would need time to build up such amplitudes. But, there are exceptions where resonance-like conditions have been seen to occur between long-distance waves and tall structures founded on deep soft soils	There is an improvement in understanding. However, site effect is not given sufficient importance

(continued)

Table 11.4 (continued)

Description	IS 1893 [19]	IS 1893 [20]	Remarks
Soil Types	Three categories of soils are considered Type 1 includes rock or hard soil, GW, GP, SW, SP or SC having N > 30. 50% increase in ABP is permitted Type 2 includes medium soil having N between 10 and 30 and SP or GP with N > 15. 25% increase in ABP is permitted Type 3 included Soft Soil. All soils other than SP having N < 10 are in this category. No increase in ABP is permitted	Four categories of soils are considered Type A includes rock or hard soil and all soils with N > 30. 50% increase in ABP is permitted Type B includes medium or stiff soil (usually SP, ML & CL) with N between 10 and 30. 25% increase in ABP is permitted Type C includes soft soil, other than SP having N < 10. No increase in ABP is permitted Type D includes unstable, collapsible and liquefiable soil Ground improvement or pile foundation up to better soil is essential	Very poor soil (Type D) considering vulnerability under earthquake is identified and ground improvement or pile foundation is suggested compulsorily on such soil
N value	Soil is classified based on N value	Soil is classified based on N value up to Type C and seismically vulnerable soils are identified as Type D	Soil type is purely based on N and not on geologic classification, soil type, groundwater conditions or liquefaction potential for most soils. No mention on how to identify Type D soil is made. Only overburden correction is mentioned for N
Depth of soil exploration	Weighted N value from depth of foundation to twice the width in shallow foundation.	Weighted N value from depth of foundation to twice the width in shallow foundation.	Clarity on weightage is not available

(continued)

Table 11.4 (continued)

Description	IS 1893 [19]	IS 1893 [20]	Remarks
Liquefaction potential	No mention is made	Loose submerged sand (SP) having N < 10 in Zone II and N < 15 in other zones are considered vulnerable. Suggested to avoid such sites, or improve the ground or provide pile foundation Simplified procedure to identify liquefaction potential as per Seed and Idriss approach is suggested	Code does not suggest the required minimum factor of safety

distance waves and tall structures founded on deep soft soils. The empirical formulae for the determination of fundamental period for different structures continue to be the same as in the previous revision.

11.9 Concluding Remarks

Seismic Site characterization is an extremely important procedure, especially in seismically active zones and for all important structures. There are many issues on this aspect which still need to be addressed to eliminate the variations in results. The following are some important issues addressed in this paper.

- Consideration of continuous variation in soil properties at the transition between layers instead of abrupt change in properties as in layered soil idealization may be useful.
- More refined procedure for computing effective strain to update dynamic properties of soil may reduce the scatter.
- The differences between actual behavior and prediction from analysis may also be attributed to the frequency effect which is normally not considered in seismic analysis of ground.
- It is important to assess the fundamental period of ground along with the structure founded on it more precisely so as to estimate the near resonance effect during earthquakes. The available methods show significant variations.
- Consideration of reduction in strength and stiffness of ground with time, especially in liquefiable ground may be essential.
- The above results emphasize the need for microzonation of at least important cities in seismically active zones. Microzonation will enable in estimating the natural

frequency of ground at a location more accurately. It will then be possible to estimate the most vulnerable structure at a particular location.

- The present IS codal provision does not address the issue of site effect thoroughly.

References

1. Beck, J.L., Hall, J.F.: Factors contributing to the catastrophe in Mexico City during the earthquake of September 19, 1985. Geophys. Res. Lett. **13**(6), 593–596 (1986)
2. Bard, P.Y., Bouchon, M.: The two-dimensional resonance of sediment filled valleys. Bull. Seismol. Soc. Am. **75**, 519–541 (1985)
3. Bray, J.D., Gookin, W.B., Riemer, M.F.: On the dynamic characterization of soils. In: Sêco e Pinto (ed) Earthquake Geotechnical Engineering. Balkema, Rotterdam (1999)
4. Cameron, W.L., Green, R.A.: Soil nonlinearity versus frequency effects. Opinion Paper, International Workshop on the Uncertainties in Nonlinear Soil Properties and their Impact on Modeling Dynamic Response, 18–19 March, PEER Headquarters, UC Berkeley (2004)
5. Carvajal, J.C., Taboada-Urtuzuástegui, V.M., Romo, M.P.: Influence of earthquake frequency content on soil dynamic properties at CAO site. Soil Dyn. Earthq. Eng. **22**, 297–308 (2002)
6. Chiang, H.H., Chang, N.: Earthquake ground motion amplification in Mexico City. In: Proceedings: Second International Conference on Recent Advances in Geotechnical Earthquake Engineering and Soil Dynamics, 11–15 Mar 1991, St. Louis, Missouri, Paper No. 8.17 (1991)
7. Dobry, R.I., Oweis, I., Urzua, A.: Simplified procedures for estimating the fundamental period of a soil profile. Bull. Seismol. Soc. Am. **66**, 1293–1321 (1976)
8. Dobry, R., Vucetic, M.: Dynamic properties and seismic response of soft clay deposits. In: Proceedings, International Symposium on Geotechnical Engineering of Soft Soils, Mexico City, vol. 2, pp. 51–87
9. Eurocode-EC8: BS EN 1998-1: design of structures for earthquake resistance-part 1: general rules, seismic actions and rules for buildings. British Standard (BS), UK (2004)
10. FEMA-P-749: Earthquake-resistant design concepts—an introduction to the NEHRP recommended seismic provisions for new buildings and other structures. Federal Emergency Management Agency Report, By the National Institute of Building Sciences Building Seismic Safety Council, Washington DC, USA (2010)
11. Gazetas, G.: Vibrational characteristics of soil deposits with variable wave velocity. Int. J. Numer. Anal. Meth. Geomech. **6**(1982), 1–20 (1982)
12. Graizer, V., Shakal, A., Hipley, P.: Recent data recorded from downhole geotechnical arrays. In: Proceedings of SMIP2000 Seminar on Utilization of Strong-Motion Data, 12 Sept, California, pp. 23–38 (2000)
13. Gookin, W.B., Bray, J., Riemer, M.: The combined effects of loading frequency and other parameters on dynamic properties of reconstituted cohesive soils. UC Berkeley Geotechnical Engineering Report No. GT/99-14 (1999)
14. Halliday, D., Resnick, R., Walker, J.: Fundamentals of Physics, Extended, 4th edn. Wiley, 1306 pp., ISBN 0-471-57578-X (1993)
15. Hanks, T.C., Brady, A.G.: The Loma Prieta earthquake, ground motion, and damage in Oakland, Treasure Island, and San Francisco. Bull. Seismol. Soc. Am. **81**(5), 2019–2047 (1991)
16. Hardin, B.O., Drnevich, V.P.: Shear modulus and damping in soils: measurement and parameter effects. J. Soil Mech. Foundation Eng. Division ASCE **98**(SM6), 603–624 (1972)
17. Idriss, I.M.: Earthquake ground motions at soft soil sites. In: Proceedings, 2nd International Conference on Recent Advances in Geotechnical Earthquake Engineering and Soil Dynamics, 11–15 March, St. Louis, Missouri, pp. 2265–2272 (1991)
18. Idriss, I.M., Sun, J.I.: Users manual for SHAKE91: a computer program for conducting equivalent linear seismic response analyses of horizontally layered soil deposits. Center for Geotechnical Modeling, Department of Civil & Environmental Engineering, University of California, Davis, California, USA (1992)

19. IS: 1893—Part 1: Criteria for earthquake resistant design of structures. General Provisions and Buildings, 5th Revision, Bureau of Indian Standards, New Delhi (2002)
20. IS: 1893—Part 1: Criteria for earthquake resistant design of structures. General provisions and buildings, 5th Revision, Bureau of Indian Standards, New Delhi (2016)
21. Isenhower, W.M., Stokoe, K.H.: Strain rate dependent shear modulus of San Francisco bay mud. In: Proceedings of International Conference on Recent Advances in Geotechnical Earthquake Engineering and Soil Dynamics, St. Louis, Missouri, vol. 2, pp. 182–191 (1981)
22. Kausel, E., Assimaki, D.: Seismic simulation of inelastic soils via frequency-dependent moduli and damping. J. Eng. Mech. Division ASCE **128**, 34–47 (2002)
23. Khan, Z.G., Cascante, M.H., Naggar, El, Lai, C.G.: Measurement of frequency dependent dynamic properties of soils using the resonant column device. J. Geotech. Geo-environ. Eng. ACSE **134**(9), 1319–1326 (2008)
24. Kramer, S.L.: Geotechnical Earthquake Engineering. Prentice Hall International Publication (1996)
25. Lin, M.L., Huang, T.H.: The effects of frequency on damping properties of sand. Soil Dyn. Earthq. Eng. **15**, 269–278 (1996)
26. Linkimer, L.: Relationship between peak ground acceleration and modified Mercalli intensity in Costa Rica. Revis. Geól. Am. Cent. **38**, 81–94 (2008)
27. Madera, G.A.: Fundamental period and peak accelerations in layered systems. Research report R70-37, Department of Civil Engineering, MIT, Cambridge, Massachusetts USA (1971)
28. Malagnini, L.: Velocity and attenuation structure of very shallow soils: evidence for frequency-dependent Q. Bull. Seismol. Soc. Am. **86**(5), 1471–1486 (1996)
29. Marino, E.M., Nakashima, M., Mosalam, K.M.: Comparison of European and Japanese seismic design of steel building structures. Eng. Struct. **27**, 827–840 (2005)
30. Matesic, L., Vucetic, M.: Strain-rate effect of soil secant shear modulus at small cyclic strains. J. Geotech. Geo-environ. Eng. ASCE **129**(6), 536–549 (2003)
31. McVerry, G.H.: Site-effect terms as continuous functions of site period and v_{s30}. In: Proceedings of the Ninth Pacific Conference on Earthquake Engineering Building an Earthquake-Resilient Society, 14–16 April, Auckland, New Zealand (2011)
32. Moehle, J.P.: Preliminary observations on the performance of concrete freeway structures. In: National Information Service for Earthquake Engineering, University of California, Berkeley, http://www.eerc.berkeley.edu/loma_prieta/moehle.html (1999)
33. Okamoto, S.: Introduction to Earthquake Engineering. University of Tokyo Press, Japan (1984)
34. Paolucci, R.: Shear resonance frequencies of alluvial valleys by Rayleigh's method. Earthq. Spectra **15**(3), 503–521 (1999)
35. Peterson, A. (ed.): Excuses on Shaky ground. Int. Constr. **29**(3):40–41. ISSN 0020-6415 (1990)
36. Prasad, S.K.: Evaluation of Deformation Characteristics of 1-G Model Ground During Shaking Using a Laminar Box. Doctoral thesis submitted to Department of civil engineering, University of Tokyo, Japan
37. Prasad, S.K., Chandradhara, G.P., Vijayendra, K.V.: Seismic ground response at Ahmedabad during Gujarat earthquake of 2001. In: Proceedings of Indian Geotechnical Conference, IGC-2001, Indore, India, 14 16 Dec, vol. 1, pp. 336–339 (2001)
38. Prasad, S.K., Vijayendra, K.V.: Relevance of site characterization in seismic studies. Ind. J. Adv. Chem. Sci. **4**(4), 1–6 (2016)
39. Prasad, S.K., Chandradhara, G.P., Nanjundaswamy, P., Revanasiddappa, K.: Seismic Bearing Capacity of Ground From Shaking Table Tests, IGC-2004, Warangal (2004)
40. Richart, F.E.: Field and laboratory measurement of dynamic soil properties. In: Proceedings, International Symposium on Dynamical Measurements in Soil and Rock Mechanics, Karlsruhe, Rotterdam, vol. 1, pp. 3–36
41. Rix, G.J.: Frequency dependence of dynamic soil properties. Opinion Paper, International Workshop on the Uncertainties in Nonlinear Soil Properties and their Impact on Modeling Dynamic Response, 18–19 Mar, PEER, UC Berkeley, USA (2004)
42. Rovithis, E.N., Parashakis, H., Mylonakis, G.E.: 1D harmonic response of layered inhomogeneous soil: analytical investigation. Soil Dyn. Earthq. Eng. **31**, 879–890 (2011)

43. Schnabel, P.B., Lysmer, J.L., Seed, H.B.: SHAKE: a computer program for earthquake response analysis of horizontally layered sites. Report No. UCB/EERC—72/12, Earthquake Engineering Research Center, University of California, Berkeley (1972)
44. Shibuya, S., Mitachi, T., Fukuda, F., Degoshi, T.: Strain rate effects on shear modulus and damping of normally consolidated clay. Geotech. Test. J. **18**(3), 365–375 (1995)
45. Sugito, M., Goda, G., Masuda, T.: Frequency dependent equi-linearized technique for seismic response analysis of multi-layered ground. J. Geotech. Eng. Proc. JSCE **493**, 49–58 (in Japanese with English abstract)
46. Towhata, I.: Seismic wave propagation in elastic soil with continuous variation of shear modulus in the vertical direction. Soils Found. **36**(1), 61–72 (1996)
47. Towhata, I.: Personal Communication on Earthquake Geotechnical Engineering. Sri Jayachamarajendra College of Engineering, Mysuru, India (2016)
48. Trifunac, M.D., Brady, A.G.: On the correlations of seismic intensity scale with peaks of recorded strong ground motion. Bull. Seismol. Soc. Am. **65**, 139–162 (1975)
49. Vankov, D.A., Sassa, K.: Dependence of pore pressure on frequency of loading at sliding surface. In: Yagi, N., Yamagami, T., Jiang, J.-C. (eds.) Slope Stability Engineering, pp. 601–606. Balkema, Rotterdam
50. Vijayendra, K.V., Prasad, S.K.: Prediction of dynamic soil properties of soil from In situ SPT test. In: International Conference on Civil Engineering, Proceedings of ICCE 2001, Bangalore, pp. 909–916 (2001)
51. Vijayendra, K.V.: Seismic Site Response of 1-D in Homogeneous Ground with Continuous Variation of Soil Properties. Ph.D. Thesis submitted to National institute of Technology Karnataka, Surathkal (2016)
52. Vijayendra, K.V., Nayak, S., Prasad, S.K.: An alternative method to estimate fundamental period of layered soil deposit. Ind. Geotech. J. (2014). https://doi.org/10.1007/s40098-014-0121-7
53. Vrettos, C.: Dynamic response of soil deposits to vertical SH waves for different rigidity depth-gradients. Soil Dyn. Earthq. Eng. **47**(4), 41–50 (2013)
54. Vucetic, M., Dobry, R.: Effect of soil plasticity on cyclic response. J. Geotech. Eng. Division ASCE **117**(1), 89–107 (1991)
55. Vucetic, M., Lanzo, G., Doroudian, M.: Damping at small strains in cyclic simple shear tests. J. Geotech. Eng. Division ASCE **124**(7), 585–594 (1998)
56. Woods, R.D.: Measurement of dynamic soil properties. In: Proceedings of Conference on Earthquake Engineering and Soil Dynamics, Specialty conference ASCE I, pp. 91–178 (1978)
57. Yashinsky, M.: Cypress Street Viaduct. US Geological Survey Professional Paper, No. 1552-8, pp. 19–26, Library of Congress catalog-card No. 92-32287 (1998)
58. Yoshida, N., Kobayashi, S., Suetomi, I., Miura, K.: Equivalent linear method considering frequency dependent characteristics of stiffness and damping. Soil Dyn. Earthq. Eng. **22**, 205–222 (2002)
59. Zhao, J.X.: Comparison between v_{s30} and site period as site parameters in ground-motion prediction equations for response spectra. In: 4th IASPEI/IAEE International Symposium on Effects of Surface Geology on Seismic Motion, 23–26 August, University of California Santa Barbara, USA (2011)
60. Zovoral, D.Z., Campanella, R.G.: Frequency effects on damping/modulus of cohesive soils. ASTM Special Technical Publication No. 1213, D (1994)

Chapter 12
Dynamic Tunnel–Soil Interaction in Soft Soils Considering Site-Specific Seismic Ground Response

Deepankar Choudhury, Milind Patil, P. G. Ranjith and Jian Zhao

12.1 Introduction

The construction of underground structure is becoming popular with an increase in the population in urban areas to fulfill the demand of public infrastructure facilities. Most of the megacities in the world constructing tunnels as a part of transportation system or other utility projects. These tunnels lie at shallow depth and pass through a soft soil zone. Tunnelling project in soft soil requires meticulous considerations to safeguard it against all possible loading conditions. Among the considerations, one of the most important aspects is to minimize damages during an earthquake. In the past, it was assumed that underground structures are less vulnerable to earthquakes compared to surface structures. Many tunnels have been constructed in seismically active regions, such as in Japan, Taiwan, Italy, the U.S., and Mexico, without considering the effect of seismic loading. These tunnels have experienced minor to major damages during past earthquake events (1989 Loma Prieta earthquake; 1995 Kobe earthquake; 1999 Chi-Chi earthquake; 1999 Duzce earthquake; 2004 Niigata earthquake; 2008 Wenchuan earthquake; 2011 Tohoku earthquake) [1–3]. Effects of earthquake on underground structures and factors influencing seismic damages are summarized in

D. Choudhury (✉)
Department of Civil Engineering, Indian Institute of Technology Bombay,
Powai, Mumbai 400076, India
e-mail: dc@civil.iitb.ac.in

M. Patil
IITB-Monash Research Academy, Powai, Mumbai 400076, India
e-mail: milindpatil@iitb.ac.in

P. G. Ranjith · J. Zhao
Department of Civil Engineering, Monash University,
Clayton, VIC 3800, Australia
e-mail: ranjith.pg@monash.edu

J. Zhao
e-mail: jian.zhao@monash.edu

M. Latha G. (ed.), *Frontiers in Geotechnical Engineering*, Developments in Geotechnical Engineering, https://doi.org/10.1007/978-981-13-5871-5_12

this article. The stresses in the tunnel lining are mainly depend on the surrounding ground movement, which makes the behavior of tunnel displacement controlled. The additional displacement that is generated during an earthquake translates the stresses in the tunnel, leading to severe damages in tunnels. Several past studies have shown a significant increment in tunnel lining forces under seismic conditions [4, 5]. Power et al. [3] and Penzien and Wu [5] gave closed-form analytical solutions to estimate the response of tunnel lining in terms of thrust and bending moment and quantified the racking and ovaling in the tunnel. Tunnels constructed in soft soil and weak rock are more vulnerable to seismic loading when compared to those that are constructed in an intact rock; nonetheless, very less studies are available explaining the seismic behavior of a tunnel in soft soil due to inadequate field observations and experimental data. Most common methods to assess the effects of ground motion on underground structures are ground deformation method and free-field shear deformation method. Owen and Scholl [6] described the various forms of tunnel deformations, such as axial deformation, curvature deformation, racking, and ovaling deformation fostered by seismic loading. The tunnel is designed to accommodate above mentioned deformations caused by an earthquake-induced ground shaking. Hashash et al. [7] explained the three-step method to quantify seismic hazard at a tunnel site and to determine magnitude of seismic loads appropriate to their location. Hashash et al. [7, 8] and Patil et al. [9] extensively reviewed the various methodologies that are used in practice of seismic analysis and design of underground structures. These methods are useful in determining stresses in tunnel lining under seismic loading conditions. Few researchers have studied the effects of various parameters such as input ground motion, ground density, tunnel flexibility, and depth on the seismic behavior of tunnels by performing centrifuge tests [10–14]. In the present state-of-the-art study, many research articles that involve experimental [15, 16], numerical [17–22] and analytical [23–25] studies, have been reviewed to understand the seismic behavior of tunnels. Various approaches that are practiced in seismic analysis of tunnel and other underground structures with similar characteristics are summarized and their significances are discussed herein. Also, a detailed comparison between seismic design philosophies based on simplified analytical methods and numerical solutions is presented. The available analytical solutions do not include effect complex ground conditions. Thus, the accuracy of prediction of lining forces is questionable [26]. Therefore, a complete analytical solution coupled with numerical modeling is needed to investigate the performance of tunnel under seismic load. It was observed that some complex issues related to soft soil tunnelling are yet to be addressed. It is important to explore them in detail, such as: (a) estimation of tunnel distortions with respect to seismic loading, (b) Evaluation of excess tunnel lining forces generated by ground shaking, (c) assessment of performance of different shapes of tunnel, and (d) effect of tunnel–soil interface conditions.

The present study investigates the complex behavior of shallow tunnels in soft soil under seismic loading conditions. A detailed finite element (FE) analysis has been carried out by using PLAXIS 2D AE.02. Numerical modeling methodology is discussed in detail and results are compared with available analytical solutions. Thereafter, a parametric study has been performed to perceive the effect of input ground motion, tunnel–soil interaction conditions, and shape of the tunnel on the response of tunnel and surrounding soil under seismic loading. Results of analyses are reported in the form of distortion of tunnel, variation of the maximum dynamic earth pressure and the residual earth pressure around the tunnel, and the distribution of thrust and bending moment in the tunnel lining. The output of the present study will help in understanding the qualitative behavior of tunnel–soil system under seismic loading conditions. In addition, it will be useful in assessing additional stresses in tunnel lining and deformations with stresses within the surrounding soil.

12.2 Effects of Earthquake on Tunnel

In general, earthquake effects on tunnel can be grouped into two categories—ground failure and ground shaking. Ground failure refers to various types of ground instability such as fault rupture, liquefaction, landslides, tectonic uplift, and subsidence. Ground shaking includes vibration of the ground produced by seismic waves propagating through the earth's crust [1, 3, 7, 27]. Various types of tunnel damages and their causes are summarized in Table 12.1.

12.2.1 Parameters Influencing Damage to Tunnels in Seismic Conditions

Hashash et al. [7] reported that damage to the tunnels decreases with increase in the overburden depth. The severity of damage depends on the peak ground acceleration, the distance from the epicenter, duration of the earthquake, and the earthquake magnitude. Damages at the near tunnel portals can be significantly attributed to slope instabilities. The large incremental lateral forces from the surrounding soil causes damage to cut-and-cover structures [6]. The important parameters that influence the damage to tunnels due to ground shaking are as follows [1, 2, 27, 29]:

Table 12.1 Various types of tunnel damage and possible causes [28]

Damage	Possible cause
Severe cracking of tunnel lining	Differential movement of ground under the tunnel
Racking of tunnel	Tunnels on skews with unbalanced soil pressures
Shifting, tilting or raising of tunnel	Liquefaction of surrounding soil
Falling rocks or tiles inside or outside the tunnel	Resulting from vibration of tunnel
Tunnel cave-in	Damage to structural members from shaking

(1) Magnitude of earthquake: severe damage occurred when typically, the earthquake magnitude is more than 6 on Richter scale.
(2) Distance to epicenter: severe damages occur when the epicentral distance is less than 25 km.
(3) Peak Ground Acceleration (PGA): severe damages were reported when the PGA is more than 0.5 g (where g is the acceleration due to gravity).
(4) Geological conditions: severe damages to tunnel during 2008 Wenchuan earthquake when the tunnel is located near the seismically active fault. Jiang et al. [2] divided geologic conditions based on soft bedrock and hard bedrock.
(5) Depth of the tunnel/Overburden.
(6) Tunnel–soil interface conditions (relative stiffness).
(7) Completion time (age of the tunnel).

12.3 Seismic Design Philosophy for Underground Structures

Wang [4] explained the difference between surface structures and underground structures with respect to their behavior under seismic loading. The seismic design approach for the underground structure is different than that of the surface structure. In surface structures, depending on their own natural frequencies, amplification of cyclic load occurs when subjected to the ground excitation. The "Force Method" analysis and design approach is used for surface structures; the seismic loads are largely expressed in terms of inertial forces. In contrast, the "Deformation Method" used in analysis and design of underground structures which is based on an approach that focuses on the displacement of the ground and the structure. Underground structures neither could move independently due to lateral confinement nor subject to vibration amplification. The simplest design method consists of pseudo-static analysis; indeed, this should be considered a kinematic approach, since the seismic action is taken equivalent to a shear strain assigned at the tunnel depth. Such an assumption is supported by field observations suggesting that the damage of the tunnel under seismic loads is strongly influenced by the deformation of the surrounding soil [30]. Wood [31] proposed the analytical solution based on the dynamic modal analysis. This solution is applicable for non-yielding rigid embedded walls and assumes the rigid behavior of structure. Therefore, mostly it is not recommended for seismic analysis of flexible underground structures. The soil nonlinearity causes the amplification of motion which is not considered in this method. The analytical methods adopt simplified closed-form expressions to determine seismic incremental lining forces, based on the maximum shear strain in the subsoil computed in free-field conditions by pseudo-static or dynamic one-dimensional site response analysis [4, 5, 32]. This assumption corresponds to an uncoupled approach, neglecting the kinematical interaction between soil and structure. In closed-form analytical solution proposed by Hashash et al. [7], the dynamic tunnel lining forces depend on the maximum shear strain of the soil in free-field conditions. Most of the methods do not consider the nonlinear properties of the soil and tunnel.

Seismic loads are considered as a cyclic, superimposed, and time-dependent loads. Determination of these seismic loads involves certain degree of uncertainty. Three kinds of earthquake events are used for the determination of seismic loads in the design or safety evaluation analysis of a tunnel, which are given as follows:

- Maximum credible earthquake (MCE).
- Maximum design earthquake (MDE) or Safety evaluation earthquake (SEE).
- Serviceability limit state earthquake (SLSE), also known as an operating basis earthquake (OBE), this can be simplified as a static live load and combined with other live loads.

In addition, the characterization and quantification of seismic hazard that tunnels may encounter are also needed. This requires definition of the damage, formation of plastic hinges, and extent of cracking, movement, and so on [33].

12.3.1 Seismic Design Steps for Underground Structures

Hashash et al. [7] suggested the following three steps for seismic analysis of tunnels.

- Definition of seismic environment
- Evaluation of the ground response to shaking
- Assessment of structure behavior due to seismic shaking.

Definition of Seismic Environment. There are two methods to determine ground motion parameters for velocities and accelerations (i) the deterministic seismic hazard analysis (DSHA), and (ii) the probabilistic seismic hazard analysis (PSHA). The later is relatively recent and gives the range of anticipated ground motions along with their probabilities of occurrence of an event. The components of ground motion can be represented by displacement, velocity or acceleration coupled with three important parameters; frequency, amplitude, and duration. The probabilities can be used to deduce the required protection level against earthquake forces [7].

Evaluation of the Ground Response to Shaking. Ground response to shaking may vary between failures to no/limited damages. Ground failure caused by seismic loading includes liquefaction, lateral spreading, landslides and fault rupture. Ground shaking, and deformation will be the most expected behavior during and after an earthquake event.

Assessment of Structural Behavior. There are several engineering approaches to compute the deformation and forces in tunnel lining induced by seismic loading. These deformation and forces depend on the type of ground, the type of structure and, the interaction between the soil and tunnel.

12.4 Response of Tunnel to Seismic Loading

Response of a tunnel to seismic load can be expressed in the form of deformations and associated strains caused by surrounding soil movement during shaking. Based on the type of waves, the direction of soil movement can be divided into longitudinal and transverse direction of tunnel based on the type of wave. Methods used to determine force and deformations in tunnel lining corresponding to ground motion are briefly described here.

12.4.1 Free-Field Deformation Method

In this method, it is assumed that the seismic ground deformation occurs in absence of tunnel. Deformations induced by the seismic event are estimated by ignoring the stiffness of the tunnel and then the tunnel is designed to sustain these free-field deformations. It is a simplified method that gives realistic results if the flexibility of tunnel is approximately equal to the surrounding soil medium. This method is suitable where ground distortions in the tunnel are small [3, 4, 7].

12.4.2 Ground Deformation Method

It is assumed that tunnels move along with surrounding soil mass during an earthquake event. The tunnels are designed to withstand the deformations imposed by the surrounding soil. Though the soil-structure interaction plays a predominant part in seismic response of tunnel, the effect of interaction condition is not considered in this method. Interaction effect should be considered in the analysis when the tunnel is surrounded by soft soil. A tunnel undergoes three major modes of deformation due to seismic shaking [6]:

- axial deformation,
- curvature deformation,
- ovaling (for circular tunnels) or racking (for rectangular tunnels).

A numerical model is necessary to understand the mechanism of all three modes of tunnel deformation as described above. Among the three types of deformations, the ovaling/racking of tunnel cross-sections caused by a shear wave propagating normal to the tunnel axis is the most critical deformation when it comes to the impact of earthquake on tunnels [34].

Axial and Curvature Deformation. Seismic waves that travel parallel or obliquely to the tunnel, develop axial and curvature deformation. Tunnel lining must accommodate the deformations in the longitudinal direction [6]. Figure 12.1 shows the schematic representation of axial and curvature deformation.

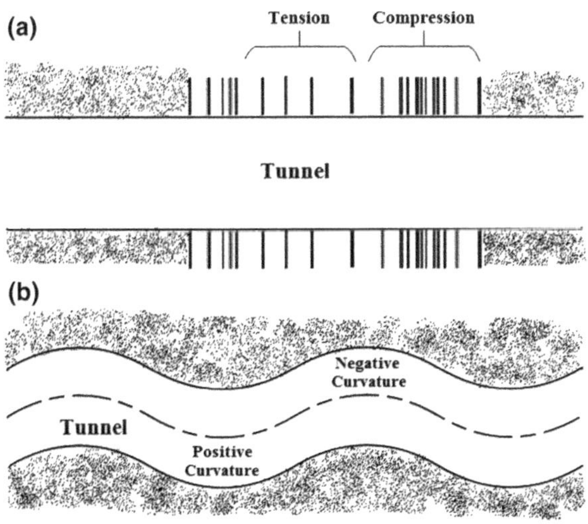

Fig. 12.1 **a** Axial deformation and **b** curvature deformation along the tunnel axis (Modified after Owen and Scholl [6])

Ovaling or Racking Deformation. The ovaling or racking deformation in tunnel occurs when waves propagate in a direction perpendicular to the tunnel longitudinal axis, thereby distorting the cross-section of tunnel lining. Deformation in the transverse direction is the most critical among all deformations that need to be considered in designing of tunnel [6]. Figure 12.2 shows the ovaling and racking deformation related with circular and rectangular tunnel, respectively.

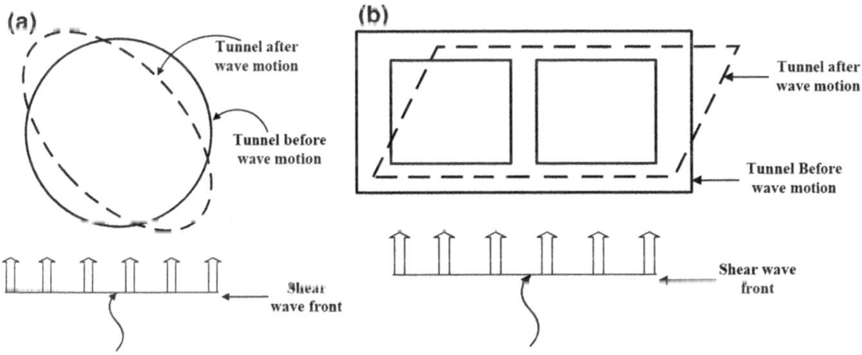

Fig. 12.2 **a** Ovaling deformation in circular tunnel and **b** racking deformation in rectangular tunnel (Modified after Owen and Scholl [6])

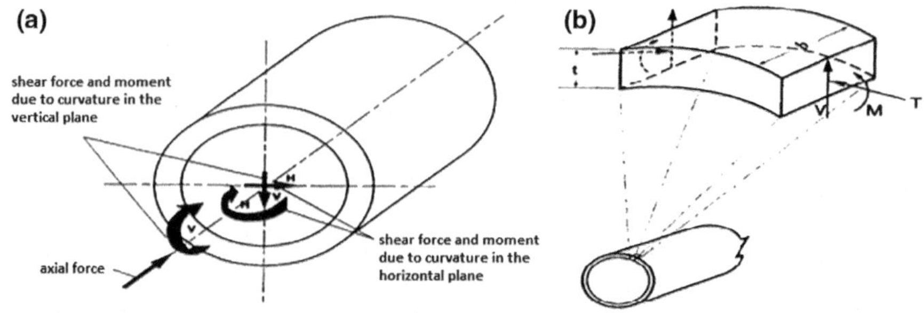

Fig. 12.3 Forces and moments in tunnel lining **a** caused by seismic waves propagating along tunnel axis. **b** Caused by seismic waves propagating perpendicular to tunnel axis [3]

12.4.3 Tunnel–Soil Interaction Approach

Based on the finite element analysis, Wang [4] reported that the seismic response of tunnel is dominated by the geometry of structure, relative stiffness of the tunnel with respect to the surrounding soil, depth of the tunnel and characteristics of the ground motion. The free-field ground deformation method is based on unrealistic assumptions. Thus, it is important to adopt a methodology which is more realistic incorporating soil-structure interaction approach. In this method, concept of the beam-on-elastic foundation is used to model the effects of soil-structure interaction. Closed-form solutions are available for circular tunnel–soil interaction. For the seismic analysis of tunnel at shallow depth, simplified tunnel–soil interaction curves were proposed by Wang [4] and Penzien [32]. Figure 12.3 shows the shear force, moment and thrust in tunnel lining caused due to seismically induced free-field axial, curvature and shear deformations.

12.4.4 Pseudo-Static Methods

Mononobe-Okabe Method. This method is useful for the design of a rectangular tunnel. Japanese Society of Civil Engineers recognized this method for designing of earthquake resistant submerged tunnels. Also, this method is used to ascertain the dynamic earth pressure on the tunnel walls, caused by the inertial force of surrounding soil [35, 36]. The Mononobe-Okabe method was primitively used for seismic analysis of above-ground structures. It is based on an assumption that the structure can move and/or tilt in such a way that an active Coulomb type soil wedge form in the backfill soil [37–50]. According to Wang [4] and Hashash et al. [7] the assumption that, the moving and/or tilting wall, is not appropriate for tunnels. This method overestimates the deformation and moment in the tunnel lining. It gives more unrealistic result with increase in depth of embedment of a tunnel. Hence, it is impractical to use in seismic analysis of tunnel. The application is limited to the typical 'U-section' type of underground construction or in case of shallow tunnel. Despite this drawback, the method was implemented in the Los Angeles Metro Project. Also, this method is commonly used for conventional design and analysis of various underground structures [37–50] by considering the effect of earthquake as equivalent static condition. Additionally, the over-simplicity of this method for seismic earth pressure estimation for a dynamic problem by converting it to equivalent static problem is found to be inappropriate as highlighted by various researchers in recent past [51–71].

Wood Method. Wood [31] derived a theoretical equation to determine dynamic earth pressure. This method gives the 1.5–2.0 times higher dynamic thrust than the Mononobe-Okabe method. This method is possibly useful for the tunnels constructed in a very stiff soil or rock. Like Mononobe-Okabe method, this method also leads to impractical results and therefore, not advisable to use in tunnel with large cover depth. It is perhaps applicable in a cut-and-cover tunnel, where the walls act as rigid shear wall that prevents the lateral movement of tunnel during earthquakes.

12.5 Finite Element Analysis

The objective of present study is to understand the response of tunnel in soft soil under seismic loading by considering in situ complex conditions. A tunnel–soil system has been modeled using finite element (FE) based specialized geotechnical computer program, PLAXIS 2D AE.02, in order to perform nonlinear plain-strain numerical analysis. The lateral extent of the numerical model, mesh size and the boundary conditions are optimized by performing sensitivity analysis. A Mohr–Coulomb model has been used to define the soil behavior, and linear elastic model is used to simulate tunnel lining. Yield strength (E_l) and Poisson's ratio (ν_l) of the concrete was taken as 32 GPa and 0.18, respectively. A similar methodology has been used earlier by many researchers [19, 22, 72] to simulate tunnel–soil system in numerical modeling. Figure 12.4 shows a schematic representation of a two-dimensional numerical model of tunnel in soft soil.

12.5.1 Problem Definition and Domain

An ideal deposit of Class D type of soil as described in NEHRP [73] is assumed as reference soil medium. Figure 12.5 shows the variation of shear wave velocity (V_s) and shear modulus (G_{max}) with depth. In the present analysis, two different acceleration–time histories were considered, 1989 Loma Prieta earthquake, and 1995 Kobe earthquake, as shown in Fig. 12.6a, b. The Fourier amplitude spectrum of the input ground motion is presented in Fig. 12.7a, b, it shows that most of the energy content of the ground motion is within the range of frequencies of 0.5–5 Hz. In nonlinear fully dynamic numerical analysis, the soil behavior is primarily dominated

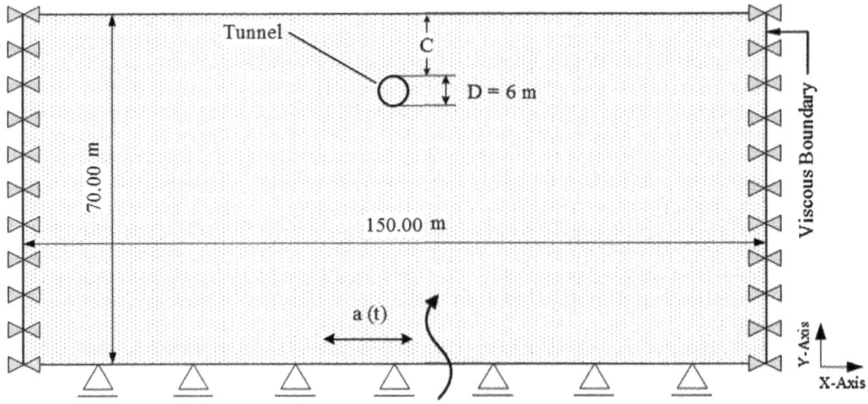

Fig. 12.4 Schematic diagram of tunnel–soil system adopted in numerical analysis

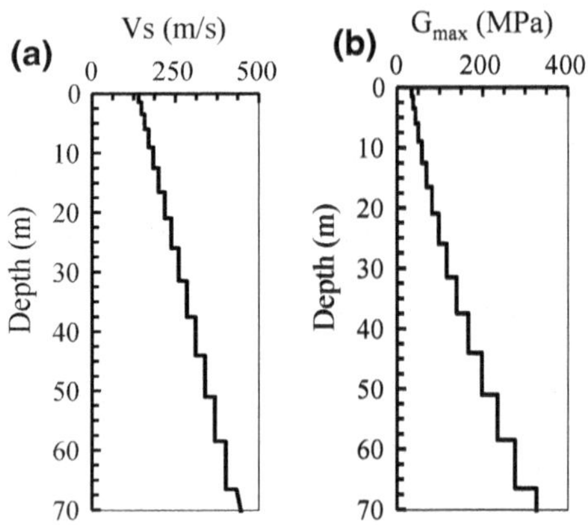

Fig. 12.5 Profile of **a** shear wave velocity V_s, and **b** small-strain shear stiffness G_{max}

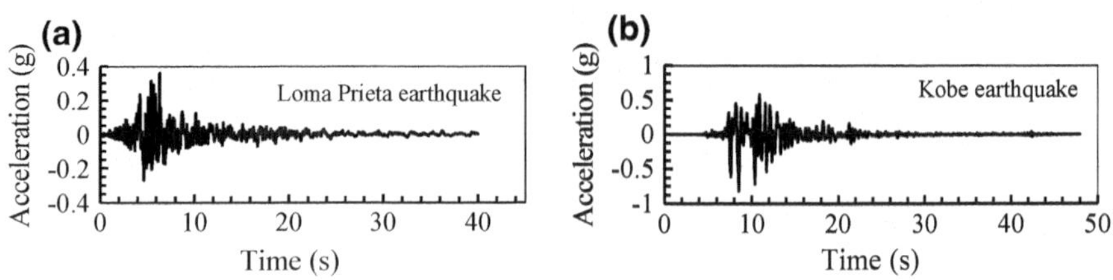

Fig. 12.6 Acceleration time histories: **a** 1989 Loma Prieta earthquake, and **b** 1995 Kobe earthquake

by dynamic properties of the soil. Therefore, shear wave velocity (V_s) is considered as a primary input parameter and strength parameter (undrained shear strength, S_u) of the soil, which is calculated from the expression given by Dickenson [74].

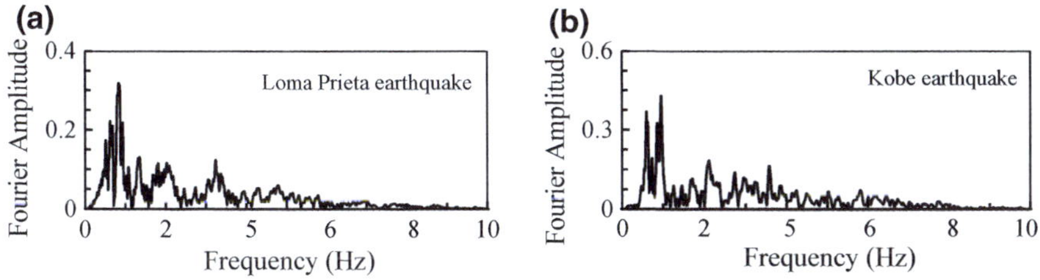

Fig. 12.7 Fourier amplitude spectra of input motions **a** 1989 Loma Prieta earthquake, and **b** 1995 Kobe earthquake

12.5.2 Calibration of Damping and Rayleigh Coefficients

During an earthquake, soil particles are subjected to oscillations, with time energy gets dissipated within a material and attenuation of seismic waves take place due to hysteretic damping. In constitutive modeling, the Mohr–coulomb has a limitation in generating hysteretic damping behavior. This was compensated using frequency-dependent Rayleigh damping formulation in terms of viscous damping. Rayleigh viscous damping is introduced by using α and β, mass and stiffness proportional damping coefficients, respectively. These coefficients are functions of target damping ratio (ξ_t) and frequency range (ω_m, ω_n) over which damping is equal to or less than ξ_t. In the present analysis, the two frequencies are identified through an iterative procedure suggested by Amorosi et al. [75] that consider the frequency interval characterized by the highest energy content in soil deposit and the amplification function between surface and base level.

Also, the equivalent linear earthquake site response analysis has been done based on the equivalent linear visco-elastic approach [76–78]. It was assumed that the results of one dimensional analysis performed in the frequency-domain using equivalent linear visco-elasticity coincide with the correspond ing two-dimensional finite element analysis performed in time-domain by taking the same constitutive behavior, provided appropriate calibration of parameters is carried out. The calibrated material parameters are utilized in subsequent numerical analyses to simulate realistic field conditions.

Equivalent Linear Earthquake Site Response Analysis. An equivalent linear seismic site response analysis for each selected input motion has been carried out with the program SHAKE2000 of Ordonez [79] to determine target damping ratio. The outcomes of the SHAKE2000 analysis plotted against depth in Fig. 12.8 for maximum shear strain (γ_{max}), normalized shear modulus (G/G_{max}), damping ratio (D), and maximum acceleration (a_{max}). The resulting damping ratio values are adopted as the target damping ratio. These damping values incorporate stiffness degradation as well as damping variation with strain for the whole soil continuum. Finally, the Rayleigh coefficients are calculated for the obtained target damping and subsequently used

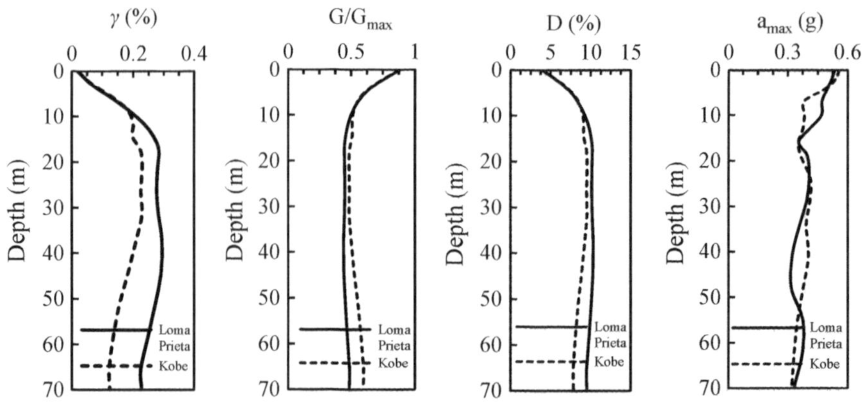

Fig. 12.8 Results of the 1D ground response analysis performed with SHAKE2000 for 1989 Loma Prieta earthquake and 1995 Kobe earthquake

in FE analysis. Similar studies on the site-specific seismic ground response analysis were considered by various researchers in recent past [80–86].

12.5.3 Parametric Study

Detailed parametric study to investigate the effects of shape of tunnel, tunnel–soil interaction condition, and input ground motion characteristic on seismic response of tunnel in soft soil has been conducted using two-dimensional (2D) finite element (FE) based computer program PLAXIS 2D AE.02 [87]. Four types of tunnel shapes (circular, square, square with rounded corners, and horseshoe) are examined. For tunnel in soft soil, the effects of interface conditions between the tunnel and soil are important to consider. Two tunnel–soil interface conditions are analyzed: (i) full-slip ($\mu = 0$, μ is the soil–tunnel interaction coefficient) condition and (ii) no-slip ($\mu = 1$) condition. Various peak ground accelerations (PGA) are considered for present analysis. Two selected earthquake motion data (a) 1989 Loma Prieta earthquake and (b) 1995 Kobe earthquake with necessary scaled down or up to get PGA of 0.2, 0.4, and 0.6 g are considered. In present study, the scaled input motion with PGA 0.4 g of Loma Prieta earthquake and Kobe earthquake are denoted as EQ1 and EQ2, respectively. The representative numerical analysis outputs for circular and rectangular tunnels are shown in Figs. 12.9 and 12.10, respectively.

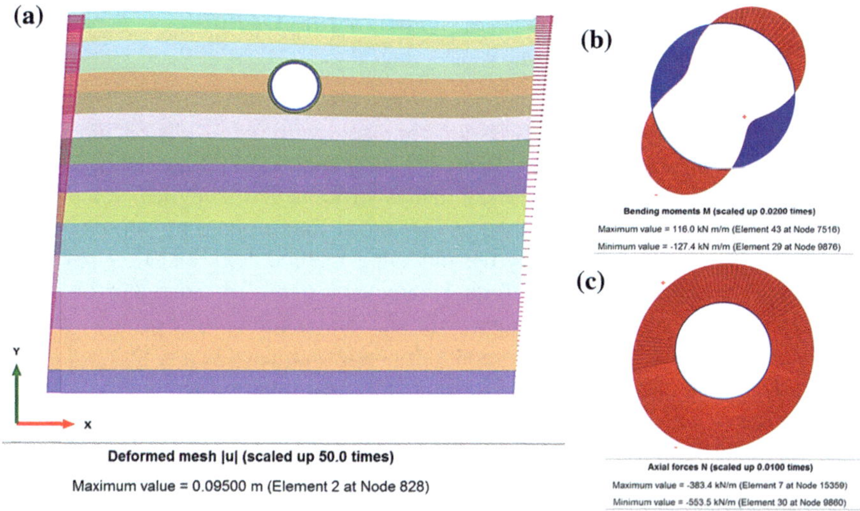

Fig. 12.9 2D FE analysis results: **a** deformed mesh, **b** total moment and **c** thrust in circular tunnel lining (earthquake: EQ1)

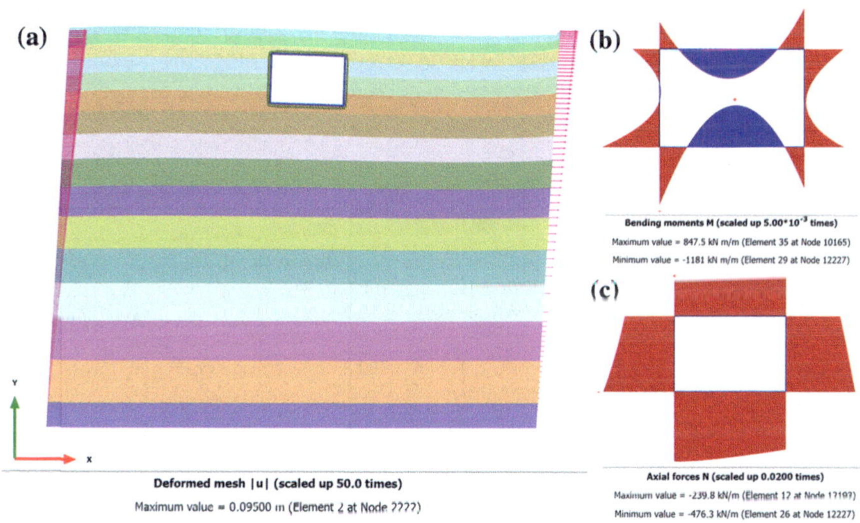

Fig. 12.10 2D FE analysis results: **a** deformed mesh, **b** total moment and **c** thrust in rectangular tunnel lining (earthquake: EQ1)

12.5.4 Comparison of Present Results with Available Solutions

The results of 2D numerical analyses are compared with the results obtained from analytical expressions provided by Wang [4] and Penzien [32]. Figure 12.11 shows the comparison between the numerical analysis results and analytical solutions in terms of bending moment and thrust in the tunnel lining under no-slip condition. The comparison is made to verify the modeling methodology and to support the assumptions made in the analysis. Tunnel diameter of 6 m is considered in the analysis with varying lining thickness to get different combinations of flexibility ratios. It was observed that analytical solutions slightly underestimate the moment and thrust in the lining when the distortion in tunnel is high. Few other researchers have made similar observations [8]. The results of numerical analysis are matching very well with the results of analytical solutions. The solution given by Penzien [32] overpredict the thrust in tunnel lining at some occasions. Overall there is a discrepancy of around 4–19% in determining maximum thrust and 2–9% in determining maximum bending moment. This variation is due to the consideration of nonlinear behavior of soil and amplification of seismic waves in the numerical model. While elastic material behavior is assumed in analytical solution and amplification is not considered. Finally, another possible source of differences is the fact that in the present analyses modulus degradation and damping variation are considered along the depth, whereas an average strain at tunnel axis level is considered in the analytical solutions.

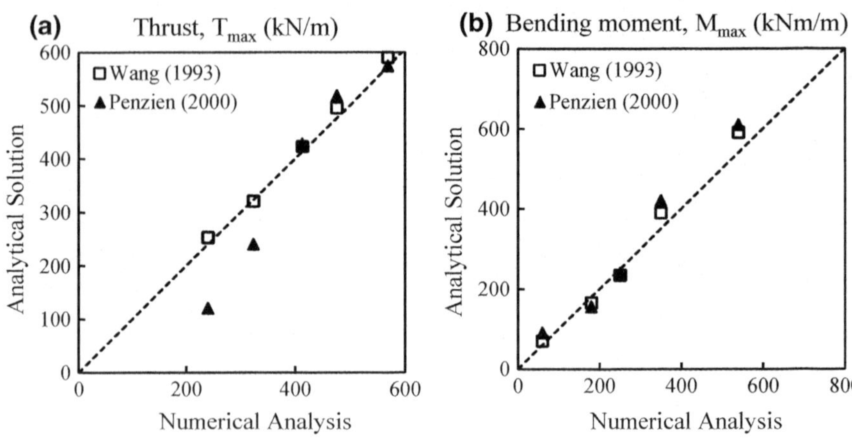

Fig. 12.11 Comparison of results on **a** thrust and **b** bending moment between numerical analysis and analytical solutions (earthquake: EQ1)

12.5.5 Results and Discussions

Comparison of Response of Different Shapes of Tunnel. The effects of shape of tunnel on seismic response are examined under earthquake EQ2. Four conventional shapes, circular, square, and horseshoe tunnel are analyzed along with newly proposed square with rounded corners. Figure 12.12a shows the schematic diagram of various shapes of tunnels that are analyzed in the present study. The equality in tunnel size and field stresses are maintained by selecting appropriate dimensions of tunnels that give the same volume of excavation in all cases. Also, all other factors such as loading conditions, tunnel embedment depth, and lining thickness are kept same in all cases. Sign convention used for representing the response of tunnel in terms of lining forces is shown in Fig. 12.12b. Figures 12.13, and 12.14 show, respectively, the maximum thrust and bending moment induced by seismic loading in tunnel lining of circular, square, square with rounded corners and horseshoe shapes. It was observed that the maximum thrust is generated in square with rounded corner tunnel lining due to the combined action of normal and hoop stress, as shown in Fig. 12.13. This type of combined action of thrusts is not observed in other shapes of tunnel. However, it can be noted from Fig. 12.14 that at the corners of square tunnel, bending moment is higher than that in the square tunnel with rounded corners. Rounded corners in the square shape subsequently reduced the bending moment, as observed. Overall, the circular tunnel lining shows the better performance compared to other shapes of tunnel in terms of resisting developed stresses under the seismic loading of EQ2. Thus, it is concluded that shape of the tunnel plays an important role in governing overall response of the tunnel–soil system for combined loading conditions.

Influence of Contact Interface Between Tunnel and Soil. A circular tunnel of 6 m diameter and 0.25 m thickness is adopted to investigate effect of interface condition on seismic response of tunnel. Two extreme interface conditions were considered

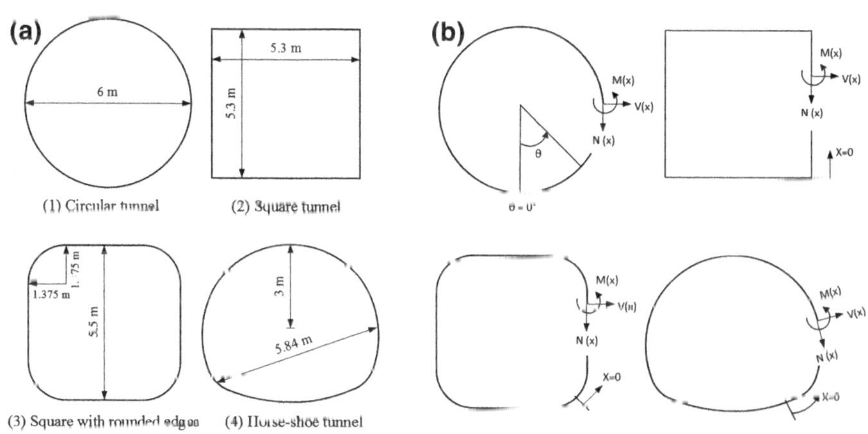

Fig. 12.12 **a** Different shapes of tunnels and **b** sign for the maximum thrust, T_{max}, and bending moment, M_{max}, in the tunnel lining

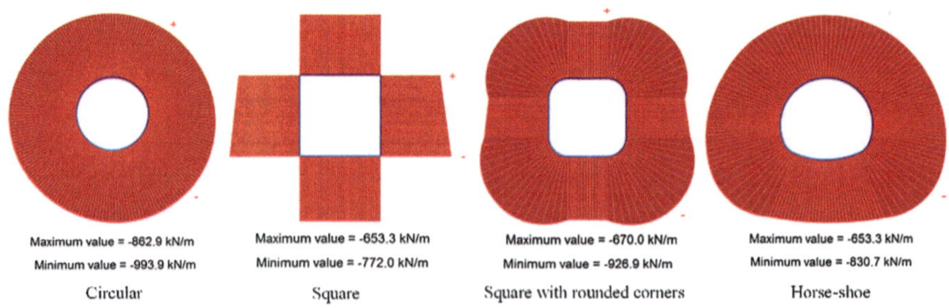

Fig. 12.13 Variation in maximum thrust, T_{max}, in the tunnel lining with different shapes (under EQ2)

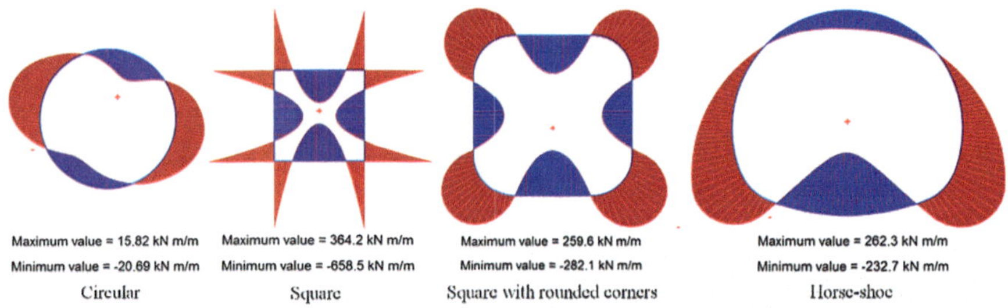

Fig. 12.14 Variation in maximum bending moment, M_{max}, in the tunnel lining with different shapes (under EQ2)

in the analysis, full-slip ($\mu = 0$) and no-slip ($\mu = 1$) interface conditions. The full-slip condition occurs in absence of tangential shear force at the interface between tunnel and soil. Full-slip assumption may underestimate the maximum thrust and overestimate the maximum bending moment. No-slip condition occurs when the contact between tunnel and soil is fully rigid. In no-slip condition, tunnel and soil move together under seismic loading conditions. Practically, interface condition lies somewhere in between full-slip and no-slip. Perfect full-slip condition is possible in tunnels that are constructed in rock with significant cover thickness. No-slip condition may exist when tunnel pass through a soft soil at shallow depth. Hence, it is important to evaluate the performance of tunnel under both conditions and more critical condition should be considered for design. Response of tunnel in transverse direction has been investigated under earthquake ground motion of EQ2. Figure 12.15 shows the lining deflections in circular tunnel. These are normalized with free-field ground deflections as function of flexibility ratio (F). Distortion in shallow tunnel is much higher compared to the deep tunnels that resulted into the ovaling deformation in shallow tunnel. In the case of a full-slip condition, the distortion in the tunnel lining increases between 5 and 21% when compared to no-slip interface condition. At a greater depth embedment ratio (C/D > 2), a negligible slip between soil and tunnel is observed that is attributed to immobilization of full shear strength at the tunnel–soil

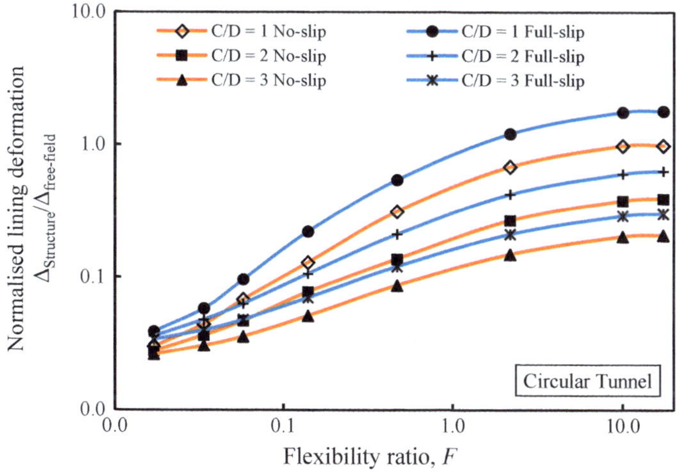

Fig. 12.15 Response of circular tunnel with full-slip ($\mu = 0$) and no-slip ($\mu = 1$) interface condition at soil–tunnel interface (D = 6 m, EQ2)

interface. Also, there is a no significant change in distortion in rigid tunnels when depth of embedment is increased.

Effect of Input Ground Motions. Earthquake induces additional earth pressure on the tunnel lining. The effect of input ground motion characteristic on tunnel and surrounding soil has been studied by varying PGA value from 0.2 to 0.6 g. The results of the parametric study are presented here for the prediction of maximum and residual dynamic earth pressures. Figure 12.16 shows the distribution of maximum dynamic earth pressure along the perimeter of the circular tunnel lining, that are obtained from FE analysis (for sign convention refer to Fig. 12.12b). The maximum dynamic earth pressure refers to the additional pressure that is generated by earthquake-induced shaking on tunnel lining. It was observed that maximum dynamic earth pressure increases with increase in amplitude of input ground motion; however, the location of maximum earth pressure remains same. The maximum dynamic earth pressure is 86.4 kPa, observed at $\theta = 225°$ at PGA of 0.6 g. The earth pressures during seismic loading are 17–32% more than earth pressure due to only static load. The residual dynamic earth pressure is an earth pressure that is generated during an earthquake and remains there at the end of the earthquake. So, this is the net permanent incremental earth pressure that tunnel must resist in addition to the parent earth pressure and is as shown in Fig. 12.17. The maximum residual earth pressure (35.02 kPa) is obtained at $\theta = 135°$ at PGA of 0.6 g. The shoulder parts of the tunnel lining in opposite combinations ($\theta = 135°$ and $\theta = 315°$) and ($\theta = 45°$ and $\theta = 225°$) experiences the maximum dynamic earth pressure due to cyclic loading, this results into ovaling deformation of tunnel under seismic loading condition. The obtained results indicate

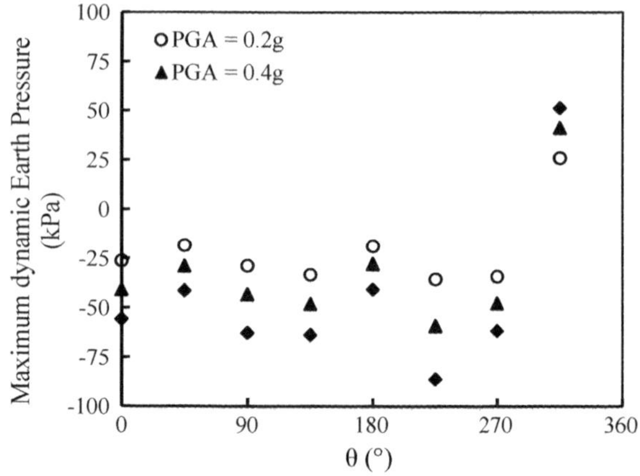

Fig. 12.16 Maximum dynamic earth pressure around the tunnel lining at PGA = 0.2, 0.4, and 0.6 g (input ground motion: Kobe earthquake)

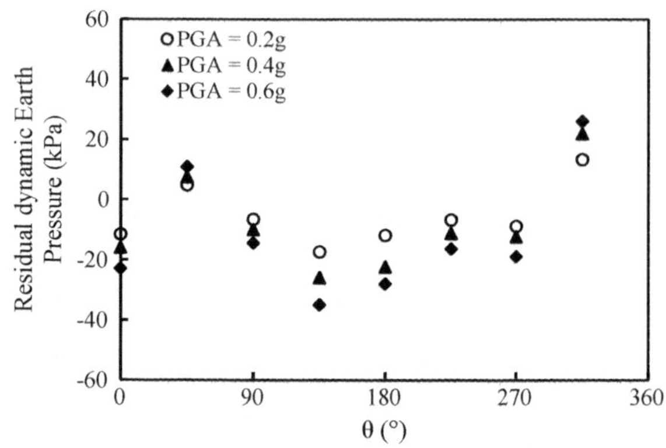

Fig. 12.17 Residual dynamic earth pressure around the tunnel lining at PGA = 0.2, 0.4, and 0.6 g (input ground motion: Kobe earthquake)

the strong influence of amplitude of input ground motion on the maximum and residual dynamic earth pressures. For higher PGA value, the earth pressure will be more.

12.6 Conclusions

A state-of-the-art review of seismic analysis of tunnel in soft soil is presented. It includes information on effect of earthquake on underground structures and parameters influencing damage to tunnel. Various approaches of estimation of ground deformation have also been discussed. Tunnels that are enclosed in soft soils suffer higher damage compared to tunnels in rock. Proper consideration of seismically induced forces in soft soil–tunnel design is essential. Few analytical methods have been developed to obtain forces in tunnel lining, but it does not incorporate effect of soil-structure interaction. The available analytical solutions are not enough to include complex ground conditions and hence a comprehensive analytical solution coupled with numerical modeling is necessary to understand the effect of seismic load on the performance of tunnel in soft soil. The effect of soil-structure interaction plays a predominant role in the seismic response of tunnel when the tunnel is enclosed in soft soil mass and therefore should be properly considered in the analysis.

The equivalent linear earthquake ground response analysis has been carried out to calibrate the material parameters that are serving as an input in finite element analysis. It was observed that selection of Rayleigh frequencies is crucial in achieving target damping in numerical modeling. A procedure for the calibration of the stiffness and Rayleigh damping parameters are discussed. An attempt has been made to study the influence of seismic loadings on tunnels in soft soil by using a FE-based program, PLAXIS 2D AE.02. A series of parametric study was performed to investigate the effect of shape of tunnel, tunnel–soil interface condition and input ground motion on response of tunnels in soft soil. Comparison between numerical analysis results and analytical solution is presented and the limitations of later were highlighted. Soil–tunnel interface conditions have a significant contribution in deciding the distortion in the tunnel lining. The distortion in the full-slip interface condition is found to be 5–21% more than that in no-slip interface condition. Interface condition significantly influence the magnitude of the bending moment in tunnel lining. Full-slip condition generated 18% higher bending moment in tunnel lining when compared no-slip condition. The shape of the tunnel plays a vital role in deciding overall response of the system under earthquake loading condition. Circular tunnel performs better than square, square with rounded corners, and horseshoe tunnel for earthquake condition. The unconventional shape of square tunnel with rounded corners shows lower bending moment than the conventional square tunnel. Input ground motion characteristics have pronounced effect on the maximum dynamic earth pressure around tunnel lining, which is observed at the shoulder region of tunnel lining, irrespective of the input motion characteristics. The present results lead to a better design and understanding of the seismic response of tunnels in soft soil considering local site effects.

References

1. Sharma, S., Judd, W.R.: Underground opening damage from earthquakes. Eng. Geol. **30**(3–4), 263–276 (1991)
2. Jiang, Y., Wang, C., Zhao, X.: Damage assessment of tunnels caused by the 2004 Mid Niigata prefecture earthquake using Hayashis quantification theory type II. Nat. Hazards **53**, 425–441 (2010)
3. Power, M.S., Rosidi, D., Kaneshiro, J.Y.: Seismic vulnerability of tunnels and underground structure revisited. In: Proceedings of the North American Tunneling Conference. Elsevier, Long Beach, CA, USA (1998)
4. Wang, J.N.: Seismic design of tunnel—a simple state of the art design approach, p. 7. Parson Brinckerhoff Quade & Douglas Inc., New York, Monograph (1993)
5. Penzien, J., Wu, C.: Stresses in linings of bored tunnels. Int. J. Earthq. Eng. Struct. Dyn. **27**, 283–300 (1998)

6. Owen, G.N., Scholl, R.E.: Earthquake engineering of large underground structures. Federal Highway Administration and National Science Foundation, Report No. FHWA/RD-80/195 (1981)
7. Hashash, Y.M.A., Hook, J.J., Schmidt, B., Yao, J.I.C.: Seismic design and analysis of underground structures. Tunn. Undergr. Space Technol. **16**, 247–293 (2001)
8. Hashash, Y.M.A., Park, D., Yao, J.I.C.: Ovaling deformations of circular tunnels under seismic loading, an update on seismic design and analysis of underground structures. Tunn. Undergr. Space Technol. **20**(5), 435–441 (2005)
9. Patil, M., Choudhury, D., Ranjith, P.G., Zhao, J.: Seismic analysis of tunnels in soft soils: a state-of-the-art review. In: Proceeding of the International Conference on Soft Ground Engineering (ICSGE 2015), 3–4 Dec 2015, Singapore, pp. 625–634 (2015)
10. Cilingir, U., Madabhushi, S.P.G.: Effect of depth on the seismic response of circular tunnels. Can. Geotech. J. **48**(1), 117–127 (2011)
11. Cilinger, U., Madabhushi, S.P.G.: A model study on the effects of input motion on the seismic behaviour of tunnels. Soil Dyn. Earthq. Eng. **31**(3), 452–462 (2011)
12. Izawa, J., Kusakabe, O., Nagatani, H., Yamada, T., Ohbo, N.: Centrifuge modelling on seismic behaviour of rectangular tunnels. Phys. Model. Geotech.—ICPMG'06, 1163–1169 (2006)
13. Tsinidis, G., Pitilakis, K., Madabhushi, G., Heron, C.: Dynamic response of flexible square tunnels: centrifuge testing and validation of existing design methodologies. Geotechnique **65**(5), 401–417 (2015)
14. Bilotta, E., Lanzano, G., Russo, G., Silvestri, F., Madabhushi, S.P.G.: Seismic analyses of shallow tunnels by dynamic centrifuge tests and finite elements. In: Proceedings of 17th International Conference on Soil Mechanics and Geotechnical Engineering, Alexandria, Egypt, pp. 474–477 (2009)
15. Lanzano, G., Bilotta, E., Russo, G., Silvestri, F., Madabhushi, S.P.G.: Centrifuge modelling of seismic loading on tunnels in sand. Geotech. Test. J. **35**(6), 854–869 (2012)
16. Tsinidis, G., Rovithis, E., Pitilakis, K., Chazelas, J.L.: Seismic response of box-type tunnels in soft soil: experimental and numerical investigation. Tunn. Undergr. Space Technol. **59**, 199–214 (2016)
17. Kontoe, S., Zdravkovic, L., Potts, D., Mentiki, C.: On the relative merits of simple and advanced constitutive models in dynamic analysis of tunnels. Geotechnique **61**(10), 815–829 (2011)
18. Kontoe, S., Avgerinos, V., Potts, D.M.: Numerical validation of analytical solutions and their use for equivalent-linear seismic analysis of circular tunnels. Soil Dyn. Earthq. Eng. **66**, 206–219 (2014)
19. Amorosi, A., Boldini, D.: Numerical modelling of the transverse dynamic behaviour of circular tunnels in clayey soils. Soil Dyn. Earthq. Eng. **29**(6), 1059–1072 (2009)
20. Debiasi, E., Gajo, A., Zonta, D.: On the seismic response of shallow-buried rectangular structures. Tunn. Undergr. Space Technol. **38**, 99–113 (2013)
21. Patil, M., Choudhury, D., Ranjith, P.G., Zhao, J. A numerical study on effects of dynamic input motion on response of tunnel-soil system. In: Proceeding of the 16th World Conference on Earthquake Engineering (16th WCEE 2017), Santiago, Chile, Paper ID: 3313 (2017)
22. Tsinidis, G.: Response characteristics of rectangular tunnels in soft soil subjected to transversal ground shaking. Tunn. Undergr. Space Technol. **62**, 1–22 (2017)
23. Huo, H., Bodet, A., Fernandez, G., Ramirez, J.: Load transfer mechanisms between underground structure and surrounding ground: evaluation of the failure of the Daikai station. J. Geotech. Geoenviron. Eng. **131**(12), 1522–1533 (2005)
24. Bobet, A., Fernandez, G., Huo, H., Ramirez, J.: A practical iterative procedure to estimate seismic-induced deformations of shallow rectangular structures. Can. Geotech. J. **45**, 923–938 (2008)
25. Bobet, A.: Drained and undrained response of deep tunnels subjected to far-field shear loading. Tunn. Undergr. Space Technol. **25**, 21–31 (2010)
26. Azadi, M.: The seismic behavior of urban tunnels in soft saturated soils. Procedia. Eng. **14**, 3069–3075 (2011)

27. Dowding, C., Rozen, A.: Damage to rock tunnel from earthquake shaking. J. of. Geotech. Eng. **104**, 175–191 (1978)
28. Bloodworth, A.G.: Three-dimensional analysis of tunneling effects on structures to develop design methods. Doctoral thesis, University of Oxford, The United Kingdom (2002)
29. Li, T.: Damage to mountain tunnels related to the Wenchuan earthquake and some suggestions for aseismic tunnel construction. Bul. Eng. Geol. Env. **71**(2), 297–308 (2012)
30. Okamato, S., Tamura, C., Kato, K., Hamada, M.: Behaviors of submerged tunnels during earthquakes. In: Proceedings of the 5th World Conference on Earthquake Engineering, vol. 1, pp. 544–553. Rome, Italy. (1973)
31. Wood, J.H.: Earthquake-induced soil pressures on structures. California Institute of Technology, Report No. EERL 73–05 (1973)
32. Penzien, J.: Seismically included racking of tunnels linings. Earthq. Eng. Struct. Dyn. **29**, 683–691 (2000)
33. Ingerslev, C., Kiyomiya, O.: Earthquake analysis. Tunn. Undergr. Space Technol. **12**(2), 157–162 (1997)
34. Kontoe, S., Zdravkovic, L., Potts, D.M., Menkiti, C.O.: Case study on seismic tunnel response. Can. Geotech. J. **45**(12), 1743–1764 (2008)
35. Okabe, S.: General theory on earth pressure and seismic stability of retaining wall and dam. J. Jpn. Soc. Civ. Eng. **12**(1), 23–134 (1926)
36. Mononobe, N., Matsuo, H.: On the determination of earth pressures during earthquakes. In: Proceedings of World Engineering Congress (1929)
37. Choudhury, D., Rao, K.S.S.: Seismic passive resistance in soils for negative wall friction. Can. Geotech. J. **39**(4), 971–981 (2002)
38. Choudhury, D., Sitharam, T.G., Rao, K.S.S.: Seismic design of earth retaining structures and foundations. Curr. Sci. **87**(10), 1417–1425 (2004)
39. Rao, K.S.S., Choudhury, D.: Seismic passive earth pressures in soils. J. Geotech. Geoenv. Eng. **131**(1), 131–135 (2005) https://doi.org/10.1061/(asce)1090-0241(2005)131:1(131)
40. Choudhury, D., Rao, K.S.S.: Seismic uplift capacity of inclined strip anchors. Can. Geotech. J. **42**(1), 263–271 (2005)
41. Choudhury, D., Singh, S.: New approach for estimation of static and seismic active earth pressure. Geotech. Geol. Eng. **24**(1), 117–127 (2006)
42. Ahmad, S.M., Choudhury, D.: Seismic internal stability analysis of waterfront reinforced-soil wall using pseudo-static approach. Ocean Eng. **52**, 83–90 (2012)
43. Choudhury, D., Savoikar, P.: Seismic stability analysis of expanded MSW landfills using pseudo-static limit equilibrium method. Waste Mgt. Res. **29**(2), 135–145 (2011)
44. Choudhury, D., Ahmad, S.M.: Pseudo-static design factors for stability of waterfront retaining wall during earthquake. J Earthq. Tsunami **4**(4), 387–400 (2010)
45. Choudhury, D., Ahmad, S.M.: External stability of waterfront reinforced soil structures under seismic conditions using a pseudo-static approach. Geosynth. Int. **16**(1), 1–10 (2009)
46. Choudhury, D., Ahmad, S.M.: Stability of waterfront retaining wall subjected to pseudo-static earthquake forces. Ocean Eng. **34**(14–15), 19471954 (2007)
47. Rangari, S.M., Choudhury, D, Dewaikar, D.M.: Pseudo-static uplift capacity of obliquely loaded horizontal strip anchor in cohesionless soil. In: Geo-Congress 2012: State of the Art and Practice in Geotechnical Engineering. Geotechnical Special Publication No. 225, pp. 185–194 (2012) https://doi.org/10.1061/9780784412121.020
48. Rangari, S., Choudhury, D., Dewaikar, D.M.: Pseudo-static uplift capacity of horizontal strip anchors. In: Geo-Frontiers 2011: Advances in Geotechnical Engineering. Geotechnical Special Publication No. 211: pp. 1821–1831 (2011) https://doi.org/10.1061/41165(397)186
49. Choudhury, D., Savoikar, P.: Seismic translational failure analysis of MSW landfills using pseudo-static approach. In: GeoFlorida 2010: Advances in Analysis, Modeling & Design. Geotechnical Special Publication No. 199: pp. 2830–2839 (2010) https://doi.org/10.1061/41095(365)288
50. Chakraborty, D., Choudhury, D.: Pseudo-static and pseudo-dynamic stability analysis of tailings dam under seismic conditions. Proc. Natl. Acad. Sci., India, Sect. A **83**(1), 63–71 (2013). https://doi.org/10.1007/s40010-013-0069-5

51. Choudhury, D., Nimbalkar, S.: Seismic passive resistance by pseudo-dynamic method. Geotechnique **55**(9), 699–702 (2005)
52. Choudhury, D., Nimbalkar, S.S.: Pseudo-dynamic approach of seismic active earth pressure behind retaining wall. Geotech. Geol. Eng. **24**(5), 1103–1113 (2006)
53. Nimbalkar, S.S., Choudhury, D., Mandal, J.N.: Seismic stability of reinforced soil-wall by pseudo-dynamic method. Geosynth. Int. **13**(3), 111–119 (2006)
54. Choudhury, D., Nimbalkar, S.: Seismic rotational displacement of gravity walls by pseudo-dynamic method: passive case. Soil Dyn. Earthq. Eng. **27**(3), 242–249 (2007)
55. Nimbalkar, S., Choudhury, D.: Sliding stability and seismic design of retaining wall by pseudo-dynamic method for passive case. Soil Dyn. Earthq. Eng. **27**(6), 497–505 (2007)
56. Ahmad, S.M., Choudhury, D.: Pseudo-dynamic approach of seismic design for waterfront reinforced soil wall. Geotext. Geomemb. **26**(4), 291–301 (2008)
57. Choudhury, D., Ahmad, S.M.: Stability of waterfront retaining wall subjected to pseudo-dynamic earthquake forces. J. Waterw. Port Coast. Ocean Eng. **134**(4), 252–260 (2008) https://doi.org/10.1061/(asce)0733-950x(2008)134:4(252) (2008)
58. Choudhury, D., Nimbalkar, S.S.: Seismic rotational displacement of gravity walls by pseudo-dynamic method. Int. J. Geomech. **8**(3), 169–175 (2008) https://doi.org/10.1061/(asce)1532-3641(2008)8:3(169)
59. Reddy, G.V.N.N., Choudhury, D., Madhav, M.R., Reddy, E.S.: Pseudo-dynamic analysis of reinforced soil wall subjected to oblique displacement. Geosynth. Int. **16**(2), 61–70 (2009)
60. Choudhury, D., Savoikar, P.: Seismic yield accelerations of MSW landfills by pseudo-dynamic approach. Nat. Hazards **56**(1), 275–297 (2011)
61. Savoikar, P., Choudhury, D.: Translational seismic failure analysis of MSW landfills using pseudo-dynamic approach. Int. J. Geomech. **12**(2), 136–146 (2012). https://doi.org/10.1061/(ASCE)GM.1943-5622.0000127
62. Rangari, S.M., Choudhury, D., Dewaikar, D.M.: Seismic uplift capacity of shallow horizontal strip anchor under oblique load using pseudo-dynamic approach. Soils Found. **53**(5), 692–707 (2013). https://doi.org/10.1016/j.sandf.2013.08.007
63. Rangari, S.M., Choudhury, D., Dewaikar, D.M.: Estimation of seismic uplift capacity of horizontal strip anchors using pseudo-dynamic approach. KSCE J. Civ. Eng. **17**(5), 989–1000 (2013). https://doi.org/10.1007/s12205-013-0046-1
64. Chakraborty, D., Choudhury, D.: Sliding stability of non-vertical waterfront retaining wall supporting inclined backfill subjected to pseudo-dynamic earthquake forces. Appl. Ocean Res. **47**, 174–182 (2014). https://doi.org/10.1016/j.apor.2014.05.004
65. Pain, A., Choudhury, D., Bhattacharyya, S.K.: Seismic stability of retaining wall-soil sliding interaction using modified pseudo-dynamic method. Geotech. Lett. **5**(1), 56–61 (2015). https://doi.org/10.1680/geolett.14.00116
66. Pain, A., Choudhury, D., Bhattacharyya, S.K.: Seismic uplift capacity of horizontal strip anchors using modified pseudo-dynamic approach. Int. J. Geomech. **16**(1), 04015025 (2016). https://doi.org/10.1061/(ASCE)GM.1943-5622.0000471
67. Pain, A., Choudhury, D., Bhattacharyya, S.K.: Seismic rotational stability of gravity retaining walls by modified pseudo-dynamic method. Soil Dyn. Earthq. Eng. **94**(3), 244–253 (2017). https://doi.org/10.1016/j.soildyn.2017.01.016
68. Rajesh, B.G., Choudhury, D.: Stability of seawalls using modified pseudo-dynamic method under earthquake conditions. Appl. Ocean Res. **65**, 154–165 (2017). https://doi.org/10.1016/j.apor.2017.04.004
69. Rajesh, B.G., Choudhury, D.: Generalized seismic active thrust on a retaining wall with submerged backfill using a modified pseudo-dynamic method. Int. J. Geomech. **17**(3), 06016023 (2017). https://doi.org/10.1061/(ASCE)GM.1943-5622.0000750
70. Rajesh, B.G., Choudhury, D.: Seismic passive earth resistance in submerged soils using modified pseudo-dynamic method with curved rupture surface. Mar Georesour. Geotech. **35**(7), 930–938 (2017). https://doi.org/10.1080/1064119X.2016.1260077
71. Pain, A., Choudhury, D., Bhattacharyya, S.K.: Seismic passive earth resistance using modified pseudo-dynamic method. Earthq. Eng. Eng. Vib. **16**(2), 263–274 (2017). https://doi.org/10.1007/s11803-017-0381-1

72. Patil, M., Choudhury, D., Ranjith, P.G., Zhao, J.: Behavior of shallow tunnel in soft soil under seismic conditions. Tunn. Undergr. Space Technol. **82**, 30–38 (2018). https://doi.org/10.1016/j.tust.2018.04.040
73. Council, B.S.S.: NEHRP Recommended Seismic Provisions for New Buildings and Other Structures (FEMA sP-750). Federal Emergency Management Agency, Washington, DC (2009)
74. Dickenson, S.E.: Dynamic response of soft and deep cohesive soils during the Loma Prieta Earthquake of October 17, 1989. Thesis dissertation, Civil Engineering, University of California. Berkeley, Californias (1994)
75. Amorosi, A., Boldini, D., Elia, G.: Parametric study on seismic ground response by finite element modelling. Comput. Geotech. **37**, 515–528 (2010)
76. Schnabel, P.B., Lysmer, J., Seed, H.B.: SHAKE: a computer program for earthquake response analysis of horizontally layered sites. Report No EERC72–12, Earthquake Engineering Research Center, University of California, Berkeley (1972)
77. Idriss, I.M., Lysmer, J., Hwang, R., Seed, H.B.: QUAD-4: a computer program for evaluating the seismic response of soil structures by variable damping finite element procedures. Report No EERC 73–16. Earthquake Engineering Research Center, University of California, Berkeley (1973)
78. Idriss, I.M., Sun, J.I.: SHAKE91: a computer program for conducting equivalent linear seismic response analyses of horizontally layered soils deposits. Center for Geotechnical Modeling, University of California, Davis (1992)
79. Ordonez, G.A.: SHAKE2000—a computer program for 1-D analysis of geotechnical earthquake problems. Geomotions, LLC, Lacey, Washington, USA (2012)
80. Choudhury, D., Savoikar, P.: Equivalent-linear seismic analyses of MSW landfills using DEEPSOIL. Eng. Geol. **107**(3–4), 98–108 (2009)
81. Phanikanth, V.S., Choudhury, D., Reddy, G.R.: Equivalent-linear seismic ground response analysis of some typical sites in Mumbai. Geotech. Geol. Eng. **29**(6), 1109–1126 (2011). https://doi.org/10.1007/s10706-011-9443-8
82. Desai, S.S., Choudhury, D.: Site-specific seismic ground response study for nuclear power plants and ports in Mumbai. Nat. Hazards Rev. **16**(4), 04015002 (2015). https://doi.org/10.1061/(ASCE)NH.1527-6996.0000177
83. Naik, N.P., Choudhury, D.: Comparative study of seismic ground responses using DEEPSOIL, SHAKE and D-MOD for soils of Goa, India. In: Geo-Congress 2014. Geotechnical Special Publication No. GSP 234, pp. 1101–1110 (2014) https://doi.org/10.1061/9780784413272.107
84. Shylamoni, P., Choudhury, D., Ghosh, S., Ghosh, A.K., Basu, P.C.: Seismic ground response analysis of KK-NPP site in the event of NCO earthquake using DEEPSOIL. In: Geo-Congress 2014. Geotechnical Special Publication No. GSP 234, pp. 840–849 (2014). https://doi.org/10.1061/9780784413272.082
85. Chatterjee, K., Choudhury, D.: Influences of local soil conditions for ground response in Kolkata city during earthquakes. In: Proceedings of the National Academy of Sciences, India, Section A: Physical Sciences, in press (2016). https://doi.org/10.1007/s40010-016-0265-1
86. Choudhury, D., Phanikanth, V.S., Mhaske, S.Y., Phule, R.R., Chatterjee, K.: Seismic liquefaction hazard and site response for design of piles in Mumbai city. Ind. Geotech. J. **45**(1), 62–78 (2015). https://doi.org/10.1007/s40098-014-0108-4
87. PLAXIS 2D AE.02 [Computer software]. PLAXIS BV, Netherlands

Chapter 13
Dynamic Characterization of Soils Using Various Methods for Seismic Site Response Studies

Pradeep Kumar Dammala and A. Murali Krishna

13.1 Introduction

Effective seismic-resistant design of structures requires thorough investigation of underlying soil's response to dynamic loading conditions. Such dynamic behavior of soils is governed by many factors and is represented in terms of strength and stiffness properties [1]. The dynamic soil stiffness is traditionally represented using the strain-dependent properties, often termed as dynamic soil properties: low-strain shear modulus (G_{max}), normalized shear modulus (G/G_{max}), and damping ratio (D) variation with shear strain (γ). The low-strain ($\leq 0.0001\%$) shear modulus (G_{max}) is the maximum dynamic shear stiffness a soil can possess and is linear elastic in nature [2]. With increase in the induced strains, the nonlinearity prevails and plastic strains are induced in the soil grains [3]. This result in the reduction of shear stiffness with shear strains, however in contrast, due to the increased work done, material damping increases [4]. Figure 13.1 illustrates the dynamic soil properties with typical strain ranges and their expected stress–strain response. In addition, the dynamic strength of saturated fine-grained soils is also represented using the ability to resist liquefaction and can be assessed by various field and laboratory equipment [5–8].

The requirement of dynamic soil behavior is application specific for earthquake geotechnical engineering studies. Some studies include liquefaction susceptibility analysis for effective mitigation applications [9]; G_{max} estimation for elastic ground response and fatigue analysis of foundations such as Offshore Wind Turbine (OWT)—[10]; comprehensive dynamic soil properties for Dynamic Soil–Structure Interaction (DSSI) analysis of deep foundations such as piles and caissons [11, 12]

P. K. Dammala · A. Murali Krishna (✉)
Department of Civil Engineering, Indian Institute of Technology Guwahati, Guwahati, India
e-mail: amurali@iitg.ac.in

P. K. Dammala
e-mail: dammala@iitg.ac.in

M. Latha G. (ed.), *Frontiers in Geotechnical Engineering*, Developments in Geotechnical Engineering, https://doi.org/10.1007/978-981-13-5871-5_13

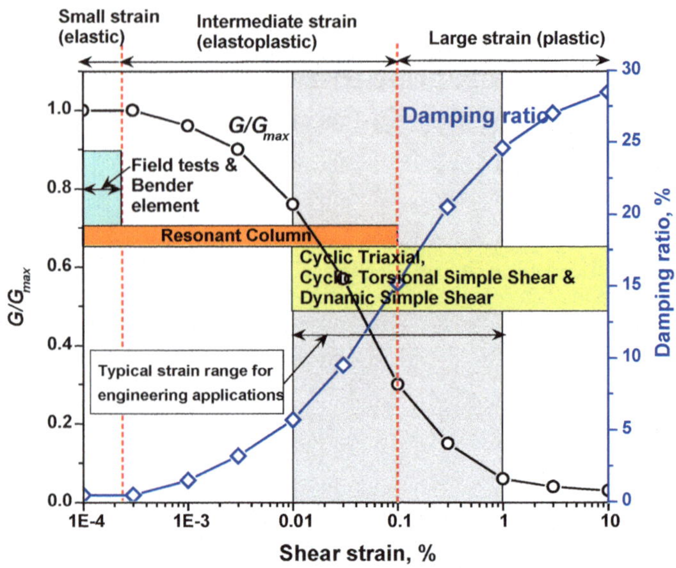

Fig. 13.1 Typical representation of strain-dependent dynamic soil properties

and studies involving nonlinear seismic ground response [13, 14]; earth retaining structures [15, 16]; seismic requalification studies [17–19].

Several field and laboratory testing methods have been developed by many researchers to investigate the dynamic behavior of soils [14, 20–24]. Field approaches involve the determination of low-strain properties (shear wave velocity, V_s) while the laboratory techniques yield the required properties over wide strain range. Some of the available field tests include seismic cross/up/down-borehole testing; Multi-channel Analysis of Surface Waves (MASW) [20]. Laboratory testing techniques include Bender element meant for G_{max} determination [22]; Resonant Column (RC) for dynamic soil properties up to a strain range of 0.1% [25]; Dynamic Simple Shear (DSS) for properties from 0.01 to 10% [26]; Cyclic Torsional Simple Shear (CTSS) apparatus for properties at strain range of 0.01–10% [27]; and Cyclic Triaxial (CTX) apparatus over strains >0.01% [28], see Fig. 13.1 for the testing techniques used to obtain the strain-dependent dynamic soil properties. Each testing technique is unique in its application with different advantages and inherent limitations [29].

The present article provides a description of some of the commonly used field and laboratory testing techniques for dynamic characterization of soils. Typical results obtained using the techniques are discussed and appropriate analytical formulations are provided. Furthermore, a seismic ground response study is also presented demonstrating the applicability of the proposed analytical formulations based on the test results.

13.2 Field Tests

Field tests generally involve the measurement of wave velocities propagating through the soil or the response of soil structure systems to dynamic excitation [29]. They can be grouped either into invasive or noninvasive techniques. Invasive methods require at least one borehole while noninvasive techniques (also called surface wave or refraction methods) are based on surface wave measurements. Invasive tests include seismic cross/up/down-borehole survey. Surface wave tests include Spectral or Multichannel Analysis of Surface Waves (SASW or MASW) and seismic refraction tests. This section describes MASW and cross-hole tests in detail with some typical results.

13.2.1 Multichannel Analysis of Surface Waves (MASW)

Multichannel Analysis of Surface Waves (MASW) is a noninvasive seismic survey for evaluating the 1D, 2D, and 3D stiffness profile of the subsurface in terms of shear wave velocity (V_s). This method has been widely used in professional practice due to the speed of implementation and is also budget friendly compared to the other seismic borehole methods. MASW test uses the dispersive characteristics of the surface waves with multi-receiver approach for the stratification of, mostly, the assumed vertically heterogeneous subsurface conditions. Active and passive MASW are different forms which are classified based on the source considered for the generation of surface waves [20]. Figure 13.2 schematically represents the active MASW test setup and instrumentation.

MASW typically consists of three-dependent stages—data acquisition, dispersion analysis, and inversion. Data acquisition is related to acquiring the response of subsurface due to the energy transmitted. Typical active MASW survey requires a sledgehammer as a source of energy, geophones (12 or more) as receivers and data acquisition system for recording and storing the data [30, 31]. Figure 13.3 shows a typical MASW survey with all the steps briefly presented. Upon acquiring the data using geophones at a specific location, data analysis (dispersion and inversion) is performed using computer programs such as SURFSEIS, EasyMASW, and Geopsy. Fixed standards have not been established in the surface wave methods due to the complexities involved in the data interpretation process and the variety of possible approaches to surface wave analysis [32]. However, reliable surface velocity stratification can be achieved by proper parameter selection during the data analysis and recommendations were proposed for efficient filtering techniques [30–32].

Two locations inside the IIT Guwahati campus were considered to provide the subsoil stratification through V_s profile. An array of 24 geophones with 2 m spacing in-between, a 10 kg sledgehammer, and 24-bit data acquisition system were used. Time sampling parameters and data processing using SURFSEIS program can be found in detail in Kashyap et al. [30]. Figure 13.4 presents the final obtained shear wave velocity profile of the chosen locations in the IIT Guwahati campus.

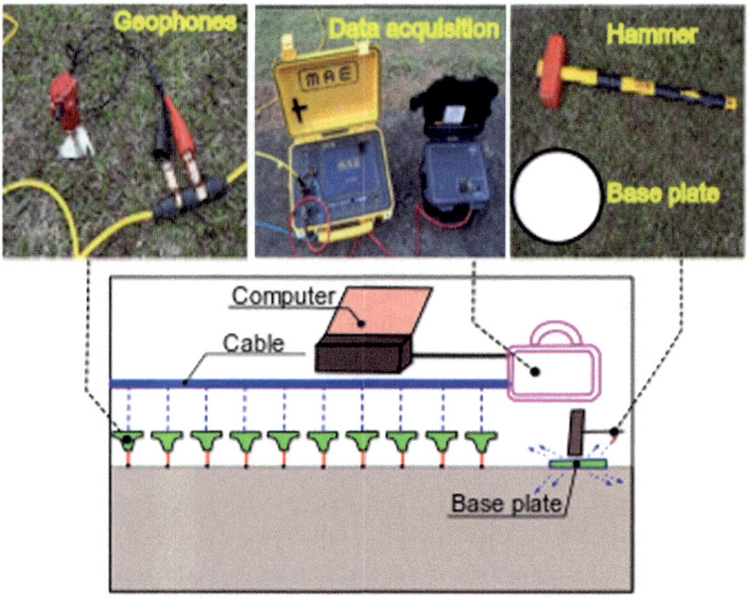

Fig. 13.2 Schematic representation of active MASW setup

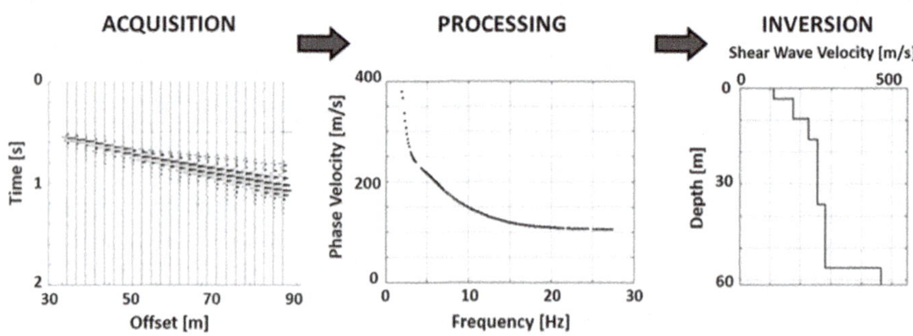

Fig. 13.3 Overall procedure of MASW survey [32]

13.2.2 Cross-Hole (CH) Test

Seismic Cross-Hole (CH) survey is an invasive test, whereby the velocity (either P wave or S wave) of the soil deposit is determined using two or more boreholes. The seismic energy is generated at the bottom of a borehole while the sensors in the adjoining borehole at same depth would act as receivers [33]. The mutual distance is measured along with the arrival times of the waves, yielding the velocity profile. Figure 13.5 a shows the essentials of the CH survey. The CH test typically requires two or more boreholes arranged in a straight line array for a better accuracy [29]. Interpretation technique in CH survey is quite straightforward and does not require

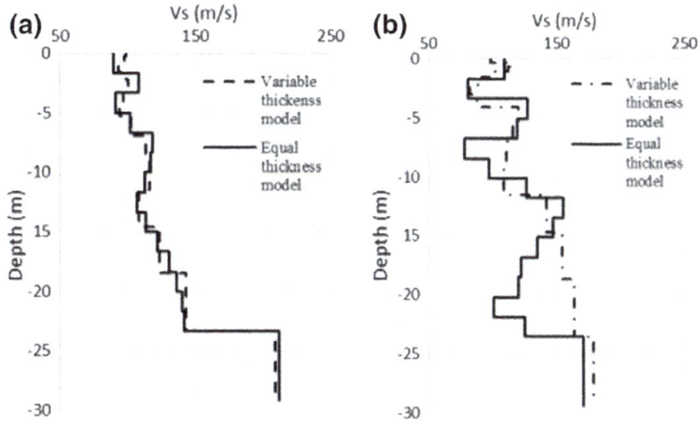

Fig. 13.4 Shear wave velocity variation with depth for the considered soil profiles [30]

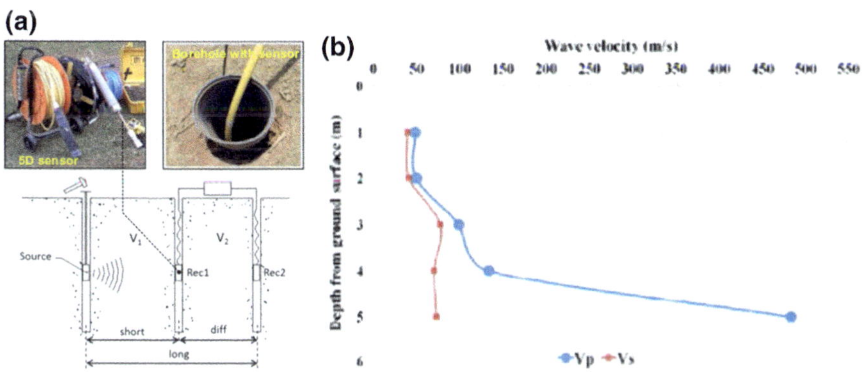

Fig. 13.5 **a** Cross-hole scheme with three boreholes and instrumentation (modified after Garofalo et al. [35]) **b** typical wave profile obtained from CH survey [36]

any complex inversion schemes unlike the surface wave methods and hence the reliability of the results mainly depends on the accuracy of the measurement and in the precision of the instrumentation [34].

Taipodia et al. [36] conducted CH tests at the IIT Guwahati campus to determine the P and S wave velocity profile. The seismic source (a Ballard shear wave generator) is installed in one borehole while the 5D sensor (geophone) is lowered to the same depth as the seismic source in the second borehole. Both the boreholes were 4 m apart and the experiments were conducted by striking the energy generator and recording the signals using the 5D sensor. The P and S wave velocities were determined by picking the "first-arrival time picking" method from the recorded signature [36]. Figure 13.5b presents the P and S wave velocities of the borehole. Further details on arrival time picking and analysis can be found in Taipodia et al. [36].

All the field tests (both invasive and noninvasive) provide the wave velocities with a certain degree of uncertainty like any other geotechnical investigations. The main drawbacks of the field tests are that only low-strain stiffness can be determined and samples for further analysis cannot be arranged. However, one or two field tests are always suggested for projects of relevant importance. The invasive test results for one-dimensional wave profiles are considered relatively reliable compared to the surface wave methods as the uncertainties involved in the testing and data analysis are less. However, the use of borehole surveys is limited in the recent days as they lead to the disturbance in the natural fabric of the soil, ineffective determination of wave velocities in lateral heterogeneity and the requirement of deep borehole for deeper wave velocity determination making it cost ineffective [20, 35].

13.3 Laboratory Element Testing Techniques

Field tests provide dynamic stiffness properties only in the low-strain range and their application is often limited due to the large scale of testing involved. However, laboratory element testing of the collected field samples would provide the desired dynamic properties over a wide strain range. In addition, different field conditions (varying void ratio, plasticity index, confining pressure, etc.) can be explored in the element testing. This section details some of the widely used laboratory element testing techniques to investigate the dynamic soil behavior.

13.3.1 Bender Element Testing

Bender element test is a low-strain test in which the maximum shear modulus (G_{max}) of the soil sample is estimated by transmitting and recording a shear/compressional wave through the sample. Lawrence [37], initially, used the shear plates for measuring the shear wave velocity in sands, and later Shirley and Hampton [38, 39] adopted bender elements for measuring shear wave velocity for marine sediments. Arulnathan et al. [40] recommended methods for effective analysis of bender test results. A bender element is an electro-mechanical transducer which can either bend by change in the induced voltage or generates a voltage as it bends [22]. Two benders are installed on the sample—one at the top plate and the latter at the bottom plate to act as transmitter and receiver, respectively. Figure 13.6 shows a pictorial view of the bender element apparatus, schematic view of the sample conditions and the piezoelectric bender elements used for the testing. These benders can be installed on triaxial, shear, or one-dimensional compressional apparatus and both the tests can be done on the same sample.

A high-frequency electrical pulse (input signal) applied to the transmitter at the bottom platen will deform the bender rapidly resulting in a stress wave, transmitting through the sample toward the receiver. Upon reaching the receiver, the stress wave

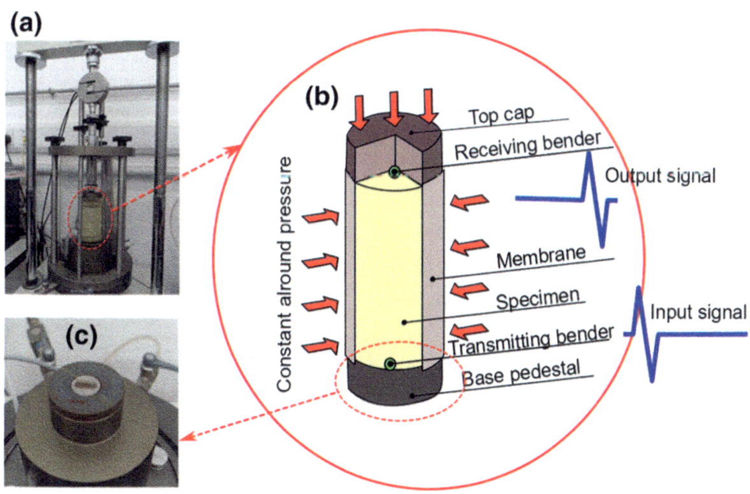

Fig. 13.6 **a** Bender element apparatus **b** schematic view of loading on the sample and **c** view of transmitting bender at the base

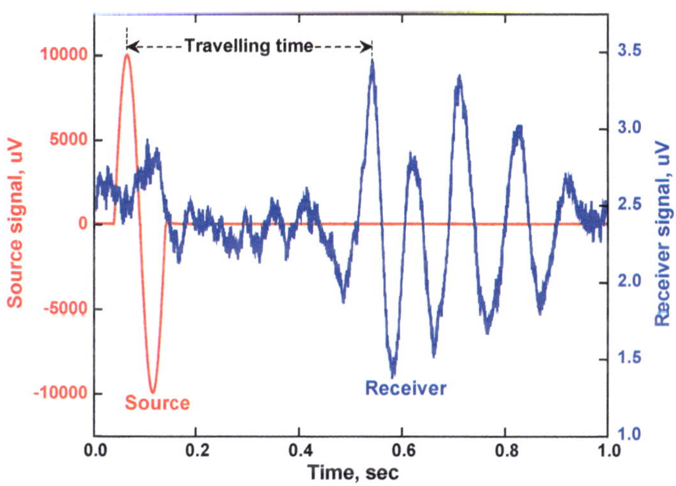

Fig. 13.7 Typical input and output signals for a bender element test

generates a voltage pulse which can be measured. Figure 13.7 illustrates the input pulse transmitted and the output signal received at the receiver. The time the stress wave travels through the sample to reach the receiver element is recorded (Fig. 13.7). The velocity (V_s) can be estimated using the effective height of the sample (distance between the two benders at transmitter and receiver—H_s) and the time the stress wave travels (T_s). Using the shear wave velocity and the density of the sample (ρ), maximum shear modulus (G_{max}) can be estimated.

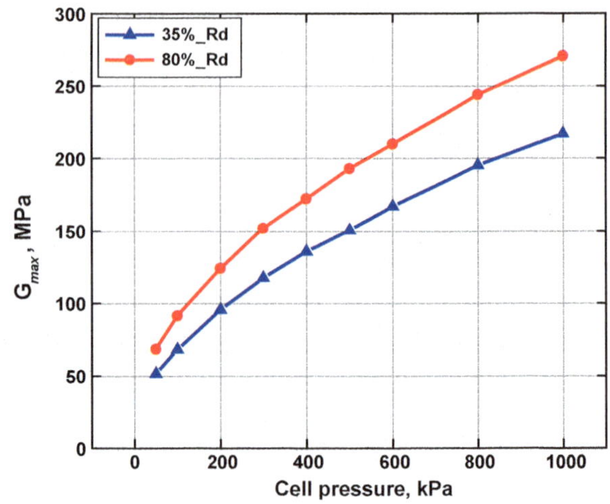

Fig. 13.8 Variation of G_{max} with confining pressure for BP sand at two relative densities

$$G\text{max} = \rho \cdot V_s^2 = \rho\left(\frac{H_s}{T_s}\right)^2 \tag{13.1}$$

Once the velocity is estimated at a particular confining pressure on the sample, the same sample can be used for determining wave velocities at other confining pressures. This is due to the low strains induced in the sample which are not expected to disturb the natural fabric [22]. Figure 13.8 presents the G_{max} results obtained from bender element test on Brahmaputra sand (BP) at two relative densities (R_d) and varying confining pressures. The index and engineering properties of BP sand can be found in the literature [14, 41]. It can be obvious that the increase in the confining pressure and relative density increases the stiffness of the sample.

13.3.2 Resonant Column (RC) Testing

Resonant column (RC) test is used to measure the shear modulus and damping characteristics of soils from low to intermediate strain levels (<0.1%). The basic principle involved in RC testing is the theory of wave propagation in prismatic rods [2], where a cylindrical soil specimen is harmonically excited till it reaches the state of resonance (peak response). The RC technique was initially used for soils by Iida [42], following which the method was further developed by many researchers [43, 44] and was also standardized in ASTM D 4015 [45].

Three different versions of RC apparatus are available, based on the end conditions to constraint the specimen. They are: (a) *Fixed-free* condition [46], in which the bottom end is fixed against rotation while the top end is free to rotate under applied torsion. A known mass is added at the free end to obtain uniform distribution of strain

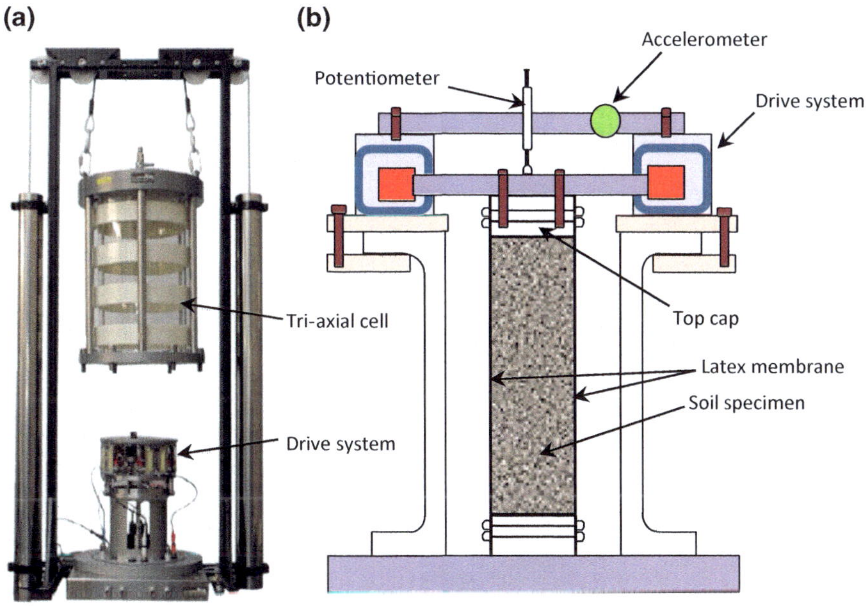

Fig. 13.9 **a** Photographic and **b** schematic view of RC apparatus [14]

throughout the length of the sample, (b) *Spring-base* [47], in which an equivalent spring of stiffness is present at the bottom end of the sample and depending on the stiffness of the spring relative to the stiffness of the soil, the base end condition can be fixed or free, (c) *Fixed-partially restrained* [48], in which the bottom end is fixed while the top cap is partially restrained by springs acting against an inertial mass.

Figure 13.9 shows a fixed-free type of RC apparatus with the instrumentation details. In brief, the soil specimen is excited under a harmonic torsional vibration, induced in the form of electric voltage through the electromagnetic drive system, consisting of four magnets (Fig. 13.9). Initially, a small amount of electric current (say 0.001 V) is passed through the magnetic coils and a broad and fine sweep is conducted to find the exact resonant frequency of the sample [14]. Using this resonant frequency (f_{nz}), height of sample (H), and the instrument constant (β) which can be obtained by calibration of the instrument [45], the shear wave velocity (V_s) and corresponding shear modulus (G) of the sample is determined (Eq. 13.2).

$$G_{\max} = \rho \cdot V_s^2 = \rho \cdot \left(\frac{2\pi f_{nz} H}{\beta} \right)^2 \tag{13.2}$$

Once the resonant frequency is obtained at a particular input voltage, the input current to the coils is switched off to perform a free vibration test. The response of the accelerometer with time is recorded from which the amplitude decay curve is obtained. The peak amplitude of each cycle (A_1—A_{n+1} with n as number of cycles of

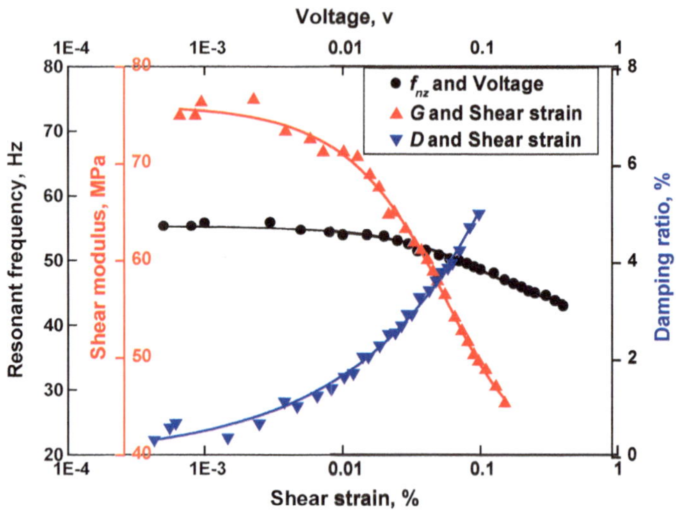

Fig. 13.10 Variation of f_{nz} with input voltage, shear modulus and damping ratio with shear strain for BP sand at 100 kPa effective confining pressure [49]

free vibration) is determined and the corresponding damping ratio (*D*) is evaluated as suggested by ASTM D 4015 [45].

$$D = \left(\frac{1}{n}\right) \ln\left(\frac{A_1}{A_{n+1}}\right) \tag{13.3}$$

Once the shear modulus and damping ratio at a particular strain (particular voltage) are obtained, then the input voltage to the system is increased to obtain resonant frequency and free vibration response at higher strains. Thus, the voltage is gradually increased at some intervals till the strains reach 0.1% to yield the variation of shear modulus and damping ratio. Figure 13.10 shows the typical variation of resonant frequency with the input voltage, shear modulus, and damping variation with shear strain for BP sand at 30% relative density and 100 kPa effective confining pressure. It can be noted that the increase in the voltage increases the strains induced in the sample resulting in reduced shear stiffness and higher damping ratio [14].

Figures 13.11 presents the variations of shear modulus with shear strain for samples at different confining pressures for BP sand at 30% relative density which were further presented in Fig. 13.12 as modulus ratio along with damping ratio curve. It is obvious that the increase in confining pressure (meaning-overburden depth) increases the stiffness of the sample while damping ratio decreases. Normalized stiffness (ratio of shear modulus to G_{max}) also increases with the confining pressure indicating that the depth of overburden decreases the rate of reduction of the stiffness of sands.

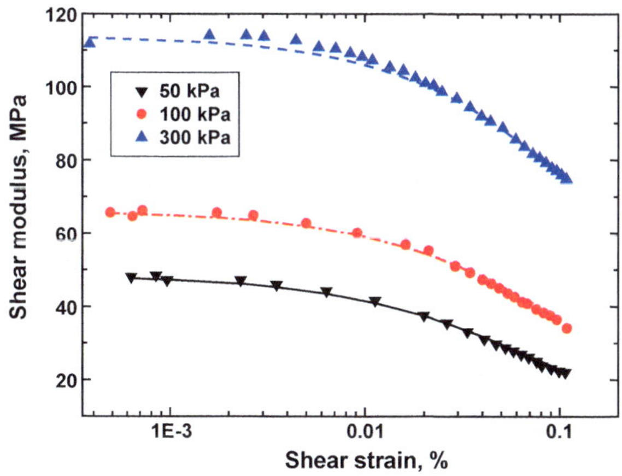

Fig. 13.11 Variation of shear modulus with shear strain for BP sand at different confining pressures for 30% relative density [14]

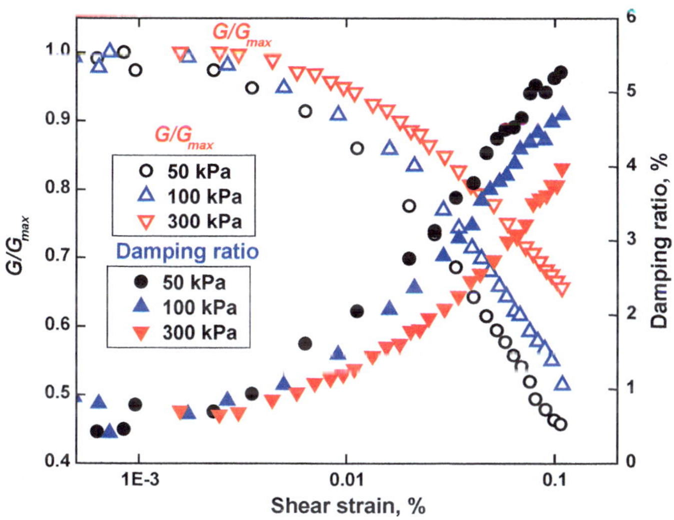

Fig. 13.12 Variation of modulus ratio and damping ratio with shear strain for BP sand at different confining pressures for 30% relative density [14]

13.3.3 Cyclic Triaxial (CTX) Testing

Cyclic Triaxial (CTX) test is a high strain test, which can be used for determining both the dynamic soil properties and liquefaction potential of soils [6, 41]. The strain range typically tested in this apparatus is 0.01–10%. Seed and Lee [6], initially, used CTX apparatus for assessing liquefaction potential of saturated sands while many researchers [28, 41, 50] adopted the apparatus for determining high-strain dynamic soil properties. Kumar et al. [51] utilized CTX apparatus for determining the dynamic behavior of northeast Indian cohesive soil and also for assessing its liquefaction potential. Lombardi et al. [52] used CTX apparatus to investigate the post-liquefaction behavior of silica sands for application in pile foundations subjected to liquefaction-induced stresses. Both the dry and saturated samples can be tested using the apparatus [53]. Irregular dynamic excitations can also be applied to investigate the effect of realistic earthquake motions [41]. The apparatus can be used for both the stress- and strain-controlled testing [54, 55].

In the CTX apparatus, the soil sample is enclosed inside a triaxial cell through which harmonic axial loads using the servo-controlled actuator at the top can be applied and necessary instrumentation is installed to measure the deformations (axial) and pressures (cell pressure and pore pressure). Figure 13.13 shows a picture of typical CTX setup and various parts associated with it. The details about sample preparation, CTX setup, and testing methodology can be found in the literature [24, 41]. Figure 13.14 presents the typical results of liquefaction potential of BP sand using the CTX system. The stress-controlled test results are presented in Fig. 13.14, which shows a cyclically varying deviatoric stress of ±20 kPa on the sample applied at 1 Hz frequency (Fig. 13.14a). Liquefaction susceptibility of a soil sample can be represented using the pore water pressure ratio (r_u) which is the ratio of excess pore pressures (Δu) to the mean effective confining pressure (σ'_c). Figure 13.14b shows the variation of axial strain and r_u with loading cycles. It can be noted that the sudden increase of axial strain was attributed to the full liquefaction condition. Also, the stress–strain response of the BP sand is presented in terms of deviatoric stress and axial strain (Fig. 13.14c) and also shear stress and shear strain (Fig. 13.14d). The possible limitation in the CTX apparatus is that the shear strains have to be estimated from the axial strains by assuming a Poisson's ratio, unlike the DSS apparatus, whereby the shear displacements can directly be applied to the specimens.

Figure 13.15 presents the variation of average pore water pressure ratio with loading cycles for BP sand at 30% relative density. The results are related to the strain-controlled tests performed at 100 kPa effective confining pressure. An increase in the pore pressure ratio can be observed with increase in the input strains and the BP sand did not show complete liquefaction for low strains (<0.075%) even up to 40 cycles of loading. However, with the strains beyond 0.075%, increased the tendency toward liquefaction has been noticed with the input strains. Liquefaction was even observed below five loading cycles for high strains (0.75%) which shows that high-intensity ground motions would lead to rapid liquefaction in the sandy deposits [41].

13.3.4 Dynamic Simple Shear (DSS) Testing

Dynamic Simple Shear (DSS) apparatus can be used to investigate the dynamic behavior of soils from intermediate (0.01%) to high strains (up to 10%). Simple Shear device dates back almost half a decade (1950s). The device was initially developed

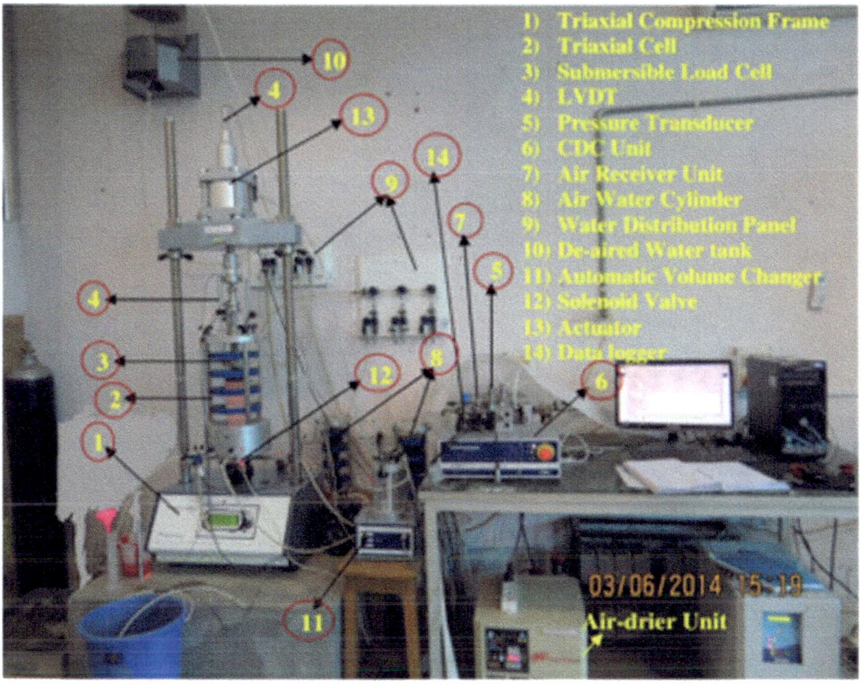

Fig. 13.13 Cyclic triaxial setup and components [41]

in the UK by Roscoe [56], Scandinavia at SGI [57], and at NGI [26]. Later, Peacock and Seed [58] extended the simple shear to assess the liquefaction potential and cyclic strength of soils and named as Dynamic Simple Shear (DSS). A pictorial view of the apparatus and a schematic view of the soil sample during loading are shown in Fig. 13.16. The device is similar to a simple shear except that the provision for cyclic loading through servo-controlled actuator. Axial and lateral displacements during the testing can be monitored using the LVDTs attached over the sample. The main advantage of DSS over the triaxial apparatus is that a shear stress/strain can be directly simulated with the two-way controlled shearing [59]. The soil specimen is contained using laminated circular rings and during shearing, the rings follow the movement of soil specimen, see Fig. 13.16 for details. The confinement is provided through the constant axial stress.

Typical results obtained from the DSS apparatus are presented briefly in this section. Figure 13.17 presents the cyclic stress strain response (also called hysteresis loops) of BP sand prepared at a medium relative density at normal stress of 100 kPa. The cyclic shear loads were applied for 1000 cycles (N), however, Fig. 13.16 shows only the first 10 cycles for brevity. Using the hysteresis loops, one can calculate the strain-dependent dynamic shear stiffness and damping ratio [41, 59]. An important aspect of hysteresis loops needs to be noted: at high shear strain or larger loading cycles, an asymmetricity in the shape of the loop is observed [41], please see

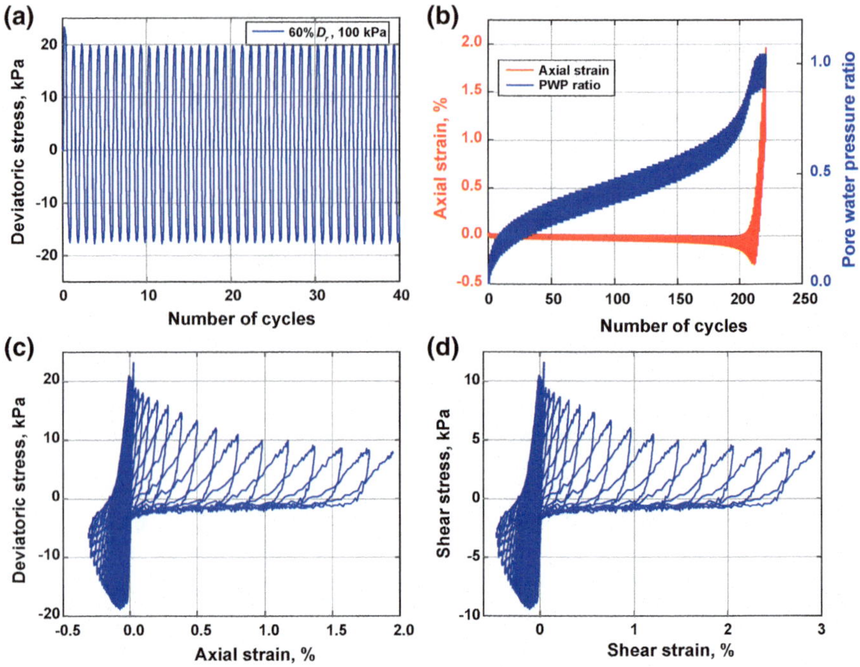

Fig. 13.14 Typical results of CTX apparatus for liquefaction assessment on BP sand

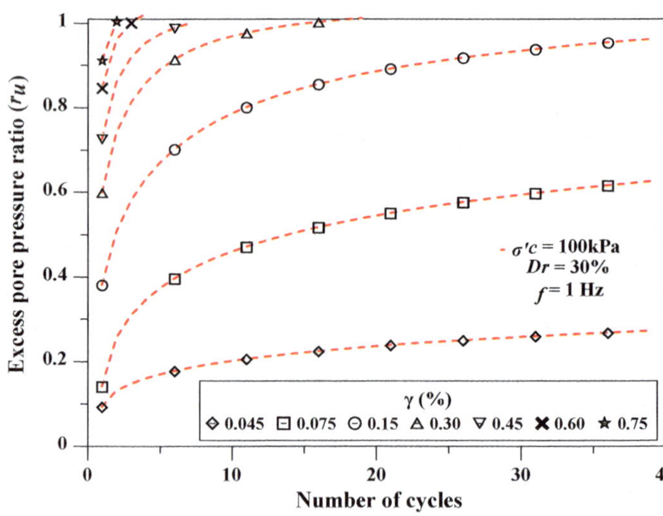

Fig. 13.15 Variation of pore water pressure ratio for BP sand at 100 kPa effective confining pressure [49]

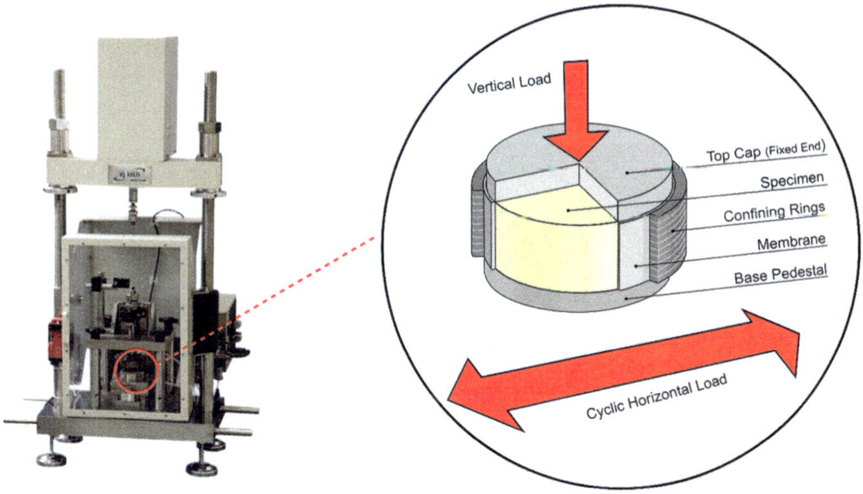

Fig. 13.16 Pictorial and schematic of DSS apparatus

Fig. 13.17 Hysteresis loops for BP sand at three confining pressures for medium density sample

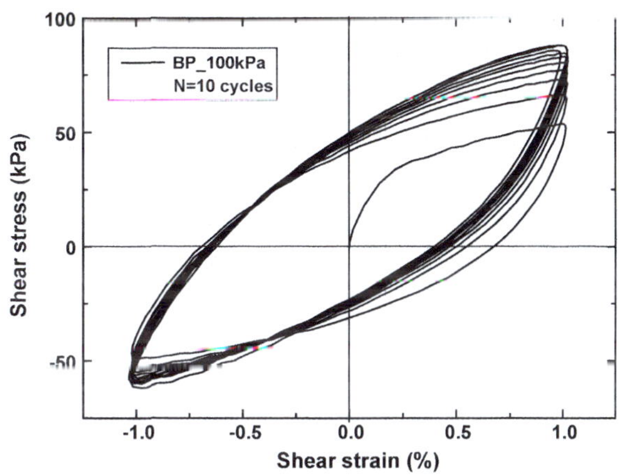

Fig. 13.17 for details. As the dynamic soil behavior is strain dependent, the dynamic soil properties should also be based on the realistic asymmetric loops rather than the traditional symmetric loop calculations. In this view, Kumar et al. [41] presented a method to calculate the strain-dependent properties using the realistic asymmetric loop. Figure 13.18 describes the methodology followed and the corresponding formulations.

Figure 13.19 presents the variation of shear modulus and damping ratio of BP sand evaluated using the asymmetric loop formulations proposed by Kumar et al. [41]. Reduction in shear modulus with shear strains is obvious with increase in shear strain, however, an untraditional trend of increase and decrease of damping ratio with

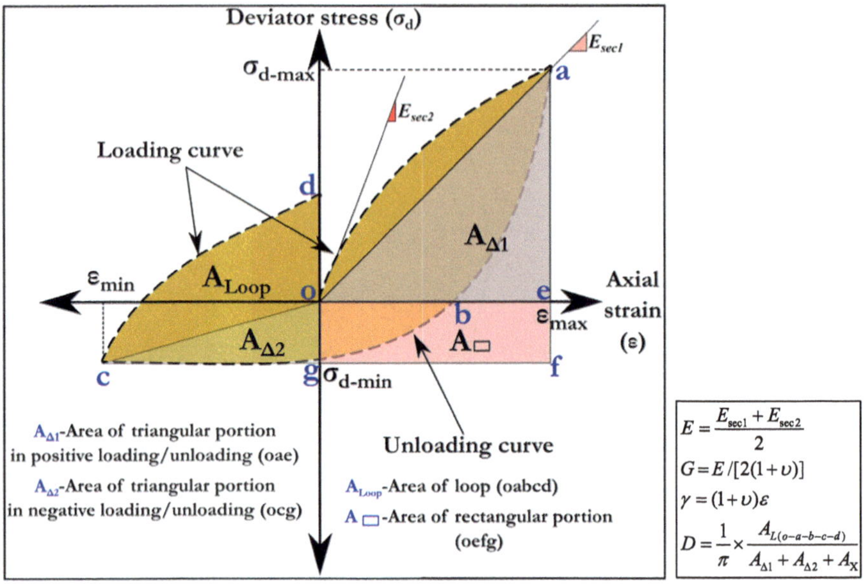

Fig. 13.18 A typical asymmetric loop with formulations for dynamic soil properties determination (modified after Kumar et al. [41])

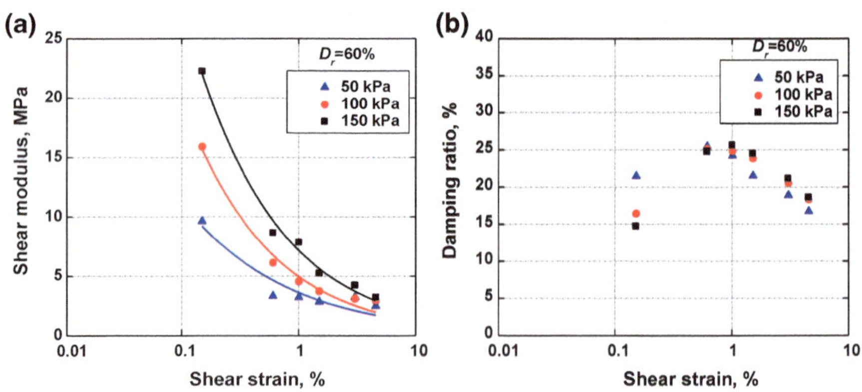

Fig. 13.19 Variation of shear modulus and D with shear strain for BP sand at 60% relative density [49]

shear strain has been observed. This uncommon damping behavior is attributed to the realistic asymmetric loops considered for the evaluation [41].

Once the dynamic soil properties at different shear strains are determined using independent apparatus, the data needs to be combined for a comprehensive data over wide strain range. Figures 13.20 and 13.21 provide the data determined using RC and DSS for BP sand at different relative densities and confining pressures, respectively. Also, the literature suggested [60, 61] ranges of modulus and damping

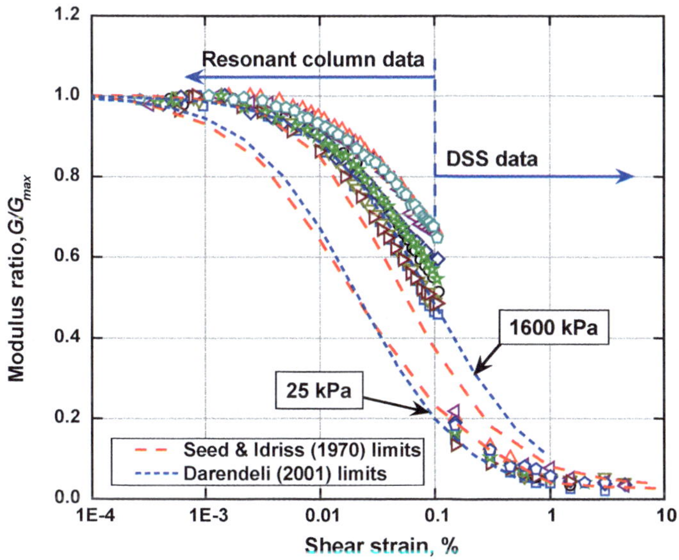

Fig. 13.20 Modulus ratio variation for BP sand determined from RC and DSS testing

curves for sands are presented (Figs. 13.20 and 13.21). It can be observed that the modulus values fall rightwards in the low-strain range than the Seed and Idriss [60] suggested ranges and follow the suggested range in the high strain levels. Similarly, damping ratio values fall below the suggested ranges, meaning-damping ratio would be overestimated if the site-specific data is not utilized for earthquake geotechnical applications [14, 62].

13.4 Analytical Comparisons

Analytical formulations are required in cases where laboratory data is insufficient for the required conditions. This section describes the analytical comparisons fitted to the laboratory data of dynamic soil properties to the literature suggested formulations.

13.4.1 G_{max}

Several forms of G_{max} estimation are available in the literature [63]. Out of the existing G_{max} formulations, the equation (Eq. 13.4) proposed by Hardin [64] is being considered most often in the earthquake geotechnical applications [63, 65, 66] due to its dimensional consistency and application even to soils of large void ratio [21].

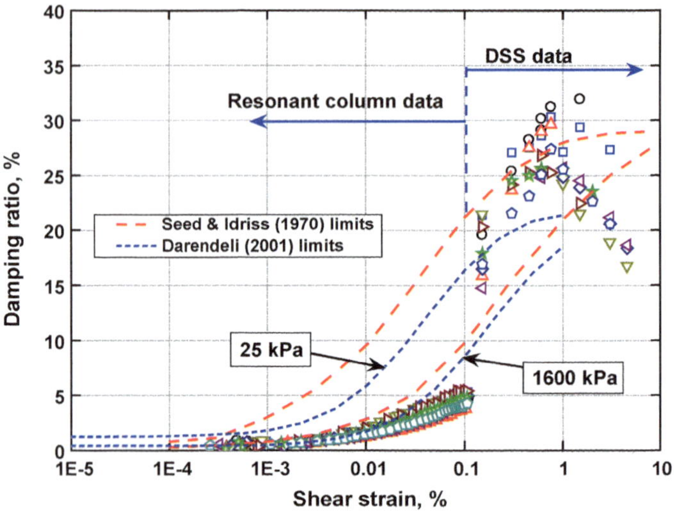

Fig. 13.21 Modulus ratio variation for BP sand determined from RC and DSS testing

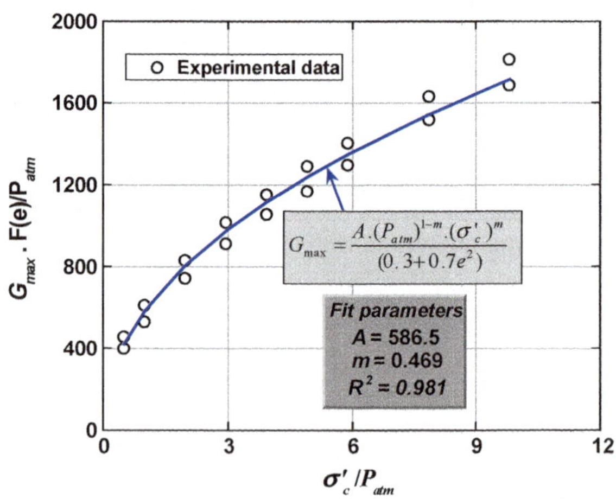

Fig. 13.22 Fitted G_{max} results in Eq. 13.4

$$G\text{max} = \frac{A \times (P_a)^{1-m} \times (\sigma_c')^m}{\left(0.3 + 0.7e^2\right)} \tag{13.4}$$

where P_a is atmospheric pressure (101 kPa), A is a constant depending on the type of soil, and m is a stress-dependent factor. The bender element tests performed on BP sand were considered for the regression analysis to obtain the best-fit parameters based on the equation (Eq. 13.4). Figure 13.22 shows the regression analysis of the bender data and the obtained best-fit parameters—$A = 586.5$, $m = 0.469$ with a correlation coefficient of $R^2 = 0.981$.

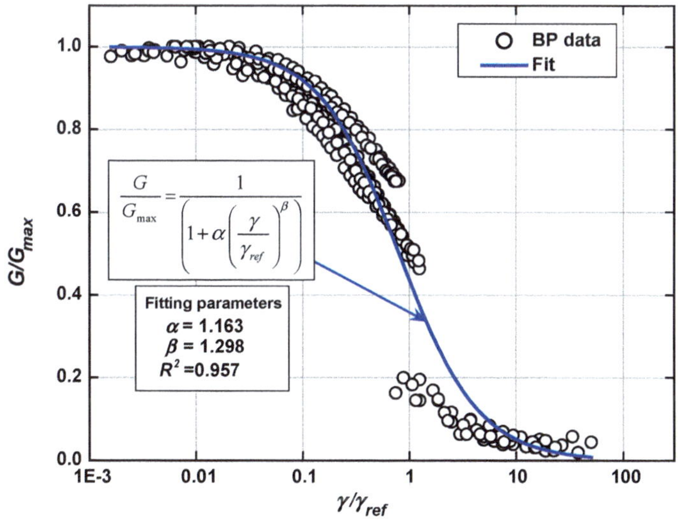

Fig. 13.23 Variation of normalized modulus to the normalized shear strain for BP sand

13.4.2 G/G$_{max}$ *and Damping*

Similarly, analytical formulations are required for modulus reduction and damping ratio variation with shear strain. Hyperbolic and modified hyperbolic stress-dependent equations were proposed by various researchers in order to model both the modulus degradation and damping variation [60, 61, 67–71]. Matasovic and Vucetic [71] formulation (Eq. 13.5) was found to be satisfactorily simulating the modulus degradation behavior.

$$\frac{G}{G_{max}} = \frac{1}{1 + \beta\left(\frac{\gamma}{\gamma_{ref}}\right)^{\alpha}} \tag{13.5}$$

where γ_{ref} is the reference shear strain (shear strain at G/G_{max} value of 0.5 according to Darendeli [61] and α and β are the curve-fitting parameters adjusting the shape of the modulus curve. As the shape of modulus curve is soil specific and hence, the corresponding α and β needs to be determined from the regression analysis of experimental data. Figure 13.23 presents the regression analysis results in terms of variation of normalized modulus to the normalized shear strain for BP sand. The obtained best-fit parameters are also mentioned—α = 1.163 and β = 1.298 with a correlation coefficient of 0.957. Similar simplified regression analysis has been performed by Dammala et al. [14]

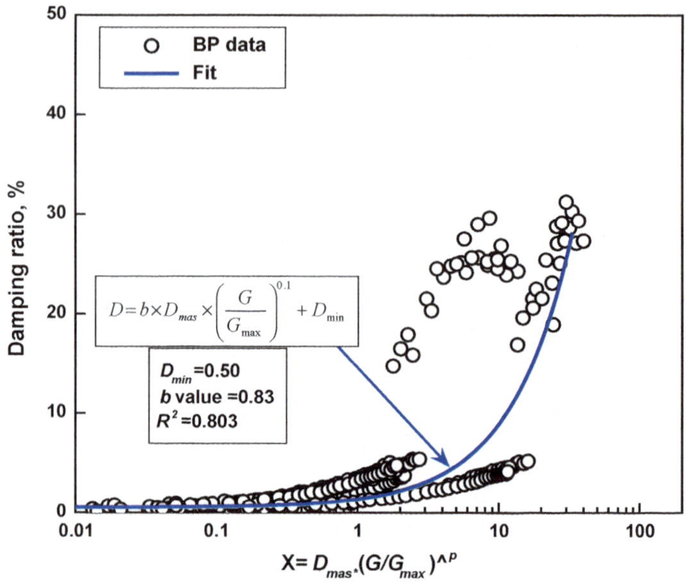

Fig. 13.24 Damping ratio variation with masing damping combined with modulus ratio

Similar to the modulus reduction, analytical expressions for evaluating damping ratio at a given shear strain was also proposed by several researchers [61, 67–69]. Darendeli [61] related damping ratio (Eq. 13.6) with the modulus reduction and achieved reasonable estimates.

$$D(\%) = b \times D_{mas} \times \left(\frac{G}{G_{max}}\right)^{p} + D_{min} \tag{13.6}$$

where b and p are the scaling coefficients adjusting the high-strain damping ratio (p considered as 0.1), D_{mas} is the masing damping (function of shear strain and the curvature coefficient (α)), and D_{min} is the minimum damping ratio at the lowest possible shear strain (0.5% based on the experimental results). A regression model is run to find the best fit values of b. A b value of 0.83 is found with an R^2 of 0.803. Figure 13.24 presents the simplified regression analysis for damping ratio.

The final stiffness curves (both modulus and damping) of BP sand can be estimated using the regression coefficients (α, β, and b) at any required σ'_c [14]. Such stiffness curves can act as a ready-made tool for design engineers, especially during the design of important structures or the seismic requalification works in the northeastern Indian region.

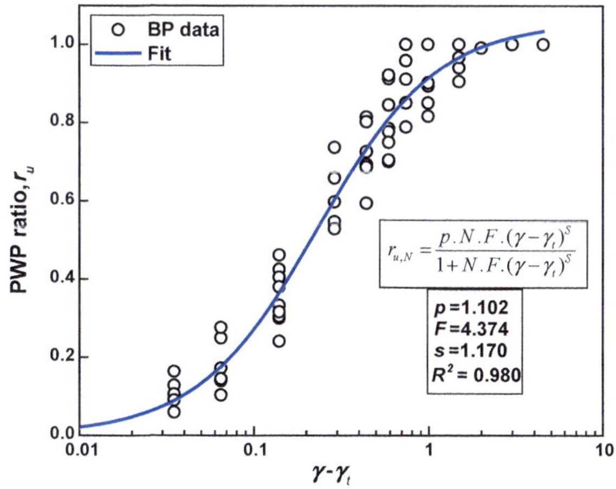

Fig. 13.25 Pore water pressure ratio variation of BP sand [49]

13.4.3 Liquefaction Potential

Liquefaction resistance or susceptibility of soils can be modeled reliably using the pore water pressure (PWP) formulations. The PWP models simulate the generation and dissipation of excess pore pressures during cyclic loading. Numerous PWP models, ranging from simple to complex nature, are available in the literature, see for example—Hashash et al. [72], Dobry et al. [73], Martin et al. [74], Byrne et al. [75]. However, to perform nonlinear effective stress analysis, some commercial programs such as DEEPSOIL [76], employ extended Dobry et al. [73] PWP model as the required input parameters can be efficiently obtained by performing stress/strain-controlled CTX/DSS tests on saturated soil samples. Based on the strain-controlled CTX tests on sandy soil, Vucetic and Dobry [77] extended the basic PWP model developed by Dobry et al. [73] (Eq. 13.7). The model emphasizes the generation of PWP with the number of cycles (N) and applied cyclic shear strain.

$$r_{u,N} = \frac{p\,.\,N\,.\,F\,.(\gamma - \gamma_t)^S}{1 + N\,.\,F.(\gamma - \gamma_t)^S} \tag{13.7}$$

where $r_{u,\,N}$ = excess PWP ratio at N number of cycles; γ = cyclic shear strain amplitude; p, F, and s are the curve-fitting parameters; and γ_t is the threshold shear strain below which no significant PWP is generated and is considered as 0.01% [3]. The fitting parameters (p, F, and s) are obtained by best fitting of the pore pressure model on the obtained experimental data of BP sand. Figure 13.25 depicts the obtained results of the PWP model with the fitting parameters. Since the dynamic properties (shear modulus and damping ratio) of BP were evaluated for the first cycle ($N = 1$), the fitting parameters of the PWP model have also been evaluated for the first loading cycle ($N = 1$).

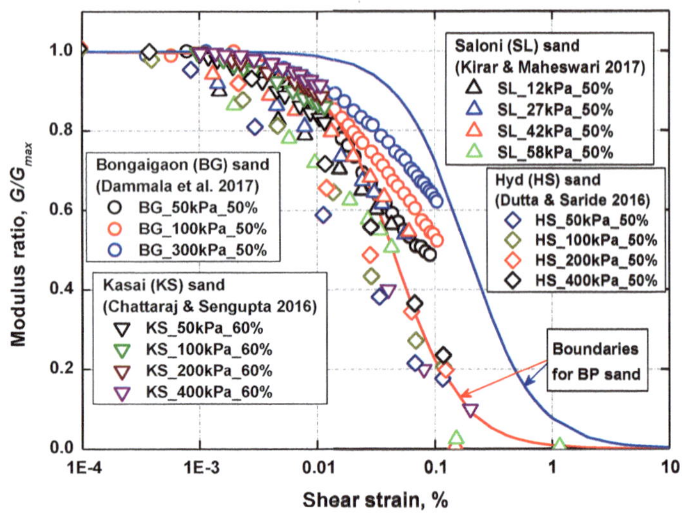

Fig. 13.26 Modulus ratio of Indian sandy soils

13.5 Indian Soil Data

Modulus and damping ratio of Indian sandy soils has been collected from the literature and is shown in Figs. 13.26 and 13.27, respectively, along with the boundaries proposed for BP sand. Data of four sandy soils determined by various laboratory element tests was presented—Bongaigaon sand from northeast India [14], Kasai sand of eastern India (Kolkata) [65], Saloni sand of northern India [78, 79], and sandy soil from southern India (Hyderabad region) [80]. Though the proposed modulus and damping boundaries for BP sand are close to the other sands data, however, there is still some gap exists in both the damping and modulus.

13.6 Application of Established Soil Properties

A one-dimensional nonlinear (NL) time-domain effective stress GRA, incorporating the PWP generation and dissipation, has been performed to demonstrate the applicability of proposed dynamic soil properties and liquefaction parameters. The computer program *DEEPSOIL* [76] has been used for the analysis as it is a widely used nonlinear time-domain site response analysis program which utilizes a discretized multi-degree-of-freedom lumped parameter model of the 1D soil column. The hysteretic soil response is captured by a pressure-dependent hyperbolic model that represents the backbone curve of the soil along with the modified extended unload–reload Masing rules [72, 81]. Further details about the theoretical back-

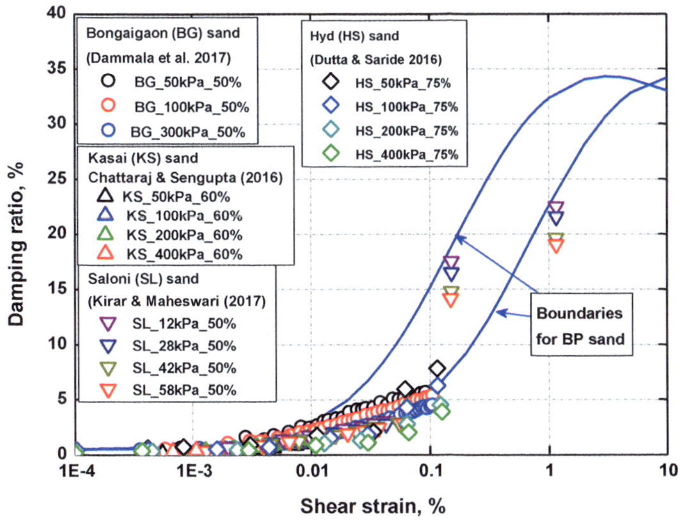

Fig. 13.27 Damping ratio of Indian sandy soils

ground of the methodology can be found in [76, 82]. Similar GRA studies for typical sites in Guwahati city were conducted by [14, 83, 84].

Soil profile considered for the present study is located near the shore of Brahmaputra river in Guwahati, which is considered as one of the most active seismic regions of the country with 0.36 g as the expected Peak Bed Rock Acceleration (PBRA) [85]. Figure 13.28 presents the location of the soil profile considered and the seismicity of the region. Dammala et al. [14, 19] details the soil stratigraphy and further details regarding the parameters considered for the analysis. In the wake of prior warnings issued by the seismologists that this region is prone to a pounding seismic event in the near future, several requalification studies were conducted in this region [19, 86]. Four recorded ground motions of varying ground motion parameters have been used for the analysis (Table 13.1).

Figure 13.29 presents the ground response analysis results in terms variation of Peak Ground Acceleration (PGA), peak shear strain, and PWP ratio along the depth of the profile. Amplification for low-intensity ground motions was observed while deamplification/attenuation for high-intensity motions was noticed. This pattern of amplification/attenuation can be justified through the damping characteristics of the soil sediments, whereby the surficial stratum which is loose is expected to experience high shear strains (Fig. 13.29b) leading to high damping, thereby resulting in attenuation of the incoming waves. In contrast, the low-intensity motions induce less strain for which the damping is also low leading to high amplifications [88, 89]. In addition, the ground motions with PBRA > 0.10g liquefy the surficial stratum which is also justified through the results obtained using the semi-empirical approach proposed by Idriss and Boulanger [7].

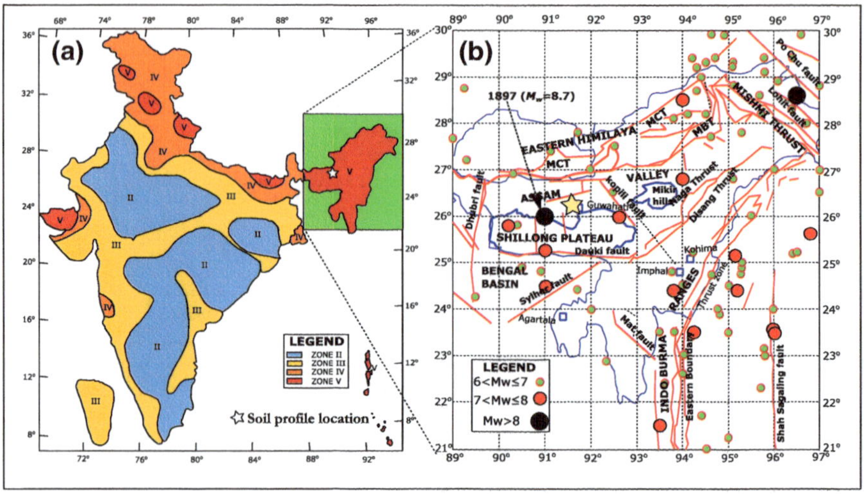

Fig. 13.28 **a** Seismic zonation map of India [85] **b** seismotectonic faults in the northeastern India [87]

Table 13.1 Strong motion parameters of the considered ground motions (modified from Dammala et al. [49])

Earthquake parameters	Sikkim 2011 earthquake	Sonitpur 2009 earthquake	Indo-Burma 1988 earthquake	Indo-Burma 1988 earthquake
Short name	SKM 0.02g	SONIT 0.03g	INDBUR 0.18g	INDBUR 0.33g
Date of occurrence	18-09-2011	19-08-2009	06-08-1988	06-08-1988
Moment magnitude	6.8	4.9	7.2	7.2
Epicenter	27.52° N 88.04° E	26.60° N 92.50° E	25.15° N 95.13° E	25.15° N 95.13° E
Recording station	IIT Guwahati	Guwahati	Diphu	Diphu
Distance from source (km)	378	424	193	193
PBRA (g)	0.02	0.031	0.18	0.330
Depth (km)	20	20	91	91

13.7 Summary and Concluding Remarks

Determining the dynamic behavior of soil is a prerequisite for an efficient earthquake-resistant foundation design. Dynamic behavior of soil can be represented using low-strain shear modulus, modulus, and damping variation with shear strain along with the liquefaction potential. Several laboratory and field techniques are available and this article describes some of the widely used field and laboratory tests for assessing

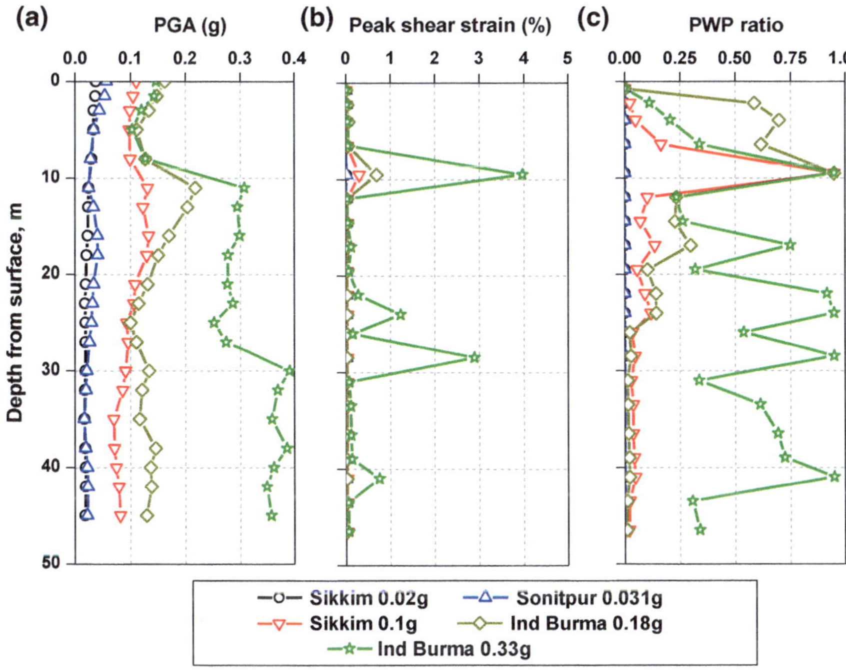

Fig. 13.29 Profiles of **a** PGA **b** peak shear strain and **c** PWP ratio [49]

the dynamic behavior of soils. Typical results using each test are also presented for a comprehensive understanding of the reader. As the field test can only provide the low-strain dynamic stiffness characteristics, it is therefore, necessary to perform laboratory element tests to completely understand the dynamic behavior of soils over wide strain range. Analytical formulations were proposed based on the laboratory test results and the data is compared to the data collected from literature for Indian sandy soils. Finally, a seismic ground response study has been conducted to demonstrate the applicability of proposed formulations.

References

1. Kumar, S.S., Murali Krishna, A., Dey, A.: Parameters influencing dynamic soil properties: a review treatise. Int. J. Innov. Res. Sci. **3**, 47–60 (2014)
2. Hardin, B.O., Richart, F.E.: Elastic wave velocities in granular soils. J. Soil Mech. Found. Div. **89**, 33–65 (1963)
3. Vucetic, M.: Cyclic threshold shear strains in soils. J. Geotech. Geoenviron. Eng. **120**, 2208–2228 (1995)
4. Kramer, S.L.: Geotechnical Earthquake Engineering, 1st edn. Prentice-Hall, New Jersey (1996)
5. Seed, H.B., Idriss, I.M.: Simplified procedure for evaluating soil liquefaction potential. J. Soil Mech. Found. Div. **97**, 1249–1273 (1971)

6. Seed, H.B., Lee, K.L.: Liquefaction of saturated sands during cyclic loading. J. Soil Mech. Found. Div. **92**, 105–134 (1966)
7. Idriss, I.M., Boulanger, R.W.: Semi-empirical procedures for evaluating liquefaction potential during earthquakes. Soil Dyn. Earthq. Eng. **26**, 115–130 (2006). https://doi.org/10.1016/j.soildyn.2004.11.023
8. Andrus, R., Stokoe, K.H.: Liquefaction resistance of soils from shearwave velocity. J. Geotech. Geoenviron. Eng. ASCE **126**, 1015–1025 (2000)
9. Krishna, A.M., Madhav, M.R.: Engineering of ground for liquefaction mitigation using granular columnar inclusions: recent developments. Am. J. Eng. Appl. Sci. **2**, 526–536 (2009). https://doi.org/10.3844/ajeassp.2009.526.536
10. Arany, L., Bhattacharya, S., Macdonald, J.H.G., Hogan, S.J.: Closed form solution of Eigen frequency of monopile supported offshore wind turbines in deeper waters incorporating stiffness of substructure and SSI. Soil Dyn. Earthq. Eng. **83**, 18–32 (2016). https://doi.org/10.1016/j.soildyn.2015.12.011
11. Wu, G., Finn, W.D.L.: Dynamic nonlinear analysis of pile foundations using finite element method in the time domain. **3**, 44–52 (1997)
12. Mondal, G., Prashant, A., Jain, S.K.: Simplified seismic analysis of soil-well-pier system for bridges. Soil Dyn. Earthq. Eng. **32**, 42–55 (2012). https://doi.org/10.1016/j.soildyn.2011.08.002
13. Phillips, C., Hashash, Y.M.A.: Damping formulation for nonlinear 1D site response analyses. Soil Dyn. Earthq. Eng. **29**, 1143–1158 (2009). https://doi.org/10.1016/j.soildyn.2009.01.004
14. Dammala, P.K., Krishna, A.M., Bhattacharya, S., et al.: Dynamic soil properties for seismic ground response studies in Northeastern India. Soil Dyn. Earthq. Eng. **100**, 357–370 (2017). https://doi.org/10.1016/j.soildyn.2017.06.003
15. Pain, A., Choudhury, D., Bhattacharyya, S.K.: Effect of dynamic soil properties and frequency content of harmonic excitation on the internal stability of reinforced soil retaining structure. Geotext. Geomembr. **45**, 471–486 (2017). https://doi.org/10.1016/j.geotexmem.2017.07.003
16. Krishna, A., Bhattacharjee, A.: Behavior of rigid-faced reinforced soil-retaining walls subjected to different earthquake ground motions. Int. J. Geomech. **17**, 6016007 (2017). https://doi.org/10.1061/(ASCE)GM.1943-5622.0000668
17. Dash, S.R., Govindaraju, L., Bhattacharya, S.: A case study of damages of the Kandla Port and Customs Office tower supported on a mat-pile foundation in liquefied soils under the 2001 Bhuj earthquake. Soil Dyn. Earthq. Eng. **29**, 333–346 (2009). https://doi.org/10.1016/j.soildyn.2008.03.004
18. Sarkar, R., Bhattacharya, S., Maheshwari, B.K.: Seismic requalification of pile foundations in liquefiable soils. **44**, 183–195 (2014). https://doi.org/10.1007/s40098-014-0112-8
19. Dammala, P.K., Bhattacharya, S., Krishna, A.M., et al.: Scenario based seismic re-qualification of caisson supported major bridges? A case study of Saraighat Bridge. Soil Dyn. Earthq. Eng. **100**, 270–275 (2017). https://doi.org/10.1016/j.soildyn.2017.06.005
20. Park, C.B., Miller, R.D., Xia, J.: Multichannel analysis of surface waves. Geophysics **64**, 800–808 (1999). https://doi.org/10.1190/1.1444590
21. Chung, R., Yokel, F., Drnevich, V.: Evaluation of dynamic properties of sands by resonant column testing. Geotech. Test. J. **7**, 60 (1984). https://doi.org/10.1520/GTJ10594J
22. Viggiani, G., Atkinson, J.H.: Interpretation of bender element tests. Geotechnique 149–154 (1995)
23. El Mohtar, C.S., Drnevich, V.P., Santagata, M., Bobet, A.: Combined resonant column and cyclic triaxial tests for measuring undrained shear modulus reduction of sand with plastic fines. Geotech. Test. J. **36**, 1–9 (2013). https://doi.org/10.1520/GTJ20120129
24. Kumar, S.S., Krishna, A.M., Dey, A.: Dynamic properties and liquefaction behaviour of cohesive soil in northeast India under staged cyclic loading. J. Rock Mech. Geotech. Eng. 1–10 (2018). https://doi.org/10.1016/j.jrmge.2018.04.004
25. Drnevich, V.P., Hardin, B.O., Shippy, D.J.: Modulus and damping of soils by the resonant column method. Dyn. Geotech. Test. Am. Soc. Test. Mater. Spec. Tech. Publ. **654**(654), 91–125 (1978)

26. Peacock, W., Seed, H.B.: Sand liquefaction under cyclic loading simple shear conditions. J. Soil Mech. Found. Div. **94**, 689–708 (1968)
27. Ishihara, K., Li, S.: Liquefaction of saturated sand in triaxial torsion shear test. Soils Found. **12**, 19–39 (1972)
28. Kokusho, T.: Cyclic triaxial test of dynamic soil properties for wide strain range. Soils Found. **20**, 45–59 (1980)
29. Saran, S.: Soil Dynamics and Machine Foundations. Galhotia Publication, New Delhi, India (1999)
30. Kashyap, S.S., Krishna, A.M., Dey, A.: Analysis of active MASW test data for a convergent shear wave velocity profile. In: Lehane, A.-M.K. (ed.) Geotechnical and Geophysical Site Characterisation, Sydney, Australia, pp. 951–956 (2016)
31. Taipodia, J., Dey, A., Baglari, D.: Influence of data acquisition and signal preprocessing parameters on the resolution of dispersion image from active MASW survey. J. Geophys. Eng. **15**, 1310–1326 (2018). https://doi.org/10.1088/1742-2140/aaaf4c
32. Foti, S., Hollender, F., Garofalo, F., et al.: Guidelines for the good practice of surface wave analysis: a product of the InterPACIFIC project. Bull. Earthq. Eng. **16**, 2367–2420 (2018). https://doi.org/10.1007/s10518-017-0206-7
33. Stokoe, K.H., Woods, R.D.: In situ shear wave velocity by cross-hole method. J. Soil Mech. Found. Div. **98**, 443–460 (1972)
34. Garofalo, F., Foti, S., Hollender, F., et al.: InterPACIFIC project: comparison of invasive and non-invasive methods for seismic site characterization. Part I: Intra-comparison of surface wave methods. Soil Dyn. Earthq. Eng. **82**, 222–240 (2016). https://doi.org/10.1016/j.soildyn.2015.12.010
35. Garofalo, F., Foti, S., Hollender, F., et al.: InterPACIFIC project: comparison of invasive and non-invasive methods for seismic site characterization. Part II: Inter-comparison between surface-wave and borehole methods. Soil Dyn. Earthq. Eng. **82**, 241–254 (2016). https://doi.org/10.1016/j.soildyn.2015.12.009
36. Taipodia, J., Ram, B., Baglari, D., et al.: Geophysical investigations for identification of subsurface. In: Indian Geotechnical Conference. JNTU, Kakinada (2014)
37. Lawrence, F.J.: Propagation velocity of ultrasonic waves through sand (1963)
38. Shirley, D.J., Hampton, L.D., Shirley, D.J., Hampton, L.D.: Shear-wave measurements in laboratory sediments. J. Accoustical Soc. Am. **63**, 607–613 (1978). https://doi.org/10.1121/1.381760
39. Shirley, D.J.: An improved shear wave transducer. J. Accoustical Soc. Am. **63**, 163–1645 (1978). https://doi.org/10.1121/1.381866
40. Arulnathan, R., Boulanger, R.W., Riemer, M.: Analysis of bender element tests. Geotech. Test. J. **21**, 120–131 (1998)
41. Kumar, S.S., Krishna, A.M., Dey, A.: Evaluation of dynamic properties of sandy soil at high cyclic strains. Soil Dyn. Earthq. Eng. **99**, 157–167 (2017). https://doi.org/10.1016/j.soildyn.2017.05.016
42. Iida, K.: The velocity of elastic waves in sand. Bull. Earthq. Res. Inst. **16**, 131–145 (1938)
43. Hardin, B.O.: Suggested Methods of Test for Shear Modulus and Damping of Soils by the Resonant Column, pp. 516–529 (1970)
44. Selig, E., Chung, R., Yokel, F., Drnevich, V.: Evaluation of dynamic properties of sands by resonant column testing. Geotech. Test. J. **7**, 60 (1984). https://doi.org/10.1520/GTJ10594J
45. ASTM D4015: Standard Test Methods for Modulus and Damping of Soils by Resonant-Column, pp. 1–22 (2014). https://doi.org/10.1520/d4015-07.1
46. Hall, J.R., Richart Jr., F.: Dissipation of elastic wave energy in granular soils. J. Soil Mech. Found. Div. **89**, 27–56 (1963)
47. Woods, R.D.: Measurement of dynamic soil properties: a state of the art. In: Proceedings of American Society of Civil Engineers Special Conference on Earthquake Engineering and Soil Dynamics, Pasadena, pp. 91–180 (1978)
48. Hardin, B.O., Music, J.: Apparatus for vibration of soil specimens during the triaxial test. ASTM STP **392**, 55–74 (1965)

49. Dammala, P.K., Kumar, S.S., Krishna, A.M., Bhattacharya, S.: Dynamic soil properties and liquefaction potential of northeast Indian soil for effective stress analysis. Bull. Earthq. Eng. (under review)
50. Sitharam, T., Govindaraju, L., Sridharan, A.: Dynamic properties and liquefaction potential of soils. Curr. Sci. 87:1370–1378 (2004)
51. Lombardi, D., Bhattacharya, S., Hyodo, M., Kaneko, T.: Undrained behaviour of two silica sands and practical implications for modelling SSI in liquefiable soils. Soil Dyn. Earthq. Eng. **66**, 293–304 (2014). https://doi.org/10.1016/j.soildyn.2014.07.010
52. Kumar, S., Krishna, A., Dey, A.: High strain dynamic properties of perfectly dry and saturated cohesionless soil. Indian Geotech. J. 1–9 (2017). https://doi.org/10.1007/s40098-017-0255-5
53. Kumar, S.S., Dey, A., Krishna, A.M.: Response of saturated cohesionless soil subjected to irregular seismic excitations. Nat. Hazards **93**, 509–529 (2018). https://doi.org/10.1007/s11069-018-3312-1
54. ASTM D3999: Standard test methods for the determination of the modulus and damping properties of soils using the cyclic triaxial apparatus. Am. Soc. Test. Mater. **91**, 1–16 (2003). https://doi.org/10.1520/d3999-11e01.1.6
55. ASTM D5311: Standard test method for load controlled cyclic triaxial strength of soil. Astm D5311 **92**, 1–11 (1996). https://doi.org/10.1520/d5311
56. Kjellman, W.: Testing the shear strength of clay in Sweden. Geotechnique **2**, 225–232 (1951)
57. Bjerrum, L., Landva, A.: Direct simple shear tests on a Norwegian quick clay. Geotechnique **16**, 1–20 (1966)
58. Lanzo, G., Vucetic, M., Doroudian, M.: Reduction of shear modulus at small strains in simple shear. J. Geotech. Geoenviron. Eng. **123**, 1035–1042 (1997). https://doi.org/10.1061/(asce)1090-0241(1997)123:11(1035)
59. Dammala, P.K., Adapa, M.K., Bhattacharya, S., Aingaran, S.: Cyclic response of cohesionless soil using cyclic simple shear testing. In: 6th International Conference on Recent Advances in Geotechnical Earthquake Engineering, ICRAGEE, pp. 1–10 (2016)
60. Seed, H.B., Idriss, I.M.: Soil moduli and damping factors for dynamic response analysis (1970)
61. Darendeli, M.: Development of a New Family of Normalized Modulus Reduction and Material Damping. University of Texas (2001)
62. Kumar, S.S., Adapa, M.K., Dey, A.: Importance of site-specific dynamic soil properties for seismic ground response studies. Int. J. Geotech. Earthq. Eng. (2018) (In press)
63. Bai, L.: Preloading Effects on Dynamic Sand Behavior by Resonant Column Tests. Technischen Universität, Berlin (2011)
64. Hardin, B.O.: The nature of stress-strain behavior of soils. In: Earthquake Engineering and Soil Dynamics. ASCE, Pasadena, CA, pp. 3–90 (1978)
65. Chattaraj, R., Sengupta, A.: Liquefaction potential and strain dependent dynamic properties of Kasai River sand. Soil Dyn. Earthq. Eng. **90**, 467–475 (2016). https://doi.org/10.1016/j.soildyn.2016.07.023
66. Saxena, S.K., Reddy, K.R.: Dynamic moduli and damping ratios for Monterey No.0 sand by resonant column tests. Soils Found. **29**, 37–51 (1989). https://doi.org/10.3208/sandf1972.29.2_37
67. Hardin, B.O., Drnevich, V.P.: Shear modulus and damping in soils: design equations and curves. J. Soil Mech. Found. Div. **SM7**, 667–692 (1972)
68. Ishibashi, I., Zhang, X.: Unified dynamic shear moduli and damping ratios of sand and clay. Soils Found. Jpn. Soc. Soil Mech. Found. Eng. **33**, 182–191 (1993). https://doi.org/10.1248/cpb.37.3229
69. Zhang, J., Andrus, R.D., Juang, C.H.: Normalized shear modulus and material damping ratio relationships. J. Geotech. Geoenviron. Eng. **131**, 453–464 (2005). https://doi.org/10.1061/(asce)1090-0241(2005)131:4(453)
70. Vardanega, P.J., Ph, D., Asce, M., et al.: Stiffness of clays and silts: normalizing shear modulus and shear strain. J. Geotech. Geoenviron. Eng. **9**, 1575–1589 (2013). https://doi.org/10.1061/(asce)gt.1943-5606.0000887

71. Matasovic, N., Vucetic, M.: Cyclic characterization of liquefiable sands. J. Geotech. Geoenviron. Eng. **119**, 1805–1822 (1994)
72. Hashash, Y.M., Dashti, S., Romero, M.I., et al.: Evaluation of 1-D seismic site response modeling of sand using centrifuge experiments. Soil Dyn. Earthq. Eng. **78**, 19–31 (2015)
73. Dobry, R., Pierce, W., Dyvik, R., et al.: Pore Pressure Model for Cyclic Straining of Sand. Troy, NY (1985)
74. Martin, G., Finn, W., Seed, H.: Fundamentals of liquefaction under cyclic loading. J. Geotech. Div. **101**, 423–438 (1975)
75. Byrne, P.: A cyclic shear-volume coupling and pore-pressure model for sand. In: Second International Conference on Recent Advances in Geotechnical Earthquake Engineering and Soil Dynamics, St. Louis, Missouri, pp. 47–55 (1991)
76. Hashash, Y.M.A., Musgrouve, M.I., Harmon, J.A., et al.: DEEPSOIL 6.1, User Manual (2016)
77. Vucetic, M., Dobry, R.: Cyclic triaxial strain-controlled testing of liquefiable sands. In: Advanced Triaxial Testing of Soil and Rock, pp. 475–485 (1988). https://doi.org/10.1520/stp29093s
78. Kirar, B., Maheshwari, B.K.: Dynamic properties of soils at large strains in Roorkee region using field and laboratory tests. Indian Geotech. J. **48**, 125–141 (2017). https://doi.org/10.1007/s40098-017-0258-2
79. Maheshwari, B.K., Kirar, B.: Dynamic properties of soils at low strains in Roorkee region using resonant column tests. Int. J. Geotech. Eng. **6362**, 1–12 (2017). https://doi.org/10.1080/19386362.2017.1365474
80. Dutta, T.T., Saride, S.: Influence of shear strain on the Poisson's ratio of clean sands. Geotech. Geol. Eng. **34**, 1359–1373 (2016). https://doi.org/10.1007/s10706-016-0047-1
81. Park, D., Hashash, Y.M.A.: Rate-dependent soil behavior in seismic site response analysis. Can. Geotech. J. **45**, 454–469 (2008). https://doi.org/10.1139/T07-090
82. Hashash, Y.M., Groholski, D.R.: Recent advances in non-linear site response analysis. In: Fifth International Conference on Recent Advances in Geotechnical Earthquake Engineering and Soil Dynamics Symposium, Honor Profr IM Idriss, vol. 29, pp. 1–22 (2010). https://doi.org/10.1016/j.soildyn.2008.12.004
83. Basu, D., Dey, A., Kumar, S.S.: One-dimensional effective stress non-Masing nonlinear ground response analysis of IIT Guwahati. Int. J. Geotech. Earthq. Eng. **8**, 1–27 (2017)
84. Kumar, S.S., Krishna, A.M.: Importance of site-specific dynamic soil properties for seismic ground response studies. Importance of site-specific dynamic soil properties for seismic ground response studies. Int. J. Geotech. Earthq. Eng. **9**, 1–21 (2018). https://doi.org/10.4018/IJGEE.2018010105
85. IS:1893: Criteria for earthquake resistant design of structures. Indian Stand. 1–44 (2002)
86. Ajom, B.E., Bhattacharjee, A.: Seismic re-qualification of Caisson supported Dhansiri River Bridge. In: Tunneling in Soft Ground, Ground Conditioning and Modification Techniques, pp 187–203 (2019)
87. Raghu Kanth, S.T.G., Sreelatha, S., Dash, S.K.: Ground motion estimation at Guwahati city for an Mw 8.1 earthquake in the Shillong plateau. Tectonophysics **448**, 98–114 (2008). https://doi.org/10.1016/j.tecto.2007.11.028
88. Kumar, A., Harinarayan, N.H., Baro, O.: High amplification factor for low amplitude ground motion: assessment for Delhi. Disaster Adv. **8**, 1–11 (2015)
89. Romero, S.M., Rix, G.: Ground Motion Amplification of Soils in the Upper Mississippi Embayment. MAE Center CD Release 05-01 (2005)

Chapter 14
Cyclic Behaviour of Scrap-Tyre Soil Mixtures

J. S. Vinod, M. Neaz Sheikh and Soledad Mashiri

14.1 Introduction

The increased seismic activity over the last few decades, particularly in areas where structures are not designed for protection against earthquakes, has motivated researchers to investigate a new application of sand–scrap tyres (e.g. [25]). The innovative civil engineering application of sand–scrap tyres mixtures is an alternative to sophisticated seismic isolation devices (e.g. rubber pads) which are expensive and need expertise for their design and installation. In the last two decades, a significant number of research studies devoted to explore practical solutions of the recycle and reuse of scrap tyres for major civil engineering projects. Despite the amount of research carried out on tyre soil mixtures (e.g. [3, 5, 6, 8–10, 17, 19, 23, 24, 28, 29]) and the discovery of their beneficial engineering properties, environmentally sustainable use of scrap tyres in civil engineering applications is still limited. It has been reported that the addition of tyre crumbs (granulated rubber) in sand reduced the shear strength of sand (e.g. [14, 17, 23, 28, 29]) highlighted significant shear strength improvement together with the reduction of dilatancy of sand with the inclusion of tyre chips, especially in sand–rubber behaviour zone where both sand and tyre chips form the skeleton of the matrix material. Even though a lot of research studies were carried out on the monotonic behaviour of scrap-tyre soil mixtures, only very limited information is available on the liquefaction and dynamic properties of sand–tyre chip mixtures [1, 7, 11, 16, 18, 22, 26, 27], Qi et al. [21]. Turer and Özden [27] reported that for shear strain (γ) < 0.1%, sand–tyre mixtures had lower shear modulus and higher damping ratio compared to sand. Hyodo et al. [12] and Kaneko et al. [13] reported that sand–tyre crumb mixtures showed a reduced liquefaction potential compared to sand. However, in a separate study carried out by Promputthangkoon and Hyde [20] reported that the addition of tyre crumbs increased the

J. S. Vinod (✉) · M. Neaz Sheikh · S. Mashiri
Faculty of Engineering and Information Sciences, University of Wollongong, Wollongong, NSW 2522, Australia
e-mail: vinod@uow.edu.au

M. Latha G. (ed.), *Frontiers in Geotechnical Engineering*, Developments in Geotechnical Engineering, https://doi.org/10.1007/978-981-13-5871-5_14

liquefaction potential of sand. Remarkable damping and seismic isolation properties of sand–tyre crumb mixtures was reported by [13]. The enhanced damping and seismic isolation properties of STCh mixture are mainly achieved from the reduced acceleration amplitudes at the ground surface due to the low stiffness of tyre crumbs. [16] carried out strain-controlled cyclic triaxial test and reported that the addition of TCh significantly reduced the liquefaction potential of sand. In addition, the shear modulus reduction curve reduced and damping ratio increased with the addition of scrap tyres. Qi et al. [21] investigated the effect of scrap tyre rubber on steel furnace slag (SFS) and coal wash (CW) mixtures. It was reported that that inclusion or rubber increased the damping ratio and energy-absorbing properties of SFS and CW mixtures.

This study evaluates the liquefaction and dynamic properties of sand–tyre chip (STCh) mixtures. Strain-controlled cyclic triaxial laboratory experiments were carried out on sand–tyre chip mixture. The effect of different parameters such as amount of TCh, confining pressure, relative density, number of cycles and cyclic strain amplitude on the monotonic and cyclic behaviour has been investigated and reported.

14.2 Laboratory Experimental Program

A series of strain-controlled cyclic triaxial experiments were conducted on sand–tyre chip (STCh) mixture specimens of 100 mm diameter and 200 mm height. Figure 14.1 shows the particle size distribution of the sand used in this study. The sand is classified as poorly graded (SP). The specific gravity, minimum and maximum dry unit weight of the sand was 2.67 kN/m^3, 14.60 kN/m^3 and 16.41 kN/m^3, respectively. The scrap tyres in the form of tyre chips (TCh) are used for the testing program (see inset of Fig. 14.1). According to ASTM D 6270-12 tyre chips (TCh) refer to pieces of scrap tyre having a basic geometric shape with nominal dimension ranging from

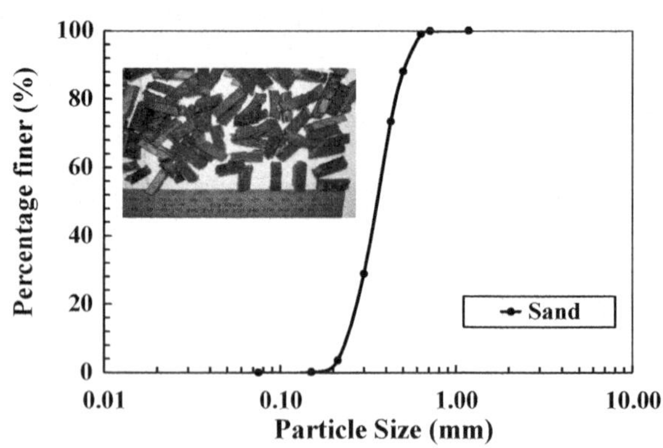

Fig. 14.1 Particle-size distribution of sand [inset Tyre chips (TCh)] (after Mashiri et al. [18])

12 to 50 mm. The maximum and minimum void ratio of TCh was 1.3 and 0.83, respectively. In this study, TCh without any steel belt were cut into rectangular shapes for uniform thickness of approximately 5–6 mm with an aspect ratio of TCh was 2.8. The maximum width of TCh was 8 mm. The STCh mixtures were prepared using dry deposition method (Ishihara, 1996). The STCh mixtures were placed in three layers and each layer was compacted by tamping the walls of the mould to achieve the required relative density of 50%. All the samples were then saturated to a B-value >0.95 using back pressure technique. All the cyclic tests were conducted at a constant frequency of 1 Hz. The cyclic tests were carried out on STCh mixtures with different gravimetric proportion of TCh, $\chi = 0$, 10, 20, 30 and 40%.

14.3 Cyclic Behaviour of Sand–Tyre Chip Mixtures

Typical results of the cyclic behaviour of STCh mixture (STCh (0%) and STCh (30%)) are presented in Fig. 14.2. Figure 14.2 shows the stress paths, hysteresis loop and pore pressure ratio with number of cycles for STCh (0%) and STCh (30%). It is evident from Fig. 14.2 that the maximum shear stress decreases with the inclusion of TCh. The process of liquefaction is clearly presented in Fig. 14.2a, b. Deviatoric stress decreases with mean p and reaches zero indicating liquefaction (Fig. 14.2a). The decrease in the deviatoric stress is mainly due to the increase in the pore water pressure (Fig. 14.2c) in the STCh mixtures. Moreover, the number of cycles for liquefaction increases with the addition of TCh (Fig. 14.2c).

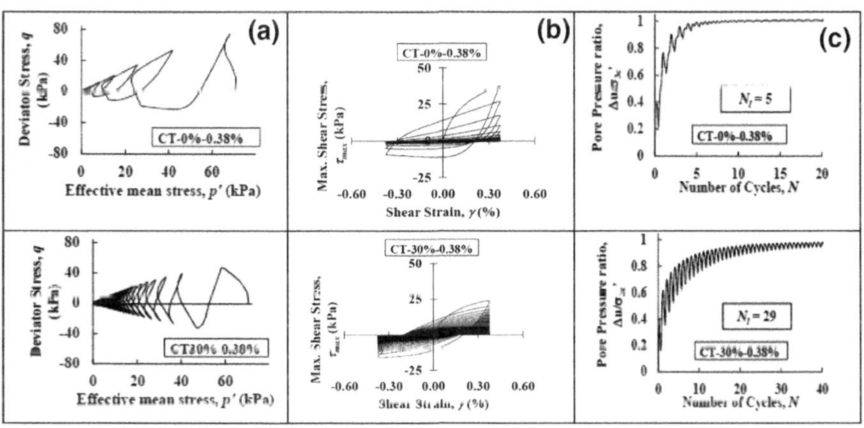

Fig. 14.2 Cyclic behaviour of STCh mixtures at $\gamma = 0.38\%$

14.4 Liquefaction Behaviour of Sand–Tyre Chip Mixtures

The variation of the amplitude of shear strain with number of cycles to liquefaction for different percentages of STCh mixtures is presented in Fig. 14.3. It is evident from Fig. 14.3 that the amplitude of shear strain has a significant influence on the liquefaction behaviour of STCH mixtures. Number of cycles to liquefaction increases with the decrease in the amplitude of shear strain. The amplitude of shear strain for liquefaction potential increase with the increase in the proportion of STCh (up to 30%). However, the liquefaction potential decreases for STCh = 40%. Moreover, STCh (10%) exhibits the same behaviour as STCh (0%). This is may be due to the sand-like behaviour exhibited by TCh (10%) mixtures during the cyclic loading [16].

Figure 14.4 presents the variation of number of cycles for liquefaction with gravimetric proportion of TCh for a confining pressure of 69 kPa. It is shown that irrespective of the amplitude of shear strain, the number of cycles for liquefaction increases nonlinearly with increase in the gravimetric proportion of TCh. It is also evident that the amount of TCh, between 30 and 33%, shows the maximum number of cycles required for liquefaction. In addition, a reduction in the number of cycles for liquefaction was observed after 33% of TCh, which is mainly due to the higher void ratio exhibited by STCh (40%).

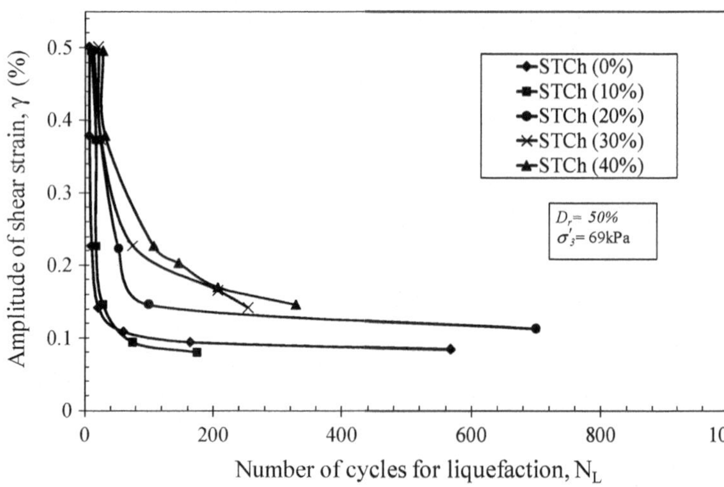

Fig. 14.3 Variation of amplitude of shear strain with number of cycles for liquefaction (after Mashiri et al. [16])

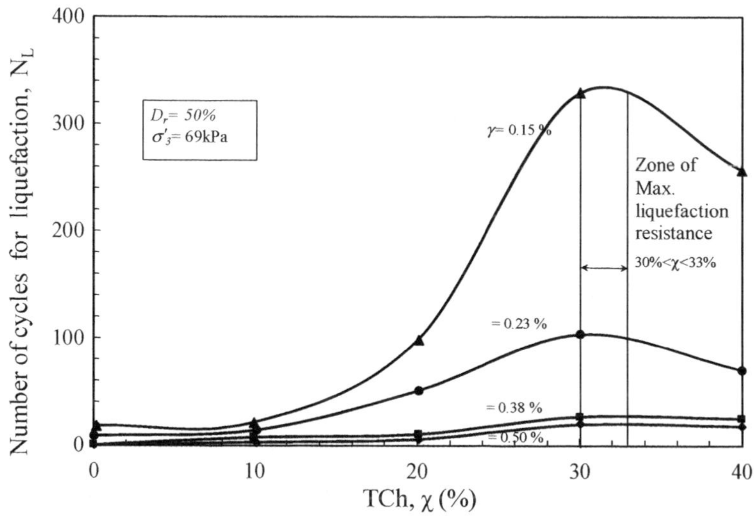

Fig. 14.4 Number of cycles for liquefaction versus TCh (after Mashiri et al. [16])

14.5 Dynamic Properties of Scrap-Tyre Soil Mixtures

Figure 14.5 shows the variation of maximum shear stress with shear strain (hysteresis loops) for different proportion of STCh mixtures. As anticipated, the maximum shear stress decreases during the strain-controlled cyclic loading. The decrease in the maximum shear stress is due to the development of pore water pressure (see Fig. 14.2). For $\gamma = 0.38\%$ the maximum shear stress (first cycle) decreases with the increase in the proportion of STCh mixtures.

Figure 14.6 shows the variation of shear modulus (G) with number of cycles (N) for different proportion of STCh mixtures. Irrespective of TCh, the shear modulus decreases with increase in the number of cycles. The decrease in the shear modulus is mainly due to the decrease in the shear stress during the strain controlled loading (Fig. 14.5). The shear modulus decreases significantly with number of cycles especially at low number of cycles and reaches a constant value at large values of N. However, the incorporation of TCh retards the reduction of shear modulus with N. This behaviour is mainly due to the potential of STCh mixtures to withstand cyclic shear stress for larger number of cycles compared to sand. For any particular value of N (e.g. N = 5), the shear modulus of sand increases with the proportion of tyre chips up to STCh = 30% and decreases with increase in the TCh.

The variation of damping ratio (ζ) with number of cycles (N) is presented in Fig. 14.7. It is shown that damping ratio decreases with the increase in the number of cycles irrespective of amount of TCh. As expected, STCh(0%) shows very low ζ compared to STCh mixtures. The reduction in ζ with N is mainly due to the loss of frictional energy in the sand skeleton due to the addition of TCh [4].

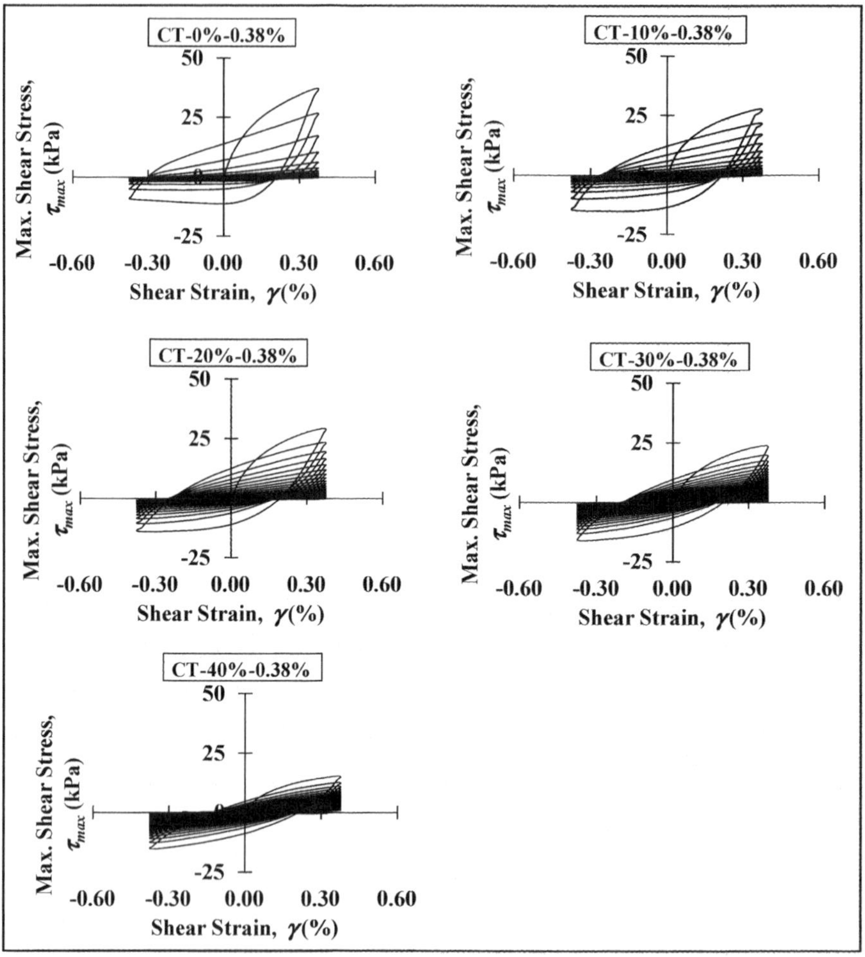

Fig. 14.5 Variation of maximum shear stress with shear strain (Hysteresis loop) for different proportions of STCh (after Mashiri et al. [17])

Fig. 14.6 Shear modulus versus number of cycles for STCh mixtures at different shear strains (after Mashiri et al. [15])

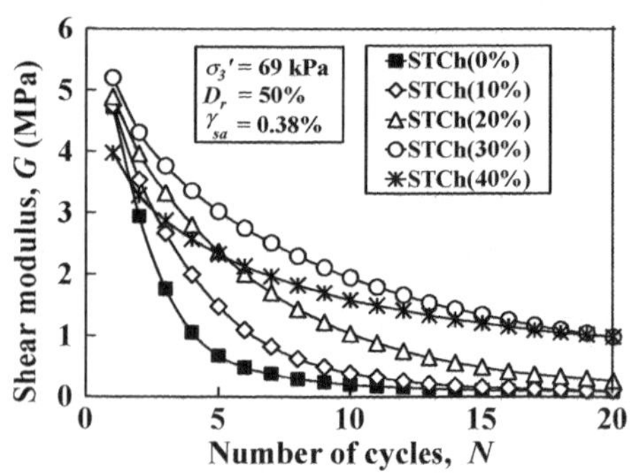

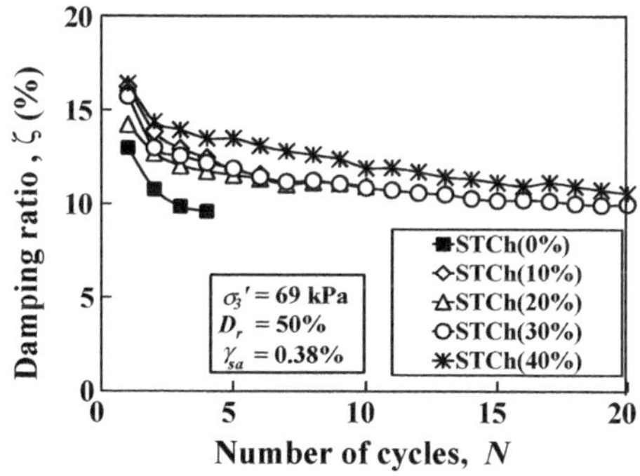

Fig. 14.7 Damping ratio versus number of cycles for STCh mixtures at different shear strains (after Mashiri et al. [15])

14.6 Shear Modulus Degradation and Damping Ratio Curves for Sand–Scrap Tyre Mixtures

The results of the shear modulus were normalised by the corresponding value of G_{max} for each STCh mixture. Detailed Bender element tests were carried out on STCh mixtures to capture the G_{max} for STCh mixtures and can be found in [18]. The normalised shear modulus (G/G_{max}) curves with the shear strain for STCh mixtures are presented in Fig. 14.8. It is evident from the figure that irrespective of TCh content the normalised shear modulus falls in a narrow band. For a particular value of shear strain, the normalised shear modulus increases with the proportion of TCh.

The variation of damping ratio (ζ) with shear strain is presented in Fig. 14.9 It is evident from the figure the damping ratio of STCh increases with shear strain. The ζ of STCh mixtures shows slightly higher value compared to sand. In addition, ζ of sand and STCh mixtures fall in a narrow band.

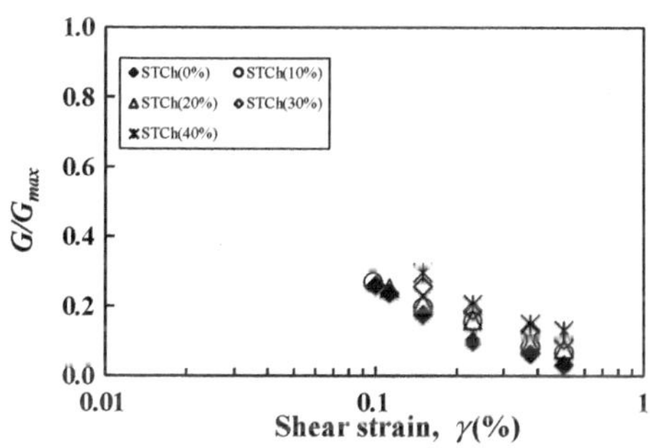

Fig. 14.8 G/G_{max} versus γ for different proportions of TCh (after Mashiri et al. [16])

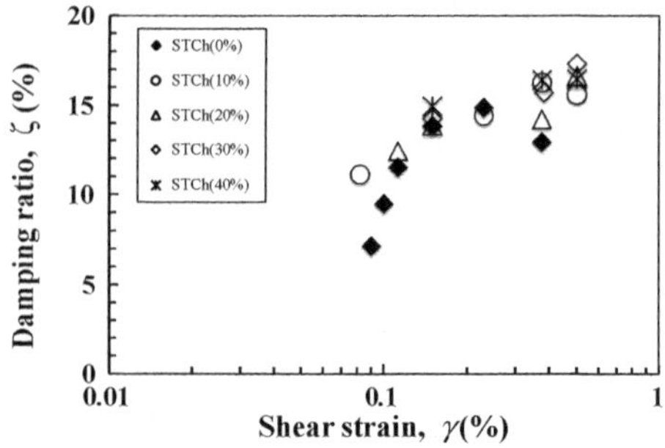

Fig. 14.9 ζ versus γ for different proportions of TCh (after Mashiri et al. [16])

14.7 Conclusions

This paper has presented the cyclic behaviour sand–tyre mixtures. A series of strain-controlled laboratory experiments were carried out on sand mixed with different proportions of tyre chips (TCh). The cyclic triaxial tests show the TCh has significant influence in controlling the liquefaction of sand. The addition of TCh reduced the liquefaction potential of sand. The optimum percentage of TCh, in this investigation, was found to be between 30 and 33%. The dynamic properties, shear modulus and damping ratio, were also influenced by the amount of TCh, number of cycles and amplitude of shear strain. The shear modulus decreases with increase in the amplitude of shear strain and number of cycles. However, damping ratio increases with the increase in shear strain amplitude and decreases with increase in number of cycles.

References

1. Anastasiadis, A., Senetakis, K., Pilitakis, K.: Small-strain shear modulus and damping ratio of sand-rubber and gravel-rubber mixtures. Geotech. Geol. Eng. **30**(2), 363–382 (2012)
2. ASTM (D6270–12) Standard Practice of Use of Scrap Tires in Civil Engineering Applications, ASTM International, West Conshohocken, PA (2012). https://doi.org/10.1520/d6270-08r12, www.astm.org
3. Balunaini, U., Varenya, Kumar D., Prezzi, M., Salgado, R.: Shear strength of tyre chip-sand and tyre shred-sand mixtures. Proc. Geotech. Eng. Instit. Civil Eng. **167**(6), 585–595 (2014)
4. Brennan, A.J., Thusyanthan, N.I., Madabhushi, S.P.G.: Evaluation of shear modulus and damping in dynamic centrifuge tests. J. Geotech. Geoenviron. Eng. **131**(12), 1488–1497 (2005)
5. Dunham-Friel, J., Carraro, J.A.H.: Shear strength and stiffness of Expansive Soil and Rubber (ESR) mixtures in undrained axisymmetric compression, Geo-Frontiers Congress 2011, Reston, VA, USA, vol. 1, pp. 1111–1120
6. Edil, T.B., Bosscher, P.J.: Engineering properties of tyre chips and soil mixtures. Geotech. Test. J. **17**(4), 453–464 (1994)
7. Feng, Z.Y., Sutter, K.G.: Dynamic Properties of Granulated Rubber/Sand Mixtures. Geotech. Test. J. **23**(3), 338–344 (2000)
8. Foose, G.J., Benson, C.H., Bosscher, P.J.: Sand reinforced with shredded waste tyres. J. Geotech. Eng. **122**(9), 760–767 (1996)

9. Ghazavi, M., Sakhi, M.A.: Influence of optimized tire shreds on shear strength parameters of sand. Int. J. Geomech. ASCE **5**(1), 58–65 (2005)
10. Hataf, N., Rahimi, M.M.: Experimental investigation of bearing capacity of sand reinforced with randomly distributed tire shreds. Constr. Build. Mater. **20**(10), 910–916 (2005)
11. Hazarika, H., Yasuhara, K., Hyodo, M., Karmokar, A.K., Mitarai, Y.: Mitigation of earthquake induced geotehcnical disasters using a smart and novel geomaterial. In: The 14th World Conference on Earthquake Engineering, October 12–17, Beijing, China (2008)
12. Hyodo, M., Yamada, S., Orense, R.P., Okamoto, M., and Hazarika, H.: Undrained cyclic shear properties of tire chip-sand mixtures. In: Proceedings of International Workshop IW-TDGM2007, Yokosuka, Japan, Taylor and Francis, 187–196
13. Kaneko, K., Orense, R.P., Hyodo, M., Yoshimoto, N.: Seismic response characteristics of saturated sand deposits mixed with tire chips. J. Geotech. Geoenviron. Eng. **139**(4), 633–643 (2013)
14. Masad, E., Taha, R., Ho, C., Papagiannakis, T.: Engineering properties of tire/soil mixtures as a lightweight fill material. Geotech. Test. J. **19**(3), 297–304 (1996)
15. Mashiri, M. S.: Monotonic and Cyclic Behaviour of Sand-Tyre Chip (STCh) Mixtures. School of Civil, Mining and Environmental Engineering, University of Wollongong, Dissertation/Thesis (2007)
16. Mashiri, M.S., Vinod, J.S., Sheikh, M.N.: Liquefaction potential and dynamic properties of Sand-Tire Chip Mixture. Geotech. Test. J. ASTM **39**(1), 69–79 (2016)
17. Mashiri, M.S., Vinod, J.S., Sheikh, M.N., Tsang, H.H.: Shear Strength and dilatancy behaviour of sand-tyre chips mixture. Soils Found. **55**(3), 517–528 (2015)
18. Mashiri, M.S., Vinod, J.S., Sheikh, M.N., Carraro, A.: Shear modulus of sand-tyre chip mixtures. Environ. Geotech. https://doi.org/10.1680/jenge.16.00016
19. Mashiri, M.S., Vinod, J.S., Sheikh, M.N.: (2015b). Liquefaction potential and dynamic properties of Sand-Tire Chip mixture. Geotech. Test. J. ASTM (in press, July 2015)
20. Promputthangkoon, P., Hyde, A.F.L.: Compressibility and liquefaction potential of rubber composite soils. In: Proceedings of International Workshop IW-TDGM2007, Yokosuka, Japan, Taylor and Francis, pp. 161–170
21. Qi, Y., Indraratna, B., Heitor, B., Vinod, J.S.: (2018). The role of rubber crumbs on the cyclic behaviour of steel furnace slag and coal wash mixtures. J. Geotech. Geoenviron. Eng. ASCE. **144**(2), 04017107
22. Senetakis, K., Anastasiadis, A., Pitilakis, K.: Dynamic properties of dry sand rubber (SRM) and gravel/rubber (GRM) mixtures in a wide range of shearing strain amplitudes. Soil Dynam. Earthquake Eng. **33**(1), 38–53 (2012)
23. Sheikh, M.N., Mashiri, M., Vinod, J.S., Tsang, H.H.: Shear and compressibility behaviours of sand-tyre crumb mixtures. J. Mater. Civil Eng. **25**(10), 1366–1374 (2013)
24. Tatlisoz, N., Edil, T.B., Benson, C.H.: Interaction between reinforcing geosynthetics and soil tire chip mixtures. J. Geotech. Geoenviron. Eng. **124**(11), 1109–1119 (1998)
25. Tsang, H.H.: Seismic isolation by rubber–soil mixtures for developing countries. Earthquake Eng. Struct. Dynam. **37**(2), 283–303 (2008)
26. Tsang, H.H., Lo, S.H., Xu, X., Sheikh, M.N.: Seismic isolation for low to medium rise buildings using granulated rubber-soil mixtures: numerical study. Earthquake Eng. Struct. Dynam. **41**(14), 2009–2024 (2012)
27. Turer, A., Özden, B.: Seismic base isolation using low-cost Strap Tire Pads (STP). Mater. Struct. **41**, 891–908 (2008)
28. Youwai, S., Bergado, D.T.: Strength and deformation characteristics of shredded rubber tire-sand mixtures. Can. Geotech. J. **40**(2), 254–264 (2003)
29. Zornberg, J.G., Viratjandr, C., Cabral, A.R.: Behaviour of tire shred—sand mixtures. Can. Geotech. J. **41**(2), 227–241 (2004)

Chapter 15
Sustainable Geotechnics: A Bio-geotechnical Perspective

Sreedeep Sekharan, Vinay Kumar Gadi, Sanandam Bordoloi, Abhisekh Saha, Himanshu Kumar, Budhaditya Hazra and Ankit Garg

15.1 Introduction

Sustainable geotechnics has emerged as an imperative branch of geotechnical engineering in response to the importance of low carbon economy, challenges of climate change, diminishing fossil fuel reserves, need for ecological conservation and environmental management [6, 14, 29]. Sustainability is the ability of preserving useful resource with minimal depletion, ensuring ecological balance and positive impact on the environment. The sustainable practices in geotechnics include the extraction of geothermal energy, waste disposal, bioengineering for slope stability, and microbial-induced calcite precipitation (MICP) for soil improvement [5, 36, 39, 52, 61]. Until recently, sustainability in geotechnics has primarily focused on reusing waste-derived materials such as rubber tires, plastic bottles, fly ash, lime, stone dust, and industrial slag for construction [5, 13, 48].

This chapter essentially focuses on bio-geotechnical interface where biomaterials are used as amendment for robust and self-sustaining green infrastructure. In this context, the question arises as to how we conceive the term sustainability in geotech-

S. Sekharan (✉) · V. K. Gadi · S. Bordoloi · A. Saha · H. Kumar · B. Hazra
Indian Institute of Technology Guwahati, Guwahati 781039, Assam, India
e-mail: srees@iitg.ac.in

V. K. Gadi
e-mail: vinay.gadi@iitg.ac.in

S. Bordoloi
e-mail: sanandam@iitg.ac.in

A. Saha
e-mail: abhisekh@iitg.ac.in

H. Kumar
e-mail: himanshu19961270@gmail.com

B. Hazra
e-mail: budhaditya.hazra@iitg.ac.in

A. Garg
Shantou University, Shantou, China
e-mail: ankit@stu.edu.in

M. Latha G. (ed.), *Frontiers in Geotechnical Engineering*, Developments in Geotechnical Engineering, https://doi.org/10.1007/978-981-13-5871-5_15

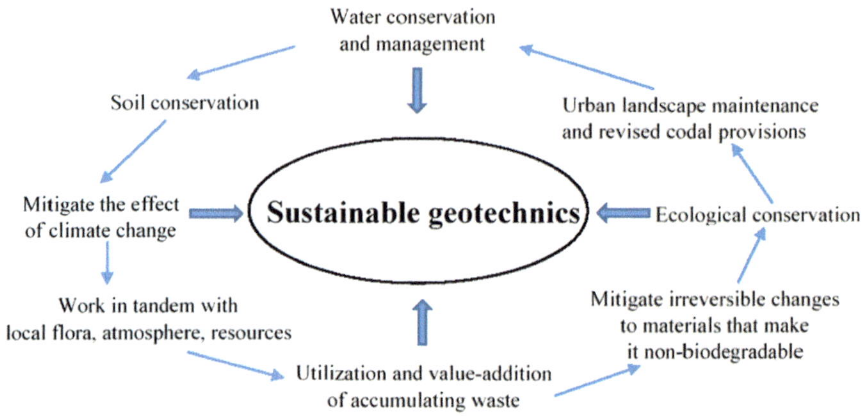

Fig. 15.1 A bio-geotechnical perspective of sustainable geotechnics

nics? The use of discrete rubber-based tires in soil reinforcement is regarded as a sustainable practice [51]. In essence, it is sustainable as it is one form of waste management. However, rubber is derived from plant sources and during its conversion to rubber becomes a nonbiodegradable entity within the soil. From this perspective, it cannot be fully perceived as eco-friendly. An ideal biomaterial should be eco-friendly and ultimately degrade while proactively helping the green infrastructure to be self-sustainable during its existence. Under this class, lignocellulose-based materials such as biochar, natural fibers, biopolymers, and compost have emerged as important bio-geo-amendment materials [43]. Their ability to retain water and nutrients in arid soils, enhance vegetation growth, induce microbial activity and ultimately degrade after a period of time makes them suitable in vegetated infrastructure.

Vegetated slope and embankment protection has emerged as one of the most sustainable shallow reinforcement measures by root reinforcement, transpiration induced suction and lowering local temperature [25, 33, 34, 42]. This form of infrastructure has gained traction under the umbrella of bio-geotechnics. The use of ideal biomaterial along with natural vegetation encompasses the interdependent attributes of sustainable geotechnics as presented in Fig. 15.1.

Incorporating vegetation and bio-enriched medium as sustainable geotechnical practice would ensure soil and water conservation, mitigate the effect of climate change and secure harmony between local flora and atmosphere [43, 46, 57]. The use of biomaterial from the abundant bio-wastes further leads to value addition and helps in enhancing ecological conservation of local flora and microorganisms. In the context of urban green infrastructures where soil needs protection along with stabilization by vegetation, there is a scope to develop a truly self-sustaining cycle of green infrastructure even in drought-affected regions by using suitable biomaterials [8, 31].

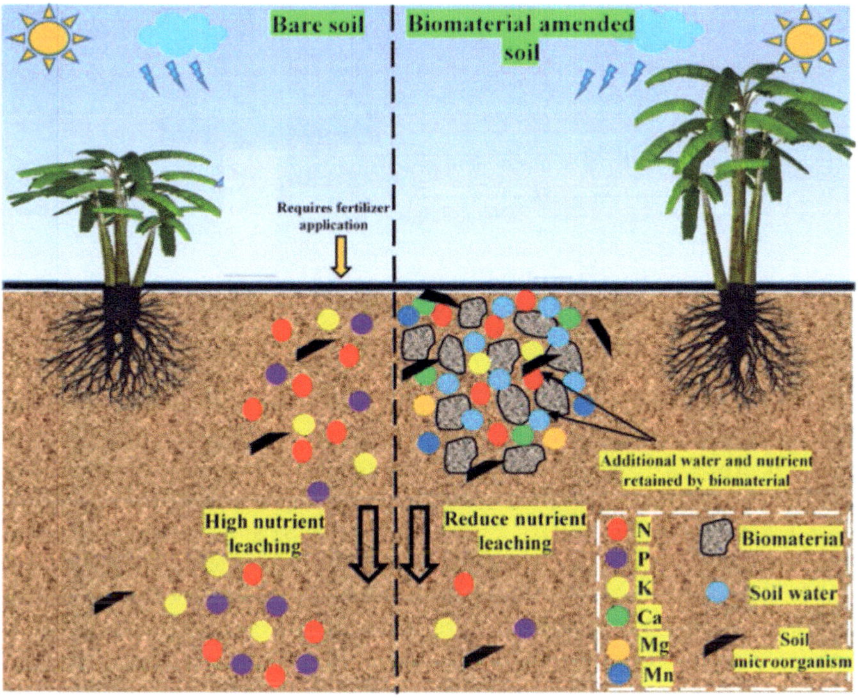

Fig. 15.2 Biomaterial amended soil as a suitable stratum for green infrastructure

15.2 Biomaterial: Origin, Usage, and Application in Green Infrastructure

One of the foremost requirements of an ideal biomaterial in vegetated infrastructures is its capacity to retain water and increase the retention capacity of the parent soil when amended [31]. This is important as vegetation development necessitates a substrate medium with adequate water availability, requisite pH and microbial activity. Figure 15.2 showcases the difference between the use of biomaterial amended soil (BAS) and fallow bare soil. BAS have high nutrient retention and low leaching, facilitates the growth of microorganisms, binds water which ultimately leads to higher plant growth. In the case of maintenance of green infrastructure, fallow soil requires regular fertilizer application to sustain growth. Biomaterial have intra-pores within individual material and suitable surface functional groups [47] that ultimately helps in retaining more water as compared to bare soil (Fig. 15.3). Some common categories of biomaterials are biochar, natural fibers, bio-absorbent polymers and compost [10, 13], which are found to be effective in increasing the water retention, plant growth, and microbial activity [10, 56]. Each of these biomaterials has been discussed in this paper as relevant to green infrastructure.

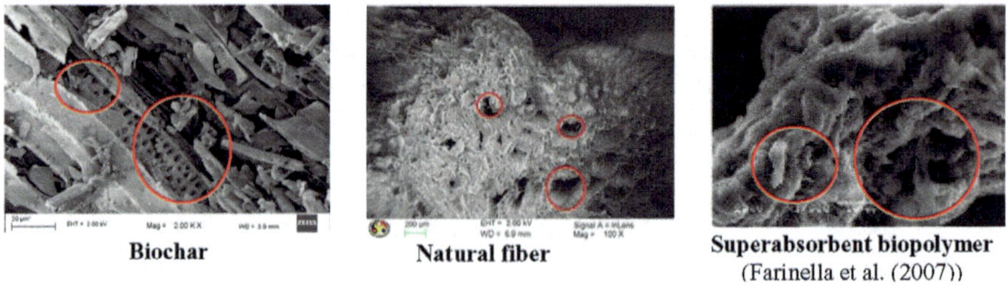

Fig. 15.3 Porous morphology of biomaterial—biochar, natural fibers, and superabsorbent biopolymer

15.2.1 Biochar

Biochar is explored as a sustainable soil amendment material in geoenvironmental infrastructures, namely landfill cover material [12] and vegetated slope [44]. Use of biochar derived from pyrolysis of biomasses is reported to decrease the carbon emission to the atmosphere by 1/4th as against direct combustion of these biomasses. Biochar amended soil from commercial plant sources was studied with respect to geotechnical applications. For this purpose, the biochar has been characterized for its hydraulic conductivity, adsorption capacity, microbial activity, vegetation potential, strength, and water retention characteristics. It was observed that biochar helps in the growth of vegetation species in compacted soil facilitating its use in bio-engineered slope stability applications [12]. A recent study by [9] have shown that addition of water hyacinth (WH) weed biochar increases the water retention of compacted soil-biochar composites and decreases the desiccation crack of soil (represented as crack intensity factor, CIF). Figure 15.4 clearly shows that water available for vegetation is more for BAS as compared to bare soil even after wilting point (1500 kPa). The biochar being finer can naturally reduce the pore size, thus increasing its retention capacity as seen in Fig. 15.4. Furthermore, biochar have suitable functional groups and intra-pores which further help in increasing available water in biochar-amended soil [47].

15.2.2 Natural Fibers

The use of natural fibers in geotechnical engineering applications has gained popularity due to the present-day need of alternate sustainable materials in contrast to conventional petroleum-based synthetic fibers. Randomly distributed fiber-reinforced soil (RDFS) using natural fibers has been extensively studied with respect to their mechanical properties. It was noted that natural fibers improve the strength of the soil 1.3–3 times based on amendment rate (Fig. 15.5a, [9]). Recently, the effect of natural lignocellulose fibers on water retention in compacted soil-fiber composite

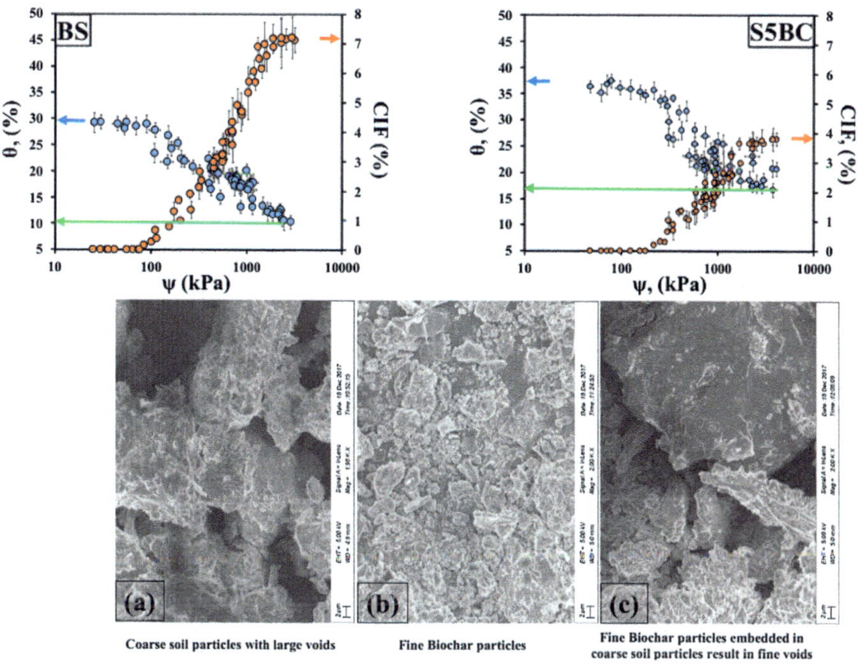

Fig. 15.4 Effect of WH biochar on water retention and desiccation cracks of soil (modified after Bordoloi et al. [9])

has been explored in context to its use in green infrastructure. Water hyacinth fiber addition of 0.75% by dry weight of soil increases the water retention of bare soil from 32.1 to 35.7% and further to 42% in case of vegetated soil ([9]; Fig. 15.5b). Natural fibers, being porous, are hydrophilic in nature and are susceptible to moisture absorption. The structural compositions of cellulose, hemicellulose, and lignin contain hydroxyl groups (–OH) (Fig. 15.5c), which allow moisture absorption from the soil matrix [10]. The fibrillary matrix (i.e., macro and microfibrils) further helps in retaining water and also increases the infiltration rate of the soil-fiber composite [44]. These fibers can be further coated with hydrophobic layer that can reduce its degradation rate and enhance strength when used for application where strength is of major concern [10]. Figure 15.6 showcases the result of CBR tests where discrete fibers were initially coated with nanoparticles. It is clearly seen in Fig. 15.6 that the surface is coated with nanoparticles, which makes it hydrophobic and increases the tensile strength of fiber which ultimately increases the CBR of the treated fiber soil composite.

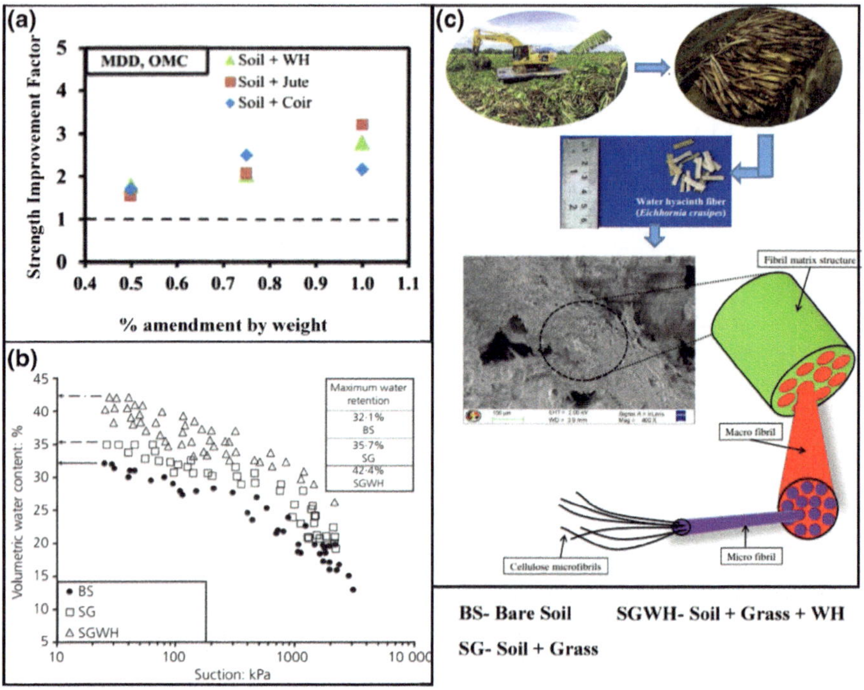

Fig. 15.5 Description of **a** strength improvement of soil due to natural fibers and **b–c** water retention of soil-natural fiber composite and hydrophilic nature (modified after Bordoloi et al. [8, 10])

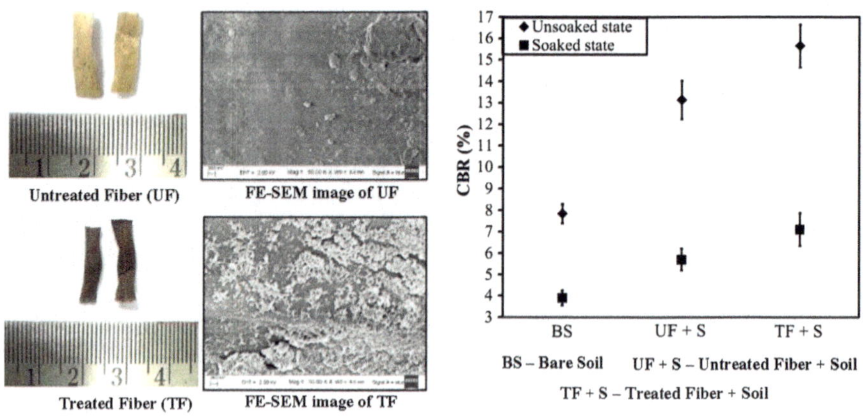

Fig. 15.6 Surface morphology due to nano-treatment and corresponding change in CBR for soil-fiber composite

15.2.3 Biopolymers

Use of biological organic matter such as biopolymers, which are produced via polymerization of biomolecules could be very promising to address some of the serious environmental issues like increasing carbon footprint, alteration in soil pH, accumulation of toxic salt, and drought management. These polymers are prepared either from the living organism or from bio-waste materials which makes it very much cost-effective. Moreover, these polymers are biodegradable with time, does not produce any harmful waste and have excellent eco-compatibility. Recent experimental studies have shown that because of its high hydrophilic nature, superabsorbent biopolymers can improve soil water holding capacity, modify soil permeability, infiltration rates; reduces soil erosion and surface water runoff [16, 30, 45, 59]. Montesano et al. [38] have highlighted that the presence of superabsorbent biopolymer in sandy soil increased the field capacity by 400% compared to the bare soil. The study also revealed that the plant-available water content (PAWC) and vegetation growth significantly increased with polymer amendment. Moreover, the saturated hydraulic conductivity of soil and the infiltration of water into deeper strata was considerably reduced due to alteration in the pore geometry with polymer addition [1, 17, 40, 51]. The total fraction of water-accessible pore space in the soil for a particular packing density decrease exponentially with the increase in polymer concentration [60].

15.3 Concept of Urban Green Infrastructure and Associated Plant Parameters

Emerging concepts like green infrastructures (i.e. bio-engineered slopes, vegetative embankments, landfill vegetation cover, green roof) address carbon footprint while conserving natural resources and minimizing waste throughout the life cycle of a project [58]. However, the establishment of these vegetated infrastructures are dependent on the substrate soil medium, which should ideally be designed such that there are minimal costs involved in maintenance. In many cases, the growth of vegetation is difficult due to arid environment or nutrient deficient soils. In such cases, especially in urban green infrastructure, there is a need to enrich the soil nutrient, microorganism and water retention. Urban forest development as a concept has been an offshoot of green infrastructure and is defined as pocket areas within the urban spaces which provide for a local ecosystem due to the high density of vegetation (trees, shrubs, grass, etc.) [2]. It ultimately develops into a self-sustainable system where nutrients are provided by shedding of leaves, water conserved by the organic humus accumulated due to shedding, evaporation is reduced due to canopy effect and ultimately reduce heat island in surrounding urban spaces [28]. For urban forestry, the use of appropriate biomaterial can help urban spaces in arid regions to develop stratum that can easily incubate vegetation and ultimately provide a micro self-sustainable ecosystem as shown in Fig. 15.7.

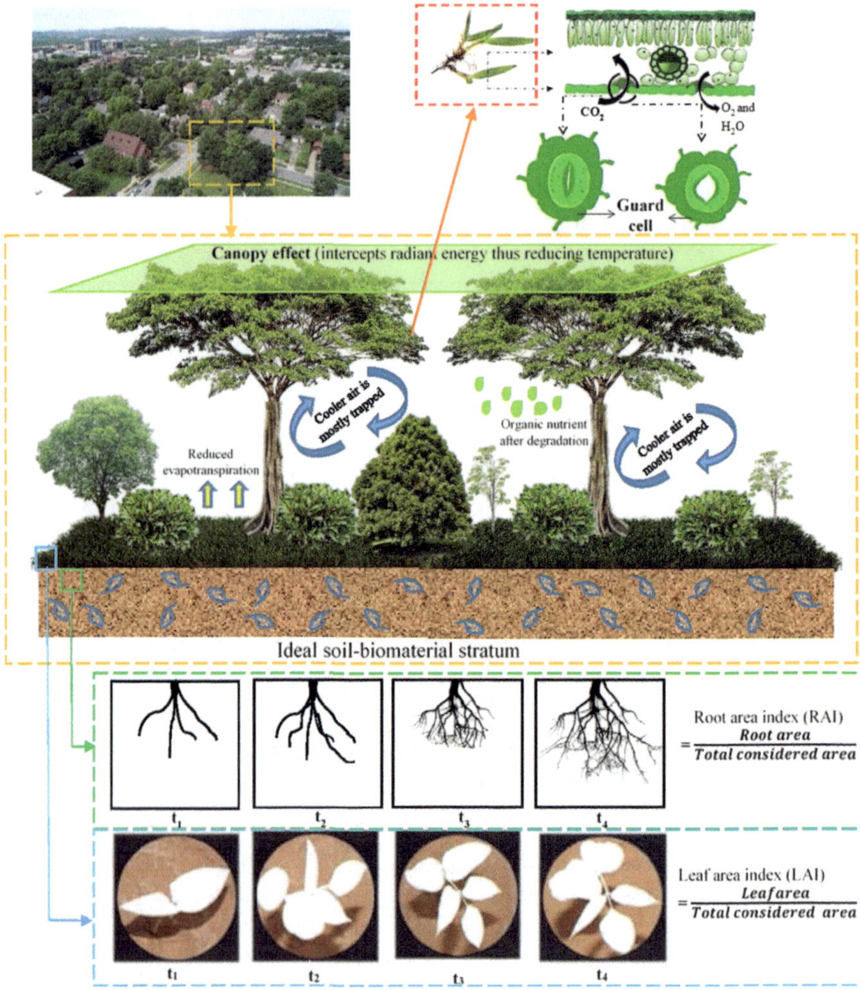

Fig. 15.7 Concept diagram of man-made urban forest and plant parameters to ascertain soil moisture conservation

Although work has been undertaken under the umbrella of bio-engineered slope stability and landfill vegetation layers using unsaturated soil mechanics in the root zone, there are not many studies that consider plant hydraulic parameters. Most studies have considered leaf area index (LAI) and root area index (RAI) to understand the plant effect on soil suction and moisture profiles [41]. LAI indicates the variation in canopy area and is dependent on the plant type. Leaf area index refers to the one-sided green leaves area per unit soil surface area and root area index is defined as the ratio of surface area of root to the soil surface area [21]. RAI indicates the root architecture, which directly induces suction and also helps in increasing soil shear strength. Root distribution is expressed in terms of root distribution function. Variation of root area index along the root depth represents the root distribution function. Leaf hydraulics is majorly dependent on plant stomatal conductance and photosynthetic yield [54], which has been neglected in understanding root zone unsaturated soil mechanics. A typical cross section of grass leaf is shown in Fig. 15.7. Stoma is the minute aperture, which occurs in leaf. Transfer rate of CO_2 (entering the leaf), H_2O and O_2 exiting through stoma during photosynthesis is termed as stomatal conductance. Transpiration occurs as long as stomata are open. Relatively large opening implies greater transpiration, whereas, small opening of stoma infers low transpiration rate. In addition, time and suction alter stomatal conductance during continuous drying period. Recent studies have shown that stomatal conductance directly effects the evolution pattern of soil suction in the root zone and influences surface desiccation cracks [11]. The authors are currently exploring inherent plant parameters such as LAI, RAI, stomatal conductance, and photosynthetic yield to understand their individual effect on root zone soil suction under controlled conditions as shown in Fig. 15.8. The effect of general plant parameters and their effect on soil unsaturated properties are discussed in the following section.

15.3.1 Leaf Area Index (LAI) and Root Area Index (RAI)

Figure 15.9a shows the various root distribution function corresponding to different LAIs, which were proposed by previous researchers. Uniform root distribution function [19] and linearly decreasing root distribution function [50] are widely adopted to understand the soil suction induced due to evapotranspiration through various canopy areas. Garg et al. [24] have investigated the variation of the root distribution function of Schefflera heptaphylla corresponding to the change in LAIs. Unlike to [19, 50], a curvilinear variation of RAI was found along the root depth of Schefflera heptaphylla. Gadi et al. [22] numerically investigated the coupled effect of canopy and root distribution on suction induced in the root zone of vegetated soil. Figure 15.9b shows computed soil suction profile of vegetated soil subjected to 10 h of drying. LAI of 1.0 was found to induce highest suction at the surface making it more effective in erosion control than higher LAIs. Significant variation of soil suction can be seen along the depth for the selected root distribution functions. This shows the importance and need of actual plant parameters to simulate the suction induced due to evapotranspiration.

A rapid increase in LAI of Eureka spring wheat was observed by [3]. LAI was found to decrease after attaining maximum value, which can be attributed to age effect. The variation in LAI with time was rarely studied for non-crop species used in bioengineering. Few researchers quantified the growth stages of forage grasses in terms of number of culms or shoot length [37, 55]. However, variation of LAI during the growth period was rarely quantified. As the radiant energy penetrate the canopy, light rays are intercepted by leaves. Transmission of light decreases with increase in LAI [4]. This can be observed from Beers law based on LAI [4] shown in Eq. 15.1.

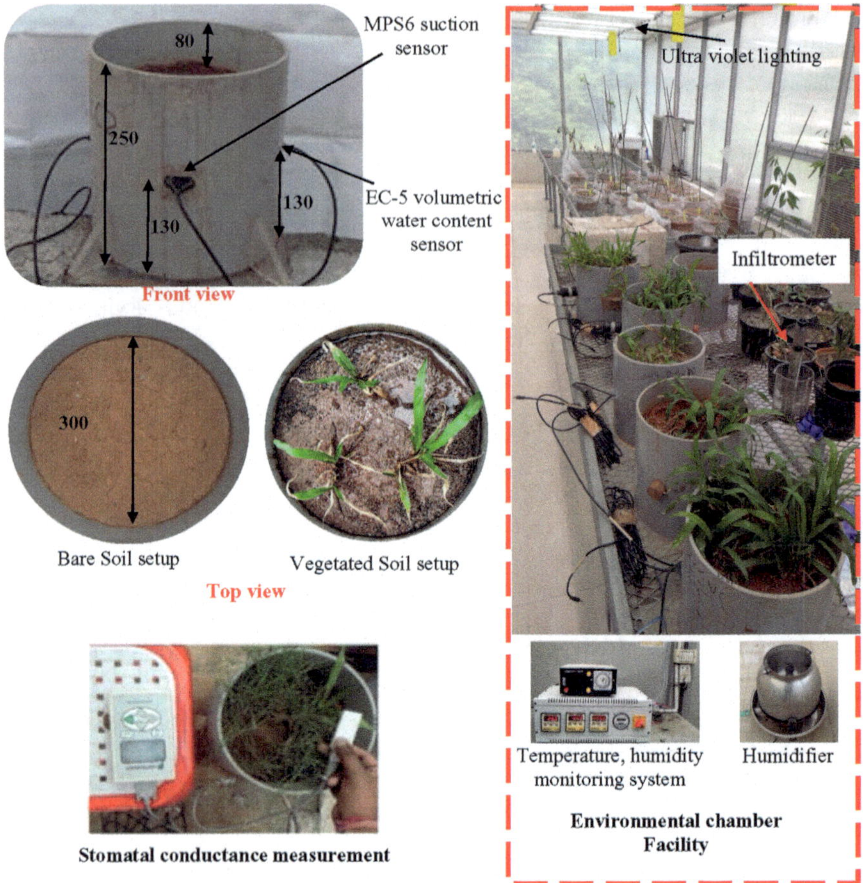

Fig. 15.8 Soil and plant parameter measurement under controlled conditions

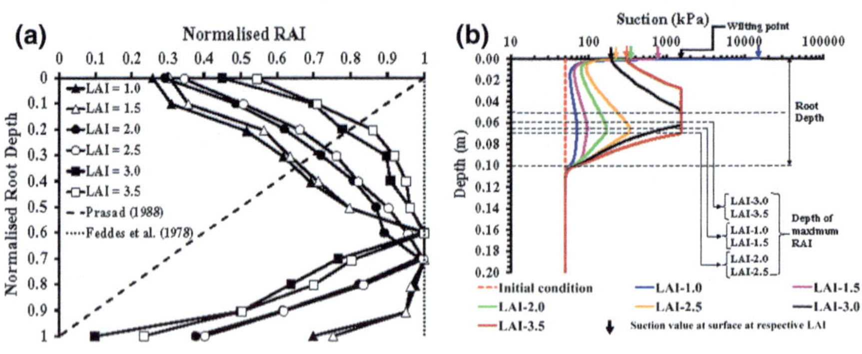

Fig. 15.9 **a** Root distribution functions corresponding to various leaf area indexes (after Garg et al. [24]) and **b** suction profiles when the vegetation is subjected to 10 h of drying (after Gadi et al. [22])

$$\frac{I}{I_0} = exp(-k \times LAI) \tag{15.1}$$

where $\frac{I}{I_0}$ is relative light transmission, LAI = Leaf area index (mm^2/mm^2), and k = proportionality constant associated with angle between leaves and sun rays.The plants are usually inclined with an angle to the slope [32]. Light transmission through the plants in plane and inclined surface would be dissimilar. This dissimilarity is rarely taken into account in previous studies [21]. Light transmission is generally quantified in terms of stomatal conductance.

15.3.2 Root Tensile Strength

Roots are categorized into two types: (i) fine roots (diameter <2 mm) and (ii) coarse roots (diameter >2 mm) [53]. De Baets et al. [15] discussed the tensile strength variation with change in root diameter for Atriplex and Anthyllis. It can be observed that tensile strength of Atriplex is 1.1–4.5 times higher than that of Anthyllis. In addition, decrease in tensile strength with increase in diameter was observed. Change in tensile strength with increase in diameter is more significant up to 3 mm. Thereafter, the tensile strength was found to be relatively constant. It is evident that root diameter varies with depth from the top surface. Furthermore, available water content and temperature alter the root diameter. In addition, root diameter changes with distance from edge of the river bank towards inner side. Root tensile strength was considered to analyze the shear strength of vegetated soil. However, changes in root diameter were rarely considered to analyze the shear strength.

15.4 Field Monitoring of Green Infrastructure

Vegetation growth in grasslands is generally responsive to meteorological parameters such as temperature, relative humidity and rainfall depth [20]. Grasslands consist of more than one type of vegetation. Figure 15.10a–e shows the overview of change in mix grass cover in the vicinity of a tree during four months (after [23]. It can be observed that area covered by grass is relatively small initially at the end of January whereas, the surface was completely covered by grass at the end of April.

15.4.1 Spatial Variation of Vegetation Density

Vegetation growth is expressed in terms of vegetation density and shoot length. Vegetation density is defined as the ratio of area covered by the vegetation to the surface area of the considered plot. Vegetation density is quantified by processing

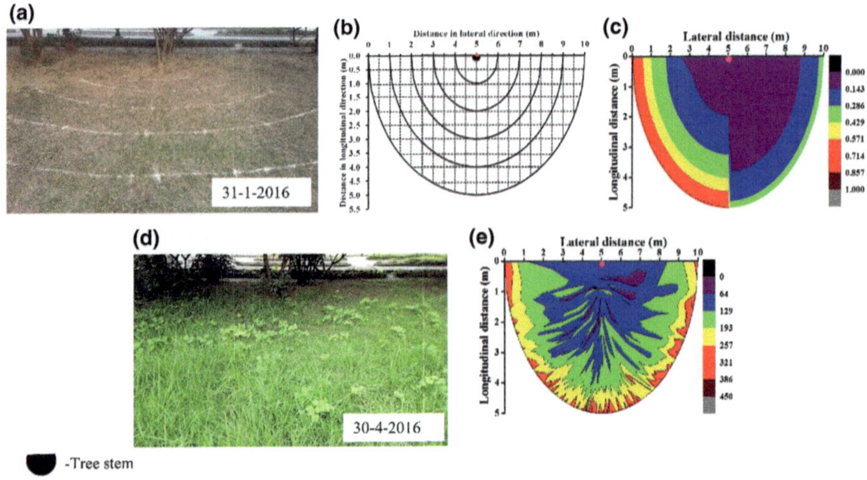

Fig. 15.10 Overview of grass cover and spatial heterogeneity of vegetation density with time (after [20, 23])

the color of vegetation cover images. ImageJ (public domain image processing tool) is used to quantify the vegetation density citerasband. Commercially available camera models are commonly used to capture the images of vegetation cover. The surface area shown in Fig. 15.10a is categorized into small zones (see Fig. 15.10b) to quantify heterogeneity in vegetation density. Trail measurements were taken to select the size of the zone. Maximum size of the grid is 0.125 m. Spatial variation of vegetation density in the selected site is shown in Fig. 15.10c. Vegetation density at higher radial distance from tree stem is around 23 times higher than that near the tree stem. One reason for this may be the difference in radiant energy inception between vegetation near and far away from tree stem. In addition, the growth competition between tree roots and grassroots may also be a reason. Grass growth is slowed due to the growth competition [7].

15.4.2 Spatial Variation of Shoot Growth

Figure 15.10d shows the overview of vegetation cover at the end of June 2016. Figure 15.10e shows the spatial variation of shoot growth rate. The surface area shown in Fig. 15.10d is categorized into small zones (see Fig. 15.10b) on the basis of trial measurements, to quantify heterogeneity in vegetation growth rate. The average growth rate of the shoots in a particular zone is assigned as the shoot growth rate of the entire zone in contours. Vegetation growth rate at higher radial distance from tree stem is around 6 times higher than that near the tree stem. This can be attributed to the difference in radiant energy inception and growth competition between vegetation near and far away from tree stem. Shoot growth is observed to be reduced due to the growth competition [7]. It must be noted that the trend of variation of shoot growth rate is dissimilar to that of vegetation density.

15.4.3 Stomatal Conductance

Soil suction in vegetated soil is influenced by evapotranspiration. Stomatal conductance is one of the crucial parameters, which governs the evapotranspiration in vegetation. Penman–Monteith equation (Eq. 15.2) is used to approximate the evapotranspiration in vegetated soil. Stomatal conductance was found to increase with time initially and attains maximum value during afternoon due to the rise in radiant energy. After attaining maximum value, it is found to decrease due to the decrease in radiant energy and aging effect. Partial closure of stomata occurs as a result of high water deficit in spite of adequate radiant energy. This shows the importance of carefully observing and analyzing stomatal conductance for bio-geotechnical research.

$$PET = \frac{\Delta(R_n - G) + \frac{\rho_a \times c_p (e_s - e_a)}{g_a}}{\lambda \times (\Delta + \gamma(1 + \frac{g_a}{g_s}))} \tag{15.2}$$

where Δ = slope of saturation vapor pressure curve (kPa k^{-1}), R_n = Radiant energy (MJ m^{-2} d^{-1}), G = Soil heat flux (MJ m^{-2} d^{-1}), ρ_a = air density (kg m^{-3}), c_p = specific heat capacity of air (MJ kg^{-1} K^{-1}), e_s-e_a = Vapor pressure deficit (kPa), e_s = saturated vapor pressure (kPa), e_a = actual vapor pressure (kPa), g_a = aerodynamic conductance (m s^{-1}), g_s = stomatal conductance (m s^{-1}), γ = psychrometric constant, ($\gamma \approx 66$ Pa K^{-1}), λ = latent heat of vaporization (MJ kg^{-1}).

15.4.4 Spatial Heterogeneity of Light Condition

It is evident that soil suction may vary spatially in the vegetated field [23]. Reason for this may be spatial heterogeneity of stomatal conductance due to nonuniform light and shade condition. The spatial heterogeneity of vegetation in field condition is generally characterized using three distinct grass covers: (1) grass cover under tree shade (GUT) (2) grass cover under self-shade (GUS), and (3) grass cover without shade (GWS). The variation of evapotranspiration induced suction (associated with stomatal conductance) is highly uncertain. This uncertainty is due to changes in precipitation and radiant energy with time. Most of the previous researchers [18, 26, 35] adopted the deterministic approach to analyze evapotranspiration induced suction. However, the deterministic approach could not predict the highly uncertain variation in suction. Therefore, the probabilistic analysis was found to be a suitable approach to analyze suction induced in vegetated soils [49]. In addition,

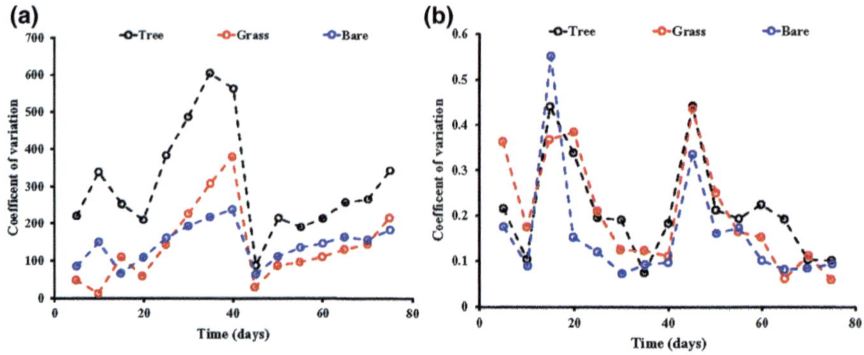

Fig. 15.11 Variation of **a** mean and **b** coefficient of variation of suction with time (after [27])

probabilistic analysis is also required to conduct reliability analysis of bio-engineered slope. A recent study by Hazra et al. [27] analyzed the suction induced in bare, grassed and treed slope at a depth of 0.1 m using probabilistic approach as shown in Fig. 15.11a, b. The result shows that the tree species induce higher suction than grass species and bare soil. Bare soil induces higher suction than grassed soil initially. Rate of decrease in suction due to wetting is higher in grassed soil as compared to that in bare soil. It is evident that changes in shoot and root parameters are also highly uncertain. Hence, further studies are required to understand variation of suction caused by the change in plant parameters using probabilistic approach.

15.5 Future Direction in Sustainable Bio-geotechnology

Figure 15.12 encapsulates the aforementioned discussion and provides a basic framework for future research direction in sustainable bio-geotechnology. Regional understanding is essential to identify suitable waste biomasses and utilize the same as soil-biomaterial composite. This can be deciphered by systematic understanding of the biomaterial in terms of physicochemical, biological and geotechnical characteristics. There need to be specific classification schemes for biomaterials akin to soil classification for recommending its use in different geotechnical and geoenvironmental projects. It is important to consider parameters such as pH, functional groups, morphology, biodegradation kinetics in addition to conventional geotechnical and hydrological properties to assess the suitability of biomaterial as soil-biomaterial composite. It is also necessary to monitor the variations in mechanical and hydraulic characteristics and hysteretic behavior and losses in desirable characteristics under climate change scenarios. The performance of the soil-biomaterial composite as a sustainable stratum needs to be studied by resorting to reduced scale tests and field monitoring. This paves way for scientific implementation of new age green infrastructure such as urban forests, green roof concepts, bio-engineered slopes, sustainable agricultural practices, and drought management.

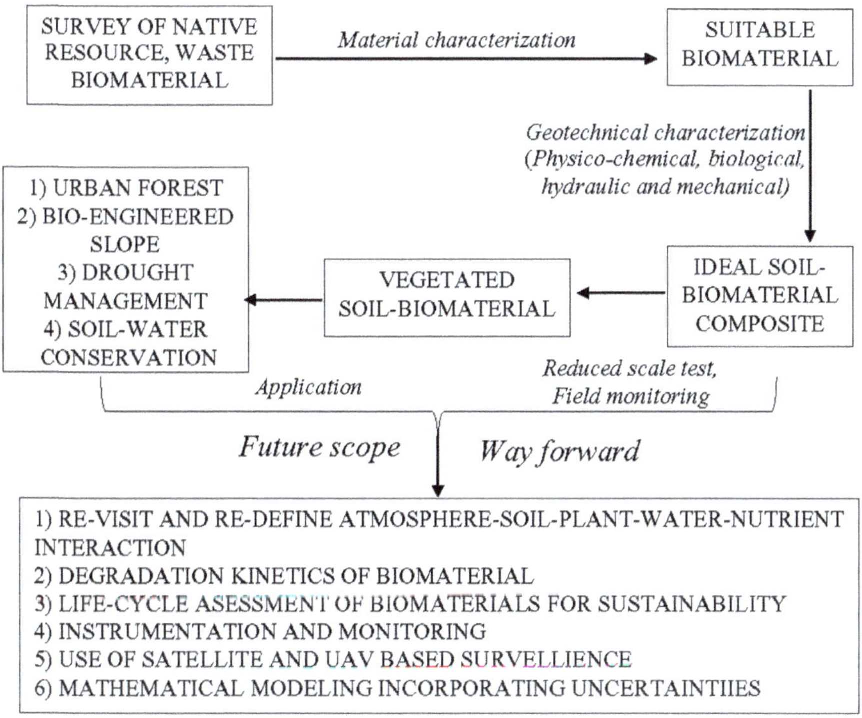

Fig. 15.12 Future direction in sustainable bio-geotechnology

The role of planned afforestation like urban forest for mitigating the effects of urban heat island, flooding, pollution and CO_2 reduction need to be explored in detail through field monitoring. A scientific approach of siting the urban forest, identifying appropriate local flora for capturing pollutants and CO_2, self-sustainability of the bio-geosystem, and life cycle assessment is challenging interdisciplinary research problem highly relevant to the socioeconomic development of the country. Green roof infrastructure is a recent concept for accommodating rainwater leading to the reduction in urban flooding, rainwater conservation, provide thermal insulation and offers a solution to urban heat island. A holistic study is needed to understand the efficiency of bio-geotechnical characteristics and soil-vegetation in green roof in achieving the above-stated goals. The concept is still in its infancy in the Indian subcontinent. The use of microbial interventions and drought-resistant species for minimizing desertification is a need of the hour for mitigating the harmful effects of climate change. Development and use of biological fertilizers and pesticides considerably prevent degradation of fertile lands leading to soil and water conservation. Sustainable practices and low energy intensive bioprocesses need to be developed for converting agricultural waste to value-added materials. This will offer solution for exorbitant pollution resulting from the burning of agricultural waste that affects the air quality of adjacent locations. Adopting the concepts of bio-geotechnology by integrating the efforts of agricultural scientists, hydrologists and geotechnologists would pave way for erosion resistance and soil conservation.

The root zone soil water dynamics is highly sensitive to species used, climatic conditions and canopy effects. The abiotic stress in plants is a function of not only plant parameters but also the soil and environmental parameters. Several definitions reported in the literature such as field capacity, wilting point, plant available water content and residual water content are vague. A well-planned interdisciplinary research involving geotechnical and biotechnological experts are needed for revisiting and redefining the concepts to facilitate effective field implementation. This necessitate short-term and long-term monitoring of soil-vegetation infrastructures using state-of-the-art low-cost sensors and instrumentation. This opens up a plethora of research requirements in terms of development and implementation of monitoring tools. Wide range of nonintrusive techniques was developed and are researched to quantify the soil-vegetation-atmosphere continua. Satellite-based space technologies have progressed manifold in the last few decades. A lot of efforts are ongoing and a lot more is needed for developing predictive and forecasting models necessary for the advancement of bio-geotechnical practices. On a smaller scale, the development and implementation of unmanned aerial vehicle (UAV) equipped with appropriate sensor for real-time monitoring of resources and parameters (agriculture, temperature, pollution, etc.) needs a focused effort.

Most of the studies dealing with soil–biomaterial interaction do not account for life cycle assessment, aging, and biodegradation of the system. The effect of gradual degradation of biomaterial and its consequent effect in root zone unsaturated parameters has not been explored in detail. Knowledge in this sphere will truly help stakeholders to invest and understand this relatively new field of geotechnics incorporating biomaterials leading to sustainability. For engineered applications, it is invariably necessary to assess and quantify the bounds of uncertainties associated with these studies. Since most of the parameters and characteristics are not well defined and are prone to several inherent and induced uncertainties, it is invariably necessary to develop mathematical models for capturing the same. Efforts need to be made to develop statistical and probabilistic models for studying bio-geo-atmosphere interface taking into account measurement fallacies related to instrument and diurnal variations of parameters. The use of trend analysis and time series modeling will be required when climate change characteristics need to be integrated with biodegradation kinetics of soil-bio-atmosphere continua.

References

1. Agaba, H., Baguma Orikiriza, L.J., Osoto Esegu, J.F., Obua, J., Kabasa, J.D., Hüttermann, A.: Effects of hydrogel amendment to different soils on plant available water and survival of trees under drought conditions. Clean-Soil Air Water **38**(4), 328–335 (2010)
2. Alvey, A.A.: Promoting and preserving biodiversity in the urban forest. Urban For. Urban Green. **5**(4), 195–201 (2006)

3. Asrar, G., Fuchs, M., Kanemasu, E., Hatfield, J.: Estimating absorbed photosynthetic radiation and leaf area index from spectral reflectance in wheat 1. Agron. J. **76**(2), 300–306 (1984)
4. Atwell, B.J.: Plants in Action: Adaptation in Nature, Performance in Cultivation. Macmillan Education AU (1999)
5. Bali Reddy, S., Pradeep Kumar, D., Murali Krishna, A.: Evaluation of the optimum mixing ratio of a sand-tire chips mixture for geoengineering applications. J. Mater. Civ. Eng. **28**(2), 06015007 (2015)
6. Basu, D., Puppala, A., Chittoori, B.: General report of tc 307 sustainability in geotechnical engineering. In: Proceedings of 18th ICSMGE (2013)
7. Belsky, A.J.: Influences of trees on savanna productivity: tests of shade, nutrients, and tree-grass competition. Ecology **75**(4), 922–932 (1994)
8. Bordoloi, S., Gadi, V., Hussain, R., Sahoo, L., Garg, A., Sreedeep, S., Mei, G., Poulsen, T.: Influence of fiber from waste weed eichhornia crassipes on water retention and cracking characteristics of vegetated soils. Géotech. Lett. 1–25 (2018)
9. Bordoloi, S., Hussain, R., Gadi, V., Bora, H., Sahoo, L., Karangat, R., Garg, A., Sreedeep, S.: Monitoring soil cracking and plant parameters for a mixed grass species. Géotech. Lett. **8**(1), 49–55 (2018)
10. Bordoloi, S., Garg, A., Sekharan, S.: A review of physio-biochemical properties of natural fibers and their application in soil reinforcement. Adv. Civ. Eng. Mater. **6**(1), 323–359 (2017)
11. Bordoloi, S., Kashyap, V., Garg, A., Sreedeep, S., Wei, L., Andriyas, S.: Measurement of mechanical characteristics of fiber from a novel invasive weed: a comprehensive comparison with fibers from agricultural crops. Measurement **113**, 62–70 (2018)
12. Chen, X.W., Wong, J.T.F., Ng, C.W.W., Wong, M.H.: Feasibility of biochar application on a landfill final covera review on balancing ecology and shallow slope stability. Environ. Sci. Pollut. Res. **23**(8), 7111–7125 (2016)
13. Consoli, N.C., Montardo, J.P., Prietto, P.D.M., Pasa, G.S.: Engineering behavior of a sand reinforced with plastic waste. J. Geotech. Geoenviron. Eng. **128**(6), 462–472 (2002)
14. Correia, A.G., Winter, M., Puppala, A.: A review of sustainable approaches in transport infrastructure geotechnics. Transp. Geotech. **7**, 21–28 (2016)
15. De Baets, S., Poesen, J., Reubens, B., Wemans, K., De Baerdemaeker, J., Muys, B.: Root tensile strength and root distribution of typical mediterranean plant species and their contribution to soil shear strength. Plant Soil **305**(1–2), 207–226 (2008)
16. Demitri, C., Scalera, F., Madaghiele, M., Sannino, A., Maffezzoli, A.: Potential of cellulose-based superabsorbent hydrogels as water reservoir in agriculture. Int. J. Polym. Sci. **2013**, (2013)
17. Ekebafe, L., Ogbeifun, D., Okieimen, F.: Polymer applications in agriculture. Biokemistri **23**(2) (2011)
18. El-Ramly, H., Morgenstern, N., Cruden, D.: Probabilistic slope stability analysis for practice. Can. Geotech. J. **39**(3), 665–683 (2002)
19. Feddes, R.A.: Simulation of Field Water Use and Crop Yield. Pudoc (1982)
20. Gadi, V., Singh, S., Singhariya, M., Garg, A.: Modeling soil-plant-water interaction: effects of canopy and root parameters on soil suction and stability of green infrastructure. Eng. Comput. **35**(3), 1543–1566 (2018)
21. Gadi, V.K., Bordoloi, S., Garg, A., Kobayashi, Y., Sahoo, L.: Improving and correcting unsaturated soil hydraulic properties with plant parameters for agriculture and bioengineered slopes. Rhizosphere **1**, 58–78 (2016)
22. Gadi, V.K., Garg, A., Prakash, S., Wei, L., Andriyas, S.: A non-intrusive image analysis technique for measurement of heterogeneity in grass species around tree vicinity in a green infrastructure. Measurement **114**, 132–143 (2018)
23. Gadi, V.K., Manogaran, I.P., Garg, A., Berretta, C., Sreedeep, S.: A novel color analysis technique for differentiation of mix grass cover under shade and without shade in green infrastructures. Adv. Civ. Eng. Mater. **6**(1), 564–582 (2017)
24. Garg, A., Leung, A.K., Ng, C.W.W.: Comparisons of soil suction induced by evapotranspiration and transpiration of S. heptaphylla. Can. Geotech. J. **52**(12), 2149–2155 (2015)

25. Garg, A., Leung, A.K., Ng, C.W.W.: Transpiration reduction and root distribution functions for a non-crop species schefflera heptaphylla. Catena **135**, 78–82 (2015)
26. Griffiths, D., Fenton, G.A.: Probabilistic slope stability analysis by finite elements. J. Geotech. Geoenviron. Eng. **130**(5), 507–518 (2004)
27. Hazra, B., Gadi, V., Garg, A., Ng, C.W.W., Das, G.K.: Probabilistic analysis of suction in homogeneously vegetated soils. Catena **149**, 394–401 (2017)
28. Heynen, N.C., Lindsey, G.: Correlates of urban forest canopy cover: implications for local public works. Public Works Manag. Policy **8**(1), 33–47 (2003)
29. Krishna, A.M., Dey, A., Sreedeep, S.: Geotechnics for Natural and Engineered Sustainable Technologies: GeoNEst. Springer (2018)
30. Lee, S.S., Chang, S.X., Chang, Y.Y., Ok, Y.S.: Commercial versus synthesized polymers for soil erosion control and growth of chinese cabbage. SpringerPlus **2**(1), 534 (2013)
31. Leung, A.K., Boldrin, D., Liang, T., Wu, Z., Kamchoom, V., Bengough, A.: Plant age effects on soil infiltration rate during early plant establishment. Géotechnique (2017) (in press)
32. Leung, A.K., Garg, A., Coo, J.L., Ng, C.W.W., Hau, B.: Effects of the roots of Cynodon dactylon and Schefflera heptaphylla on water infiltration rate and soil hydraulic conductivity. Hydrol. Process. **29**(15), 3342–3354 (2015)
33. Leung, A.K., Garg, A., Ng, C.W.W.: Effects of plant roots on soil-water retention and induced suction in vegetated soil. Eng. Geol. **193**, 183–197 (2015)
34. Leung, A.K., Ng, C.W.W.: Analyses of groundwater flow and plant evapotranspiration in a vegetated soil slope. Can. Geotech. J. **50**(12), 1204–1218 (2013)
35. Li, K., Lumb, P.: Probabilistic design of slopes. Can. Geotech. J. **24**(4), 520–535 (1987)
36. Mimouni, T., Laloui, L.: Towards a secure basis for the design of geothermal piles. Acta Geotech. **9**(3), 355–366 (2014)
37. Mitchell, R.B., Moser, L.E., Moore, K.J., Redfearn, D.D.: Tiller demographics and leaf area index of four perennial pasture grasses. Agron. J. **90**(1), 47–53 (1998)
38. Montesano, F.F., Parente, A., Santamaria, P., Sannino, A., Serio, F.: Biodegradable superabsorbent hydrogel increaseswater retention properties of growing media and plant growth. Agric. Sci. Procedia **4**, 451–458 (2015)
39. Montoya, B., DeJong, J., Boulanger, R.: Dynamic response of liquefiable sand improved by microbial-induced calcite precipitation. Géotechnique **63**(4), 302 (2013)
40. Narjary, B., Aggarwal, P., Singh, A., Chakraborty, D., Singh, R.: Water availability in different soils in relation to hydrogel application. Geoderma **187**, 94–101 (2012)
41. Ng, C.W.W., Garg, A., Leung, A.K., Hau, B.: Relationships between leaf and root area indices and soil suction induced during drying-wetting cycles. Ecol. Eng. **91**, 113–118 (2016)
42. Ng, C.W.W., Leung, A.K., Woon, K.: Effects of soil density on grass-induced suction distributions in compacted soil subjected to rainfall. Can. Geotech. J. **51**(3), 311–321 (2013)
43. Ni, J., Chen, X., Ng, C., Guo, H.: Effects of biochar on water retention and matric suction of vegetated soil. Géotech. Lett. 1–23 (2018)
44. Ni, J., Leung, A.K., Ng, C.W.W., Shao, W.: Modelling hydro-mechanical reinforcements of plants to slope stability. Comput. Geotech. **95**, 99–109 (2018)
45. Nnadi, F., Brave, C.: Environmentally friendly superabsorbent polymers for water conservation in agricultural lands. J. Soil Sci. Environ. Manag. **2**(7), 206–211 (2011)
46. Nowak, D.J.: Atmospheric carbon reduction by urban trees. J. Environ. Manag. **37**(3), 207–217 (1993)
47. Pardo, G., Sarmah, A., Orense, R.: Mechanism of improvement of biochar on shear strength and liquefaction resistance of sand. Géotechnique pp. 1–10 (2018)
48. Patidar, A., Mahiyar, H.: An experimental study on stabilization of black cotton soil using hdpe wastage fibres, stone dust and lime. Int. J. Adv. Sci. Tech. Res. **6**(4) (2014)
49. Phoon, K.K., Santoso, A., Quek, S.T.: Probabilistic analysis of soil-water characteristic curves. J. Geotech. Geoenviron. Eng. **136**(3), 445–455 (2010)
50. Prasad, R.: A linear root water uptake model. J. Hydrol. **99**(3–4), 297–306 (1988)
51. Reddy, K., Maruthi, V., Umesha, B., et al.: Influence of super absorbent polymers on infiltration characteristics of alfisols in semi-arid region. Indian J. Dryland Agric. Res. Dev. **30**(2), 11–16 (2015)

52. Renou, S., Givaudan, J., Poulain, S., Dirassouyan, F., Moulin, P.: Landfill leachate treatment: review and opportunity. J. Hazard. Mater. **150**(3), 468–493 (2008)
53. Rosado, B.H.P., Martins, A.C., Colomeu, T.C., Oliveira, R.S., Joly, C.A., Aidar, M.P.M.: Fine root biomass and root length density in a lowland and a montane tropical rain forest, SP, Brazil. Biota Neotrop. **11**(3), 203–209 (2011)
54. Scoffoni, C., Chatelet, D.S., Pasquet-kok, J., Rawls, M., Donoghue, M.J., Edwards, E.J., Sack, L.: Hydraulic basis for the evolution of photosynthetic productivity. Nat. Plants **2**(6), 16072 (2016)
55. Simon, J., Lemaire, G.: Tillering and leaf area index in grasses in the vegetative phase. Grass Forage Sci. **42**(4), 373–380 (1987)
56. Steinbeiss, S., Gleixner, G., Antonietti, M.: Effect of biochar amendment on soil carbon balance and soil microbial activity. Soil Biol. Biochem. **41**(6), 1301–1310 (2009)
57. Tyrväinen, L., Pauleit, S., Seeland, K., de Vries, S.: Benefits and uses of urban forests and trees. In: Urban Forests and Trees, pp. 81–114. Springer (2005)
58. Tzoulas, K., Korpela, K., Venn, S., Yli-Pelkonen, V., Kaźmierczak, A., Niemela, J., James, P.: Promoting ecosystem and human health in urban areas using green infrastructure: a literature review. Landsc. Urban Plan. **81**(3), 167–178 (2007)
59. Vundavalli, R., Vundavalli, S., Nakka, M., Rao, D.S.: Biodegradable nano-hydrogels in agricultural farming-alternative source for water resources. Procedia Mater. Sci. **10**, 548–554 (2015)
60. Wei, Y., Durian, D.J.: Effect of hydrogel particle additives on water-accessible pore structure of sandy soils: a custom pressure plate apparatus and capillary bundle model. Phys. Rev. E **87**(5), 053013 (2013)
61. Zhu, H., Zhang, L.M.: Evaluating suction profile in a vegetated slope considering uncertainty in transpiration. Comput. Geotech. **63**, 112–120 (2015)

Chapter 16
Recent Developments in Earth Pressure Reduction Techniques

S. M. Dasaka and V. K. Gade

16.1 Introduction

Retaining walls are structures constructed to retain the soil in steep and deep basements or in slopes where an abrupt change in elevation that exceeds the angle of repose of the soil [1]. They need to resist loads due to retaining soils, traffic movements, adjoining structures, and earthquakes, during the service life. Retaining walls may be vulnerable to catastrophic failure during earthquakes, due to sudden increase in lateral loads. Optimum design of earth retaining structures depends primarily on the total lateral force and distribution of lateral earth pressures acting on them. Lateral earth pressure on retaining walls decides the sectional dimensions of walls, and it depend the on height of wall, properties of the soil to be retained and movement of the wall with reference to the backfill. However, there are instances, such as basement walls, bridge abutments, and box culverts, where wall movements are highly restricted and they are considered as non-yielding walls. In general, these walls are designed for lateral earth pressures correspond to the at-rest pressures, which are considered to be higher than active earth pressures. Significantly higher lateral earth pressures make the retaining walls bulky and prove uneconomical. In the past, several methods had been developed to reduce lateral earth pressure acting on retaining wall. Different approaches such as reinforced retaining walls [2, 3], use of recycled tire chips [4, 5] and using compressible inclusion at the interface between the retaining wall and backfill [6–8] were found in the literature.

Any material which is readily compressible due to pressures acting on it is known as compressible inclusions. Numerous analytical, numerical, field, and laboratory

S. M. Dasaka (✉)
Department of Civil Engineering, IIT Bombay, Mumbai, India
e-mail: dasaka@civil.iitb.ac.in

V. K. Gade
Department of Civil Engineering, VRSEC, Vijayawada, India
e-mail: gadevinilkumar23@gmail.com

M. Latha G. (ed.), *Frontiers in Geotechnical Engineering*, Developments in Geotechnical Engineering, https://doi.org/10.1007/978-981-13-5871-5_16

experimental studies had been conducted and revealed that use of compressible inclusion significantly reduces the lateral earth pressure (both static and dynamic) on the rigid retaining structures, and may provide a cost-effective solution.

In recent scenario of compressible inclusions, Expanded Polystyrene Styrofoam (EPS), aptly named as geofoam, is widely used because of its unique advantages, such as lightweight, small amplitude wave damping, thermal insulation, etc., compared to conventional geosynthetic materials, such as geotextiles, geogrids, and geomembranes. Geofoam is a special kind of material, which is free from the common phenomenon observed in most geologic materials, known as dilatancy. EPS inclusions induced lesser lateral thrust when used behind retaining structures [7, 9, 10, to name a few]. There were no attempts to understand the actual deformation of compressible inclusion during loading conditions [7, 11, 12] and effect of creep of compressible inclusions on the long-term lateral earth pressure reduction needs further understanding to economize design of retaining structures.

16.2 Motivation and Objectives

In general, most of the civil engineering structures are designed for an expected lifespan of 30–100 years [13], and it is essential to predict the behavior of materials used in the structures throughout the lifespan of structures. If material properties change with time, such effects should be incorporated in the design of the structures. However, studies on the long-term behavior of geofoam and its effect on the earth pressure have not been given much attention. The aim of the study is to understand the time-dependent behavior of compressible geo-inclusion on the reduction of static earth pressures on retaining walls retaining a cohesionless backfill.

16.3 Materials Used and Their Characterization

In the present study Grade III Indian Standard Sand was used as retaining wall backfill material. From the results of the particle size analyses, it can be observed that Grade III sand is uniformly graded medium to fine sand and are classified as SP according to the Unified Soil Classification System, USCS [14]. Salient details of the backfill material adopted in the study are listed in Table 16.1. Expanded polystyrene (EPS) geofoam samples of three different densities (15 kg/m^3, 20 kg/m^3, and 25 kg/m^3, referred to as EPS15, EPS20, EPS25, respectively) were adopted for the experimental study. Geofoam panels with 600 mm in length, 300 mm in width, and 25 mm in thickness, procured locally, were used in the retaining wall model studies. The mechanical properties of geofoam are presented in Table 16.2. Detailed experimental program, testing procedure, and results of properties EPS geofoam can be obtained from Gade and Dasaka [15]. The melting temperature of the geofoam was obtained as 145 °C from thermo-gravimetric analysis. Similarly, the glass transition temperature

Table 16.1 Properties of Grade III sand used in the present study

Description	Grade III
Specific gravity of soil solids, G_s	2.65
Mean diameter of soil particles, D_{50} (mm)	0.28
Coefficient of uniformity, C_u	1.42
Coefficient of curvature, C_c	0.93
Minimum dry unit weight, γ_{dmin} (kN/m^3)	14.29
Maximum dry unit weight, γ_{dmax} (kN/m^3)	17.33

Table 16.2 Properties of EPS geofoam

Designation	Density (kg/m^3)	Elastic limit (%)	Young's modulus, E (MPa)	Compressive strength, σ_c (kPa)	Yield strength, σ_y (kPa)
EPS10	10	1.4–1.8	0.88–0.93	28–30	21.5
EPS15	15	2.6–2.8	1.99–2.28	63–70	55
EPS20	20	2.7–3	2.99–3.61	96.3 108	90
EPS25	25	2.1–3	5.99–6.46	126–142	122

for the Geofoam samples was obtained as 42 °C. Time–Temperature–Stress Superposition (TTSS) accelerated creep testing was carried out the geofoam samples to understand the long-term creep behavior of geofoam used in the present analysis. Generally, an average room temperature in Mumbai region varies in the range of 28 ± 2 °C, due this a reference temperature of 29 °C is adopted. Creep strains induced in the geofoam specimen were estimated for a period of 100 years. At the end of 100 years, geofoam samples of EPS15, EPS20, and EPS25 experienced 2.12, 2.4, and 2.11% creep strains, respectively. Figure 16.1 shows the results of the creep test carried out on EPS25 geofoam samples.

16.4 Experimental Program

Small-scale 1g model tests were carried out in the laboratory to measure lateral earth pressures on the retaining walls retaining dry cohesionless backfill, with and without the use of compressible geofoam inclusion. Air pluviation method using Mechanized Traveling Pluviator (MTP) was adopted in the study to prepare the sand beds, and miniature cone penetration testing was used to evaluate the uniformity of the prepared sand beds. All model backfills are prepared at 65% relative density, using 10 mm diameter orifice, four diffuser sieves, and 12.5 cm height of fall. More details on the pluviator (MTP) system used and the methodology adopted to prepare the sand beds can be found in Gade and Dasaka [16]. Diaphragm-based earth pressure sensors were used to measure the lateral earth pressures on the wall. All pressure

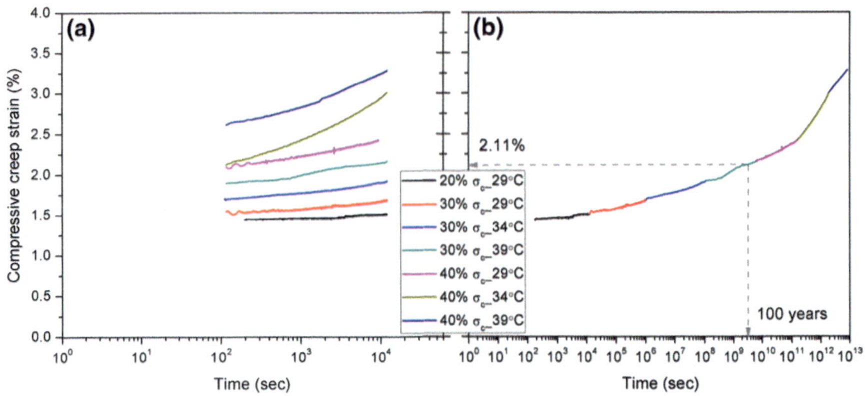

Fig. 16.1 Results of TTSS method for EPS25 geofoam **a** initial test data **b** master compressive creep strain curve

sensors were properly calibrated before using them in the model tests. In case of no-foam conditions, the earth pressure sensors were in direct contact with sand. Conversely, with geofoam inclusions at the interface between the retaining wall and the backfill, earth pressure sensors were in contact with the geofoam, but not with sand. In view of two different materials, with which the earth pressure sensors were in direct contact during the experiments, extensive calibration studies were carried out to understand the calibration response of sensors based on the type and nature of contacting material. More details on the calibration of earth pressure sensors can be found in Gade [17] and Gade and Dasaka [18].

In the present study, earth pressures were measured on rigid non-yielding walls, and the model set up consists of rigid tank, rigid stainless steel plate, earth pressure sensors, surcharge load, camera, and its fixtures, illumination set up, rigid support and plastic markers, as shown in Fig. 16.2. Stainless steel material was selected to prepare model rigid tank. However, acrylic glass material was used on one side of the tank to capture the soil and geofoam movements during the testing. The model tank has internal dimensions of 1200 mm × 305 mm × 700 mm. Additional stiffeners were placed on acrylic side of the container to enhance the rigidity of the tank. The model plate was made of stainless steel material and along the central line grooves were made to accommodate EPCs to measure the lateral earth pressures from backfill. A stainless steel plate of 700 mm × 300 mm × 18 mm thick is adopted for model tests. The model plate was hinged at bottom. Pictorial view and schematic view of model wall are shown in Fig. 16.3a and b, respectively. Total six sensors were affixed at midline of plate at 100 mm spacing. Care was exercised to protect the cables of the sensors from damage.

Majority of model tests were performed by placing surcharge loading at distance of 150 mm from the face of model plate, which is one-fourth of height of backfill. Single surcharge load of 43 kPa is applied for 10 min duration on backfill; lateral earth pressure on wall achieves steady state well within the adopted duration of loading.

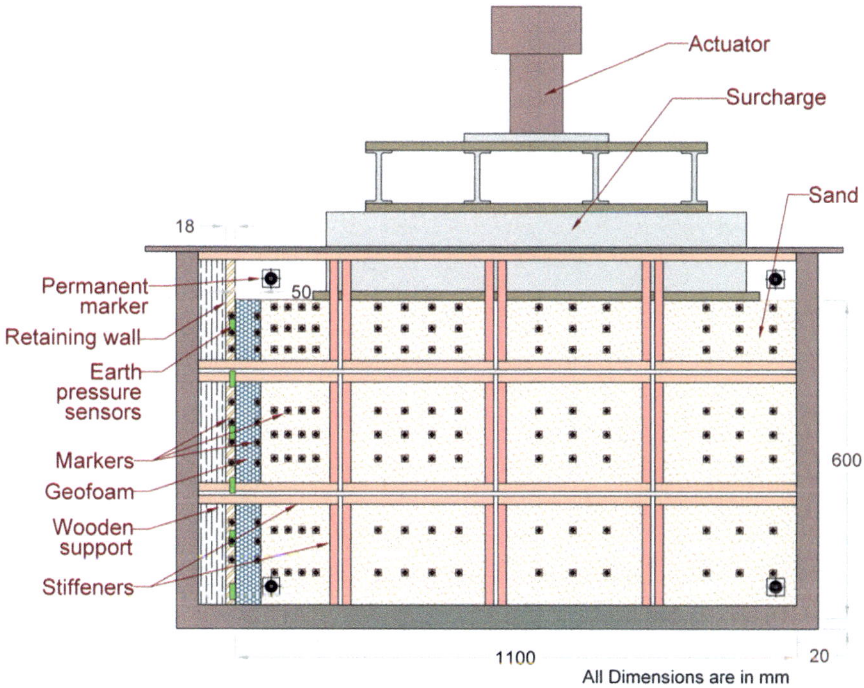

Fig. 16.2 Experimental set up including loading system used to measure lateral earth pressures on the wall

16.5 Results and Discussions

16.5.1 Creep Tests on Geofoam Samples

Servo-hydraulic actuator was used for performing the pseudo-long-term tests on 100 mm cube samples of geofoam. The test procedure comprises of four steps (i) initially required amount of load was applied to induce an desired value of compressive creep in geofoam samples in 60 s, followed by (ii) maintaining applied compressive creep for 3 h (>10^4 s), (iii) removal of external load causing compressive creep from the geofoam sample, and (iv) testing geofoam samples under compressive load at 10%/min strain rate, similar to static compression tests. During the testing, load on the sample was continuously monitored during steps 1–3 through servo data logger, as shown in Fig. 16.4a, b.

In this study, EPS15, EPS20, and EPS25 geofoam samples were subjected to compressive creep of 3% for 3 h duration. At the end of 100 years, compressive creep strain, obtained from TTSS accelerated creep tests, ranges from 2.11 to 2.4% for three densities of geofoam. A slightly higher value (3%) of compressive creep was chosen, to accommodate strain relaxation during the transit, and to evaluate the

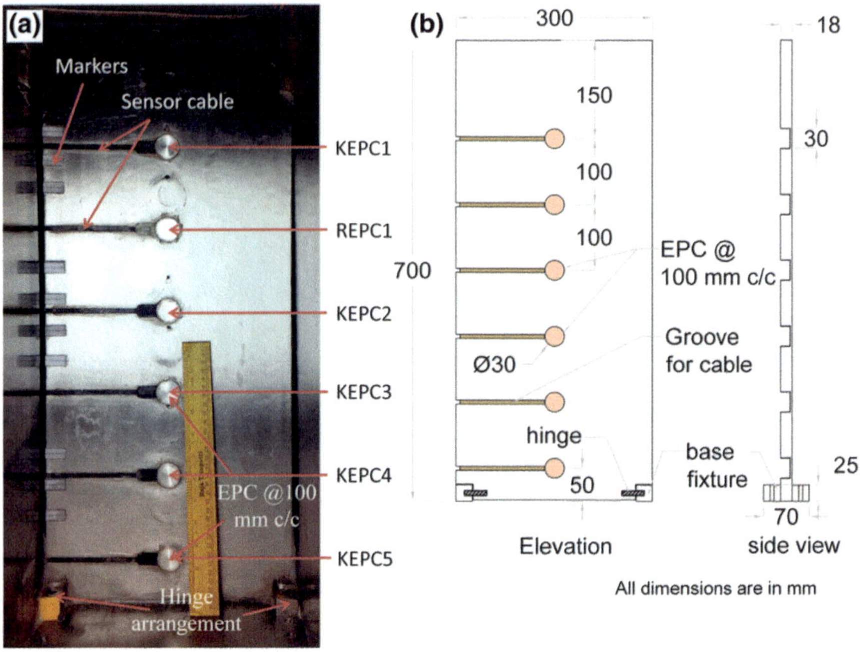

Fig. 16.3 Model retaining wall with location of earth pressure sensors **a** pictorial view **b** schematic view

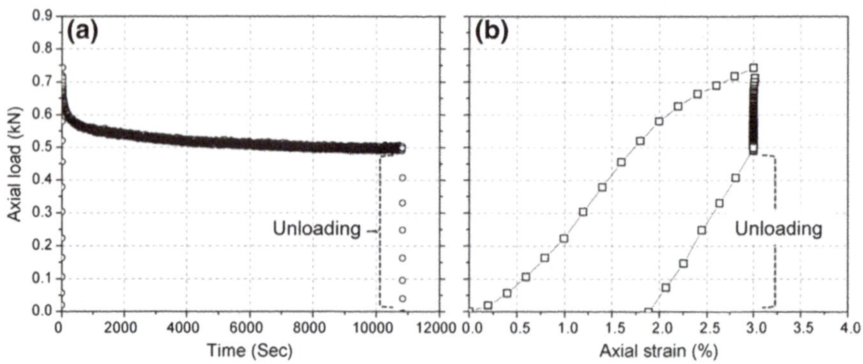

Fig. 16.4 EPS20 geofoam **a** resistance offered during holding period under 3% compressive creep and **b** axial load-strain response

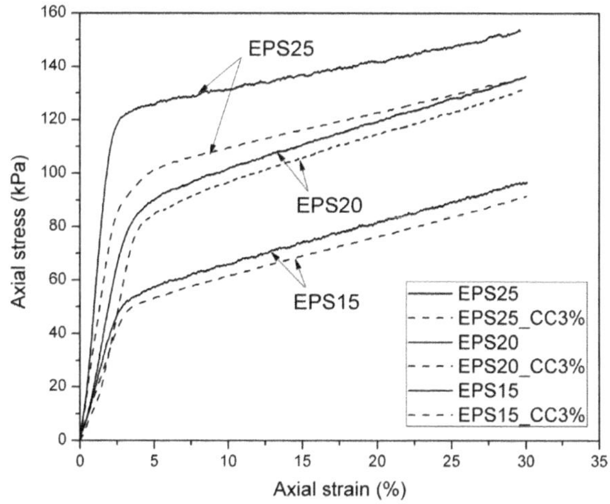

Fig. 16.5 Compressive stress–strain behavior of initial and 3% compressive creep sustained geofoam samples

long-term behavior of geofoam. The results obtained from pseudo-long-term test on EPS20 geofoam sample are shown in Fig. 16.4. It was observed that after 3 h of holding period, resistance offered by geofoam sample was more or less constant. After geofoam samples sustained required compressive creep strain, they were tested at 10% strain/minute, similar to static compression tests. Compressive creep induced and non-sustained geofoam samples stress–strain responses are shown in Fig. 16.5. It was noted that irrespective of density of geofoam, compressive creep sustained samples experienced lower stress levels at any strain values compared to original geofoam samples. This behavior leads to lower Young's modulus, compressive strength, and yield strength. Table 16.3 summarizes properties of 3% compressive creep sustained samples and the corresponding non-sustained geofoam samples. Variation of Young's modulus with increase in compressive strain is shown in Fig. 16.6 for all the three densities of geofoam. It was observed that Young's modulus of geofoam decreases with increased compressive creep.

16.5.2 *Effect of Creep of Geofoam on Long-Term Static Lateral Earth Pressure*

From the preliminary pseudo-long-term static compression studies on the 100 mm cubes, it was observed that compressive creep sustained geofoam samples exhibit lower values of mechanical properties than that of original samples. For assessing the benefit of geofoam in reducing the long-term earth pressures, creep sustained geofoam panels were used in model tests. For conducting retaining wall model tests simulating the long-term conditions, geofoam panels were initially compressed to a desired level before placing them at the interface between the model retaining

Table 16.3 Properties of geofoam samples subjected to compressive loads

Properties	Density					
	EPS15		EPS20		EPS25	
	Initial	After CC3%	Initial	After CC3%	Initial	After CC3%
Young's modulus (MPa)	3.03	2.33	5.08	3.48	6.24	4.67
Compressive strength (kPa)	66.5	61.4	102	96	130.5	109
Yield strength (kPa)	55	51.5	91	86	123	99.5

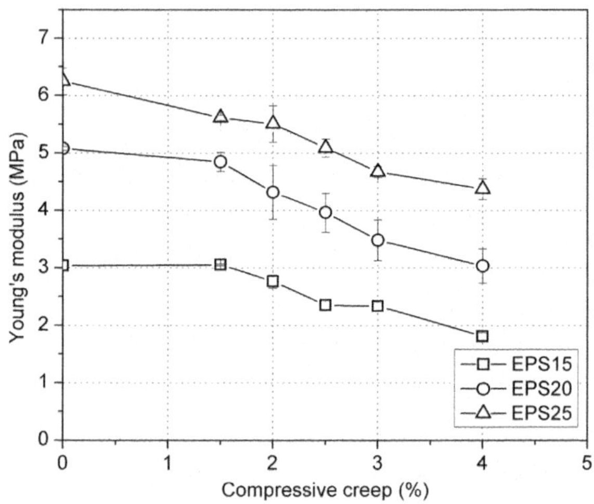

Fig. 16.6 Variation of Young's modulus of geofoam samples with compressive creep strain

wall and the backfill. Detailed schematic view of experimental set up used to induce pre-compression in geofoam panels is shown in Fig. 16.7.

Pseudo-long-term (PLT) static model tests were performed on four densities (EPS10, EPS15, EPS20, and EPS25) of geofoam, and details of PLT static tests performed in the study are listed in Table 16.4. A compressive strain of 3% was induced in geofoam panels before using them in pseudo-long-term model tests.

Results of pseudo-long-term static load tests with four geofoam densities are shown in Fig. 16.8. For comparison, measured lateral earth pressures on retaining walls with backfill alone (without foam—WOF) and with using original geofoam panels (uncompressed) at the interface between the model retaining wall and backfill are also shown in Fig. 16.8. Slightly lower earth pressures were observed for the creep induced geofoam model tests compared to that of geofoam panels. In pseudo-long-term tests, desired compressive creep was applied to geofoam samples prior to

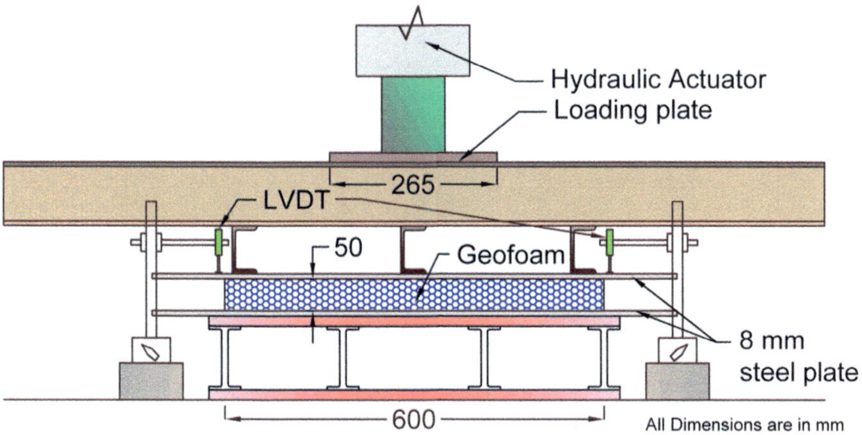

Fig. 16.7 Drawing details of geofoam panel compression set up

Table 16.4 List of pseudo-long-term model rigid retaining wall tests

S. no.	Density of foam	Thickness of geofoam (GT)	Surcharge position
1	EPS10	50	H/4
2	EPS15	50	H/4
3	EPS20	50	H/4
4	EPS25	25	H/4

model testing. In PLT tests, geofoam samples experience permanent strains, due to which much lower lateral earth pressure measure on model wall was noticed.

Image analysis of geofoam compression was carried out based on the images captured during the pseudo-long-term model tests. Geofoam compression was measured at ten selected points along the height of backfill. It was noted that irrespective of the density, creep induced geofoam panels exhibited higher geofoam compression than that of original geofoam panels.

In order to assess the extent of reduction of lateral thrust on retaining wall in the presence of geofoam, the term "isolation efficiency" is used. In the past, Ertugrul and Trandafir [19], Zarnani and Bathurst [20], and other researchers used this term for understanding the efficiency of geofoam in reducing the earth pressure acting on retaining wall. Isolation efficiency is defined here as the ratio of difference between total lateral thrust for no geofoam case (T_{wof}) and geofoam case (T_{wf}) and the total lateral thrust for no geofoam case (T_{wof}).

$$I_T = \frac{T_{wof} - T_{wf}}{T_{wof}} \times 100 \tag{16.1}$$

where I_T—isolation efficiency, %,

T_{wof}—Lateral thrust on retaining wall without geofoam, kN/m,

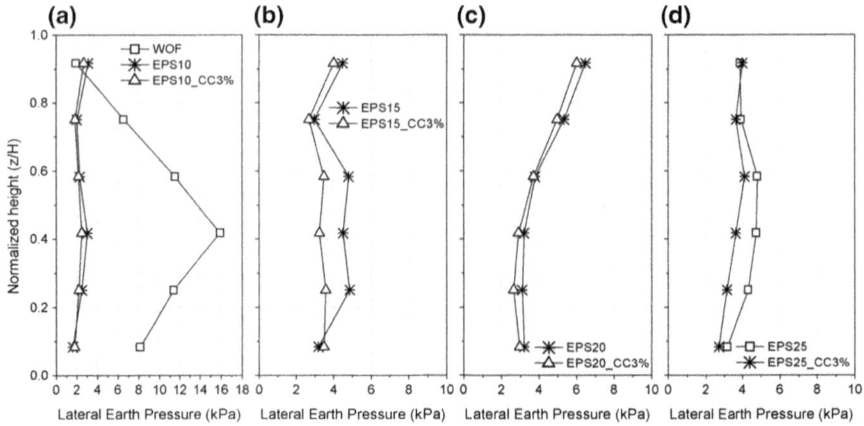

Fig. 16.8 Lateral earth pressure on retaining wall with **a** EPS10_CC3%, **b** EPS15_CC3% **c** EPS20_CC3% and **d** EPS25_CC3%

Fig. 16.9 Lateral earth pressure isolation efficiency from pseudo-long-term tests

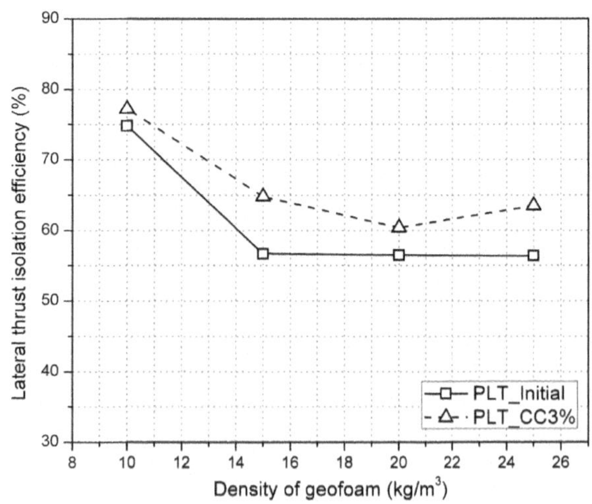

T_{wf}—Lateral thrust on retaining wall with geofoam, kN/m.

Isolation efficiency of lateral thrust was evaluated for pseudo-long-term tests using compressive creep sustained geofoam panels. For comparison, respective isolation efficiency using original geofoam panels are also plotted in Fig. 16.9. Under static loading, it was observed that lateral thrust isolation efficiency using compressive creep sustained geofoam was found to be 2.4–8.1% higher compared to that of original geofoam.

From pseudo-long-term tests on 100 mm geofoam cubes, it was observed that Young's modulus decreases with increase in compressive creep strain. Also, it can be observed that lower the density of geofoam lesser the lateral earth pressure on retaining wall, and Young's modulus directly depends on density. The above observations might have attributed to the much lower lateral earth pressure obtained from pseudo-long-term static tests, compared to respective short-term static tests. Geofoam panels were compressed to 3% compressive creep before using in pseudo-long-term model tests, which eventually lead to reduction in Young's modulus of geofoam. From these observations, it can be concluded that Young's modulus of a particular density of geofoam decreases with increase in compressive creep strains and it helps in reduction of lateral earth pressure on retaining wall.

16.6 Conclusions

Based on the extensive experimental investigation carried out on laboratory scale models to measure lateral earth pressures on the rigid non-yielding retaining wall retaining dry cohesionless backfill, with and without compressible geofoam inclusion, the following conclusions can be drawn.

- From TTSS accelerated creep testing method, EPS15, EPS20, and EPS25 geofoam samples experienced compressive creep strain of 2.12%, 2.4%, and 2.11%, respectively, at the end of 100 years.
- From pseudo-long-term tests, it was observed that geofoam samples exhibit reduced values of Young's modulus, compressive strength, and yield strength compared to that of short-term tests.
- Creep induced geofoam panels were found more efficient than original geofoam panels in decreasing the earth pressure acting on the retaining wall.
- Higher lateral thrust isolation efficiency, in the range of 2.4–8.1%, was observed from the pseudo-long-term tests compared to that of short-term studies.
- Lastly, from the series of model tests carried out in the present study, it can be observed that use of geofoam panels at the wall–backfill interface in rigid retaining walls is highly beneficial in decreasing the lateral earth pressures on the wall. In addition, through a systematic study carried out for the first time, it was revealed that use of Geofoam at the wall–backfill interface continuously reduces the in-service earth pressures on retaining walls.

References

1. Ching, F.D.K.: Building Construction Illustrated, 5th edn. Wiley (2014)
2. Saran, S., Garg, K.G., Bhandari, R.K.: Retaining wall with reinforced cohesionless backfill. J. Geotech. Eng. **118**(12), 1869–1888 (1992). https://doi.org/10.1061/(ASCE)0733-9410(1992)118:12(1869)
3. Garg, K.G.: Retaining wall with reinforced backfill—a case study. Geotext. Geomembr. **16**, 135–149 (1998). https://doi.org/10.1016/S0266-1144(98)00003-X
4. Lee, H.J., Roh, H.S.: The use of recycled tire chips to minimize dynamic earth pressure during compaction of backfill. Constr. Build. Mater. **21**, 1016–1026 (2007). https://doi.org/10.1016/j.conbuildmat.2006.02.003

5. Lee, J.H., Salgado, R., Bernal, A., Lovell, C.W.: Shredded tires and rubbersand as lightweight material. Geotech. Geoenviron. Eng. **125**(2), 132–141 (1999). https://doi.org/10.1061/(ASCE)1090-0241(1999)125:2(132)
6. Partos, A.M., Kazaniwsky, P.M.: Geoboard reduces lateral earth pressures. In: Proceedings of the Geosynthetics'87, pp. 628–639. Industrial Fabrics Association International. New Orleans, LA, USA (1987)
7. Karpurapu, R., Bathurst, R.J.: Numerical investigation of controlled yielding of soil-retaining wall structures. Geotext. Geomembr. **11**, 115–131 (1992). https://doi.org/10.1016/0266-1144(92)90040-H
8. Horvath, J.S.: Geofoam Geosynthetics. Horvath Engineering, Scarsdale, NY (1995)
9. Horvath, J.S.: An overview of the functions and applications of cellular geosynthetics. In: International e-Conference on Modern Trends in Foundation Engineering: Geotechnical Challenges and Solutions (2003)
10. Inglis, D., Macleod, Naesgaard, E., Zergoun, M.: Basement wall with seismic earth pressure and novel expanded polystyrene foam buffer layer. In: Proceedings of the 10th Annual Symposium on Earth Retention System, pp. 1–10. Canadian Geotechnical Society, Vancouver, Canada (1996)
11. Bathurst, R., Zarnani, S., Gaskin, A.: Shaking table testing of geofoam seismic buffers. Soil Dyn. Earthq. Eng. **27**, 324–332 (2007). https://doi.org/10.1016/j.soildyn.2006.08.003
12. Hazarika, H., Okuzono, S., Matsuo, Y.: Compressible geo-inclusion as seismic earth pressure reduction technique. In: Proceedings of the 38th (2003) Japan National Conference on Geotechnical Engineering, Japanese Geotechnical Society (JGS), 2–4 July 2003, Akita, Japan, pp. 1625–1626 (2003). https://doi.org/10.11512/jiban.jgs38.0.1625.0
13. Yeo, S.S.: Evaluation of creep behavior of geosythetics using accelerated and conventional methods. Ph.D. thesis, Drexel University, PA, 2007
14. ASTM D2487-11: Classification of Soils for Engineering Purposes (Unified Soil Classification System), West Conshohocken, PA (2011). https://doi.org/10.1520/d2487-11
15. Gade, V.K., Dasaka, S.M.: Short-term and long-term behavior of EPS geofoam. J. Test. Eval. (ASTM) (2018). https://doi.org/10.1520/jte20170207
16. Gade, V.K., Dasaka, S.M.: Development of a mechanized traveling pluviator to prepare reconstituted uniform sand specimens. J. Mater. Civ. Eng. (ASCE) **28**(2), 1–9 (2016). https://doi.org/10.1061/(ASCE)MT.1943-5533.0001396
17. Gade, V.K.: Reduction of earth pressure on rigid non-yielding retaining walls subjected to traffic loading using EPS Geofoam. Ph.D. thesis, Indian Institute of Technology Bombay, Mumbai, India, 2017
18. Gade, V.K., Dasaka, S.M.: Calibration of earth pressure sensors. Indian Geotech. J. **48**(1), 142–152 (2017). https://doi.org/10.1007/s40098-017-0223-0
19. Ertugrul, O.L., Trandafir, A.C.: Reduction of lateral earth forces acting on rigid non-yielding retaining walls by EPS geofoam inclusions. J. Mater. Civ. Eng. **23**(12), 1711–1718 (2011). https://doi.org/10.1061/(ASCE)MT.1943-5533.0000348
20. Zarnani, S., Bathurst, R.: Experimental investigation of EPS geofoam seismic buffers using a shaking table tests. Geosynth. Int. **14**(3), 165–177 (2007). https://doi.org/10.1680/gein.2007.14.3.165

Chapter 17
Probabilistic Study on the Geotechnical Behavior of Fiber Reinforced Soil

Arif Ali Baig Moghal, B. Munwar Basha and Mohammed Ashfaq

17.1 Introduction

Expansive soils cause millions of dollars of damage each year due to their swell/shrink behavior during moisture ingress and digress. Through a detailed review of expansive soils, the annual cost of damage due to these soils in the United States alone was estimated as $2.3 billion [1]. Petry and Armstrong [2] noted that it was more economical to perform initial stabilization of these soils before/during construction of the overlying structures rather than performing remedial treatments later with existing structures around. An efficient and stable solution in the short term has been attained by chemical stabilization of expansive soils. However, these methods are unpredictable in their long-term durability and are dependent on environmental conditions such as availability of water, mineralogy and construction methods [3, 4]. Providing reinforcement using randomly mixed synthetic fibers is a cost-effective and eco-friendly option to strengthen the soils in tension. The inclusion of Polypropylene (PP) fibers in soil showed a minor increase in optimum water content but a significant reduction in the maximum dry unit weight [5]. However, durability could be an issue when natural fibers (palm and jute fibers) are employed as they offer an improvement on only short-term basis and are beneficial for only shallow depths [6–8]. In order to circumvent the durability issue, synthetic polypropylene fibers having durable lifespan were used [9, 10]. The currently available literature suggests that behavior of fiber-reinforced clays is affected by fiber characteristics

A. A. B. Moghal (✉) · M. Ashfaq
Department of Civil Engineering, NIT, Warangal, Warangal 506004, India
e-mail: reach2arif@gmail.com

M. Ashfaq
e-mail: gmohdashfaq@gmail.com

B. M. Basha
Department of Civil Engineering, IIT Hyderabad, Kandi, Sangareddy 502285, Telangana, India
e-mail: basha@iith.ac.in

M. Latha G. (ed.), *Frontiers in Geotechnical Engineering*, Developments in Geotechnical Engineering, https://doi.org/10.1007/978-981-13-5871-5_17

(including fiber orientation, type, length, thickness, dosage, and geometry), nature of clay, and type of stabilizer [11–19].

Consoli et al. [20] studied the effect of fiber reinforcement (monofilament polypropylene fibers) on the UCS of cemented soils. The study revealed that, for the whole range of cement tested, the addition of fiber resulted in an increase in UCS for both fibers reinforced as well as unreinforced samples. In an another extended study by Consoli [09], reported that porosity and the cement content play a crucial role in improving the q_u value for different fiber contents. It was observed that an explicit relation between q_u and cement/porosity ratio has been established (C_{iv} = η) by considering the change in q_u values on increasing porosity and cement values. In a study on six soil samples at various combinations of lime and fiber contents, it was observed that the compressive strength of samples increased with an increase in fiber content [21].

On studying the wetting and drying cycles of subgrade material, it was observed by Mirzaii and Negahban [22] that CBR is a function of nature of soil-water characteristic curve (SWCC) and the initial compaction dry density and matric suction. Yideti et al. [23] demonstrated a relation between the amount of load-carrying aggregate particles and increasing CBR values after considering the influence of aggregate packing configuration on the CBR values. CBR behavior of natural soils is generally dependent on nature and type of soil, optimum moisture content, compacting density, method of compaction, Los Angeles abrasion value and heavily reliant on the concentration of fibers [24]. It was observed by previous researchers that the CBR values were higher in case of multifilament fibers compared with fibrillated fiber mainly due to the harder texture of fibrillated polypropylene fiber, Also, they observed that the CBR values started to decrease when fiber doses were greater than 0.5% for both types of fibers [25–27].

In landfill lining systems, compacted clay liners form the integral part. The essential purpose of a compacted clay liner is to inhibit flow and subsequently, hydraulic conductivity becomes the most significant factor affecting it [28, 29]. It was also observed that linear shrinkage values are greatly affected by fiber inclusion in the presence of lime [30]. In this regard, stabilized soil in combination with randomly oriented fibers may be considered in combination with stabilized soil. In this research, the effect of randomly oriented polypropylene fiber inclusion on the hydraulic conductivity, strength performance and CBR of a lime stabilized expansive soil was studied.

17.2 Testing Program

17.2.1 Materials

In the present study, high plastic clay from Al-Ghat area has been selected. The soil exhibited a natural moisture content of 3.2%. The liquid limit, plastic limit and

Table 17.1 Properties of tested fiber materials

Property	FC	FM
Specific gravity	0.91	0.91
Electrical conductivity	Low	Low
Tensile strength (N/mm^2)	440	330
Ignition point	1100 °F (593 °C)	1100 °F (593 °C)
Acid and salt resistance	High	High
Thermal conductivity	Low	Low
Melt point	324 °F (162 °C)	324 °F (162 °C)
Alkali resistance	Alkali proof	Alkali proof
Water absorption	Nil	Nil

Note "FC" and "FM" refer to "Fiber Cast® 500" and "Fiber Mesh® 300" respectively

plasticity index values were found to be 66, 32 and 34% respectively, whereas the shrinkage limit and bar linear shrinkage limit values were found to be 15 and 31% respectively. As per unified soil classification system, the soil has been classified as clay of high plasticity, i.e., "CH." The maximum dry density and optimum moisture content values were found to be 1.64 g/cm^3 and 25% respectively. Lime is a proposed binder for both unreinforced and fiber-reinforced soils. Its addition to expansive soils leads to the cementation bonds formation due to the flocculation of clay particles [31]. Lime dosage was fixed at 6% by dry weight of the soil, satisfying initial lime consumption and optimum lime requirements [32, 33]. Two types of fibers were studied in this research, the FIBERCAST® 500(FC) and FIBERMESH® 300(FM). These fibers of nominal 6 and 12 mm lengths were procured from a UK based company Propex, LLC. The fiber amount was fixed at 0.6% by dry weight of soil based on previous studies [22, 34, 35] and the physicochemical properties of fibers are presented in Table 17.1.

17.2.2 Methodology

The standard proctor tests in accordance with ASTM code [36] were performed to determine the optimum moisture content (OMC) and maximum dry density (MDD) for each mix.

Hydraulic Conductivity Test. The compacted specimen along with the perspex hydraulic conductivity mold was kept in a desiccator and thereby maintaining a relative humidity more than 95%. Hydraulic conductivity of 7, 14 and 28 days cured specimen was determined as per ASTM code [37]. The variation in Hydraulic conductivity was found to be very less on changing hydraulic gradient from 5 to 40 [38].

Unconfined Compressive Strength Test. The compacted soil samples were wrapped in plastic wraps and preserved in a 100% humidity room to avoid any moisture movements. These samples were cured for 1.7, 28, 60, 120, 180, and 360 days. The weight of the cured specimens to be tested were compared to their original weight, the specimens which showed a reduction of more than 5% of its weight due to heat of hydration were rejected. Further, the samples whose diameter deviated by more than ±0.5 mm and length by more than ±1 mm were discarded and the samples whose dry density to degree of compaction ratio was falling outside the range of 99–101% were discarded.

California Bearing Ratio (CBR) Test. To determine the force required for a penetration up to 12.5 mm, the samples were tested at their respective optimum moisture content and maximum dry density as per ASTM [39]. Lime addition to dry soil was done before mixing with fibers and was cured for a period of 14 days in a 100% humidity room. Further, care was taken to prevent any moisture movements from the samples by wrapping them in plastic wraps with the help of adhesive tape.

17.3 Results and Discussions

The results obtained from hydraulic conductivity, UCS and CBR tests performed on Al-Ghat soil are discussed in the following subsections. The untreated UCS, and CBR for Al-Ghat soil was 589.1 kPa, 5.97, respectively, and the hydraulic conductivity tests results are presented in Table 17.2. Based on the results, in the following sections, the effect of amount of fiber, fiber type and curing period on the lime-treated expansive soil is discussed. In this study, UCS tests were extended for curing periods up to 360 days to establish the length of time beyond which no additional strength gain is possible.

17.3.1 Effect of Fiber Type

It is clear from Table 17.2, that at higher dosages of fiber (0.6%), the FM is found to be more effective in increasing the hydraulic conductivity and at lower dosage levels, their performance is like one another. It can also be observed that as the dosage increased from 0 to 0.6%, the hydraulic conductivity of the soil showed an increasing trend for both 6 and 12 mm lengths.

Relatively higher secant moduli are noticed for FC fibers compared to FM type. This is attributed to higher friction mobilization in FM type along with crystallization of pozzolanic compounds which makes the sample more brittle compared to FC type

Table 17.2 Hydraulic conductivity test results

Length	LI			L2		
Dosage (%)	D1	D2	D3	D1	D2	D3
Case 1						
FC	0.589×10^{-7}	0.477×10^{-7}	0.314×10^{-6}	0.715×10^{-7}	0.622×10^{-6}	0.289×10^{-5}
FM	0.614×10^{-7}	0.815×10^{-7}	0.722×10^{-6}	0.844×10^{-7}	0.466×10^{-6}	0.617×10^{-5}
Case 2						
FC	0.516×10^{-7}	0.288×10^{-7}	0.119×10^{-6}	6.79×10^{-7}	0.317×10^{-6}	0.102×10^{-5}
FM	0.455×10^{-7}	0.419×10^{-7}	0.216×10^{-6}	0.322×10^{-7}	0.289×10^{-6}	0.114×10^{-5}
Case 3						
FC	0.621×10^{-8}	8.23×10^{-8}	0.314×10^{-7}	0.415×10^{-7}	0.371×10^{-7}	0.185×10^{-6}
FM	0.324×10^{-8}	0.687×10^{-8}	0.117×10^{-7}	0.267×10^{-7}	0.233×10^{-7}	0.138×10^{-6}
Case 4						
FC	0.255×10^{-9}	0.672×10^{-9}	0.827×10^{-8}	0.366×10^{-8}	0.528×10^{-8}	0.837×10^{-7}
FM	0.244×10^{-9}	0.586×10^{-9}	0.862×10^{-8}	0.277×10^{-8}	0.466×10^{-8}	0.655×10^{-7}
Case 5						
FC	0.169×10^{-9}	0.563×10^{-9}	0.477×10^{-8}	0.263×10^{-8}	0.422×10^{-8}	0.386×10^{-7}
FM	0.144×10^{-9}	0.927×10^{-9}	0.321×10^{-8}	0.186×10^{-8}	0.566×10^{-8}	0.244×10^{-7}

Note L1 and L2 refer to 6 mm and 12 mm fiber lengths respectively; D1, D2 and D3 refer to 0.2, 0.4 and 0.6% dosages (by weight of soil) respectively; Case 1 refers to "untreated case and immediate"; Case 2 refers to "6% lime treated soil and immediate"; Case 3, 4 and 5 refer to "6% lime treated after 7, 14 and 28 days curing period" respectively. All values reported herein are in "cm/s"

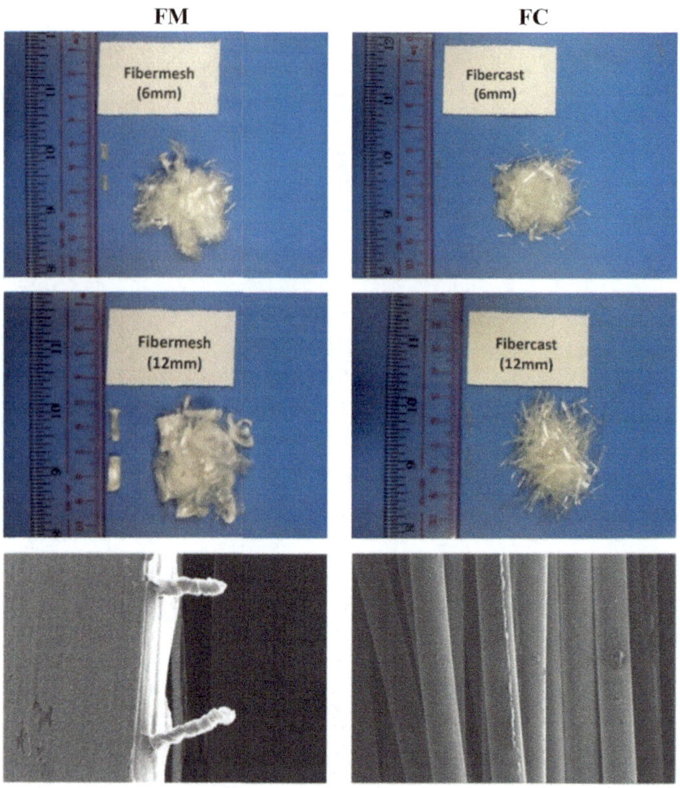

Fig. 17.1 Fibers used in the study and their SEM images

where these friction mobilization levels are low owing to smooth morphological texture features (Fig. 17.1).

It was observed that CBR of FM was higher than FC by about 20% and 60% in the case of 1 day and 14 days cured samples respectively. This higher value of CBR for FM can be attributed to their rough micro surface texture compared to FC fiber.

17.3.2 *Effect of Fiber Length and Amount of Fiber*

From the results, it was evident that hydraulic conductivity of the soil was affected by the length of the fiber used. Table 17.2 shows that for both the type of fibers, the hydraulic conductivity values were found to be higher for longer fiber lengths. This might be due to the action of fibers as drainage conduits facilitating the flow of water, longer the fiber the longer the conduit.

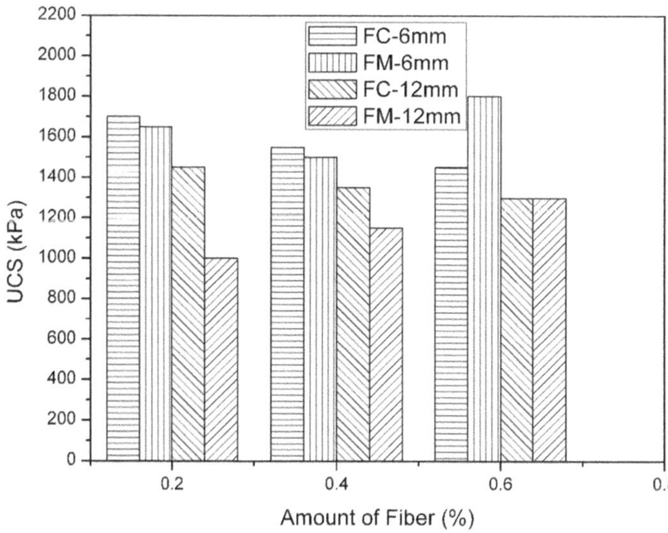

Fig. 17.2 Variation in UCS data for amount and length of fiber with lime treatments at 28-day curing

Figure 17.2 demonstrates the effect of the amount and length of fiber on the UCS values for the selected soil. It can be observed that in case of FC fiber as the amount of fibers increased from 0.2% to 0.6% the UCS values reduced by 32% and 39% for 6 mm and 12 mm lengths, respectively. This reduction is due to the fact that smooth textured fibers (FC) (Fig. 17.1) do not allow the soil to compact upon reinforcement and they reduce the cohesion of the soil by increasing the distance between particles. The increase in friction induced by the smooth FC fibers does not compensate this loss in cohesion. On the contrary, in the case of FM fiber, for a similar change in fiber amount, the UCS values for samples mixed with 6 mm fiber length reduced by 30% while those of 12 mm fiber lengths showed an increase of 12%. The reduction in strength with increase in the case of FC compared to FM is predominantly due to difference in the respective surface morphological properties as seen from Fig. 17.1. The surface of FC is relatively smooth compared to that of FM with elongated protrusions enabling greater mobilization of friction levels. The rate of increase in UCS values with increase in fiber length for FM fiber is only marginal 12%, which reinforces the fact that higher dosage of shorter fibers is preferred over longer fibers. This is in agreement with the findings of Consoli [09], wherein the authors claim that for cement treated fiber-reinforced soils, the adjusted cement/porosity ratio does not depend heavily on the fiber content.

From Fig. 17.3, For FC-type fibers (6 and 12 mm), the rate of increase in secant moduli values (at 50% peak strength) compared to untreated soil, reduced with increase in fiber dosage. This reduction is much higher in FM type fibers compared to FC type fibers. In fact, FM fibers owing to their distinct surface morphological

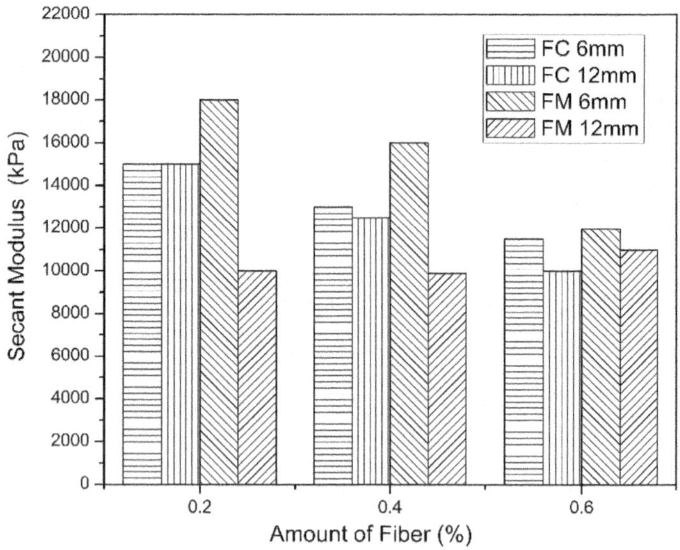

Fig. 17.3 Effect of fiber dosage and length on soil secant modulus

properties (Fig. 17.1) endure higher peak stress values at lower strain levels compared to FC fibers and this drastically affects their secant modulus values. Further, this effect is more pronounced for shorter 6 mm fibers compared to longer 12 mm fibers with increase in dosage levels for FM type fibers.

The higher increase in CBR value was noted for longer fibers (12 mm). This may be due to their greater contact surface with soil particles and pozzolanic compounds, enabling them to mobilize greater friction compared to shorter fibers (6 mm). However, for both 1 day and 14 days cured samples, the increase in CBR values between both 6 and 12 mm fibers was found to marginal (6–7%) (see Fig. 17.4).

17.3.3 Effect of Curing Period

From Table 17.2 it can be noted that hydraulic conductivity of zero-day cured sample increased on addition of fibers when compared to the untreated soil. However, with this value decreased with progress in curing period. This behavior is due to the formation of pozzolanic compounds by lime, eventually making the soil less conductive. Further, it was also noted that even after 28 days of curing, the hydraulic conductivity is still lower than the untreated soil.

It was observed that rate of gain of UCS was greatly influenced by the combined addition of lime and fiber It should also be noted that the UCS values increased with curing periods up to 60 days and relatively stayed constant after that. Prima facie, it appears that up to 60 days curing period, the release of reactive silica and

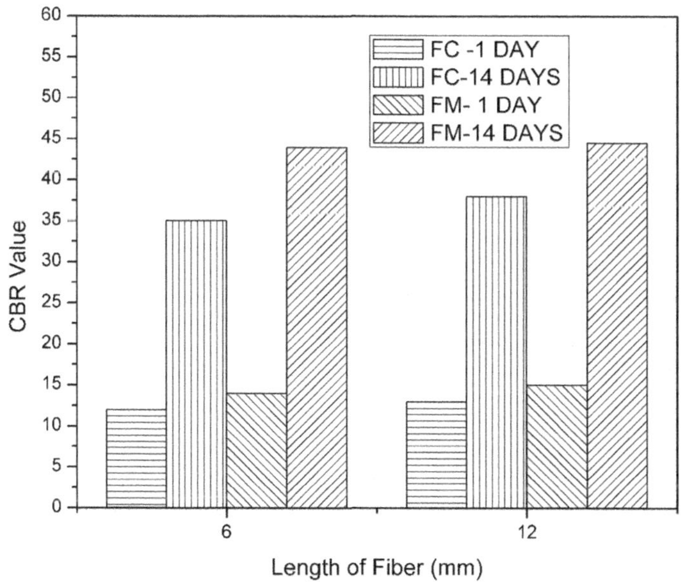

Fig. 17.4 Effect of curing period and fiber length of both fiber types on CBR values at a dosage of 0.6% after 1 day and 14 days curing period

alumina from inert clay phase proceeds without any hindrance owing to smooth solubility of silica. At higher curing periods (above 60 days), the readily available reactive silica is consumed and the pH of the resultant soil-system drops inhibiting any further release of amorphous reactive silica [40]. Further, the addition of fibers disturbs the alteration of water film surrounding the clay minerals and does not allow the pozzolanic materials to progress after reaching equilibrium levels (in this study equilibrium levels can be taken as 60 days). Accordingly, the rate of gain in UCS at higher curing periods proceeds at a slower rate (Fig. 17.5).

It can be noted from Fig. 17.6 that the secant modulus values increased with curing periods up to 60 days and stayed approximately constant until 180 days and this increase is proportional to the increase in UCS values (Figs. 17.2 and 17.3). Up to 180 days, the amorphous nature of pozzolanic compounds formed allows the samples to endure higher strain levels, which resulted in proportional increase in secant modulus values on par with UCS values. The secant modulus values increased considerably for the 360 day cured samples for all mixes, irrespective of the type (i.e., FC and FM) and length (i.e., 6 or 12 mm) of fiber. This is attributed to the transformation of amorphous pozzolanic compounds to crystalline form, which endures relatively lower strain levels even at higher peak stress levels as seen from Figs. 17.2 and 17.3 [40].

From Fig. 17.7 it can be observed that CBR characteristics have been highly affected by curing and showed an increase of about 310%. Lime could stabilize the soils through cementation and this is due to the reaction between lime and reac-

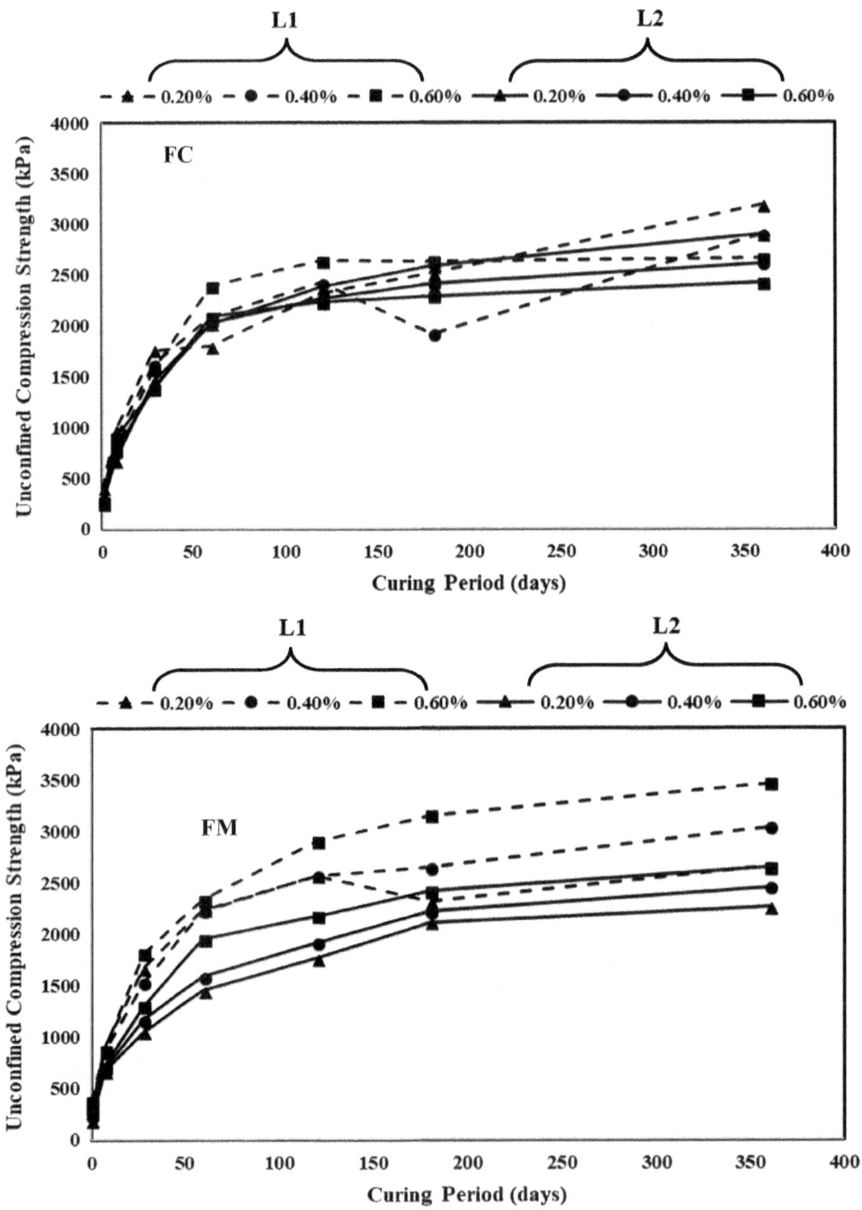

Fig. 17.5 Variation in UCS strength with curing period for FC and FM-treated soil

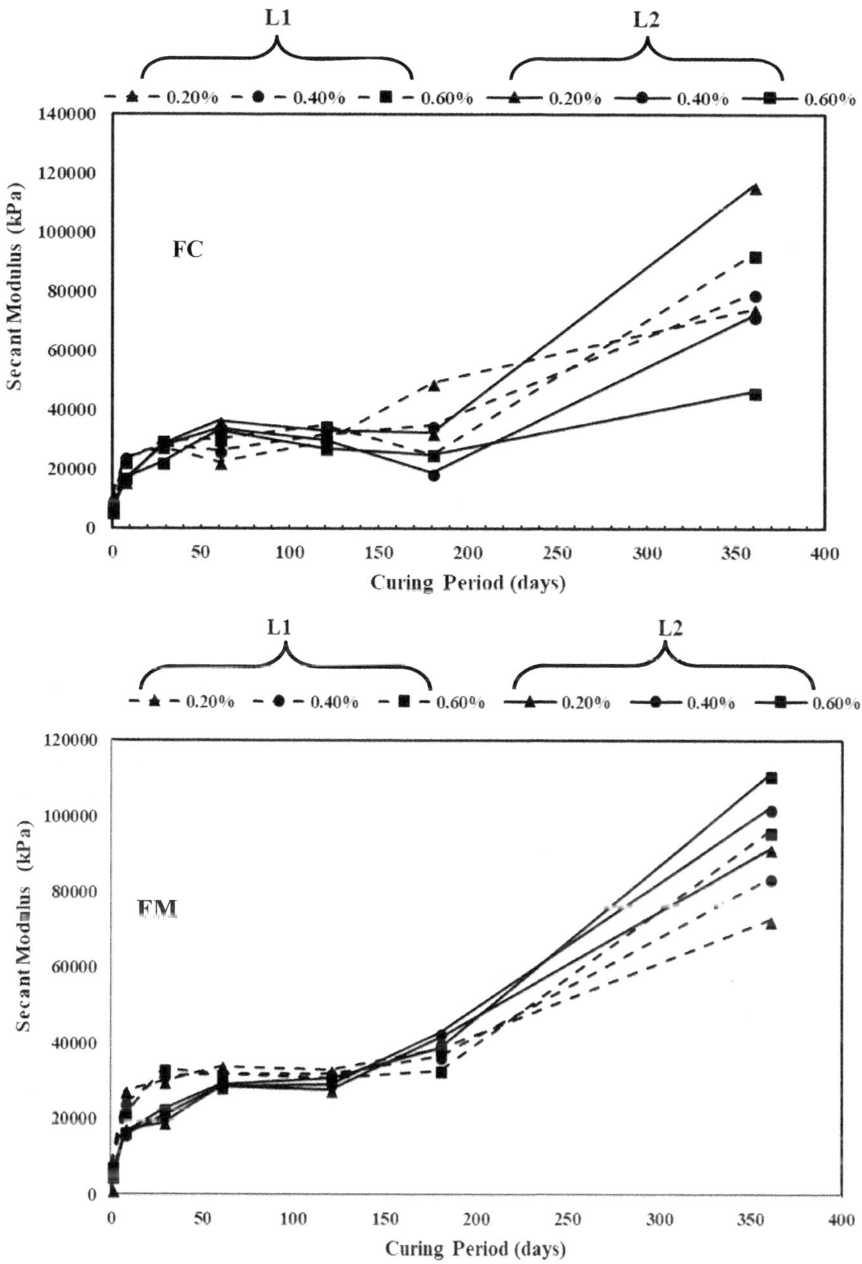

Fig. 17.6 Variation in FC and FM secant modulus with curing period

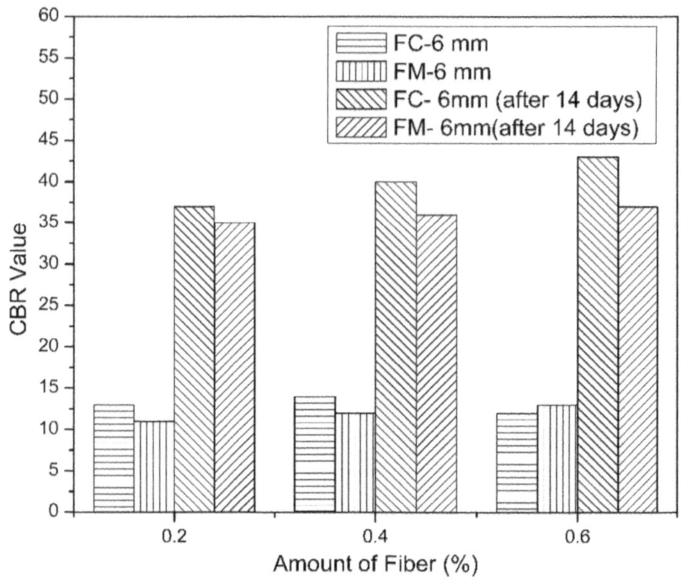

Fig. 17.7 Effect of curing time on CBR values for FC and FM fiber reinforcement

tive alumina and silica present in the soil, eventually leading to the formation of pozzolanic compounds [41, 42].

From Fig. 17.7 it can be observed that CBR characteristics have been highly affected by curing and showed an increase of about 310%. Lime could stabilize the soils through cementation and this is due to the reaction between lime and reactive alumina and silica present in the soil, eventually leading to the formation of pozzolanic compounds [41, 42].

17.4 Nonlinear Regression Fitting

Nonlinear regression models are chosen to represent the experimental values of hydraulic conductivity, UCS and CBR. The proposed equations for hydraulic conductivity with FC and FM reinforcements are provided below:

$$K_{Fit\_FC\_AG1} = 3.941 \times 10^{-12}(\mathrm{L_{FC}})^{3.016}(36324.088)^{D_{FC}} \tag{17.1}$$

$$K_{Fit\_FM\_AG1} = 2.486 \times 10^{-11}(\mathrm{L_{FM}})^{2.910}(1215.103)^{D_{FM}} \tag{17.2}$$

The proposed equations for CBR of the reinforced expansive soil with FC and FM reinforcements are provided below:

$$CBR_{Fit\_FC} = 19.94 + 0.453L_{FC} + 11.325D_{FC} + 2.7720 \times 10^{-14}(D_{FC})^2 \quad (17.3)$$

$$CBR_{Fit\_FM} = 21.29 + 0.837L_{FM} + 11.303D_{FM} - 0.075(D_{FM})^2 \quad (17.4)$$

The best-fit equations for the estimation of UCS of the reinforced expansive soil with FC and FM reinforcements are given below:

$$UCS_{Fit\_FC\_28} = 2137.79 - 39.86(L_{FC}) - 1006.95(D_{FC}) + 839.75\,(D_{FC})^2 \quad (17.5)$$

$$UCS_{Fit\_FM\_28} = 2329.16 - 82.74(L_{FM}) - 1662.39(D_{FM}) + 2688.31(D_{FM})^2 \quad (17.6)$$

17.5 Reliability-Based Design Optimization

The problematic soils may be stabilized by using discrete fibers with subgrade soils. The fiber reinforcement prohibits the formation of weak planes parallel to the oriented reinforcement. The effect of fibers on the dynamic and mechanical properties was reported by Ozkul and Baykal [43], Cai et al. [44], and Tang et al. [38]. It is also important to consider the high degree of variability associated with synthetic fibers due to the type of fiber material, configuration of the fiber reinforcements, the environmental factors, saturation of fibers because of percolation of rainwater, the magnitude of stress, and biological effects of additives and stabilizers [31]. Therefore, the frictional resistance between soil particles and fibers and tensile strength over the period of time reduces substantially. Mirzaii and Negahban [22] reported that the CBR value depends on matric suction, initial compaction dry density and soil-water characteristic curve (SWCC). Therefore, the hydraulic conductivity, UCS and CBR of the subgrade soil is considerably affected. It is essential to evaluate the safety of pavements with suitable approaches that consider variability and uncertainty in a rational way.

Chou [45] and Divinsky et al. [46], Chua et al. [47], Timm et al. [23], and Kenis and Wang [48], AASHTO [49], Kim et al. [50], Kim and Buch [51] and Retherford and McDonald [52] proposed reliability-based design of pavements using CBR method. Sani et al. [53] reported the first-order reliability method (FORM) for the probabilistic analysis of treated black cotton soil for UCS, CBR, and resistance to loss in strength. Recently, Moghal et al. [54] proposed reliability analysis to study the influence of fiber reinforcement on the hydraulic conductivity behavior of lime blended expansive soils. Moreover, the influence of fiber reinforcement on UCS behavior of lime blended semiarid soil is proposed by Moghal et al. [30] using the target reliability approach. However, guidelines need to be proposed for the design of pavements. In view of the above discussion, present paper reports on reliability-based design of pavements considering variability associated with hydraulic conductivity, UCS and CBR of subgrade, fiber type, fiber content, and fiber length.

The present work reports a framework for target reliability-based approach (TRA) for ensuring adequate safety against liner material failure, unconfined compressive strength failure, CBR strength failure of lime blended expansive subgrade soil considering the variability associated with the coefficient of permeability, UCS, CBR, and characteristics of fiber reinforcement (type, length and dosage). Nonlinear equations are presented for computing the hydraulic conductivity, UCS, and CBR of treated and fiber-reinforced expansive soils in terms of the dosage and length of fiber reinforcement. The optimum dosage and lengths of fiber reinforcement needed to maintain stability against hydraulic conductivity, UCS, and CBR failure modes by targeting various reliability indices.

17.5.1 Performance Functions for Hydraulic Conductivity, CBR, and UCS Strength

The reliability of liner material using Al-Ghat soil characterized by hydraulic conductivity (K) can be defined

$$\text{Reliability} = P(K < K_{\max}) \tag{17.7}$$

where K_{max} = the maximum specified hydraulic conductivity (i.e., 1×10^{-7} cm/s in the present study). Failure occurs when hydraulic conductivity $\geq$ the maximum K_{max} of 1×10^{-7} cm/s.

$$g_1(x) = \frac{K_{\max}}{K_{Fit\_FC\_AG1}} - 1 \tag{17.8}$$

$$g_2(x) = \frac{K_{\max}}{K_{Fit\_FM\_AG1}} - 1 \tag{17.9}$$

A reliability measure of assurance of performance of subgrade material characterized by CBR is defined as

$$\text{Reliability} = P\left(CBR \geq CBR_{\min}\right) \tag{17.10}$$

where $CBR_{\min}$ is the minimum CBR value to prohibit subgrade soil failure. The CBR failure of lime blended and reinforced expansive soil occurs when CBR < $CBR_{\min}$. Therefore, the limit state functions can be written as

$$g_3(x) = \frac{CBR_{FIT\_FC}}{CBR_{\min}} - 1 \tag{17.11}$$

$$g_4(x) = \frac{CBR_{Fit\_FM}}{CBR_{\min}} - 1 \tag{17.12}$$

The UCS of pavement subgrade material failure occurs when UCS < $UCS_{\min}$ as shown below.

$$P_f = P\left(UCS < UCS_{\min}\right) \tag{17.13}$$

where $UCS_{\min}$ is the minimum UCS value to prohibit failure of subgrade soil. The limit state functions for UCS failure for FC and FM reinforcements with 28 days curing period are given by

$$g_5(x) = \frac{UCS_{Fit\_FC\_28}}{UCS_{\min}} - 1 \tag{17.14}$$

$$g_6(x) = \frac{UCS_{Fit\_FM\_28}}{UCS_{\min}} - 1 \tag{17.15}$$

17.5.2 Estimation of Length of Fiber Cast (L_{FC}), Length of Fiber Mesh (L_{FM}), Amount of Fiber Cast (D_{FC}), Amount of Fiber Mesh (D_{FM})

The following nonlinear constrained optimization can be performed in standard normal space as illustrated by Basha and Babu [21, 55]

$$\text{Find } L_{FC} \text{ and } D_{FC} \text{ which} \begin{cases} \text{minimizes} & g_1(u) \\ \text{subject to} & \sqrt{u^{\mathrm{T}}u} = \beta_{K\_FC} \end{cases} \tag{17.16}$$

$$\text{Find } L_{FM} \text{ and } D_{FM} \text{ which} \begin{cases} \text{minimizes} & g_2(u) \\ \text{subject to} & \sqrt{u^{\mathrm{T}}u} = \beta_{K\_FM} \end{cases} \tag{17.17}$$

$$\text{Find } L_{FC} \text{ and } D_{FC} \text{ which} \begin{cases} \text{minimizes} & g_3(u) \\ \text{subject to} & \sqrt{u^{\mathrm{T}}u} = \beta_{UCS\_FC\_28} \end{cases} \tag{17.18}$$

$$\text{Find } L_{FM} \text{ and } D_{FM} \text{ which} \begin{cases} \text{minimizes} & g_4(u) \\ \text{subject to} & \sqrt{u^{\mathrm{T}}u} = \beta_{UCS\_FM\_28} \end{cases} \tag{17.19}$$

$$\text{Find } L_{FC} \text{ and } D_{FC} \text{ which} \begin{cases} \text{minimizes} & g_5(u) \\ \text{subject to} & \sqrt{u^{\mathrm{T}}u} = \beta_{CBR\_FC} \end{cases} \tag{17.20}$$

$$\text{Find } L_{FM} \text{ and } D_{FM} \text{ which} \begin{cases} \text{minimizes} & g_6(u) \\ \text{subject to} & \sqrt{u^{\mathrm{T}}u} = \beta_{CBR\_FM} \end{cases} \tag{17.21}$$

where β_{K_FC} = reliability index against liner material treated with FC reinforcement, β_{K_FM} = reliability index against liner material treated with FM reinforcement, $\beta_{UCS\_FC\_28}$ = reliability index against UCS failure when treated with FC reinforcement, $\beta_{UCS\_FM\_28}$ = reliability index against UCS failure when treated with FM reinforcement, β_{CBR_FC} = reliability index against CBR failure when treated with

FC reinforcement, and β_{CBR_FM} = reliability index against CBR failure when treated with FM reinforcement.

17.5.3 Optimum Length and Dosage of Fiber Cast and Fiber Mesh for Al-Ghat Soil

The length and dosage of FC and FM for lime blended expansive soil have a considerable effect on the safety against liner material failure, UCS failure, CBR strength failure. Hence, an adequate dosage and length of the fiber need to be provided to maintain the desired performance. Figure 17.8a–c show the optimum values of length (L_{FC}) and dosage (D_{FC}) of FC for various target values of target reliability indices, β_{K_FC}, $\beta_{UCS\_FC\_28}$ and β_{CBR_FC} for $UCS_{\min} = 1103$ kPa, $CBR_{\min} = 20\%$, dosage of FC reinforcement (D_{FC}) = 0.2%, 0.4%, and 0.6%, length of FC reinforcement (L_{FC})

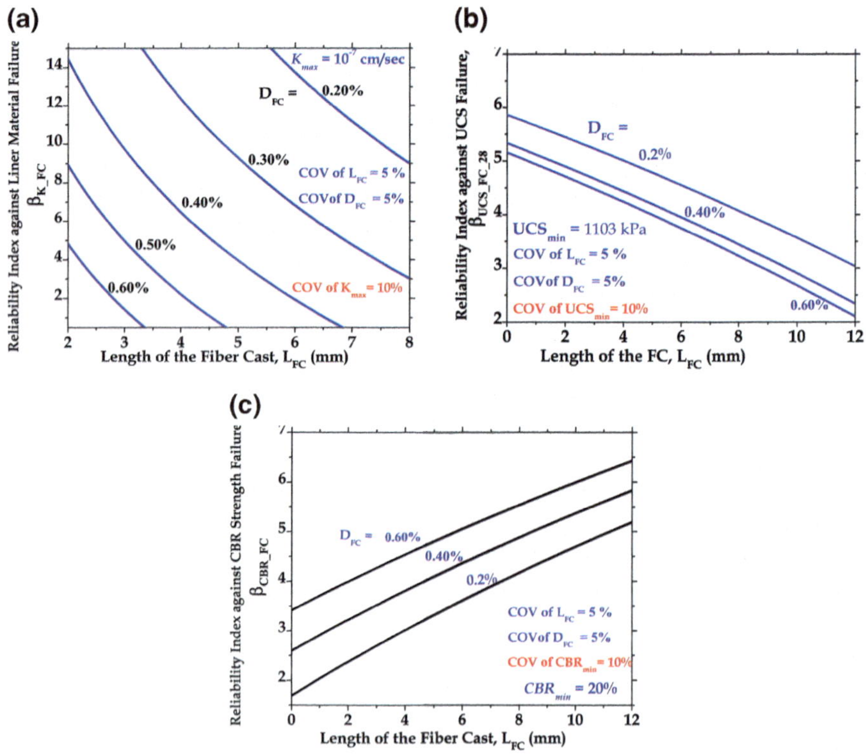

Fig. 17.8 Effect of L_{FC} with D_{FC} on **a** β_{K_FM}. **b** $\beta_{UCS\_FM\_28}$ and **c** β_{CBR_FM} for Al-Ghat soil with fiber cast for COV of $UCS_{\min}$ = 10%, COV of $CBR_{\min}$ = 10%, and COV of $K_{\max}$ = 10%, respectively

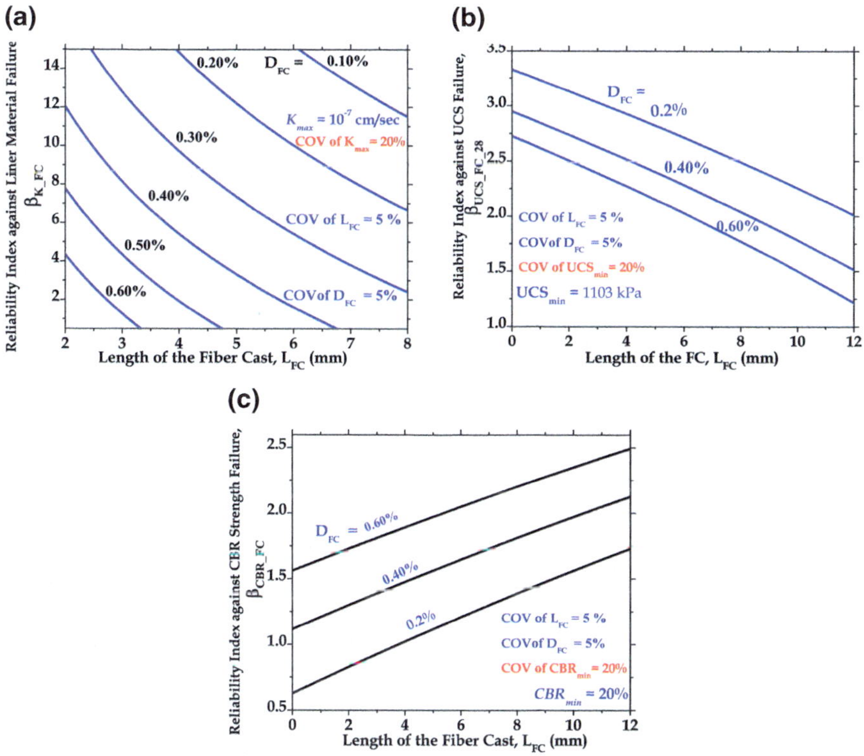

Fig. 17.9 Effect of L_{FC} with D_{FC} on **a** β_{K_FM}. **b** $\beta_{UCS\_FM\_28}$ and **c** β_{CBR_FM} for Al-Ghat soil with fiber cast for COV of UCS_{min} = 20%, COV of CBR_{min} = 20%, and COV of K_{max} = 20%, respectively

ranges from 0 to 12 mm, COV of K_{max}= 10%, COV of UCS_{min}= 10%, COV of CBR_{min} = 10%, respectively. Similarly, the results presented in Fig. 17.9a–c show the optimum values of length (L_{FC}) and dosage (D_{FC}) of FC reinforcement for COV of K_{max}= 20%, COV of UCS_{min}= 20%, and COV of CBR_{min} = 20%.

Figure 17.10a–c show the optimum values of length (L_{FM}) and dosage (D_{FM}) of FM for various target values of reliability indices, β_{K_FM}, $\beta_{UCS\_FM\_28}$ and β_{CBR_FM} for UCS_{min} = 1103 kPa, CBR_{min} = 20%, dosage of FM reinforcement (D_{FM}) = 0.2%, 0.4%, and 0.6%, length of FM reinforcement (L_{FC}) ranges from 0 to 12 mm, COV of K_{max}= 10%, COV of UCS_{min}= 10%, COV of CBR_{min} = 10% respectively.

On a similar note, the results presented in Fig. 17.11a–c show the optimum length (L_{FM}) and optimum dosage (D_{FM}) of FM reinforcement for COV of K_{max}= 20%, COV of UCS_{min}= 20%, and COV of CBR_{min} = 20%. An observation that may be made from Figs. 17.8a and 17.9a that for constant values of the reliability index (β_{K_FM}) and COV of K_{max} = 10 and 20%, the length of the FC reinforcement (L_{FC}) should be decreased from 8 to 2 mm as the fiber dosage (D_{FC}) increases from 0.1 to 0.6%. Similar observation can be made from Figs. 17.10a and 17.11a for fiber

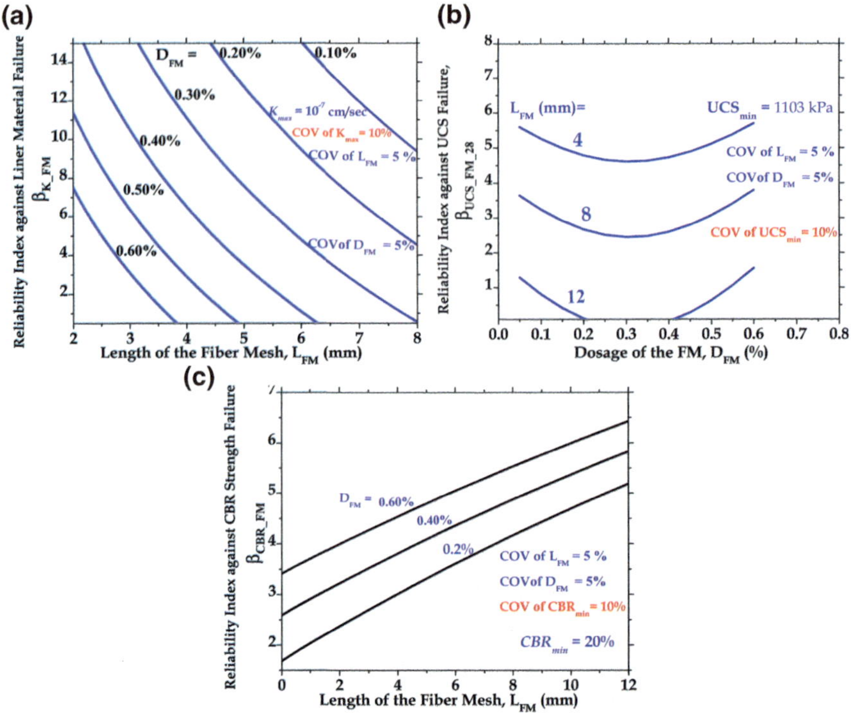

Fig. 17.10 Effect of L_{FM} with D_{FM} on **a** β_{K_FM}. **b** $\beta_{UCS\_FM\_28}$ and **c** β_{CBR_FM} for Al-Ghat soil with fiber mesh for COV of $UCS_{min} = 10\%$, COV of $CBR_{min} = 10\%$, and COV of $K_{max} = 10\%$, respectively

mesh. Figures 17.10a and 17.11a show that for a constant value of COV of K_{max}, reliability indices, β_{K_FC} and β_{K_FM} show considerable decrease with increase in values of L_{FC}, L_{FM}, D_{FC} and D_{FM}.

The results presented in Figs. 17.8b and 17.9b show the effect of adding the length and dosage of FC reinforcement on the reliability index against UCS failure ($\beta_{UCS\_FC\_28}$). An important observation that can be made from Figs. 17.8b and 17.9b that the reliability index, $\beta_{UCS\_FC\_28}$ decreases with increase in magnitudes of L_{FC} from 0 to 12 mm and D_{FC} from 0.2 to 0.6%. The more dosage of FC reinforcement reduces the UCS value because of smooth fibers do not allow the soil to compact when reinforced the lime blended expansive soil. Furthermore, the cohesion of the soil is reduced by the fibers due to increase in the distance between particles.

On contrary, it can be noted from Figs. 17.10b and 17.11b that the reliability index ($\beta_{UCS\_FM\_28}$) reduces with the increase in FM content up to 0.3%. Whereas, the reliability index, $\beta_{UCS\_FM\_28}$ increases when the FM content increases beyond 0.3% because of the rough fibers decrease the cohesion of soil by increasing the distance between particles, but the increase in friction induced by the rough fibers does compensate the loss in cohesion.

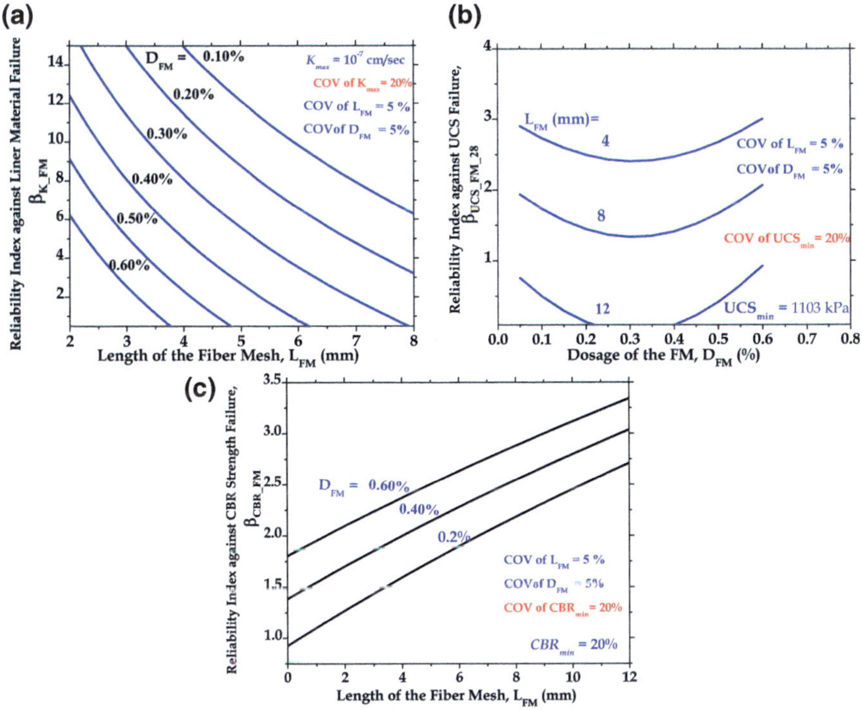

Fig. 17.11 Effect of L_{FM} with D_{FM} on **a** β_{K_FM}. **b** $\beta_{UCS\_FM\_28}$ and **c** β_{CBR_FM} for Al-Ghat soil with fiber mesh for COV of $UCS_{\min} = 20\%$, COV of $CBR_{\min} = 20\%$, and COV of $K_{\max} = 20\%$, respectively

Figures 17.8c, 17.9c, 17.10c, and 17.11c show that the reliability indices, β_{CBR_FC} and β_{CBR_FM} increase with increase in amounts of L_{FC} from 0 to 12 mm and D_{FC} from 0.2 to 0.6%. The higher FC and FM content enhances the CBR strength further due to the more mobilized frictional resistance around the fibers, and therefore, higher tensile stresses are developed in the FC and FM reinforcements.

17.6 Conclusions

1. Hydraulic conductivity of the soil increases with increase in fiber content and longer fibers gave higher increase in CBR value.
2. Fiber mesh type fibers with extended protrusions have distinct surface morphological advantage over fiber cast type fibers enabling them in mobilizing higher friction levels and also in increasing hydraulic conductivity. FM fibers outperformed the FC fibers with about 20% higher CBR increment.

3. Higher concentration of fibers resulted in higher increase in CBR values and, the CBR value increased with increasing fiber amount, irrespective of the type of the fibers.
4. The combined addition of lime and fiber has a significant effect on the rate of gain of UCS. UCS behavior is directly dependent on the nature and type of fiber. For FC fibers, with increase in fiber dosage from 0.2% to 0.6%, the UCS values reduced by 32% and 39% for 6 mm and 12 mm lengths, respectively. On the contrary, for FM fibers, for a similar change in fiber amount, the UCS values for samples mixed with 6 mm fiber length reduced by 30% while those of 12 mm fiber lengths showed an increase of 12.
5. The curing has considerable effect on the UCS, Hydraulic conductivity and CBR data. For a Curing period up to 60 days, Hydraulic conductivity kept decreasing and UCS showed an increasing trend.
6. The increase in CBR value, when compared to untreated expansive soils, stayed constant with increasing fiber content for FM fibers while the same for FC fibers increased with fiber content.
7. The variability associated with the maximum value of hydraulic conductivity, and minimum values of UCS and CBR have significant influence on the reliability index against hydraulic conductivity failure, UCS and CBR strength failure modes.
8. The more dosage of FC reinforcement reduces the reliability index ($\beta_{UCS\_FC\_28}$) value as smooth fibers do not allow the soil to compact. The reliability index ($\beta_{UCS\_FM\_28}$) reduces with the increase in FM content up to 0.3%. However, the increase in fiber mesh beyond 0.3% increases the reliability index, $\beta_{UCS\_FM\_28}$ due to the rough fibers decrease the cohesion of soil by increasing the distance between particles. Moreover, the increase in friction induced by the rough fibers does compensate the loss in cohesion.
9. The adequate lengths and dosage of fiber cast and fiber mesh reinforcements are needed for a target reliability index based on the hydraulic conductivity, UCS, and CBR performance of natural and lime stabilized soils as pavement subgrade.

References

1. Gromko, G.J.: Review of expansive soils. J. Geotech. Eng. Div. **100**(6), 667–687 (1974)
2. Petry, T.M., Armstrong, J.C.: Stabilization of expansive clay soils. Transp. Res. Record (1219) (1989)
3. Chittoori, B., Puppala, A., Pedarla, A., Vanga, D.: Durability studies on native soil-based controlled low strength materials. Ground improvement and geosynthetics. Geotechnical Special Publication No. 238, pp. 249–257 (2014)
4. Chittoori, B.S., Puppala, A.J., Saride, S., Nazarian, S., Hoyos, L.R.: Durability Studies of lime stabilized clayey soils. In the Proceedings of 17th International Conference on Soil Mechanics and Geotechnical Engineering, ICMGE 2009, pp. 2208–2211 (2009)
5. Shukla, S.K., Shahin, M.A., Abu-Taleb, H.: A note on void ratio of fibre-reinforced soils. Int. J. Geosynth. Ground Eng. Switz **1**(3), 29.1–29.5 (2015)

6. Anggraini, V., Asadi, A., Huat, B.B.K., Nahazanan, H.: Effects of coir fibers on tensile and compressive strength of lime treated soft soil. Measurement **59**, 372–381 (2015)
7. Babu, G.L.S., Vasudevan, A.K.: Strength and stiffness response of coir fiber-reinforced tropical soil. J. Mater. Civ. Eng. **20**(9), 571–577 (2008)
8. Şenol, A.: Effect of fly ash and polypropylene fibers content on the soft soils. Bull. Eng. Geol. Env. **71**(2), 379–387 (2012)
9. Consoli, N.C., Zortéa, F., De Souza, M., Festugato, L.: Studies on the dosage of fiber-reinforced cemented soils. J. Mater. Civ. Eng. **23**(12), 1624–1632 (2011)
10. Correia, A.A.S., Oliveira, P.J.V., Custódio, D.G.: Effect of polypropylene fibers on the compressive and tensile strength of a soft soil, artificially stabilised with binders. Geotex. Geomem. **43**(2), 97–106 (2015)
11. Ahmad, F., Bateni, F., Azmi, M.: Performance evaluation of silty sand reinforced with fibers. Geotext. Geomembr. **28**(1), 93–99 (2010)
12. Consoli, N.C., Bassani, M.A.A., Festugato, L.: Effect of fiber-reinforcement on the strength of cemented soils. Geotex. Geomem. **28**, 344–351 (2010)
13. Dafalla, M.A., Moghal, A.A.B.: Effect of fibercast and fibermesh inclusion on the direct shear and linear shrinkage response of clay. Arab. J. Geosci. **9**, 555 (2016)
14. Gray, D.H.: Role of woody vegetation in reinforcing soils and stabilizing slopes. In: Proceedings, Symposium on Soil Reinforcement and Stabilizing Techniques. Sydney, Australia, pp. 253–306 (1970)
15. Gray, D.H., Maher, M.H.: Admixture stabilization of sand with discrete randomly distributed fibers. In: Proceedings XII International Conference on Soil Mechanics and Foundation Engineering, vol. 2. Rio de Janeiro, Brazil, pp. 1363–1366 (1989)
16. Maher, M.H., Ho, Y.C.: Mechanical-properties of kaolinite fiber soil composite. J. Geotech. Eng. **120**(8), 1381–1393 (1994)
17. Moghal, A.A.B, Chittoori, B., Basha, B.M., Al-Shamrani, M.A.: Target reliability approach to study the effect of fiber reinforcement on UCS behavior of lime treated semi-arid soil. J. Mater. Civil Eng. **29**(6), 04017014. https://doi.org/10.1061/(asce)mt.1943-5533.0001835 (2017)
18. Petry, T., Little, D.: Review of stabilization of clays and expansive soils in pavements and lightly loaded structures—history, practice, and future. J. Mater. Civ. Eng. **14**(6), 447–460 (2002)
19. Zhu, H.H., Zhang, C.C., Tang, C.S., Shi, B., Wang, B.J.: Modeling the pullout behavior of short fiber in reinforced soil. Geotext. Geomembr. **42**(4), 329–338 (2014)
20. Consoli, N.C., Lopes, L.S. Jr., Foppa, D., Heineck, K.S.: Key parameters dictating strength of lime/cement treated soil. Proc. Inst. Civ. Eng. – Geotech. Eng. **162**(2), 111–118 (2009)
21. Basha, B.M., Babu, G.L.S.: Target reliability based design optimization of anchored cantilever sheet pile walls. Can. Geotech. J. **45**(3), 535–548 (2008)
22. Mirzaii, A., Negahban, M.: California bearing ratio of an unsaturated deformable pavement material along drying and wetting paths. Road Mater. Pav. Des. **17**(1), 261–269 (2016)
23. Yideti, T.F., Birgisson, B., Jelagin, D.: Influence of aggregate packing structure on California bearing ratio values of unbound granular materials. Road Mater. Pav. Des. **15**(1), 102–113 (2014)
24. Moghal, A.A.B, Chittoori, B., and Basha, B.M.: Effect of fiber reinforcement on CBR behaviour of lime-blended expansive soils: reliability approach. Road Mater. Pav. Des. **19**(3), 690–709 (2018)
25. Alawi, M.H., Rajab, M.I.: Prediction of California bearing ratio of subbase layer using multiple linear regression models. Road Mater. Pav. Des. **14**(1), 211–219 (2013)
26. Timm, D.H., Newcomb, D.E., Birgisson, B., Galambos, T.V.: Incorporation of reliability into the Minnesota mechanistic-empirical pavement design method. In: Final Report Prepared to Minnesota Department of Transportation, Minnesota Univ., Department of Civil Engineering, Minneapolis (1999)
27. Sivapullaiah, P.V., Moghal, A.A.B.: CBR and strength behavior of stabilised low lime fly ashes. Int. J. Geotech. Eng. **5**, 121–130 (2011)
28. Benson, C.H., Trast, J.M.: Hydraulic conductivity of thirteen compacted clays. Clays Clay Miner. **43**(6), 669–681 (1995)

29. Daniel, D.: Earthen Liners for waste disposal facilities. ASCE Geotechnical Special Publication No. 13, pp. 21–39 (1987)
30. Moghal, A.A.B, Chittoori, B., Basha, B.M., Al-Mahbashi A.M.: Effect of polypropylene fibre reinforcement on the consolidation, swell and shrinkage behaviour of lime-blended expansive soil. Int. J. Geotech. Eng. **12**(5), 462–471 (2018)
31. Abu-Sharkh, B.F., Hamid, H.: Degradation study of date palm fiber/polypropylene composites in natural and artificial weathering: mechanical and thermal analysis. Polym. Degrad. Stab. **85**(3), 967–973 (2004)
32. Eades, J.L., Grim, R.E.: Reaction of hydrated lime with pure clay minerals in soil stabilization, vol. 262. Highway Research Board, Washington DC, Bulletin, pp. 51–63 (1960)
33. Moghal, A.A.B., Al-Obaid, A.K., Al-Refeai, T.O., Al-Shamrani, M.A.: Compressibility and durability characteristics of lime treated expansive semiarid soils. ASTM J. Test. Eval. **43**(2), 254–263 (2015)
34. Malekzadeh, M., Bilsel, H.: Swell and compressibility of fiber reinforced expansive soils. Int. J. Adv. Technol. Civil Eng. **1**(2), 42–45 (2012)
35. Puppala, A.J., Wattanasanticharoen, E., Porbaha, A.: Combined lime and polypropylene fiber stabilization for modification of expansive soils. In: Ali Al-Rawas, A., Zeynal, F.A. (eds.) Goosen Book Chapter: Expansive Soils Recent Advances in Characterization and Treatment. Taylor & Francis, https://doi.org/10.1201/9780203968079.ch24 (2006)
36. ASTM D698–07e1: Standard Test Methods for Laboratory Compaction Characteristics of Soil Using Standard Effort. West Conshohocken, PA (2007)
37. ASTM. D5856-07: Standard test method for measurement of hydraulic conductivity of porous material using a rigid-wall, compaction-mold permeameter. West Conshohocken, PA (2007)
38. Tang, C., Wang, D., Cui, Y., Shi, B., Li, J.: Tensile strength of fiber-reinforced Soil. J. Mater. Civ. Eng. https://doi.org/10.1061/(asce)mt.1943-5533.0001546 (2016)
39. ASTM D1883—14: Standard test method for California bearing ratio (CBR) of laboratory-compacted soils. ASTM International, West Conshohocken, PA (2014)
40. Moghal, A.A.B., Al-Obaid, A.K., Al-Refeai, T.O.: Effect of accelerated loading on the compressibility characteristics of lime treated semi arid soils. J. Mater. Civ. Eng. **26**(5), 1009–1016 (2014)
41. Dash, S., Hussain, M.: Lime stabilization of soils: Reappraisal. J. Mater. Civil Eng. 707–714. https://doi.org/10.1061/(asce)mt.1943-5533.0000431 (2012)
42. Qu, J., Zhao, D.: Stabilising the cohesive soil with palm fiber sheath strip. Road Mater. Pav. Des. **17**(1), 87–103 (2016)
43. Ozkul, Z.H., Baykal, G.: Shear strength of clay with rubber fiber inclusions. Geosyn. Int. **13**(5), 173–180 (2006)
44. Cai, Y., Shi, B., Ng, C.W.W., Tang, C.: Effect of polypropylene fiber and lime admixture on engineering properties of clayey soil. Eng. Geol. **87**(3–4), 230–240 (2006)
45. Chou, Y.T.: Reliability design procedures for flexible pavements. J. Transp. Eng. **116**(5), 602–614 (1990)
46. Divinsky, M., Ishai, I., Livneh, M.: Simplified generalized California bearing ratio pavement design equation. Transportation Research Record 1539, Transportation Research Board, National Research Council, Washington DC, pp. 44–50 (1996)
47. Chua, K.H., Kiuredhian, A.D., Monismith, C.L.: Stochastic model for pavement design. J. Transp. Eng. **118**(6), 769–785 (1992)
48. Kenis, W., Wang, W.: Pavement variability and reliability. In: International Symposium on Heavy Vehicle Weights and Dimensions, Maroochydore, Queensland, Australia, Part 3, pp. 213–231 (1998)
49. AASHTO: Guide for Design of Pavement Structures. American Association of State Highway and Transportation Officials, Washington DC (1993)
50. Kim, H.B., Harichandran, R.S., Buch, N.: Development of load and resistance factor design format for flexible pavements. Can. J. Civ. Eng. **25**, 880–885 (1998)
51. Kim, H.B., Buch, N.: Reliability-based pavement design model accounting for inherent variability of design parameters. In: Transportation Research Board, 82nd Annual Meeting, and Washington DC (2003)

52. Retherford, J.Q., McDonald, M.: Reliability methods applicable to mechanistic-empirical pavement design method. Transportation Research Record No. 2154. J. Transp. Res. Board, 130–137 (2010)
53. Sani, J.E., Bello, A.O., Nwadiogbu, C.P.: Reliability estimate of strength characteristics of black cotton soil pavement sub-base stabilized with bagasse ash and cement kiln dust. Civil Env. Res. **6**(11), 115–135 (2014)
54. Moghal, A.A.B., Basha, B.M., Chittoori, B., Al-Shamrani, M.A.: Effect of fiber reinforcement on the hydraulic conductivity behavior of lime-treated expansive soil—reliability-based optimization perspective. Geo-China **2016**, 25–34 (2016). https://doi.org/10.1061/9780784480069.004
55. Basha, B.M., Babu, G.L.S.: Seismic reliability assessment of external stability of reinforced soil walls using pseudo-dynamic method. Geosynth. Int. **16**(3), 197–215 (2009)

Chapter 18
Foundations for Tall Buildings on Alluvial Deposits—Geotechnical Aspects

Ravi Sundaram, Sanjay Gupta and Sorabh Gupa

18.1 Introduction

With rapid economic development, the urban land cost is skyrocketing. Consequently, construction of tall multi-storeyed buildings has become common in metropolitan cities of India.

As of today, the structural engineer is the primary decision maker in the design process of such buildings in India. Typically, the geotechnical design is done by the structural engineer and there is no involvement of the geotechnical engineer beyond the site investigation stage. Therefore, the approach on geotechnical aspects still remains conventional whereas structural design has seen several innovations.

Geotechnical engineers need to rise to the challenge and come out with innovative solutions that not only ensure safety of the building but also places emphasis on reliability and economy.

The paper discusses the importance of conducting a thorough geotechnical investigation and use of advanced in situ tests to assess geotechnical parameters, particularly for the design of foundations for tall buildings in the Indo-Gangetic Alluvium. Load testing and quality assurance tests shall enhance the reliability of the predictions. Using the soil parameters and pile-load-settlement behavior as inputs into a soil–structure interaction analysis, the design may be optimized to minimize the number of piles and account for contribution of the intervening soil between the piles.

R. Sundaram (✉) · S. Gupta · S. Gupa
Cengrs Geotechnica Pvt. Ltd., Noida, Uttar Pradesh, India
e-mail: ravi@cengrs.com

S. Gupta
e-mail: sanjay@cengrs.com

S. Gupa
e-mail: sorabh@cengrs.com

M. Latha G. (ed.), *Frontiers in Geotechnical Engineering*, Developments in Geotechnical Engineering, https://doi.org/10.1007/978-981-13-5871-5_18

18.2 Design Approach

18.2.1 Definition of Tall Building

IS: 16700-2017 [1] defines "tall building" as a building with height exceeding 50 m. The code is applicable for building height not exceeding 250 m.

The Council of Tall Buildings and Urban Habitat [2] definitions are as follows:

- Building height > 100 m: Skyscraper
- Building Height > 300 m: Super tall
- Building Height > 600 m: Mega tall.

18.2.2 Factors Influencing Foundation Behavior

Foundation behavior is governed by the following aspects:

- strata characteristics,
- foundation type,
- magnitude of the load, and
- distribution of loads.

To limit the extent of geotechnical uncertainty in the foundation design and construction, a realistic and reliable geotechnical model of the stratigraphy, soil strength, and stiffness characteristics should be established.

18.2.3 Design Criteria

The design of the foundations should satisfy the following criteria (Quick et al. [3]):

Ultimate Limit State (*ULS*): The factor of safety against failure of foundation and the supporting soils should be adequate.

Serviceability Limit State (*SLS*): Total and differential settlement of the foundation under the working loads should not affect the serviceability of the building.

In addition, safety and stability of nearby buildings and services should not be put at risk during the construction stage or in the long-term (post-construction)—ultimate limit state and serviceability limit state.

The foundation design of high-rise buildings should be done considering performance-based soil–structure interaction (SSI). It should not limited to traditional empirically based design methods such as a bearing capacity approach with an applied factor of safety (Poulos and Badelow [4]).

18.2.4 From Concept to Construction

The various steps that govern the design and construction of tall buildings include the following:

- Preliminary studies, planning, and data collection.
- Conceptual design.
- Geological and geological investigation—may be in two stages, preliminary and detailed.
- Preliminary foundation analysis based on the geotechnical investigation and selection of suitable foundation type.
- Detailed foundation design.
- Foundation construction
  - In situ testing and
  - pile-load tests and pile integrity tests (if piles are planned) and/or footing load tests.
- Review of design based on the test results and assessment of foundation performance.

The flowchart in Fig. 18.1 describes the steps involved in initial and detailed stages of the design. The geotechnical engineer should be a part of the design team during each of these stages. Presently, for the detailed design, the role of the geotechnical engineer is usually performed by the structural engineer.

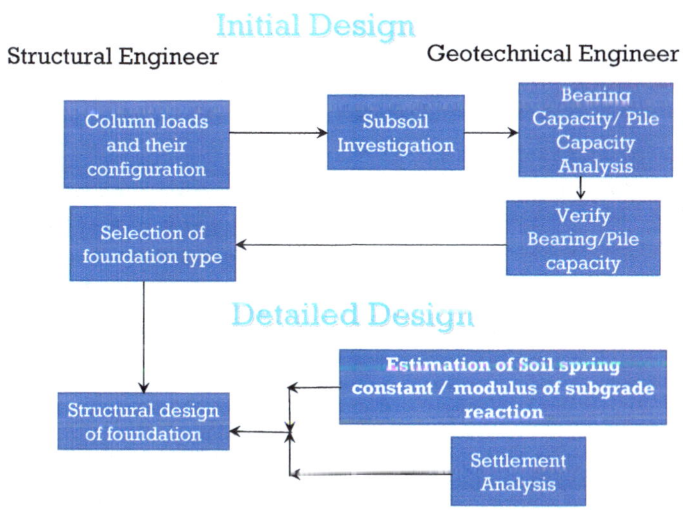

Fig. 18.1 Flowchart explaining the role of structural engineer and geotechnical engineer

18.3 Geotechnical Data

18.3.1 Initial Studies

The geological features at the site that may influence the design and performance of the foundations should be assessed so as to identify any specific measures to be taken. A desk study of the published literature and Internet search is the first step. This should be followed up with site visits to observe the following:

- topography,
- soil type, any rock exposures, and geological mapping (if required),
- groundwater level in wells in the locality, presence of nearby water bodies,
- performance of foundation of nearby buildings, details of any building failures in the vicinity, and
- local experience, etc.

These and any other information that can be obtained can provide valuable pointers that could guide the process of characterizing the ground conditions and quantifying the relevant geotechnical parameters required for foundation design.

18.3.2 Levels of Geotechnical Input

The geotechnical input that the owner/designer gets may be categorized into four levels (Sundaram [5]) as explained in Fig. 18.2.

As one moves up along the pyramid from bronze toward platinum, the factor of ignorance reduces and the reliability of design is enhanced. It also results in the reduction in foundation cost and construction time.

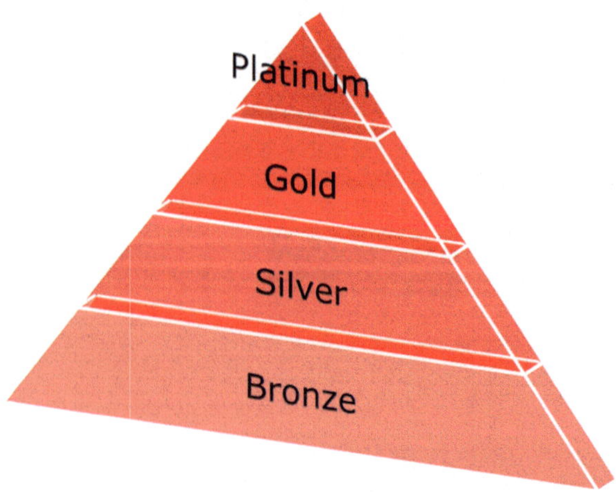

Fig. 18.2 Geotechnical input pyramid

For the design of tall buildings, particularly in thick alluvial deposits that are not underlain by rock within depth range of 50–100 m, it is vital to move toward Gold/Platinum level of investigations, while avoiding bronze level investigations. The soils at shallow depth of recent origin may be loose or prone to liquefaction during earthquakes. The properties of the deeper soils are vital for a good assessment of foundation settlement.

This requires detailed geotechnical and geophysical investigation and in situ testing, developing site-specific design parameters, using advanced design methods, load tests as well as construction monitoring.

18.3.3 Site Investigation

The site investigation should include a comprehensive borehole drilling and in situ testing program. For tall buildings in thick alluvial deposits, the investigation should extend to significant depths, to at least 50–100 m depth. The geotechnical data obtained should be reviewed carefully to select realistic design parameters.

In alluvial deposits, liquefaction analysis is important. It is usually done using SPT data, cone tip resistance and shear wave velocities (IS: 1893-2016[6], Youd and Idriss [7] NCEER Report). The highest level of groundwater should be carefully assessed so as to perform the liquefaction analysis for the worst condition.

SPT and SCPT. Two in situ tests are commonly used in geotechnical investigations are the Standard Penetration Tests (SPT) and the Static Cone Penetration Tests (SCPT). SPT conducted using an automatic trip hammer can give useful results in alluvial soils. SCPT gives a continuous profile of soil resistance with depth and may be used for bearing capacity and settlement analysis.

In the Indo-Gangetic alluvium, refusal (SPT N-value > 100) is usually met below 30–40 m depth. These high SPT values in the refusal stratum do not yield a fair assessment of the soil stiffness. Therefore, the designer may have to conservatively select modulus of elasticity (E-values) for settlement analysis.

SCPT conducted using a 20-tonne capacity penetrometer also encounters refusal, usually around 20–30 m depth. The test cannot be used to assess E-values in the refusal stratum.

Pressure meter Tests. Good quality pressure meter data in soils can provide very useful results which can effectively assess the E values. The test provides deformation properties at strain levels which are commensurate with those of the ground when subjected to service loads from the building (Haberfield [8]).

However, in sands below water table that may collapse during drilling, the ground may get disturbed and result in oversized diameter of the hole. This could result in reporting lower values of deformation modulus.

Seismic Tests. Cross-hole or down-hole seismic test usually gives a good assessment of shear wave velocities with depth. This may be supplemented with seismic refraction or SASW/MASW tests to assess the lateral variation of the ground characteristics.

But E-value for small strain cannot be applied directly to foundation analysis since ground strains under dead, live and wind/earthquake loads are significantly higher than those experienced during seismic testing. The influence of the strain level should be taken into account in the test interpretation. Haberfield [8] suggests dividing the E-values from cross-hole seismic test by a factor of 5 to obtain the static E-value for settlement analysis.

In Situ Permeability. In situ permeability tests may be required in areas of shallow water table. Since tall buildings may usually have at least 2–3 basements, substantial dewatering could be required in areas of shallow water table. In sands, dewatering could be a challenge due to the high inflow.

Field permeability tests are usually done in boreholes by falling head method or constant head method. Pump-out test can give a more realistic assessment of the hydraulic parameters for the design of the dewatering system.

Experience has shown (Gupta and Sundaram [9]) that the coefficient of permeability and transmissivity measured from pump-out tests in the Indo-Gangetic Alluvium could be 1–2 orders of magnitude higher than that determined from a borehole in situ permeability test.

Laboratory Tests. A suite of laboratory tests to assess the soil classification and index properties is essential to develop a representative soil profile. This should be supported by tests to characterize strength and stiffness properties as well as consolidation properties. Conventional unconsolidated undrained triaxial tests and consolidated undrained triaxial tests (with pore pressure measurement) are usually performed in cohesive soils whereas in granular soils, consolidated drained triaxial or direct shear tests are conducted. Sufficient tests should be done so as to develop a design profile for foundation analysis.

Laboratory tests should also be done to assess the concentration of harmful salts like sulfates and chlorides in soil and groundwater.

18.4 Geotechnical Interpretative Studies

18.4.1 Selection of Foundation Type

The interpretative aspect of the geotechnical data generated is an important overlap zone between the structural engineer and the geotechnical engineer. Selection of appropriate foundation type and depth based on the loading conditions and soil characteristics sets the tone for the foundation design and construction.

The commonly used foundation systems in alluvial deposits in the Indian scenario are:

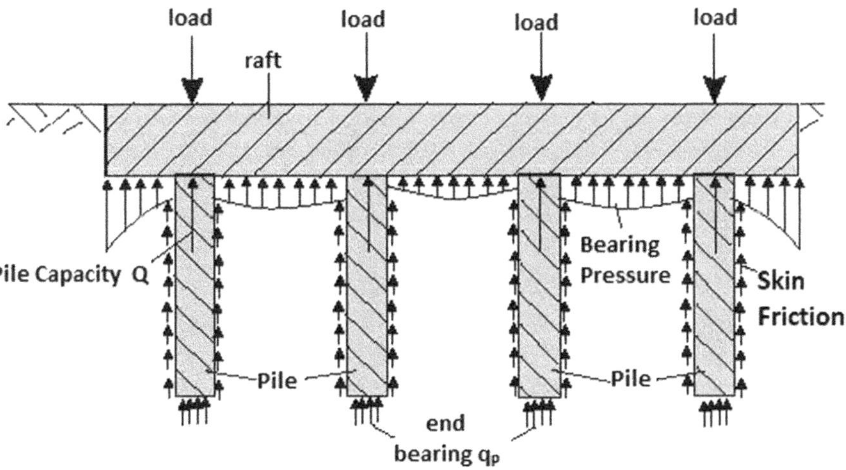

Fig. 18.3 Load transfer mechanism in piled raft

- *Raft foundations*: Loads are transferred to the ground through a raft slab covering the whole footprint of the building.
- *Pile foundations*: Piles below the columns can transfer the loads to the deeper more competent stratum. The pile caps cast over each pile group are usually connected through a beam to give rigidity to the structure.
- *Hybrid piled-raft system*: This is an economical solution for tall buildings with high foundation loads.

The hybrid piled raft is increasingly being advocated and adopted in the design of tall buildings all over the world (Kachzenbach et al. [10], Poulos [11], de Sanctis and Mandolini [12]).

It may be used to transfer the loads to the piles with the intervening soils below the raft also contributing to the load transfer (see Fig. 18.3). As a result, it has potential cost-saving and better control of differential settlement (Amornfa et al. [13]).

The contribution of the intervening soils could be substantial in stiff to hard clays and medium dense to dense sands. However, it may not be significant where the soils immediately below the raft are loose sands prone to liquefaction during earthquakes and soft to firm clays with undrained shear strength less than 50–60 kPa that may undergo consolidation/long-term settlement.

18.4.2 Piled-Raft Foundation

The piled-raft soil–structure interaction study should be done using actual column locations and loads. The single pile capacity and load-settlement behavior should be assessed from initial load tests on test piles. Zoned spring constants should be used and the settlement analysis of the piled raft should be done using PLAXIS 3D or other appropriate software to assess the load share between the piles and raft. This can be effectively used to optimize the number of piles.

In alluvial deposits, total and differential settlement and horizontal displacement, as well as stiffness of soil and pile (serviceability limit state), are the issues of primary concern that govern the design.

Shear strength and bearing capacity (ultimate limit state) should be checked although it may not govern the design.

The advantages of the piled-raft system are:

- The piles act as settlement reducers and limit the total and differential settlement.
- The piles may be loaded beyond the computed safe pile capacity (up to 70–80% of the ultimate pile capacity) since it the behavior as a piled-raft governs.
- The tilt due to eccentric loading and inhomogeneous soil conditions reduces.
- The number of piles required reduces in comparison with the pile system without raft contribution.

Amornfa et al. [13] demonstrated that in Bangkok clays, that with increasing depth of the raft, the load share of the piles reduces. For a three basement case with the foundation at about 15 m depth, the load shared by the piles reduces down to 72% of the total building loads. For rafts at shallow depth however, the contribution of the soils immediately below the raft may be less with the piles transferring substantial part of the load.

18.4.3 Foundation Design Considerations

Based on a detailed geotechnical investigation, a geotechnical model should be prepared. A proper understanding of the constitutive behavior of the soil is required to select a design profile for foundation analysis. The design considerations (Haberfield [8]) in a layered alluvial deposit are illustrated in Fig. 18.4.

Design considerations and parameters of interest in the design of piled raft are summarized in Table 18.1.

The piles are usually used up to 70–80% of the ultimate bearing capacity which is higher than the permissible safe design value for a comparable single pile. Therefore, a thorough understanding of soil–structure interaction is necessary.

18.4.4 Load Tests and Integrity Tests

To validate the design, load tests are important. Initial load tests should be performed on sufficient piles to model the pile behavior and to determine its stiffness. Static load tests traditionally performed by the maintained load method (step loading or cyclic)

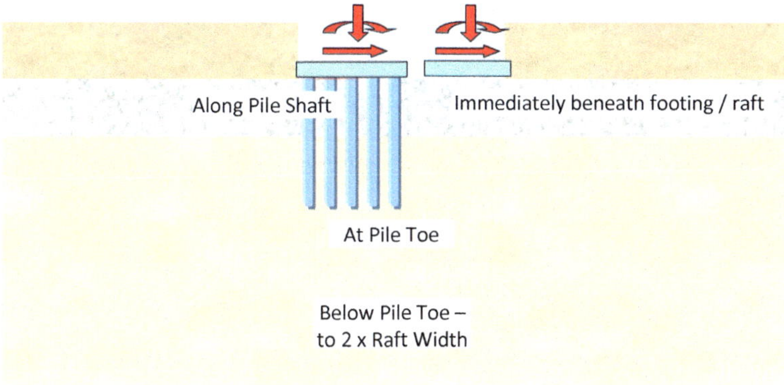

Fig. 18.4 Design considerations for raft and piled raft

Table 18.1 Geotechnical design considerations for piled raft

Zone	Parameters of interest
Immediately below raft	• Strength: bearing capacity • Settlement
Along the pile shaft	• Soil characteristics, c-ϕ values • Piles shaft resistance • Possibility of negative skin friction • Stiffness/modulus of subgrade reaction • Pile installation/drivability issues
At the pile tip	• Presence of hard strata for good end bearing • Soil characteristics, c-ϕ values • Stiffness/modulus of subgrade reaction • Pile installation/drivability issues
Below pile tip to 2 times raft width	• Soil characteristics, c-ϕ values • Stiffness/modulus of subgrade reaction

are a preferred method. But for piles very high ultimate capacity, often exceeding 1000 tonnes, it could be very cumbersome, expensive and time consuming.

High-strain dynamic load test has gained popularity in India as a cost-effective option and is now an acceptable test method in the construction industry. The test can assess the structural integrity as well as load-settlement behavior using CAPWAP analysis (Vaidya and Likins [14]).

Bi-directional static load tests using Osterberg cell has been used on some projects in India and its application is rising. It involves casting a sacrificial jack near the pile tip and can be used to test piles to high loads (Osterberg [15], George et al. [16]).

Footing load tests on 1.5–2 m wide footings can be used to realistically assess the modulus of subgrade reaction of the soil immediately below the raft and also to validate the theoretical settlement estimates (Gupta et al. [17]).

18.5 Provisions of IS: 16700-2017

The Bureau of Indian Standards has recently brought out a new code of practice, IS: 16700-2017 [1] outlining the criteria for structural safety of tall concrete buildings. Section 9.0 of this code covers foundations and has several requirements regarding geotechnical aspects.

18.5.1 Geotechnical Investigation

As per Clause 9.3.1 of IS: 16070-2017 [1], geotechnical investigation for tall buildings should comply with the following requirements:

- Geotechnical investigation should establish the safety of the building and should include liquefaction analysis and estimation of spring constants/modulus of subgrade reaction.
- Boreholes for tall buildings should be spaced about 30 m apart. Minimum of three boreholes should be drilled per tower.
- The depth of the investigation should be at least 1.5 times the raft width in soil and 30 min rock.

The authors advocate the use of static cone penetration tests and pressure meter tests for better assessment of E-values and modulus of subgrade reaction. Cross-hole seismic tests can also give a valuable input for static analysis as well as to assess liquefaction potential.

18.5.2 Foundation Depth

The embedded depth of the building shall be at least 1/15 of the height of the building for raft foundations and 1/20 of the height in case of piled raft excluding pile length (Clause 9.4 of IS: 16700 [1]). Some relaxations are available for foundations on rock and for the case of no uplift loads.

18.5.3 Soil Modeling

While modeling raft foundation through modulus of subgrade reaction, the code recommends that zoned spring constants should be used. For buildings taller than 150 m, soil–structure interaction study should be done to obtain the zoned spring constants (Clause 9.7 of IS: 16700 [1]).

Haberfield [18] highlights the importance of selection of proper values of spring constant which is a function not only of the soil but also the loading conditions and geometry.

18.5.4 Settlement

Clause 9.8 of IS: 16700-2017 refers to IS: 1904-1986 RA 2006 [19] for permissible settlement of foundations on soil and to IS: 12070-1987 RA 2010 [20] for foundations on rock.

Table 18.1 of IS: 1904-1986 RA 2006 [19] specifies permissible total settlement of 75 mm for raft foundations bearing on sands and hard clays. It also specifies a permissible differential settlement of 0.0021L (L = center-to-center distance between columns) and permissible angular distortion of 1/500 for concrete buildings.

The code permits relaxation of the permissible total settlement to 125 mm for rafts and piled rafts provided that angular distortion of the raft does not exceed 1/500.

The authors advise caution in implementing the relaxed settlement criterion and suggest that it should be backed up by the following:

- Detailed and thorough geotechnical investigation is an essential prerequisite.
- It should be ensured that the foundation is safe in shear (bearing capacity).
- Soil–structure interaction using FEM software such as PLAXIS 3D to assess settlement, differential settlement, stresses, etc. should be done.
- Monitoring settlement during the construction period will not only enhance the reliability of prediction but also give advance warning that can help prevent potential failure before it occurs.

On the basis of the review of over 52 case studies, Zhang and Ng [21] suggest limiting tolerable total settlement to 106 mm and angular distortion to 1/500.

18.6 Quality Assurance

The best of designs can fail if the requisite quality is not maintained during construction. Routine static load tests are usually done as per IS: 2011 (Part 4)-2013 [22] to 1.5 times the design safe load. Bi directional static load test provides time saving and economical option, particularly for high load-carrying-capacity piles.

The current industry practice in India on all major projects is to supplement this with low-strain pile integrity tests (PIT) on 100% of the piles (Likins et al. [23]). Cross-hole sonic logging (CSL) is also being used to locate construction defects in piles.

The results of PIT and CSL should be used to select appropriate piles for static load test and/or high-strain dynamic load tests (HSDLT). This will form the basis of acceptance of the piles and ensure quality assurance.

If any piles fail during load test or HSDLT, coring through the pile is a good option (Gupta et al. [24]) to reconfirm the pile quality. In such case, redesign of the piling system shall be required based on the actual pile stiffness and lower capacities.

18.7 Case Study

18.7.1 Project Details

This case study gives details of foundation design for a 38-storeyed building with two basements under construction in Noida (UP) located east of Delhi, not very far from the River Yamuna. The project, described as mixed land use, offers high-end commercial, retail and residential units.

18.7.2 Regional Geology

The deposits in the area belong to the Indo-Gangetic Alluvium and are river deposits of the River Yamuna and its tributaries. The Pleistocene and Recent deposits in the project area are composed primarily of sands and silts.

The Indo-Gangetic alluvial tract in the nature of a synclinal basin formed concomitantly with the elevation of the Himalayas to its north (Krishnan [25]). It was formed during the later stages of the Himalayan Orogeny by the buckling down of the northern border of the peninsular shield beneath the sediments thrust over it from the north.

The newer Alluvium, locally called *Khadar*, consists primarily of fine sand that is often loose in condition at shallow depths. The older alluvium, locally called *Bhanger*, consists of compact sands and silts and is generally, rich in concretions or nodules of impure calcium carbonate (kankars).

18.7.3 Initial Investigation

The builder got an investigation done consisting of seven boreholes to 30 m depth and three boreholes to 40 m depth. These boreholes were drilled after excavating to 6 m depth.

The boreholes revealed the presence of alluvial sands to 40 m depth with occasional sandy silt zones in some boreholes. Groundwater was met about 1.5 m depth below the excavated level. Typical borehole profiles are presented in Fig. 18.5.

Based on the borehole data, the soil investigation agency had recommended safe axial compressive capacities of 300 and 350 tonnes, respectively, for 30 and 35 m

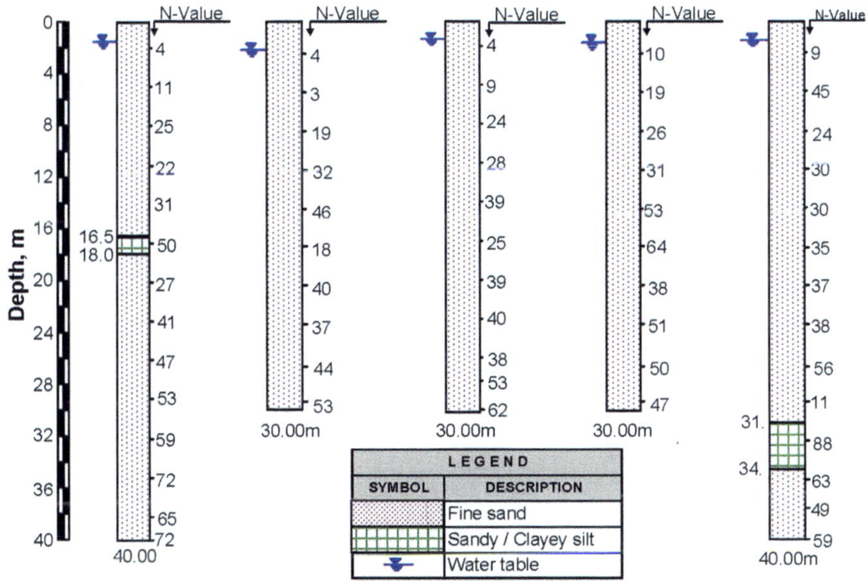

Fig. 18.5 Typical profiles of boreholes drilled for initial investigation

long 1000 mm diameter RCC bored piles. The corresponding values given for a 1200 mm diameter pile were 420 and 480 tonnes, respectively. These values are typical of the capacities used on most other sites in the surrounding areas within a 4–5 km radius of the project.

18.7.4 Initial Load Tests

Initial load tests were performed on 1000 and 1200 mm diameter piles of lengths 30–35 m. A photograph of the load test in progress is presented in Fig. 18.6. Test results are summarized in Table 18.2.

As illustrated in Fig. 18.7, the pile-load test results are fairly inconsistent/scattered with no clear trend. Many piles (especially TP-1, 2, and 3 of 1200 mm diameter piles; and TP-1 of 1000 mm diameter) significantly underperformed, indicating possible structural defects in the pile and/or poor bottom cleaning.

On the contrary, some of the piles (such as TP-2, 3, and 4 of 1000 mm diameter) performed rather well. The safe pile capacities substantially exceeded the computed theoretical capacities.

Fig. 18.6 Pile load test in progress

Table 18.2 Results of static load tests on initial piles

Pile no.	Pile dia (mm)	Pile length (m)	Interpreted safe capacity (MT)[1]	Type of slurry used
TP-1	1200	40.0	317†	Bentonite
TP-2		35.0	280†	Bentonite
TP-3		34.0	613	Composite*
TP-4		34.0	1000	Composite*
TP-1A	1000	35.0	353	Polymer
TP-2A		30.0	686	Polymer
TP-3A		30.0‡	860	Polymer
TP-4A		30.0‡	>890	Polymer

‡Gravel pad provided at pile tip
†Possible construction defects resulted in lower capacity
*Polymud + Alfabond (a thickening agent) and bentonite
[1]Safe pile capacity interpreted in accordance with IS: 2911 (Part 4)-2013 [22]

18.7.5 Detailed Geotechnical Investigation

Having gained confidence that well-constructed piles shall be able to carry loads substantially higher than that computed based on the initial geotechnical investigation, it was decided to perform an additional geotechnical investigation which included

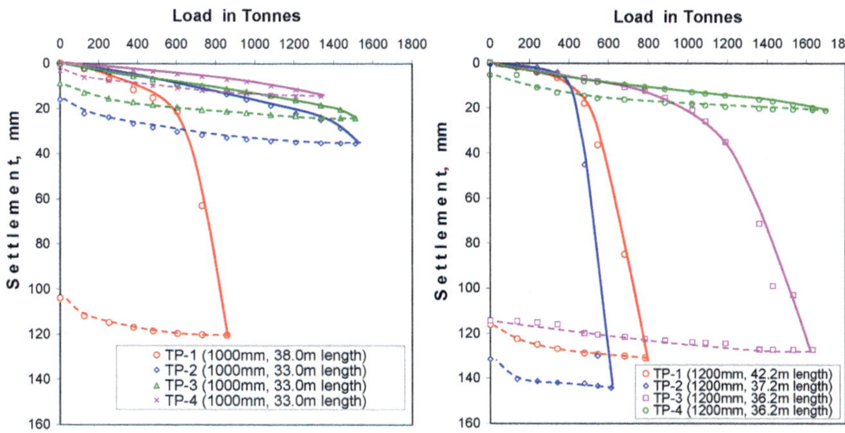

(a) 1000 mm diameter pile **(b)** 1200 mm diameter pile

Fig. 18.7 Results of pile load tests

Fig. 18.8 Boreholes and pressure meter tests in progress

three boreholes to 60 m depth, pressure meter tests in two boreholes and a cross-hole seismic test (Fig. 18.8).

Figure 18.9 presents typical borehole data.

Figure 18.10 presents typical pressure meter test data showing the profile of limit pressure and deformation modulus with depth.

Primary and shear wave velocities and dynamic shear modulus and dynamic Young's modulus from cross-hole seismic test are presented in Fig. 18.11.

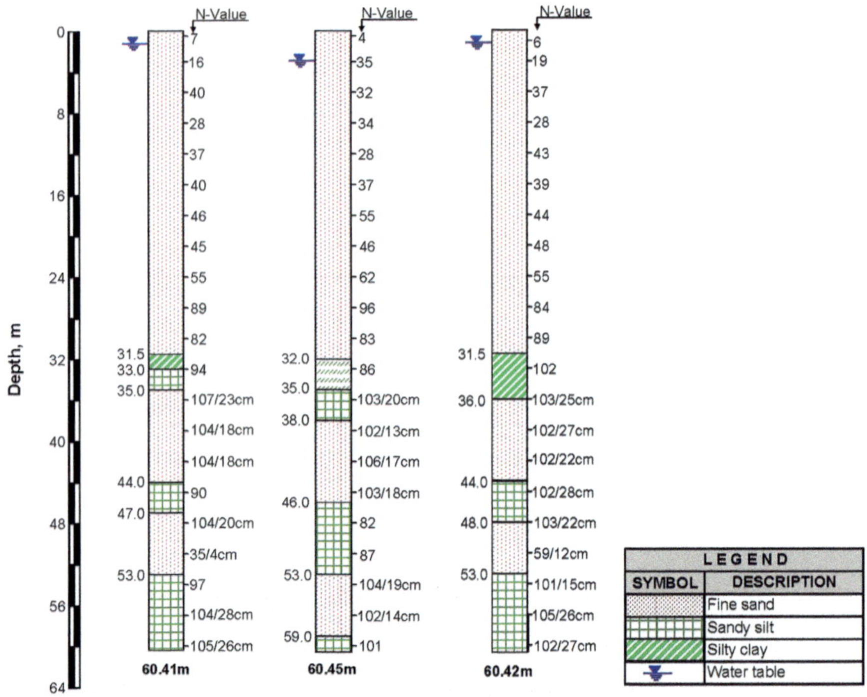

Fig. 18.9 Data from deep boreholes drilled for piled-raft analysis

18.7.6 Design Profile

Reviewing the soil characteristics, SPT values, pressure meter modulus, laboratory tests, etc., soil parameters selected for analysis of statically loaded foundations are presented in Table 18.3.

The design groundwater level was considered at the existing ground level for the worst condition. Poisson's ratio, μ, for the soil was taken as 0.33.

18.7.7 Computed Pile Capacities

Pile capacities under compression loading has been computed based on the c-ϕ values as per IS: 2911 (Part 1 Section 2) 2010 [26]. Pile capacities were also computed using the pressure meter design rules (Clarke [27]). Analysis was done for 1000 and 1200 mm diameter bored piles with cutoff level at 8.0 m depth below average ground level. The computed pile capacities are presented in Table 18.4.

The above values include a factor of safety of 2.5 in accordance with IS: 2911 (Part 1 Section 2) 2010 [26].

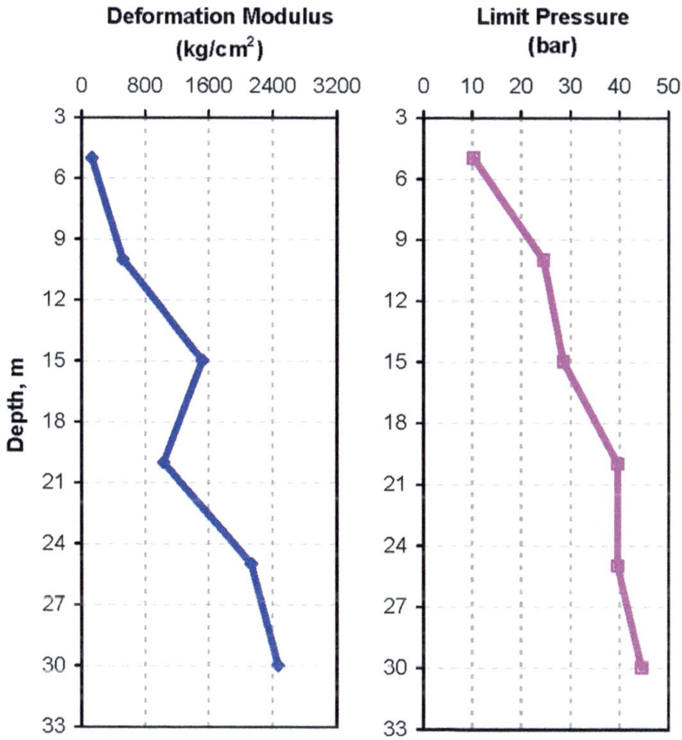

Fig. 18.10 Pressure meter modulus and limit pressure versus depth—typical results from one borehole

18.7.8 Piled Raft

Piled-raft foundation was used to optimize and economize the design. The analysis was done considering 1000 mm diameter pile of length 35 m installed below cutoff level of 8 m. The raft thickness was taken as 3.5 m in consultation with the structural engineer.

Piles were introduced at strategic locations based on the column loads. The pile locations were finalized after several hit and trials to achieve maximum economy ensuring that the load on any individual pile does not exceed 900 tonnes (2/3 of the ultimate pile compressive capacity of 1350 tonnes).

Figure 18.12 illustrates the pile layout finalized for the project.

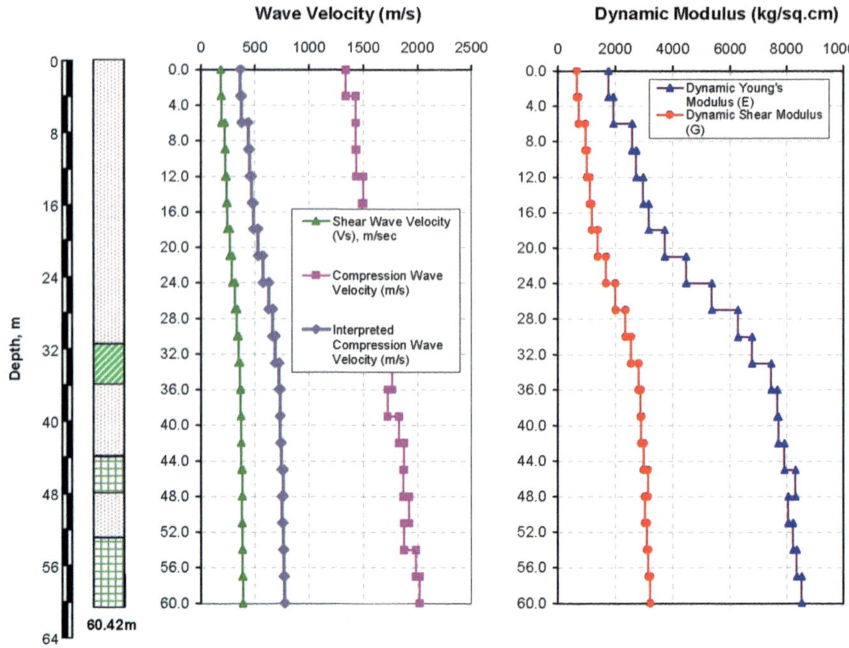

Fig. 18.11 Wave velocities and dynamic moduli from cross-hole seismic test

Table 18.3 Design profile

Depth (m)		Soil classification	γ (kN/m^3)	c (kN/m^2)	$\phi°$	E (kN/m^2)	E_{PMT} (kN/m^2)	l_m (kN/m^2)
From	To							
0.0	2.0	Silty sand/fine sand	18.0	0	30	3000	10,000	500
2.0	10.5		19.0	0	31	6200	70,000	2000
10.5	13.0		19.5	0	32	29,600	100,000	2500
13.0	23.0		20.0	0	33	43,700	160,000	3600
23.0	33.0		20.0	0	33	62,400	220,000	4000
33.0	36.0	Clayey silt	21.0	0	33	90,000		
36.0	44.0	Silty sand	21.0	0	33	124,800		
44.0	47.0	Clayey silt	21.0	90	6	124,800		
47.0	53.0	Fine sand	21.0	0	34	124,800		
53.0	60.0	Sandy silt	21.0	110	5	124,800		

where
γ = Bulk density of soil
c = cohesion intercept
ϕ = angle of internal friction
E = Modulus of elasticity of soil
E_{PMT} = Pressure meter (deformation) modulus
l_m = Limit Pressure

Table 18.4 Pile capacities under compression loading

Pile diameter (mm)	Pile length below COL	Safe compression pile capacity (kN)		
		c-ϕ values	Pressure meter	Selected design value
1000	30	4640	4800	4700
	32	4900	5100	5000
	35	5290	5560	5400
	40	5940	6310	6000
1200	30	6960	7390	7000
	32	7340	7750	7500
	35	7910	8540	8100

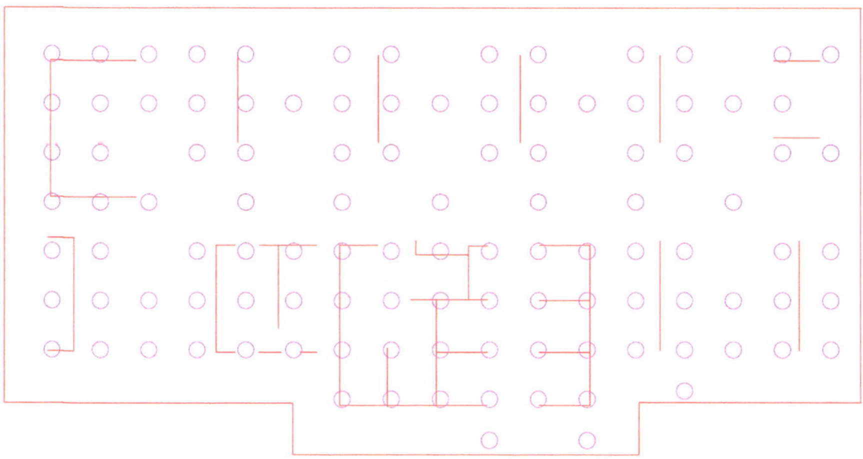

Fig. 18.12 Pile layout

18.7.9 PLAXIS Modeling

Initial analysis using the PLAXIS 3D software was done for raft without piles. However, settlement of a raft foundation was found to be about 330 mm which is higher than the permissible 75 mm total settlement specified in IS: 1904-1986 RA 2006 [19] for raft foundation. Differential settlements were also beyond permissible limits and the angular distortion was about L/400.

Thus, the analysis confirmed that piles are required to be installed as settlement reducers. Piles were therefore introduced in the finite element model below the raft at strategic locations within the high settlement zones.

Pile-load tests conducted at the site were simulated in the FEM model for calibration of the model, using soil profile and soil E-values given in Table 18.3. It was observed that soil E-values had to be reduced by 40% based on single pile simulation,

Table 18.5 Summary of piled raft settlements and pile loads

Load combination	DL + LL
Pile diameter	1000 mm
Pile length	35 m
Maximum raft total settlement	74.3 mm
Maximum pile load	8900 KN
Maximum angular distortion	L/1200

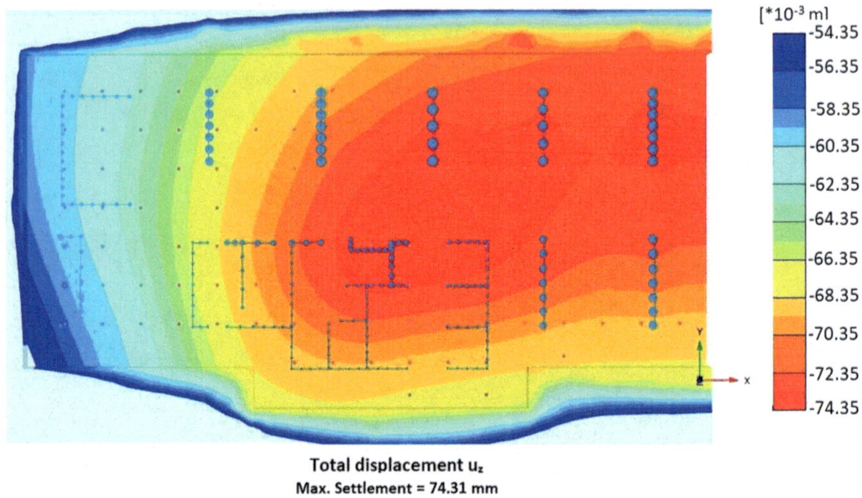

Fig. 18.13 Settlement contours

for the pile settlements obtained by the software to match the field pile settlements. Loads distribution and settlements were assessed to optimize the foundation design.

18.7.10 Results of Analysis

Based on detailed analysis, the proposed building can be supported on a piled-raft foundation as per the configuration in Fig. 18.12. Maximum settlements of the piled-raft foundation and pile loads are summarized in Table 18.5.

The maximum total settlement is less than the permissible. Differential settlement/angular distortion is also less than permissible value of L/500. Figure 18.13 presents the settlement contours. Figure 18.14 presents the pressure distribution at the soil–raft interface.

The spring constants for soil and for piles, computed as the ratio of contact pressure and settlement are presented in Figs. 18.15 and 18.16 which were used by the structural engineer for the design of foundation.

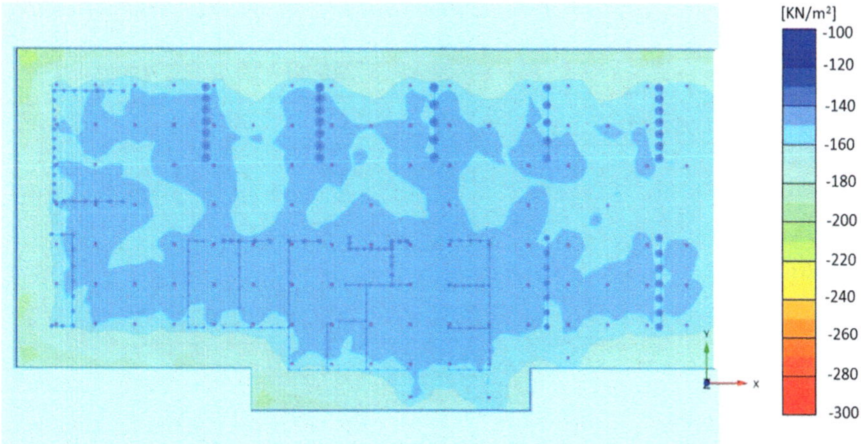

Fig. 18.14 Pressure contours at soil–raft interface

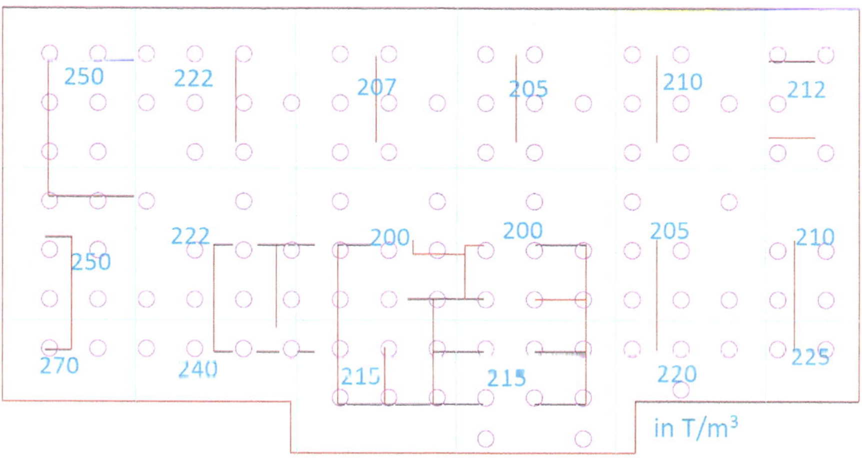

Fig. 18.15 Soil spring constant

18.7.11 Cost-Saving Realized

To illustrate the financial saving achieved by thorough geotechnical investigation and soil–structure interaction analysis, the foundation costs is summarized in Table 18.6.

It may be seen that the cost of the detailed geotechnical investigation and the SSI is significantly less than the savings in the foundation cost. While the number of piles decreases, the reliability of the design increases substantially.

The authors of the opinion that further saving is possible if the following in situ tests are performed prior to starting work on the routine piles:

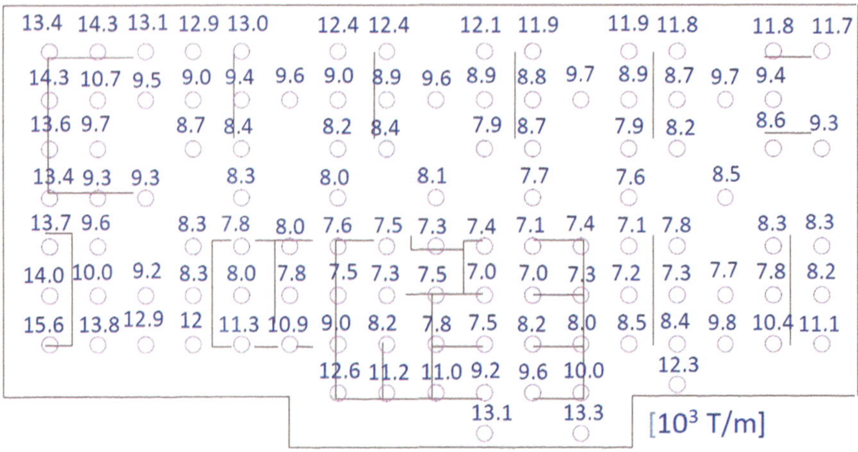

Fig. 18.16 Pile spring constant

Table 18.6 Cost-savings analysis

Design stage	Load on pile* (MT)	No. of piles	Estimated piling cost (in million rupees)	Savings† (in million rupees)	Consultancy cost (in million rupees)
Based on limited geotechnical investigation	300	550	440.0		₹ 0.5
Based on detailed geotechnical investigation	536	500	400.0	₹ 40	₹ 2.0
Piled raft (based on assumed spring constant)	536	358	286.4	₹ 153.6	–
Piled raft system—SSI using PLAXIS	900#	221	176.8	₹ 263.2	₹ 0.25
Direct project cost saving				₹ 263.2	₹ 2.75

*as per the design based on the safe pile capacity
#approximately 67% of ultimate load carrying capacity of pile
†Savings in cost of piling in comparison to that based on limited geotechnical investigation without SSI

- Footing load tests on 1.5–2 m size footing may be used to assess the E-values and spring constant of soil effectively.
- Sufficient number of high-strain pile integrity tests and/or bi-directional static load test on initial test piles used in conjunction with the static pile-load tests can give a better assessment of load-settlement behavior of piles that can be input into the PLAXIS model for the piled-raft analysis.

18.8 Concluding Remarks

In the Indo-Gangetic alluvium in northern India, sands and silts are encountered to substantial depths. These soils are usually loose to medium dense at shallow depths and very dense (SPT N > 100) below 30–40 m depth. Construction of tall buildings in such soils needs a thorough geotechnical investigation. This should be followed up by soil–structure interaction using PLAXIS or other software.

Piled-raft foundation is increasingly being used for the design of tall buildings. It takes advantage of the load sharing of the intervening soils and thus can optimize the foundation design.

With IS 16025-2017 permitting up to 125 mm total settlement for raft foundations and piled rafts, the following issues are critical:

- Detailed geotechnical investigation should be performed.
- Liquefaction potential assessment should be done for highest water table expected.
- Assessment of modulus of subgrade reaction is an important aspect.
- Differential settlement/tilt should be restricted to 1/500.

For piled rafts, the piles are usually loaded to 70–80% of the ultimate pile capacity. Therefore, quality assurance of pile construction is essential to ensure that the piles behave as designed. Conventional static load tests should be supplemented with high-strain dynamic pile-load tests and bi-directional static load test (O-cell) for a realistic assessment of pile-load settlement behavior.

Quality assurance tests such as low-strain pile integrity tests and cross-hole sonic logging should be an integral part of field testing of the piles. This will assist in optimizing the design and ensuring that the piles are capable of supporting the applied loads.

A case study of a 38-storeyed building in Noida presented in this paper demonstrates the effectiveness of the design and the advantage achieved by conducting a thorough geotechnical investigation and soil–structure interaction analysis. Savings of nearly Rs. 260 million were realized on the piling costs in comparison to what the owner would have spent based on a simplistic limited geotechnical investigation.

References

1. IS: 16700: Criteria for structural safety of tall concrete buildings. Bureau of Indian Standards, New Delhi (2017)
2. Council on Tall Buildings and Urban Habitat. http://www.ctbuh.org/TallBuildings/HeightStatistics/Criteria/tabid/446/language/en-US/Default.aspx
3. Quick, H., Meissner, S., Michael, J.: Innovative foundation systems for high-rise buildings. In: Proceedings, 1st Intelligent Building Middle East Conference, Bahrain (2015)
4. Poulos, H.G., Badelow, F.: Geotechnical parameter assessment for tall building design. Int. J. High-Rise Build. **4**(4), 227–239 (2015)
5. Sundaram, R.: Geotechnical investigations in difficult ground conditions—Indian experiences. In: Sixth Madhav Lecture, Proceedings, GeoPractices-2017, IGS Hyderabad (2017)
6. IS: 1893 Part-1: Criteria for earthquake resistant design of structures. Bureau of Indian Standards, New Delhi (2016)
7. Youd, T.L., Idriss, I.M.: Liquefaction resistance of soils: Summary report from the 1996 NCEER and 1998 NCEER/NSF workshops on evaluation of liquefaction resistance of soils. J. Geotech. Geo-environ. Eng. 297–313 (2001)
8. Haberfield, C.M.: Practical experience in piled-raft design for tall buildings. In: Proceedings, 18th ICSMGE, Paris (2013)
9. Gupta, S., Sundaram, R.: In-situ testing-current trends in geotechnical investigations. In: Proceedings, Indian Geotechnical Conference, IGC-2003. Roorkee, pp. 635–645 (2003)
10. Katzenbach, R., Arslan, U., Moormann, C.: Piled raft foundation in Germany. In: Hemsley, J.A. (ed.), Design Applications of Raft Foundations. London, Thomas Telford Ltd, pp. 323–391 (2000)
11. Poulos, H.G.: Piled raft foundations: design and application. Geotechnique, **51**(2), 95–113 (2001)
12. de Sanctis, L., Mandolini, A.: Bearing capacity of piled rafts on soft clay soils. J. Geotech. Geoenviron. Eng. **132**(12), 1600–1610 (2006)
13. Amornfa, K., Phienwej, N., Kitpayuck, P.: Current practice on foundation design of high rise buildings in Bangkok, Thailand. J. Lowland Technol. Int. **14**(2), 70–83 (2012)
14. Vaidya, R., Likins, G.: Guidelines for successful high-strain dynamic load tests and low strain integrity tests for bored piles. In: Proceedings, DFI-India-2013, Mumbai (2013)
15. Osterberg, J.: New device for load testing driven and drilled shafts separates friction and end bearing. In: Proceedings International Conference Piling and Deep Foundations, London, pp. 421–427 (1989)
16. George, J.S., Sundaram, R., Gupta, S.: Bi-directional static load tests—case studies. In: Proceedings DFI-India 2018, Gandhinagar, paper communicated (2018)
17. Gupta, S., Sundaram, R., Gupta, S.: Footing load tests on sand: validating theoretical predictions. In: Proceedings, Indian Geotechnical Conference-IGC-2016, Chennai (2016)
18. Haberfield, C.M.: Practical application of soil structure interaction analysis. In: Gregory Tschebotarioff Lecture, Proceedings, 19th ICSMGE, vol. 1. Seoul, Korea, pp. 83–101 (2017)
19. IS: 1904-1986 Reaffirmed: Design and construction of foundations in soils: general requirements. Bureau of Indian Standards, New Delhi (2006)
20. IS: 12070-1987 Reaffirmed: Design and construction of shallow foundations on rock. Bureau of Indian Standards, New Delhi (2010)
21. Zhang, L., Ng, A.M.Y.: Limiting tolerable settlement and angular distortion for building foundations. Geotechnical Special Publication No. 170, Probabilistic Applications in Geotechnical Engineering, ASCE (2006)
22. IS: 2911 Part 4: Design and construction of pile foundations—code of practice part 4 load test on piles. Bureau of Indian Standards, New Delhi (2013)
23. Likins, G., Robinson, B., Piscsalko, G.: A brief overview of testing of deep foundations. In: Proceedings, Testing and Design Methods for Deep Foundations. Kanazawa, Japan, pp. 97–104 (2012)

24. Gupta, S., Sundaram, R., Gupta, S.: Identifying defects in bored piles—case studies from India. In: Proceedings, 19th ICSMGE. Seoul, Korea, pp. 2397–2400 (2017)
25. Krishnan, M.S.: Geology of India & Burma. CBS Publishers, New Delhi (1986)
26. IS: 2911 Part 1 Section-2: Design and construction of pile foundations—code of practice part 1 concrete piles section-2 bored cast in-situ concrete piles. Bureau of Indian Standards, New Delhi (2010)
27. Clarke, B.G.: Pressuremeters in Geotechnical Design. Blackie Academic and Professional, London (1995)

Chapter 19
Sustainable Landslide Mitigation Measures-Demonstration Through a Live Case Study in Uttarakhand

Minimol Korulla, Rudra Budhbhatti, P. S. Meenu and Vikalp Kamal

19.1 Introduction

En route from Nandprayag to Govindghart, a landslide prone zone, Birahiin Chamoli District of Uttarakhand State on the riverbanks of Alaknanda and Birahi Ganga at NH 58, has been active since the Chamoli Earthquake (1999). Many landslides have since then tormented the locals and created inconvenience for the travelers. Moreover, the 2013 cloudburst in Uttarakhand made it more susceptible.

This route has distinct importance because of a variety of destinations of pilgrim and tourists interest like Badrinath, Hemkund Sahib, Valley of Flowers, and Auli. Chamoli occupies the northeastern corner of the Garhwal tract and lies in the central or mid-Himalayas in the very heart of the landslide and rockfall prone region.

To mitigate this phenomenon, MoRTH along with PIU Dehradun decided to undertake measures for the rehabilitation and treatment of the landslide-prone area. The unstable zone in this area stretches around 100 m and the solution with brief details on the project will be described here forth. Moreover, this paper discusses in detail about the analysis and design method adopted for the mitigation measures for landslide, rockfall and drainage improvement with appropriate and relevant codal referencing. Utmost preference was given for use of sustainable and environmental-friendly solutions adopted for Birahi Landslide. Moreover, the necessity of post-construction instrumentation and monitoring is briefly described.

M. Korulla (✉) · R. Budhbhatti · P. S. Meenu · V. Kamal
Maccaferri Environmental Solutions Pvt. Ltd., Gurgaon, Haryana, India
e-mail: minikorulla@maccaferri-india.com

M. Latha G. (ed.), *Frontiers in Geotechnical Engineering*, Developments in Geotechnical Engineering, https://doi.org/10.1007/978-981-13-5871-5_19

Fig. 19.1 Landslide zone at Birahi

19.1.1 Location and Geology

Birahi is a valley region situated 8 kms from Chamoli District of Uttarakhand State, India at an altitude of 1200 m above the sea level. The landslide prone region stretches up to100 m and is most susceptible near a hairpin bend at location 30°24'35.18" N 79°23'22.33" as shown in Fig. 19.1 depicted.

Through the geological investigation, it was found that there are very high folds and faults with fault zone passing toward NE of the landslide location. The exposed rocks in the region are highly jointed, weathered and fractured quarzitic schist belonging to Garhwal group. Surficial material of slope in between the roads consists of soil and boulders and is prone to sliding. Birahi Fault is in the vicinity of the area; the landslide zone is seismically disturbed. The geological cross section of the Birahi region is depicted in Fig. 19.2.

19.2 Problem Description

The 100 m stretch of the Birahi region is highly susceptible to landslides due to steeper slopes at both hill and valley side, with instability in the topsoil over the rock. Moreover, the faults in the hill and presence of boulders make it vulnerable to rockfall as shown in Fig. 19.3. Constant roadblocks due to the landslides and rockfall have caused distress among the local and the tourists. In addition, there is no proper

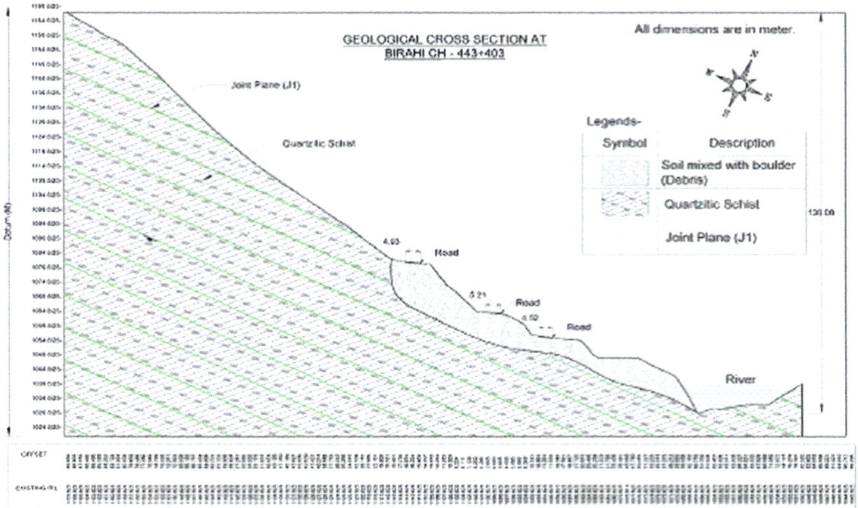

Fig. 19.2 Geological cross section at Chainage 443+403, Birahi

Fig. 19.3 Landslide at Birahi hairpin bend

drainage system to reduce the effect of surface runoff leading to increased pore water pressure, which causes destabilization of slope which results in landslide. Moreover, this region lies in the seismic zone V.

19.2.1 Causative Forces

a. **Anthropogenic Activities**: Construction operations during the widening of road are one of the major destabilizing forces in Birahi. Cutting of hillside vegetative slope results in release of the in-situ stresses in the rock mass and disturbed the equilibrium of the forces. This equilibrium forces remains untreated without any proper support system which makes the hill slope vulnerable to slide.
b. **Tectonic Settings**: The region of Birahi falls under seismic V region and the presence of fault lines have made the area more vulnerable to landslide.
c. **Surface Erosion**: Accumulation of colluviums on the valley side slope starts the erosion when it encountered surface runoff water. This surface erosion disturbed the stability of road.

19.3 Geotechncial and Topographical Investigation

To mitigate the problem of landslide, a comprehensive geotechnical investigation was performed to properly understand the terrain and proceed with the most sustainable and economic solution. Various boreholes at many desired locations were proposed and investigations as per the relevant codes were carried out up to a depth ranging from 20 m to 26 m, to obtain suitable geotechnical design data.

19.3.1 Rotary Drilling

Rotary drilling was performed using heavy-duty, hydraulic-feed, and skid-mounted rotary drill rig. The drilling rig has a hydraulic feed and is driven by a bevel gear system run by a 28 HP Perkins engine. The drill chuck has four jaws to accommodate NW size drill rod. Four numbers of boreholes were performed (designated as BH-1 to BH-4). One of such drilling is depicted in Fig. 19.4.

Drilling and sampling of the rock were performed using an NX size double tube core barrel. A diamond impregnated bit was used to drill through rock strata/refusal. The strata observed at the site generally comprise of overburden of cobbles/boulders underlain by Rock (Quartz Mica Schist). However, the exposed rock was also visible at some locations. Sampling of rock core can be seen in Fig. 19.5.

19.3.2 Seismic Reflection Test

Seismic refraction test was conducted in order to characterize the subsurface geologic conditions and geological structure.

Fig. 19.4 Borehole 4 at Birahi Site

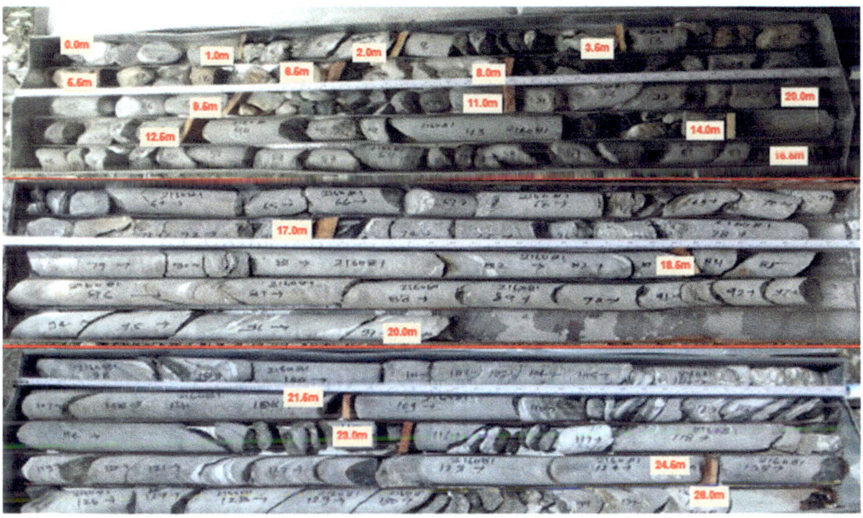

Fig. 19.5 Rock samples from Borehole 1

A stress applied at the surface of an elastic medium creates conditions for associated strains to propagate as elastic waves (P&S) in the subsurface material. The waves travel as a pattern of particle deformation, with velocities that are dependent upon the elastic properties and densities of media through which it travels. SRT tests were performed at two locations and designated as 3A and 3B. The test uses the travel time of first-arrival wave component only regardless of its travel path. The success of the refraction test depends highly on the properties of layer interface. An interface can only be found if the layer below is shown a significantly higher velocity. The travel time curve was developed using WinSims-16 based on first-arrival picking from the raw data. Velocity then computed using linear regression between selected receiver and results are shown in Figs. 19.6 and 19.7.

19.3.3 Laboratory Tests

Laboratory tests were conducted on selected rock samples to determine their physical and engineering properties. The testing procedures were in accordance with current applicable IS specifications. The following tests were conducted on selected rock samples recovered from the test locations which is tabulated in Table 19.1 and general properties of the samples is tabulated in Table 19.2.

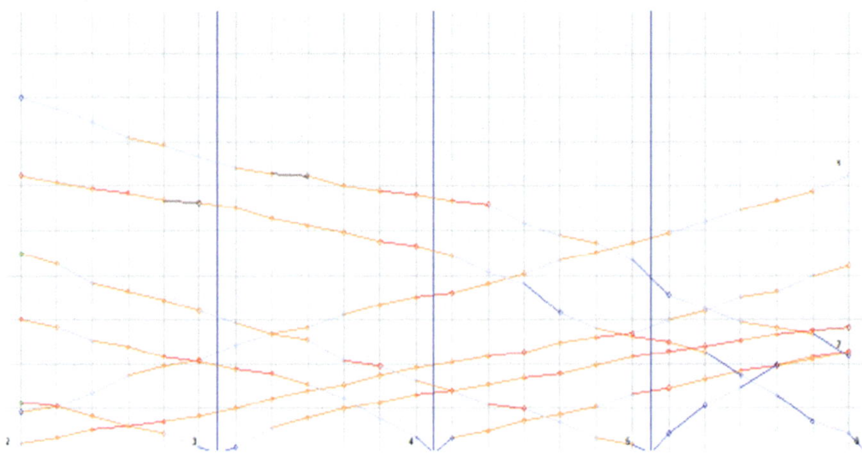

Fig. 19.6 Travel time Curve SRT-location-3A

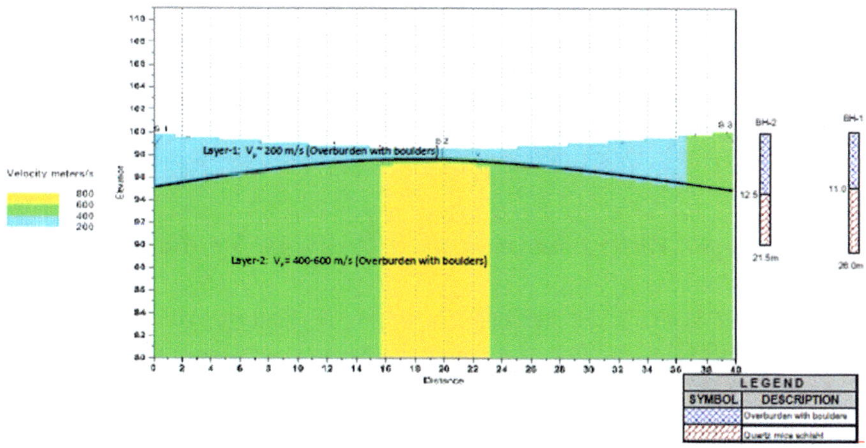

Fig. 19.7 Seismic profile for Birahi site location-3B

Table 19.1 Laboratory tests on samples

Name of test	Is code no
Porosity and void ration	By calculations
Specific gravity	IS: 2720 (PART 3)-1980
Moisture content, density and water absorption	IS: 13030-1991
Unconfined compressive strength	IS: 9143-1979, RA-2006
Point load strength index	IS: 8764-1998, RA-2008

Table 19.2 General properties of the sample from investigation report

Index	Range	Unit
Bore hole depth	20–26	m
Unconfined compressive strength of rock (IS: 9143-1979)	151.1–551.7	kg/cm^2
Density of rock (IS: 13030-1991)	2.71–2.79	gm/cc
Moisture content (IS: 13030-1991)	0.11–0.64	%
Water absorption (IS: 13030-1991)	0.38–0.76	%
Void ratio	0.01–0.04	
Porosity	0.01–0.04	
Point index (IS: 8764-1998, RA-2008)	12.5–32	kg/cm^2
Unconfined compressive strength (IS: 9143-1979, RA-2006)	151.1–551.7	kg/cm^2

19.4 Selecting Advanced Solution and Design of System

Following are the proposed rehabilitation and treatment measures adopted to counter the destabilizing forces.

19.4.1 Proper Water Channelization/Drainage Solution

It was observed from the site that there are no proper channels to divert the surface runoff to a nearby culvert. Hence, a proper drainage channel has been provided on hill as well as on valley side including cascade, side drains, and culverts (already existing). Water has been channelized from the hill slope and is safely discharged into the culvert through cascade and roadside drain. After entering inside the culvert, water is directed to the river through chute drain. Chute drain shall be extended until the hard strata or rock is encountered on the valley side so that water shall be safely discharged into the river. This proper drainage network will reduce the danger of further accumulation of water in the in-situ soil. The details of the proposed drainage network can be seen in Figs. 19.8 and 19.9.

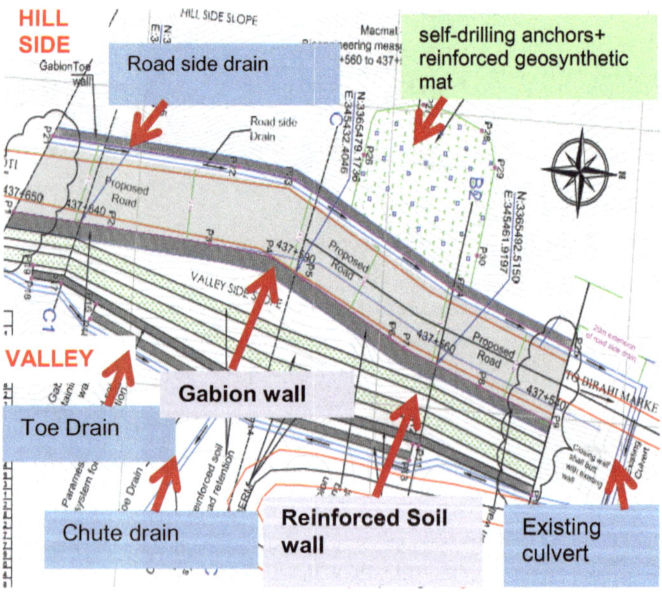

Fig. 19.8 Proposed plan for drainage and protection work

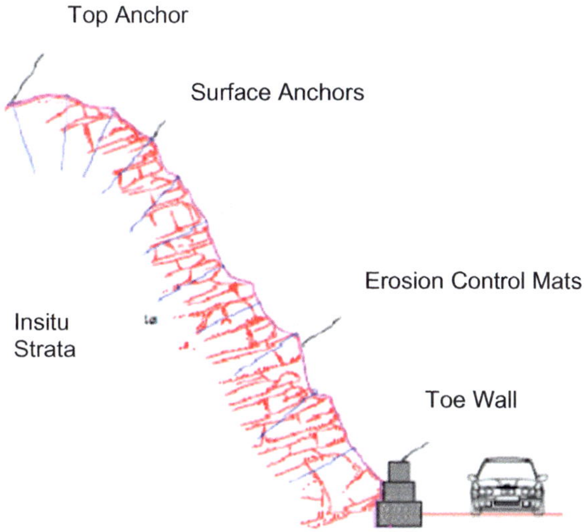

Fig. 19.9 Cross section of hillside road protection works

19.4.1.1 Design for Drainage

To provide above drainage solution, rainfall intensity of 14.58 cm/h was considered for a 100-year period as per isopluvial maps of Uttarakhand. All the discharge calculations were done using the empirical relations and parameters mentioned in IRC: SP 42 (guidelines for road drainage) and highest discharge was calculated as 1.633 m^3/s. Trapezoidal section of the drain was hence proposed to facilitate the required self-cleansing velocity and allowable discharge from the section was duly checked with the peak runoff generated using Manning's Formula. Similarly, chute drains were also designed to safely guide the discharge of water into the river.

19.4.2 Solution for Hillside Erosion Control and Rockfall Mitigation

On the hillside, the slope was left untreated after the cutting during the construction of road portion where the slide has occurred. It was stabilized by providing a secured drapery comprising of erosion control geomats as facia system along with self-drilling anchors of 25 mm diameter at 3 m spacing in longitudinal and transverse direction. Geomat is a three-dimensional erosion control mat to take care of surficial erosion by preventing splashing effect of rainfall on exposed soil surface. Also, it helps to provide strength to the base of stem or root of the grass. In order to facilitate the growth of vegetation, soil amendments along with special hydraulically applied erosion control mulchin combination with seeds has been applied. These measures are environmentally friendly and sustainable. At the toe of the slope, for the affected stretch, 3 m high gabion wall was provided as per IS 16014:2012 for toe protection (refer Figs. 19.8 and 19.9).

19.4.2.1 Design for Rockfall Mitigation and Erosion Control

For the design of the above-advanced solution, IRC: HRB-23 was referred and limit equilibrium method was adopted. The system was checked for static and seismic conditions. Further, to control the surface erosion, suitable erosion control mat was provided as per the strength requirement in line with MoRTH section 700.

19.4.3 Solution for Valley Side Retention System

On valley side, for retention of proposed 12 m wide road, reinforced soil system of 25 m maximum height has been provided. In this location, the retention is provided by the combination of reinforced soil system and gabion wall (Fig. 19.10).

The system is based on principles of soil reinforcement wherein, tension elements, i.e., high strength polymeric grids are introduced in the soil mass as reinforcement to retain the soil vertically or at steep slope by virtue of interaction between soil and reinforcing elements [6], arrangement can be shown in Fig. 19.11.

Considering the high seismicity of the region, heavy rainfall, drainage requirement, large amount of cutting and filling work, etc., a composite reinforced soil system with high strength geo-grid as primary reinforcement has been selected as the

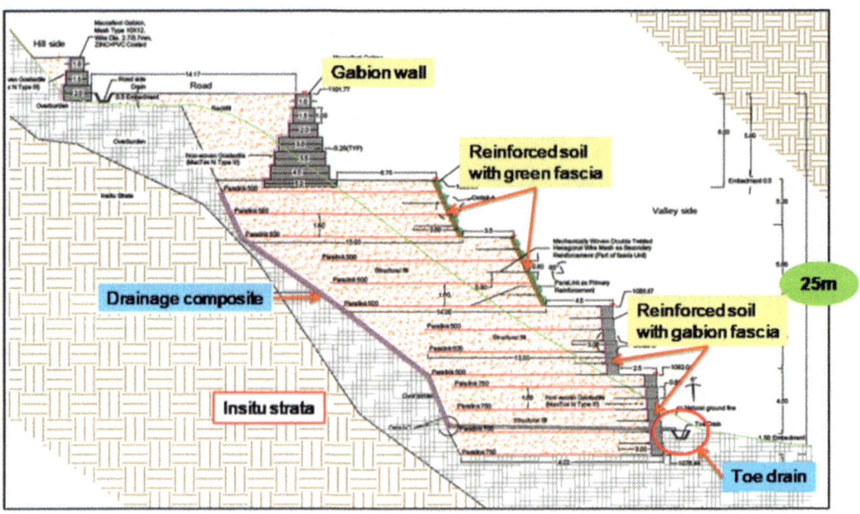

Fig. 19.10 Typical section of valley side retention system

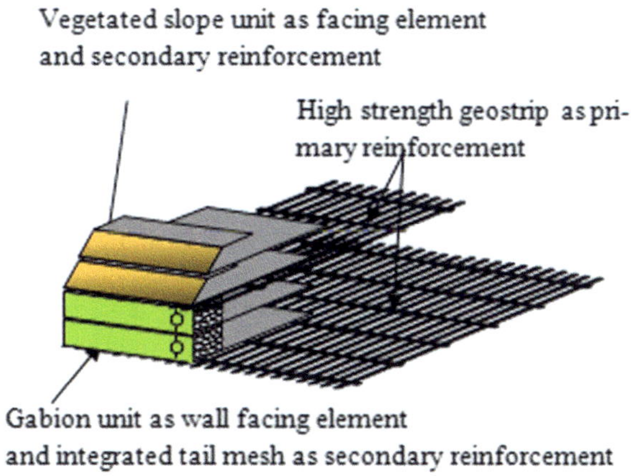

Fig. 19.11 Schematic arrangement of composite reinforced soil system

best possible solution for retention system. On the top of this composite reinforced soil structure, a gabion retaining the wall of maximum 6.5 m height is provided. Further, at the interface of in-situ strata and structural soil, a drainage composite has been provided to dissipate the pore water pressure developed and to assist the flow of seeping water. As the topography is steep and variable, the height of valley side retaining reinforced soil structures is varied throughout the stretch. The stability analysis is carried on representative section depends upon the wall height and other parameters.

19.4.3.1 Design of Reinforced Soil Structures

For the composite reinforced soil system, the static design was done as per British Standards (BS: 8006-1:2010) and seismic design as per AASHTO Standards (FHWA-NHI-10-024). Seismic coefficients were taken from Indian Standard (IS: 1893:2014-Part-1). All stability analyses including, global stability, internal stability, and bearing capacity checks were done for static as well as seismic conditions. The fill material was adopted as per clause 3014 of MoRTH. Reinforcing element specifications follows MoRTH 3013.7. The design live load was adopted as per IRC 6: 2014. Moreover, the system was provided with geotextile at the back of the facia and at the hill slope for proper drainage of the percolating water, to diminish the effect of pore water pressure. All the required checks such as input flow, retention criteria, and clogging resistance were checked as per FHWA-HI-95-038.

19.5 Design Methodology

All the above solutions were designed using limit state equilibrium method and serviceability method. For the design of reinforced soil structure as described in Sect. 19.4.3 of this paper, in-house built software was used. A typical output of the analysis is presented in Fig. 19.12. The software has the ability to perform slope stability analysis for static as well as seismic conditions.

The summarized factors of safety for stability checks for static as well as seismic condition are given in Tables 19.3 and 19.4.

Similarly, for the design of secured drapery, in-house built software was used. The analysis for the design for hillside slope protection and rockfall mitigation as described in Sect. 19.4.2 was performed using this software. A typical output file of the analysis can be seen in Fig. 19.13.

The summary of the factor of safety used in this analysis for seismic conditions is given in the Table 19.5.

Moreover, to check the settlement and deformation of the retention structure finite element analysis was conducted using Plaxis 2D (plane strain analysis) was done. A typical output of settlement analysis for the problem as described in Sect. 19.4.3 is depicted in Fig. 19.14.

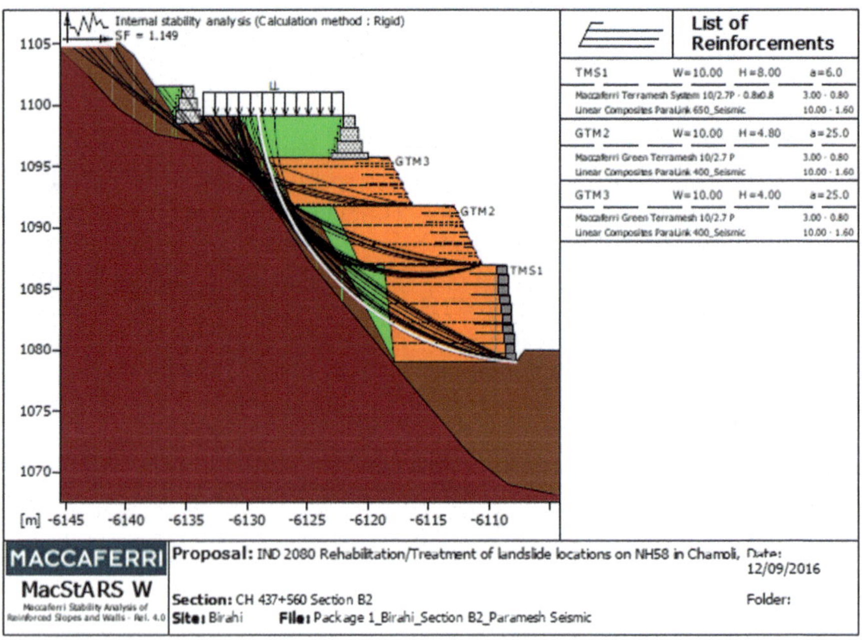

Fig. 19.12 Slope stability analysis of typical cross section

Table 19.3 Summarized factors of safety for soil reinforced wall (static case design as per BS 8006-1:2010)

Stability checks	Target FOS	FOS for B2 Section	FOS for CC Section	FOS for C1 Section
Global stability	1	1.346	1.480	1.886
Sliding stability	1.3[a]	3.023	4.042	>1.3
Bearing Capacity	1.35[a]	1.829	3.380	3.611
Internal sliding	1	1.230	1.494	1.274

[a]The software used for the analysis applies the minimum required safety factors in static case, as mentioned in above table, to the resisting forces for respective stability checks and hence the targeted safety factor is unity in all checks of static analysis

Table 19.4 Summarized factors of safety for soil reinforced wall (seismic case design as per FHWA-NHI-00-043)

Stability checks	Target FOS	FOS for B2 Section	FOS for CC Section	FOS for C1 Section
Global stability	1	1.258	1.193	2.292
Sliding stability	1.125	1.353	1.421	>1.125
Bearing Capacity	1.875	1.969	2.813	9.625
Internal sliding	1.125	1.149	1.128	1.128

Table 19.5 Summarized factor of safety for rockfall mitigation (seismic case design)

Stability checks	Target FOS	FOS for hillside section
Bar design	1	1.41
Mesh design	1	1.20
Serviceability design	1	1.55

Moreover, the drainage flow checks through the drainage composite were designed with another in-house built software, and the analysis of the seepage was done using Slide Software. The inputs from Slide seepage analysis was used to further compute the desirable grade of drainage composite to be used in the system. The typical result of analysis can be seen in Fig. 19.15.

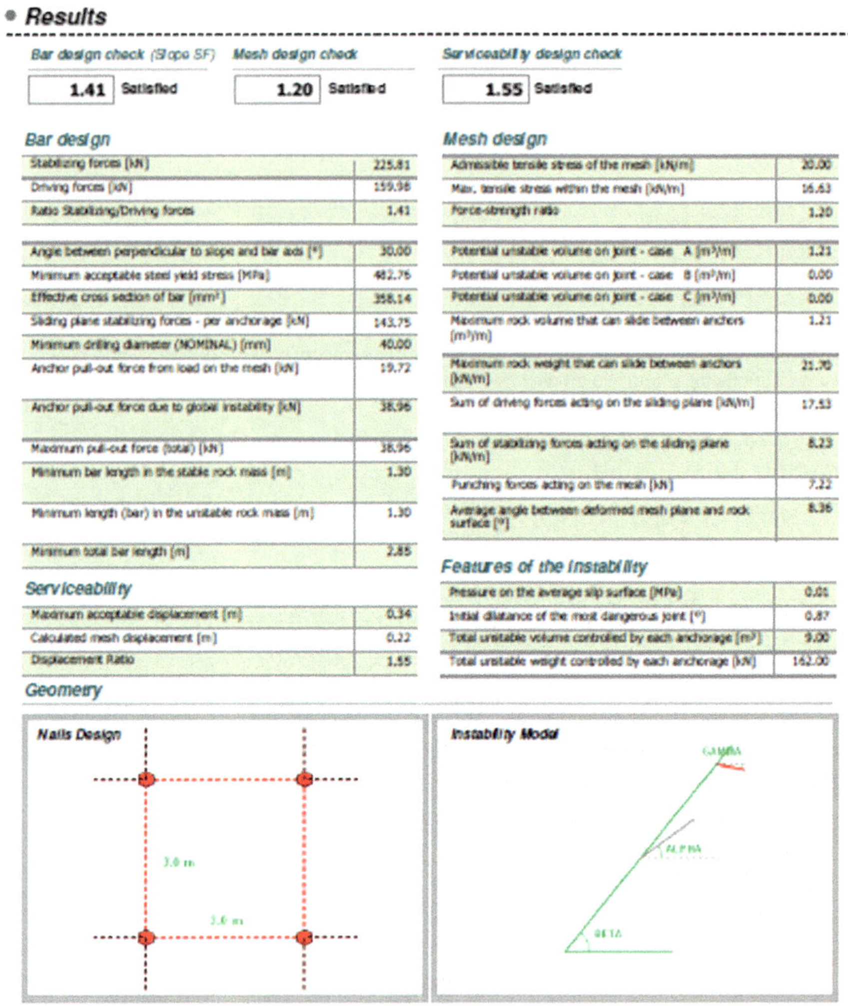

Fig. 19.13 MacRO analysis for rockfall mitigation

19.6 Advantages of Solution Adopted

Composite reinforced soil system was selected due to its flexibility, seismic performance, cost-effectiveness, speed of construction, and suitability for use with the local materials. Other than these major advantages, this system also has the following general and technical advantages in comparison to the conventional structure like reinforced cement concrete. Flexibility of the system within acceptable limits helps to accommodate differential ground settlement without any compromise in structural integrity and help to stand stable during seismic effect.

Fig. 19.14 Settlement analysis using Plaxis (2D) software

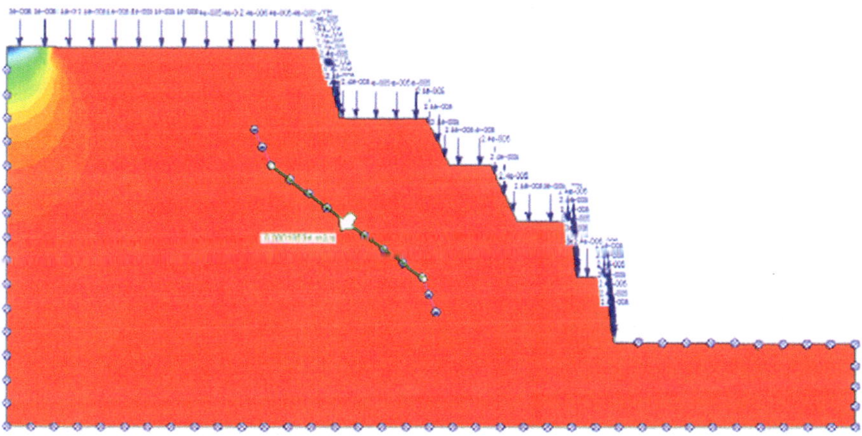

Fig. 19.15 Seepage analysis using slopes software for retentions system

For rigid structures like RCC, failure is spontaneous, and it is highly sensitive to even slightest settlement in foundations and hence extensive ground improvement techniques may be required or wall needs to be founded on hard strata. Also, there is no limitation in terms of height of retention to which the composite reinforced soil systems can be constructed. As on date case study exists for more than 70 m height retention.

Fig. 19.16 Current status of retention system at Birahi

In the case of composite soil reinforcement system, fascia system options like gabions with integrated tail and composite system of coir, welded mesh, and woven wire mesh that allows growth of vegetation are available. Moreover, social and economic advantages are other vital aspects to be considered. As the technology suggested is simple and the construction does not require highly skilled labor force or special equipment, it allows recruitment of local workforce usually hired on site which is sometimes a much more important social benefit than that represented by the final work.

19.7 Construction Progress

The project is currently under construction. However, the retentions system for most of the stretch of the road has been developed and drainage system has been installed. This can be seen in Fig. 19.16. Moreover, the slope protection works has been done with proper anchorage for rockfall mitigation. This can be seen in Fig. 19.17. Similarly, a roadside gabion toe wall for hillside protection can also be seen in Fig. 19.17. Few more site photographs of the site are depicted in Figs. 19.18 and 19.19.

Fig..19.17 Rockfall mitigation and erosion control at Birahi

Fig. 19.18 Retention works at hairpin bend at Birahi site

Fig. 19.19 Site photograph at Birahi site

19.8 Instrumentation and Post-Construction Monitoring

It is quite essential to investigate any structure of this magnitude via proper instrumentation and monitoring. Moreover, due to the challenging site location of Birahi where the rainfall intensity is high, and which falls in seismic zone V with high vulnerability of surface erosion, makes instrumentation and monitoring work more essential. Hence to observe and safeguard the integrity of the structure, suitable contraption for landslide monitoring were implemented on the site during construction. Major parameters that needed to be monitored are as follows: (1) Drainage behavior of the ground. (2) Polymeric reinforcement and nail force magnitude. (3) Deterioration of facing and other soil nailing elements. (4) Performance of any structure supported by reinforced valley slopes. (5) Horizontal and vertical movement of slope surface and overall structure. Some of the function and instruments are given in Table 19.6.

19.8.1 Site Instrumentation

Various strain gauges at many desired locations under the retention system at the valley site were put to observe the deformations at different layers at the site. Moreover, to detect the water pressure, piezometers were also placed in a PVC pipe of diameter 4 inches which reached till the bottom of the valley side retention system.

Table 19.6 Instruments and their functions

Instrument	Function
Reflectometer	Deflection of facia elements in all three directions
Strain gauge	Strain developed in primary reinforcement
Slope inclinometer	Wall deformation
Earth pressure cells	Total pressure in earthfills and embankments
Piezometers	Pore water pressure measurement

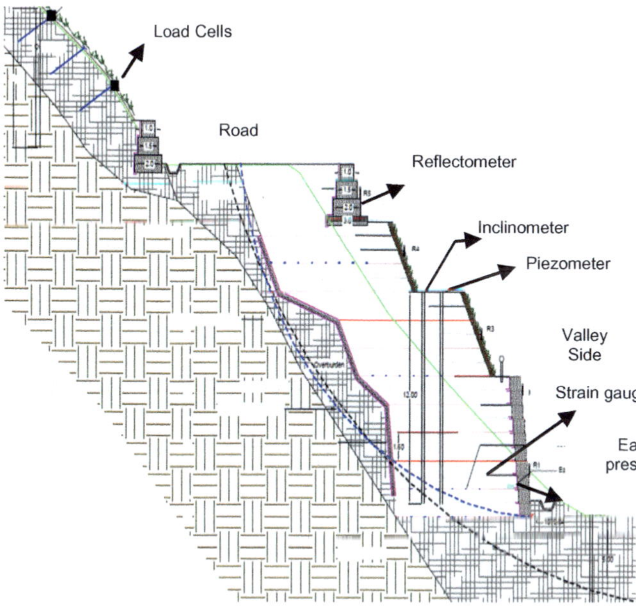

Fig. 19.20 Instrumentation scheme for Birahi site

An earth pressure cell at the bottommost retentions system at the back side of the facia was positioned to monitor the variation of the stresses where the shear failure is most anticipated. Apart from these, reflectometers were also placed at the valley side retention system to examine the tilt of the structure with time. As for the hill-side monitoring, five load cells were positioned to detect the rock slide. Moreover, inclinometers were put at the valley side. The typical cross-sectional view of the instrument scheme can be seen in Fig. 19.20. Moreover, Figs. 19.21 and 19.22 show the implication of the strain gauge and reflectometer at the site.

Fig. 19.21 Implementation of strain gauge at the site

Fig. 19.22 Placement of reflectometer at the site

References

1. British Standard Institute Publication BS 8006-1:2010 'Code of Practice for strengthened/reinforced soils and other fills
2. IS 1893 part1 (2002) Criteria for earthquake resistant design of structures
3. IRC-SP-042—Guidelines of Road Drainage
4. IRC-HRB-Special Report-23. State of the Art: Design and Construction of Rockfall Mitigation Systems. New Delhi:s.n. (2014)
5. U. S. Department of Transportation Federal Highway Administration, Publication No. FHWA-NHI-00-043, 'Mechanically stabilized earth walls and reinforced soil slopes design & construction guidelines
6. Di Pietro, P.: Design and construction of soil reinforced structures using composite reinforcement systems: modern and cost-effective alternatives for high walls and slopes. In: Proceedings of the Seventh International Conference on Geosynthetics, 7-ICG (2002)
7. IS: 2720 (Part 3)-1980, Method of tests for soils, Part III: Determination of specific gravity

8. IS: 13030-1991, Method of test for laboratory determination of water content, porosity, density and related properties of rock material
9. IS: 9143-1979, RA-2006, Method for the determination of unconfined compressive strength of rock materials
10. IS: 8764-1998, RA-2008, Method for determination of point load strength index of rocks

Chapter 20
Geotechnical Education from Field

I. V. Anirudhan

20.1 Introduction

The formal teaching of soil mechanics started in India more than six decades ago. Terzaghi and Peck [1], and Taylor [2] were taught in classes, while the actual application in the field was limited since the civil engineers were engaged in other matters of design and construction. Foundation design and construction continued without much application of soil mechanics. Some specific problems encountered in the field were looked into by the professors and solutions were offered using their limited time allotted for such engagements. Special foundations like driven cast-in-situ piles, bored cast-in-situ piles, well foundations and caissons were being constructed by specialist contractors collaborating with specialists from other countries. These technologies brought along the design procedures. Presently, there are different foundation and ground improvement systems and earth retention systems being adopted in India. Design of all these systems is based on the empirical approaches developed elsewhere, either by the specialist contractors or by the research institutions.

Most of the systems being adopted in India have standards specified by Bureau of Indian Standards. These standards were initially drawn from the accepted practices elsewhere based on the successful execution of the system in some projects. Very rarely research is pointing towards the improvement or understanding of these design procedures because the procedures are taken for granted. As a result, some of the geotechnical design codes are closer to theory without addressing the field-related variabilities and boundary conditions.

India witnessed several advancements in the foundation construction methods, deep excavation procedures, and ground improvement systems, all borrowed from specialist contractors. Three critical aspects of a foundation system, vis-à-vis, design, construction and performance, are not well connected and there are no serious efforts

I. V. Anirudhan (✉)
Geotechnical Solutions, Chennai, India
e-mail: anirudhen@gmail.com

M. Latha G. (ed.), *Frontiers in Geotechnical Engineering*, Developments in Geotechnical Engineering, https://doi.org/10.1007/978-981-13-5871-5_20

of the academic institutions to bridge this gap. However, such efforts are the only way to equip a geotechnical graduate student to foresee problems and then design a solution for execution. Involvement of the students in such practical research programs shall improve their problem-solving capabilities.

Terzaghi and Peck [1] developed their theories based on the problems they faced and solved in the field, and their books and papers present the interlacing between the theory and the practice. The curriculum, however, does not attempt to relate the theory to its background while teaching from these books.

Many institutions do not understand the whole ambit of research and execute theoretical projects. One such example is microzonation of cities in which the research is not bringing out any authentication of an application. Several modifications are being done on the commonly used Indian Standard Codes on Geotechnical Engineering, but rarely have the inputs from well-documented case histories.

This paper attempts to bring in the ways to reduce the gap between the design and practice citing some selected field experience.

20.2 Literature

Some of the textbooks, like the Terzaghi and Peck [1], use the field situations to explain the theory behind an application and take back the reader to the practical world of soil mechanics and the geotechnical engineering. One classic example is the bearing capacity theory developed by Terzaghi from the observation of 'punching' of the shallow footing into the ground. Unfortunately, the visualisation of the bearing capacity failure in soil is only in our imagination because actual ones seldom occur as represented in theory. Centrifuge model studies and 1g model studies help us to reaffirm our imaginations if the boundary conditions are adequately captured.

Bolton [3] developed his book 'A guide to Soil Mechanics' narrating three case histories, taking the reader from the field to a classroom.

The understanding of anisotropy in clay deposits by observing the mobilisation of shearing resistance along the failure plane embedded in the soft clay foundation supporting embankment [4] is a classic example of learning from the field.

The limiting frictional resistance along the length of a drilled shaft measured by instrumentation of several such foundations lead to the development of empirical expressions [5] instead of classical static formulations and later modified and being practised by FHWA [6, 7]. The success of such empirical expressions lies in the fact that they follow the accepted theoretical outlines. The classroom lectures may not be sufficient to convince a student about such deviations from the theoretical formulations.

Lateral squeezing of very soft to soft clays under embankment loading is a common occurrence in the field. However, such a possibility is not taught in a classroom where classical theories of slip failures are taught. The literature report several such cases. Radhakrishnan and Anirudhan [8] illustrate an example of lateral squeeze and heave of soft clay under platform fill loading. High water table conditions or standing water reduces the resistance to lateral squeezing. Anirudhan and Ramaswamy [9] discuss a similar case in which a full raft foundation experiences large settlement and tilt partly due to lateral squeezing of soft clay under insufficient overburden conditions. These lessons from the field should advise a geotechnical student to check for the possible lateral squeezing when the design of an embankment or foundation over soft clay is undertaken. Ramaswamy and Anirudhan [10] present how the case histories can aid teaching the foundations on expansive soils.

These citations are only a fraction of the valuable case studies and field observations available in the literature. The classrooms shall introduce such well-documented case studies as part of the teaching process.

20.3 Linear and Nonlinear Load-Settlement Behaviour

Geotechnical design often stresses the serviceability criterion to define the design load-carrying capacities of any foundation element. This criterion relates to the structural capacity of the foundation system also since the structures are not very flexible. Making a structure more and more rigid or more flexible to tolerate any amount of differential movement is neither rational nor practical. Hence, the serviceability criterion relates to a sustainable differential movement between different components of the structure and also of the foundation. There is a difficulty in estimating the possible differential movement because of the unknown variables. The starting point of defining the differential movement is the estimation of the total expected movement of the foundation, and then the differential movement is suggested as a percentage of the total movement.

From the elastic theory, the movement of a rigid foundation is

$$S_i = qB\frac{(1-\mu^2)}{E}I, \qquad (20.1)$$

where S_i is the immediate elastic settlement, q is the incremental load, B is the foundation width, E is the elastic modulus μ is the Poisson's ratio and I is influence factor that is a function of other foundation dimensions, depth and the Poisson's ratio.

In all practical purposes, this immediate elastic settlement is taken as the total settlement when cohesionless (free draining) soil is forming the foundation soil.

There are other expressions for estimating the compression of a soil layer encountered below the foundation element once the incremental stress in the layer is known.

$$S_f = 2.303\frac{H_i}{C}log_{10}\left[\frac{(\bar{p}_0 + \Delta p)}{\bar{p}_0}\right], \qquad (20.2)$$

where S_f is the compression of the layer of thickness H_i, C is the compressibility coefficient as a function of static cone resistance and the effective overburden pressure p_0' and Δp is the incremental stress due to the load on the foundation.

$$S_f = \frac{H_i}{1+e_0} c_c log_{10} \left[\frac{(\bar{p}_0 + \Delta p)}{\bar{p}_0} \right], \quad (20.3)$$

where e_0 is the initial void ratio and c_c is the compression index.

$$S_f = \Delta p \, m_v H_i, \quad (20.4)$$

where m_v is the coefficient of volume compressibility and H_i is the thickness of the interested layer.

Equations 20.1–20.4 are from IS 8009 Part 1 [11]. Equations 20.1 and 20.4 are linear to the load increment resulting proportionally increasing settlement as the applied load increment increases. Equations 20.2 and 20.3 may somewhat converge and produce a mildly increasing rate of settlement as the incremental load increases.

20.3.1 Load-Settlement Response from Plate Load Test

Figure 20.1a is illustrating a load-settlement curve from a plate load test in the field. The plate size is 450 mm × 450 mm and placed on the surface. There is no overburden effect around the plate for some distance. Figure 20.1b is presenting the hyperbolic representation of the same curve.

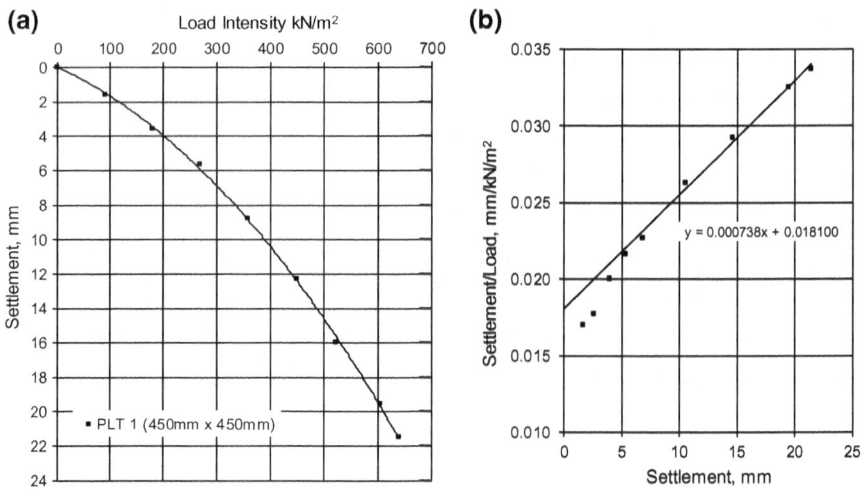

Fig. 20.1 **a** Load intensity-settlement response of a plate load test **b** hyperbolic representation of the load-settlement response in figure (**a**)

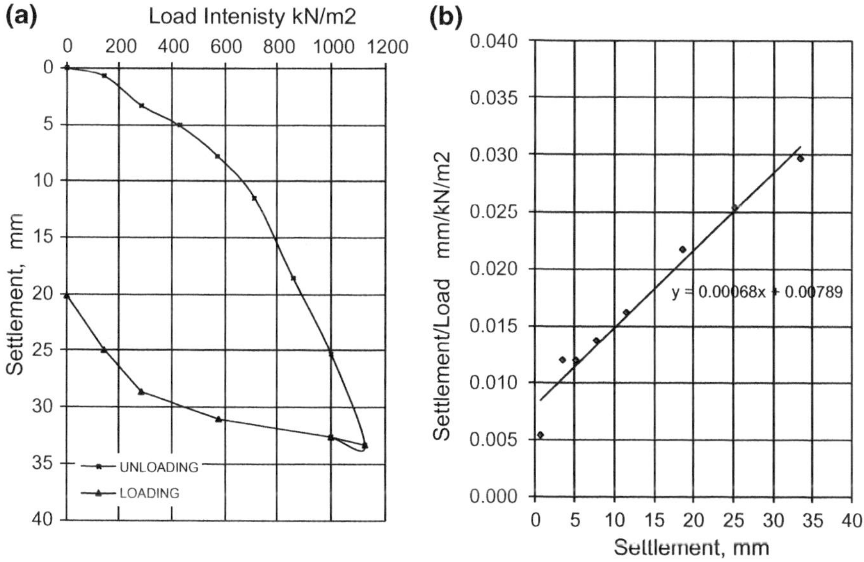

Fig. 20.2 **a** Load-settlement response of a 1.50 m × 1.50 m footing at surface **b** hyperbolic representation of the load-settlement response in figure (**a**)

20.3.2 Load-Settlement Response from Footing Load Test

Results from a 1.5 m × 1.5 m size footing load test conducted over fully weathered friable rock formation are presented in graphical form in Fig. 20.2a. The hyperbolic fitting of the same is shown in Fig. 20.2b.

20.3.3 Load-Settlement Response of Driven Cast-In-Situ Pile

The response of a vertically driven cast-in-situ pile of 500 mm diameter under compression loading is shown in Fig. 20.3a. The load-settlement response is not linear. The hyperbolic representation shown in Fig. 20.3b illustrates that the load-settlement response is a combination of two hyperbolic curves.

20.3.4 Load-Settlement Response of a Large Pile Group

The load-settlement response of a piled raft system (pile used as a settlement reducer) is not much different from the above-referred responses. Figure 20.4 is presenting an estimation made using Eqs. 20.2 and 20.3 for a 33 m diameter ammonia storage

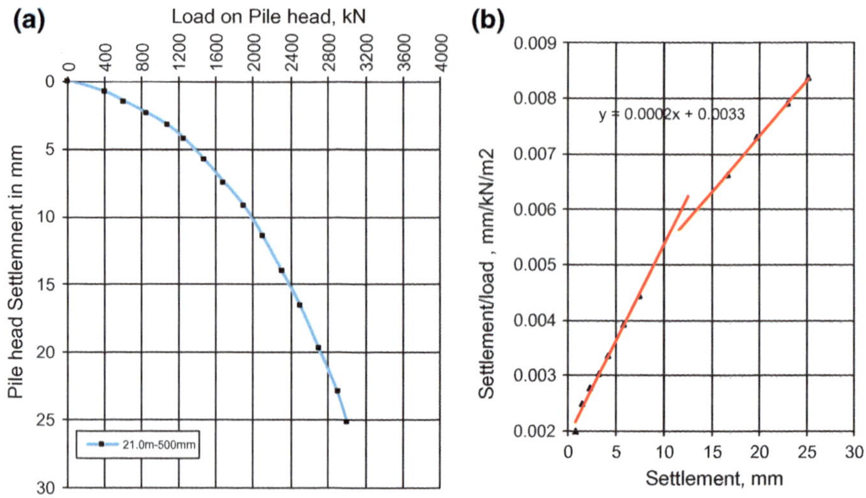

Fig. 20.3 **a** Load-settlement response of a 21-m-long, 500-mm-diameter-driven cast-in-situ pile resting in sand (mostly passing through sand) **b** hyperbolic representation of the load–displacement response shown in Fig. 20.3a

tank supported on short-driven cast-in-situ piles. The piles are terminated at about 13.0 m depth passing through medium dense sand and soft clay layers and resting in the medium dense sand. The estimated settlement at different loading is somewhat linear to the load increments. The actual measured settlement of the tank under hydro-testing is also shown in Fig. 20.4. The actual load-settlement response is not linear as the rate of settlement is increasing as the load increases. More details of this case study are available in Anirudhan and Balakumar [12].

20.3.5 Discussion

All the above responses from the field confirm that the load-settlement reaction of a foundation is not linear as many equations suggest. The nonlinear response is due to the nonlinear behaviour of one of the variables in all these estimations. The variables here are the incremental load and the stiffness of soil layers. The stiffness and probably the Poisson's ratio are the only nonlinear parameters here.

The classrooms talk about nonlinearity in the stiffness, i.e. its variability under different stress levels of the soil. Different soil models incorporate this nonlinear behaviour, and a visit to the field responses helps the student to understand it better. The hyperbolic representation is one of the nonlinear response curve fitting technique that has broad applications in the geotechnical problems [13].

The nonlinearity in its simplest form may be expressed as

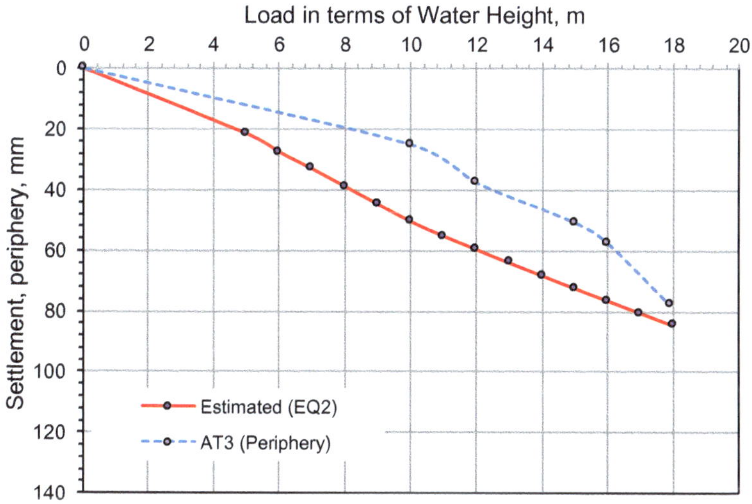

Fig. 20.4 Load-settlement response of a flexible raft on piles under hydro test (33 m diameter tank on 450-mm-diameter-driven cast-in-situ piles resting in medium dense sand)

$$\Delta H = \Delta p(a + k_a \Delta H), \tag{20.5}$$

where ΔH is the displacement, Δp is the incremental load, a is a constant and k_a is the stiffness coefficient. ΔH is appearing on both the sides. Kota et al. [14] attempted a solution for the load-settlement behaviour of piles, and partial solutions are reported.

20.3.6 Summary Points

1. The theoretical estimation of the load–displacement behaviour of a foundation system is generally linear.
2. The real behaviour is nonlinear even in the usual working range, and this non-linearity is emerging from the elasto-plastic nature of the soil.
3. The nonlinear stiffness factors shall be introduced for a realistic estimation of the load–displacement response.

20.4 Construction Procedures and Related Variabilities

In many geotechnical construction procedures, the boundary conditions and the natural state of the soil are altered by varying degree. These changes affect the performance of the structure thus constructed. Construction of earth retention systems, deep foundations, ground improvement programs and tunnels is influenced by construction-induced variables. Theoretical postulations alone are not sufficient to address the design of such geotechnical elements. Various soil models try to incorporate such variabilities. Often an empirical or semi-empirical approach is needed to address the variabilities resulting from the construction procedures. One typical example is the design of bored cast-in-situ piles (drilled shafts). The following sections discuss the design, both by theoretical and empirical approaches, and the field performance of deep-bored cast-in-situ piles in an infrastructure project.

20.4.1 Theoretical Design of Bored Cast-In-Situ Piles

Most textbooks refer to static theories for estimating the load-carrying capacity of a deep foundation. The end-bearing component is determined using bearing capacity equations, while the shaft friction (skin friction) is determined from the shearing resistance along the pile–soil interface as derived from the horizontal soil pressure acting on the pile surface. The IS code for the design and practice of the construction of pile foundations [15] uses these recommendations. The ultimate vertical compression load-carrying capacity Q_u of a uniform diameter bored cast-in-situ pile in granular soil is estimated using the expression,

$$Q_u = A_p\left(1/2D\gamma N_\gamma + P_D N_q\right) + \sum_{i=1}^{n} K_i P_{Di} \tan \delta_i A_{si} \tag{20.6}$$

where A_p is the pile base area, D is the pile diameter, γ is the effective unit weight of the soil at the pile base, N_γ and N_q are the bearing capacity factors depending upon the angle of shearing resistance of the soil around the pile base, P_D is the effective overburden pressure at the pile tip, $\sum_{i=1}^{n}$ is the summation for layers 1 to n contributing positive skin friction, K_i is the coefficient of earth pressure applicable to the *i*th layer, P_{Di} is the effective overburden pressure at the mid-depth of the *i*th layer, δ_i is the wall friction between the soil and the pile in the *i*th layer taken as 0.75–1.00 times the angle of shearing resistance ϕ of the soil and A_{si} is the surface area of the pile shaft in the *i*th layer.

The value of P_D is limited to a value equal to the effective overburden pressure from the soil thickness equal to 15 to 20 times the pile diameter above the pile tip.

The ultimate compression capacities of 35-m-long and 55-m-long piles in a site are estimated using the above expression, and the results are shown in Table 20.1.

The above-estimated values are by strictly following the parameters recommended by the code and textbooks. Often, the design practice is to reduce the skin friction and end bearing in the case of bored cast-in-situ piles citing the possibility of disturbances during the boring operations, the effect of the drilling mud, etc.

Table 20.1 Estimated compression capacities of 35- and 55-m-long piles using IS 2911 (Part 1, Section 2) procedure

Resistances/pile length	35 m long (TP5)	55 m long (TP8)
Ultimate friction (kN)	6251	23,743
Ultimate end bearing (kN)	5120	5120
Total ultimate compression (kN)	11,371	28,862

20.4.2 *Empirical Design of Bored Cast-In-Situ Piles*

Reese and O'Neill [6] recommended an empirical approach limiting the side friction as the depth increases and by estimating a limiting end-bearing resistance based on the SPT N value at the pile base when the bearing soil is sand. This procedure is adopted by FHWA [7] and subsequently adopted by AASHTO.

According to this procedure, the base resistance (kN) in cohesionless soil is

$$Q_e = A_p(57.5N_{60}),\ \ N_{60} \text{ limited } 50, \tag{20.7}$$

where A_p is the cross-section area at the pile base in m^2, N is the SPT blows for 300 mm penetration corrected to SPT N_{60}; the factor 57.5 is modified proportionately for pile length less than 10 m.

The side resistance (kN) in cohesionless soil is estimated as

$$Q_s = \sum_{k=1}^{n} \beta P_{Di} A_{si} \tag{20.8}$$

where A_{si} is the surface area of the pile length in the respective soil layer, in m^2, β is estimated from the empirical expression $\beta = 1.5–0.245\ \sqrt{z}(m)$ for corrected SPT $N > 15$, β is reduced by a factor $N_{60}/15$ when the corrected SPT N_{60} is ≤ 15. The unit side friction resistance is limited to 200 kN/m^2.

The ultimate compression capacities of 35-m-long and 55-m-long piles are estimated using the above expressions suggested by O'Neill and Reese and the results are shown in Table 20.2.

Table 20.2 Estimated compression capacities of 35- and 55-m-long piles using O'Neill and Reese procedures

Resistances/pile length	35 m long (TP5)	55 m long (TP8)
Ultimate friction (kN)	3365	6315
Ultimate end bearing (kN)	2608	2608
Total ultimate compression (kN)	5973	8923

20.4.3 Performance of the Pile Under Load Testing and Discussion

Figure 20.5a, b is presenting the load-settlement response of these 35-m-long, 55-m-long and 760-mm-diameter-bored cast-in-situ piles when load tested. The ultimate capacities as defined by the load corresponding to a settlement equal to 10% pile diameter are determined by hyperbolic extrapolation.

The estimated values are compared with the actual test results and presented as the percentage of the actual capacity. The IS2911 value is compared with the ultimate value, and the Reese values are compared with the load corresponding to the settlement equal to 5% pile diameter. The comparisons are made with respect to different base values considering the basis of the two approaches. IS 2911 predicts the ultimate resistance corresponding to pile movement equal to 10% pile diameter, while Reese procedure predicts a resistance corresponding to a pile tip movement of 5% of the pile diameter.

The summary of this comparison is presented in Table 20.3. The ultimate value is reported from the inverse slope of the hyperbolic curve.

The estimations using the IS code procedure is overpredicting the capacities by a considerable margin (+48% to +142%) mainly because of the theoretical approach without considering the construction procedures. The values determined by O'Neill and Reese procedure are comparable (−3% to +6%) with the test results because of the adaptation from field results. The method from IS 2911 was used for comparison because of its closeness to most analytical methods.

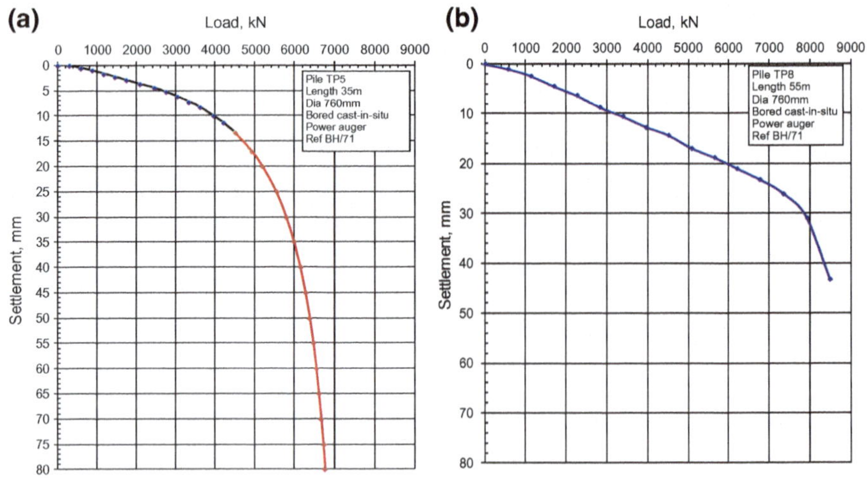

Fig. 20.5 Load-settlement response of 760-mm-diameter-bored cast-in-situ piles under compression loading. **a** 35-m-long pile and **b** 55-m-long pile

Table 20.3 Comparison of compression capacities of 35- and 55-m-long piles as obtained from load test and the estimations using O'Neill and Reese procedures and IS 2911 procedure

Resistances/pile length	35 m long (TP5)	55 m long (TP8)
O'Neill and Reese procedure (kN)	5973 (−3%)	8923 (+6%)
IS 2911 procedure (kN)	11,371 (48%)	28,862 (142%)
Ultimate from the load tests (kN)	7692	11,905
From the load tests @10% pile diameter (kN)	6840	9851
From the load tests @5% pile diameter (kN)	6157	8401

20.4.4 Summary Points

1. The theoretical postulations and expressions used in the design of different geotechnical components are not incorporating the variations occurring during the construction.
2. The nature of variations differ between the systems and suitable modifications in the design procedures are necessary to provide a safe design.
3. The empirical design procedures developed based on the real-field response studies produce more economical and safe designs.
4. The design codes shall incorporate such empirical design procedures.

20.5 Stage Loading and the Risk of Plastic Flow

Soft clay layers are often encountered at shallow depths in and around coastal regions and are susceptible to large consolidation settlement. Special ground improvement techniques are in practice to control the settlement. Among the methods, precompression, otherwise, known as the pre consolidation procedure is quite often used in the construction of road embankments, large storage tanks, etc., to minimise post-construction settlement. Preloading is, generally, done in stages, allowing the weak soil to gain strength to accept more loads. The principles of preloading are well established, but this technique is typically combined with vertical drains (sand drains/band drains) to accelerate the consolidation. However, in situations where adequate time is available for the construction and commencement of the project, preloading alone is sufficient to complete the required amount of consolidation improving the shear strength.

Preloading of soil is, generally, carried out before the actual construction. But in specific construction works preloading is possible even after the installation and one such example is large storage tanks wherein relatively large settlements are allowed. Successful examples of such preloading are described by Penman and Watson [16] and Darragh [17]. However, there have also been unsuccessful instances, where the foundation soil could not be stabilised under the scheduled time, and the rapid loading leads to a base failure [18]. A slow filling is successful in consolidating and strengthening the clay and for subsequent satisfactory performance.

20.5.1 Case of Stage Loading of Surge Tanks

Three surge tanks of 12.5 m diameter and 8.0 m water column height were constructed on soft clay foundation overlain by roughly 3.0 m thick medium dense to dense granular soil. Even though the thickness of soft clay with liquidity index 0.55 is 4.5–6.0 m, the ground improvement using standard procedures was not considered. Instead, very slow incremental loading of the total 8.0 m water over a period of 3 years was agreed upon as the time for the commissioning was not critical.

In the present case, the undrained shear strength of the soft clay at 2.50 m depth was 15–17 kN/m^2. The average coefficient of consolidation c_v was determined as 2 m^2/year from the laboratory investigation. The maximum pre-consolidation pressure is about 64 kN/m^2. The soft clay is followed by very stiff to hard clay and weathered rock, almost incompressible under the expected load increment.

The effective overburden pressure at the onset of the soft clay is 45 kN/m^2. The pre-consolidation pressure being 64 kN/m^2, roughly 20 kN/m^2 additional loads can be applied without any significant settlement. Consolidation begins when the applied load exceeds 20 kN/m^2. Hence, the load increments were recommended as 35, 45 kN/m^2, etc. with steps of 10 kN/m^2 after that. Based on the estimations, each additional load was to be maintained for 3–4 months for improving the shear strength of clay making it strong enough to support the next load increment.

The hydro test was carried out as designed until load increment of 45 kN/m^2. Figure 20.6 presents the response under the hydro test of one of the tanks. OB-1 to OB-8 are the observation points along the tank periphery.

20.5.2 Rapid Loading and Plastic Flow

There was a request to speed up the filling, and holding rate and the site management decided to load the tank beyond 55 kN/m^2 load after waiting for only 5 days. The rate of settlement suddenly increased as seen in the load-settlement responses at eight observation points around the tank. The load was immediately reduced to 6 m filling height to prevent any possible failure. The revised estimation considering the increased rate of loading is also presented in Fig. 20.6. The rate of settlement was

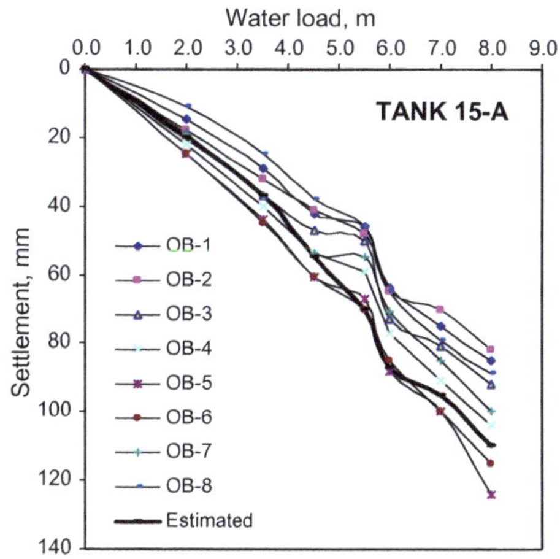

Fig. 20.6 Load-settlement response of the surge tank under hydro test at a rate of 10 kN/m^2 at every 3–4 months

suggesting a possible plastic flow in the soft clay when the tank was loaded at a faster rate. There is a standard assumption that the clay may undergo plastic flow when the vertical incremental stress exceeds 4 c_u. Four times the undrained shear strength is 4×16 kN/m^2 $= 64$ kN/m^2. There was a possibility of plastic flow in the soft clay below 2.50 m when the water load exceeded 6.5 m height. More details of this case are available in Kalaiarasi and Anirudhan [19].

20.5.3 Summary Points

1. The risk of plastic flow shall be thoroughly studied when constructions over soft clay are planned.
2. Since there is no definite solution for estimating the shear strength of soil involved in plastic flow, a conservative approach is necessary in most cases.
3. A plastic flow may be expected when the incremental vertical stress exceeds four times the undrained shear strength, i.e. 4 c_u.
4. Stage loading procedures, with or without vertical drains, shall address the issue of plastic flow and the corresponding rate of loading.

20.6 Pile-Driving Formulae for Driven Cast-In-Situ Piles

During the development of driven cast-in-situ piling system, various dynamic formulae were also developed for confirming the pile seating depth and for verifying the compression capacity. The basis for such equations is the compatibility between the input energy used for driving the pile and the work done in terms of the pile penetration into the ground. One such method developed by Hiley [20] became very popular because of its simplest form.

$$R_u = \frac{Whn}{S + C/2}, \tag{20.9}$$

where R_u is the ultimate resistance, W is the weight of the driving hammer, h is the effective falling height, n is the energy efficiency of the driving system that is a function of the ratio of the driving weight and the pile casing weight and the type of cushion used over the casing top, S is the permanent pile penetration under one hammer blow and C is the temporary compressions of the casing pipe, wooden cushion and related parts of the system and the soil.

The construction of driven cast-in-situ pile is achieved by driving a temporary casing made of seamless pipe using a heavy hammer. A dispensable shoe made of cast iron or mild steel is provided at the bottom preventing the soil from entering into the casing. The driving is thus displacing the soil sideways increasing its density when drained type soil is present. Once the driving reaches the desired depth, the reinforcement is lowered into the casing, and then the concrete of high workability is poured to fill up the tube. The casing tube is immediately withdrawn during which the casing is tapped to release the concrete.

This system of piling became very popular in India, and several projects were executed. IS 2911 (Part 1, Section 1) [21] was developed for this piling procedure that incorporated the Hiley dynamic pile-driving formula. However, there was disagreement over the use of this driving formula because of the severe mismatch between the predictions from the equation and the actual load test results. Subsequently, the IS code in its latest revision withdrew the Hiley formula citing the reason that it was not scientific.

The driving formula cannot be used to estimate the pile capacity during the design stage since the driving records are not available at this stage. There are several cases in which very dense soil is not present for a considerable depth, whereas the piles could be terminated before reaching the very hard stratum. It is also because of the limitations in the length of driven cast-in-situ piles that could be installed using a single piece casing. The pile-driving formula becomes handy in such cases, wherein the pile-driving record is used to ensure the desired capacity once it reaches a predetermined depth.

The difficulties in the appropriate use of pile-driving formula are deriving from the inadequate understanding of the pile-driving system and its response during driving. Problem in the prediction of recoverable penetration, C, used in the formula and the presence of the dispensable shoe that is larger than the pile casing are the primary sources of errors in the use of a pile-driving formula. The following sections describe the difficulties in using the dynamic driving formula in the case of driven cast-in-situ piles in a project.

20.6.1 Soil Profile, Pile Capacity Estimation and Pile-Driving Records

Driven cast-in-situ piles of 500 mm diameter and 21.0 m long with design capacity 120 tonnes were recommended in one of the sites in Manali, Chennai, based on a detailed geotechnical investigation. Figure 20.7 is presenting the typical soil profile, the shear strength profile and the pile capacity estimations. The subsoil profile comprises weak clay and sandy clay layers up to 5.0 m followed by fine-to-medium sand with medium-to high-relative density extending up to about 28 m. Profile up to 20.0 m is only presented in Fig. 20.7. The estimated ultimate pile capacity was 3090 kN (309 tonnes) using static procedures given in the standard books. 500 mm diameter piles were installed using an 'A' frame-piling machine supported on roller pipes. 4.30-tonne drop hammer guided through the casing guide frame was dropped from an average height of 1.40 m for pile driving. Several trial drivings, EB 1 to EB 12, were executed on the site to confirm the drivability and also to confirm the desired capacity for 21-m-long piles. Uniform pile penetration profiles were derived from these trial drivings as illustrated in Fig. 20.8. It presents four trial-driving details up to 28 m depth.

20.6.2 Pile Capacity Estimation Using Driving Records

Estimation of safe compression capacity of 18-m-long and 21-m-long piles was made using the pile-driving records and the Hiley formula. The summary of the analysis is presented in Table 20.4. Standard values of elastic compression C are used in the estimations.

The estimated safe capacities are far below the capacities derived from the soil properties using static formula. There were discussions on reducing the safe capacities or increasing the depth to achieve the desired safe capacities. However, the trial-driving records show more or less uniform diving resistance between 21 m and 26 m that would suggest practically no improvement in the safe capacity if analysed using the driving records.

20.6.3 Pile Capacity from Static Maintained Load Tests and Comparison with the Estimations

Static maintained load tests were conducted on 18.0-m-long and 21-m-long piles and the corresponding load-settlement responses are presented in Fig. 20.9.

The load-settlement response curves are analysed for arriving at the safe compression capacities based on the shear criteria and settlement criteria as recommended in IS 2911 Part 4 [22]. The summary of this review is presented in Table 20.5.

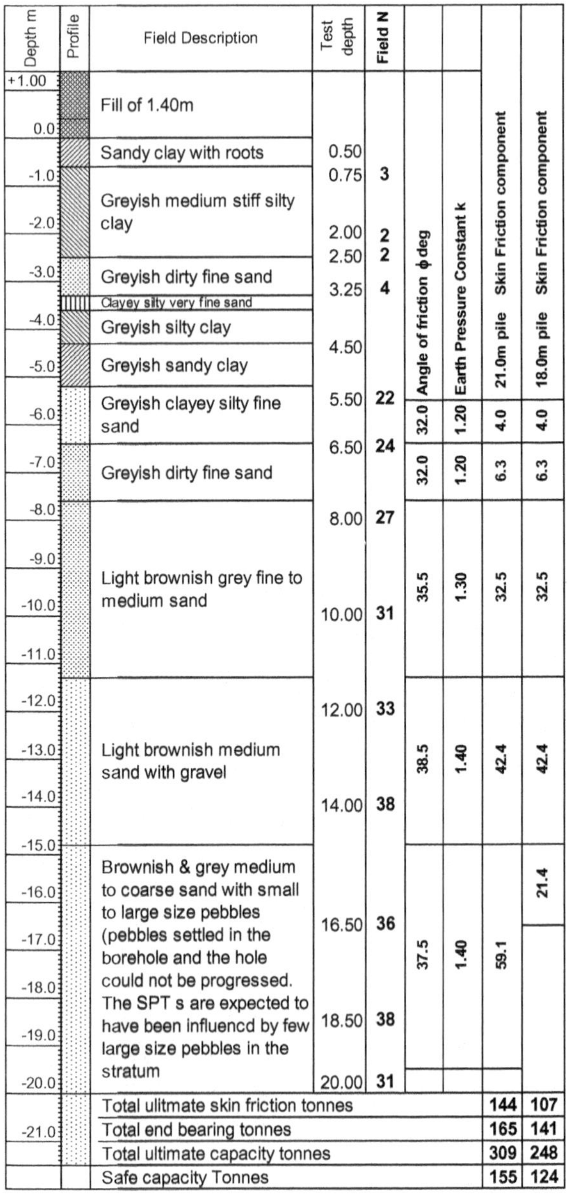

Fig. 20.7 Subsoil profile and shear strength parameters and pile capacities

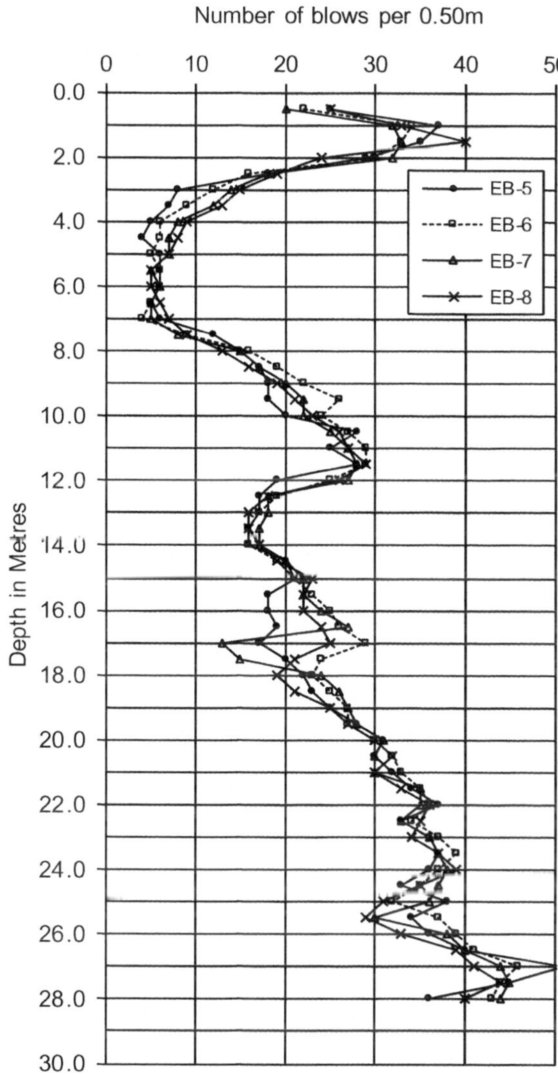

Fig. 20.8 Pile-driving profile in terms of the number of blows for every 500 mm penetration

The estimated ultimate capacities using the standard static formula are comparable with the actual capacities confirmed by the static load tests. Marginally higher resistance may attribute to the additional skin friction from the top weak soil layers. The estimations using dynamic pile formula are only about 38% of the actual ultimate capacities derived from the static load tests.

Table 20.4 Estimation of safe compression capacities using Hiley formula and the driving records

Trial driving no.	EB-5	EB-6	EB-7	EB-8
18.0-m-long piles				
W tonnes	4.3	4.3	4.3	4.3
H m	1.40	1.40	1.40	1.40
Np (blows between 6 and 18 m)	432	493	458	464
Blows for last 500 mm	22	23	24	19
S cm	2.27	2.17	2.08	2.63
Ultimate capacity using Hiley formula	98 t	103 t	105 t	88 t
21.0-m-long piles				
Np (blows between 6 and 21 m)	601	667	630	629
Blows for last 500 mm	32	33	30	30
S cm	1.56	1.52	1.67	1.67
Ultimate capacity using Hiley formula	128 t	130 t	123 t	123 t

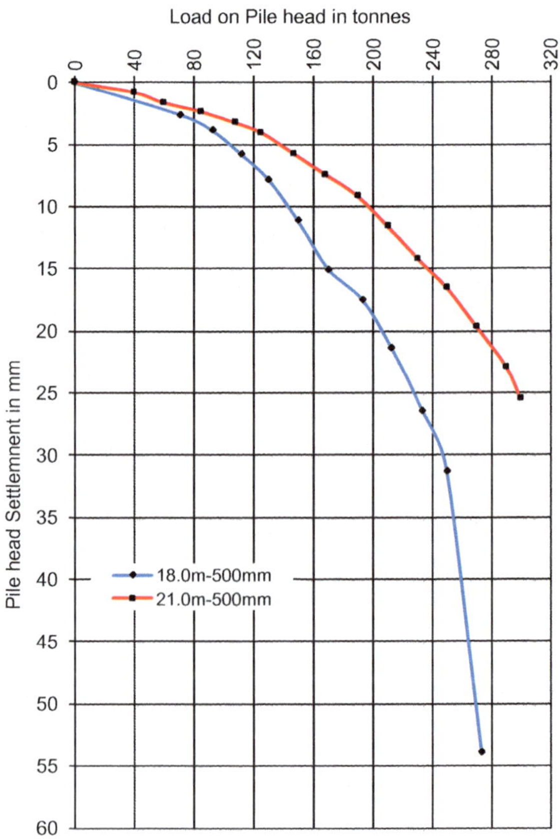

Fig. 20.9 Load-settlement response from static maintained load test

Table 20.5 Estimation of safe compression capacities from the load tests

Pile length	18.0 m	21.0 m
Load corresponding to 12 mm settlement P_{12}	154 t	210 t
Safe load as per IS code provisions = $P_{12} \div 1.5$	103 t	140 t
Ultimate load from load–settlement curve	270 t	
Load corresponding to 50 mm settlement as per hyperbolic curve	270 t	340 t
Safe load from shear capacity $P_{50} \div 2$	135 t	170 t

20.6.4 A Relook into the Use of Hiley Formula and the Driving Records

Figure 20.8 is presenting a very consistent pile-driving record for all the four trials reported. Such consistency cannot be considered accidental. In spite of the general opinion that the pile-driving formulae are erratic or not dependable, the uniformity in the driving profile in a given soil suggests its usefulness if appropriately analysed.

The basis of different dynamic formulae for pile driving is not known to many in the field, because the primary data from which these formulae were derived and validated are not readily available. Whitaker [23] reviewed ten such formulae, including the one based on wave equation, and finally suggested that 'If possible, avoid using a driving formula except for piles that are to be end bearing in sand or gravel'. He, however, acknowledged the fact that a driving formula is of more practical use as a control where the ground conditions are substantially uniform to ensure that piles of one kind are driven to approximately the same resistance.

With this limited purpose, a rational approach to the most common dynamic formulae practised in India is attempted here. Further perusal of driving records in the present instance suggests that the computations using Hiley formula do not reflect any significant increase in the safe capacity from 15 to 21 m. It cannot be true as the soil between 15 and 21 m is reasonably dense sand adding substantial skin friction.

Mostly, driven cast-in-situ piles use flat dispensable MS shoes. The diameter of these shoes is 40–60 mm more than the pile diameter creating an annular spacing between the soil and the casing pipes. The Hiley formula is not accounting the side friction along the casing pipe during pile driving, partly because of an oversize shoe and partly because of lubrication along the smooth casing. It is evident from the trial driving made on this site also in which the blow counts only marginally increase with increasing depth in spite of good sand layers.

Many researchers have experimentally proved that the pile derives its compression capacity by friction gradually developing from the top segments and the end-bearing mobilisation in the initial stages is marginal. A substantial percentage of the total friction is mobilised before a sizable amount of end bearing is realised [24]. They have concluded that all the piles with a reasonable amount of skin friction carry almost the

Table 20.6 Revised summary of pile capacity estimation from static formula

Pile length	18.0 m	21.0 m
Ultimate end bearing	141 t	165 t
Ultimate skin friction	107 t	144 t
Total ultimate capacity	243 t	305 t
Safe skin friction for FS = 1.75	61.3 t	82 t
Safe end-bearing component with FS 2.5	56.5	66 t
Design safe capacity tonnes	118 t	148 t

entire design load by friction that too with a small amount of relative displacement between pile and soil. Hence, it is justifiable to assume the partial factor of safeties such as 1.50–1.75 for skin friction and about 2.50 for end bearing while arriving at the safe capacity from the load tests. It is also pertinent here that the mobilisation of more than 30% of the end bearing requires large movement of the pile base. In most of the cases, where both the components are prominent, the overall safety factor will be close to 2.0–2.5. Tomlinson [25] also points out the predominant presence of the skin friction component under working load.

Two issues arise from the above discussions. (1) The 'set' measured at the end of pile driving reflects mostly the end conditions with a negligible presence of side friction and (2) the significant contribution to the working capacity of a pile (except for strictly end-bearing piles) is from skin friction.

A renewed look as summarised in Table 20.6 at the end-bearing and the skin friction component as estimated from the static formula is relevant at this juncture.

The driving records show that the pile penetration per blow of 4.30 m hammer falling from 1.40 m height is $500 \div 21 = 23.8$ mm/blow at about 18.0 m depth and $500 \div 31 = 16$ mm/blow at approximately 21.0 m (Ref Fig. 20.8). The ultimate capacities for these rates of penetration estimated using Hiley formula after applying standard elastic compression parameters are 115 tonnes and 145 tonnes, respectively, for the 18-m-long and 21-m-long piles.

20.6.5 Discussion

This static capacity derived from the dynamic response under specific impact energy is not expected to be the same for different types of soils. Thus, the static resistance estimated above shall be subjected to suitable modification taking into account the dynamic–static response of the given type of soil.

Various studies on the correlation between SPT N and static cone resistance q_c revealed that the ratio between q_c (static response) and field N (dynamic response) increases when the average particle size increases [24, 26]. The ratio may be 110 for clay and 600 for coarse sand when q_c is expressed in kN/m^2. Many have equated the static end bearing of a pile to the resistance of the cone [24]. The pile driving resembles SPT, while the static resistance of the pile is similar to the resistance of the cone pushed into the ground by static loading.

Even while the elastic compression of the soil (a component in the temporary penetration C in the dynamic formula) at the pile tip is selected based on the type of the soil, there is also a need to choose a suitable 'conversion' factor for estimating the static resistance from the dynamic resistance. In the present case of load testing, the ultimate static resistances of 18- and 21-m-long piles from end bearing are roughly 140 and 165 tonnes. The ratio between the real load and the computed load is about 1.20 for this given soil and assuming that the dynamic response is fully realised by end bearing only.

Whitaker [23] quotes Terzaghi's observation after analysing the performance of several timber, steel and concrete piles, the ratio of the real load to the computed load covered the range of 0.25–4.0. This range is expected to cover plastic clays to weathered rock formations. Accordingly, the resistance to driving into different soils with comparable static resistance differs considerably. Resistance from clay to driving will be considerable compared to that from sand having similar static resistance. The factor of safety applied to ultimate dynamic resistance for deriving the safe static resistance shall reflect such varying resistance to driving. The range of the factor of safety can be 1.5 for coarse sand and 3.5 or more for clay with intermediate values for different effective particle sizes. Selection of the most suitable factor of safety is a difficult task because of high empiricism. The soil below 8.0 m in the present case is fine-to-medium and medium sand and factor safety 2.0 is found appropriate for arriving at the safe end-bearing component.

The dynamic pile-driving formulae can be a handy tool in the decision on the termination level of driven piles. However, a suitable calibration mechanism and more sophisticated ways of measuring the quake in the soil, recoverable penetration due to elastic responses of the pile and soil, contribution from the skin friction resistance during driving, etc. are necessary to make it work for a given soil condition and driving mechanism. Modern pile drivability analysis systems take care of this to a reasonable extent.

In spite of the shortcomings in the dynamic pile-driving formulae, the pile-driving records are effectively used in deciding the pile termination depths for a given safe design capacity in a given soil condition. Anirudhan [12] describes one such case.

20.6.6 Summary Points

1. The driven pile system is relatively economical, and one of the advantages is the possibility of individual assessment of its performance during construction.
2. Various dynamic pile-driving formulae do not produce a consistent prediction under different soil conditions. There are several factors influencing the driving mechanism other than the simple energy work done relation.

3. The use of dynamic pile-driving formulae in the decision of pile termination depth in a given soil condition requires careful understanding of the equivalent static response.

20.7 Field Exposure to Geotechnical Student

The internship programs for the postgraduate students in geotechnical engineering implemented by many institutions are not very useful because of its short duration and the absence of periodical reviews by the institute. The reviews are not possible as the training period is 4 to 5 weeks only. It is not an exaggeration that there is a lack of interest in the students and their teachers in this matter.

A productive and fruitful practical training is possible only if the postgraduate student spends 3 to 6 months in a design and construction organisation. The period of training shall be split between the field and design office. Relatively large duration of training shall provide the opportunity for periodical reviews by the parent institute making the training useful.

There is also a lack of knowledge and understanding of the desired outcome of such training. It shall be understood that the student is not expected to learn a new topic during this training, but required to gain experience of what was taught in the classroom. The practical application of a procedure or theory explained in the classroom can be understood when similar problems are solved in the field or a design office. It means that the student shall need an initial preparation to undergo a useful training program. Interaction with the organisation offering such practical training by the student supervisor is necessary to make the student prepared for the training. The time spent on the training can be more effective if the training program is outlined with the help of such interaction.

Often the postgraduate project work is not on a practical application. It is mainly because of the short duration available for the project work and the necessity of certain 'completeness' of the work carried out. The field problems require more time, and this could delay the project report preparation. This shortcoming can quickly be resolved if the teachers are committed to producing a resourceful geotechnical engineer.

Students can seek internships in prestigious organisations and specialist contracting firms outside their curriculum requirements. Resourceful companies shall come forward to offer such facilities that would immensely gain not only the student but also the geotechnical profession as a whole.

Some premier institutions are offering courses on design and practices developed from case studies. Experts from the field are conducting such classes, and the approach is advantageous. This program shall be developed further to make such courses to cover more areas of geotechnical applications. Some of the pertinent areas may be,

1. Design of foundations in varying ground conditions.
2. Development of ground improvement system for given site conditions and the construction proposals.
3. Comparison between different foundation systems to achieve more carbon footprint gains.
4. Underground constructions including deep excavations, basement constructions and tunnels.
5. Ground and rock anchors, micropiles and uplift anchors.
6. Field investigation procedures and subsurface characterisation.
7. Safety in geotechnical constructions such as deep foundations, deep excavation, grouting and tunnelling.

20.8 Concluding Remarks

This paper attempts to explain the necessity for a geotechnical engineering student to get familiarised with the field problems and the possible ways of relating it to the theory taught in the classrooms using some case studies.

The variations in the field due to the complex nature of the soil are not often represented by the classical theories. There are also the difficulties in incorporating different influences imparted by the field implementation procedures in the theoretical formulations. The evolving of empirical methods, keeping them as close as possible to the theory was possible by constant interaction between the classrooms and the field.

Relatively long duration practical training programs suitably structured by the interaction between the institute and the organisation offering the training shall improve the quality of a geotechnical engineering graduate.

References

1. Terzaghi, K., Peck, R.B.: Soil Mechanics in Engineering Practice, 2nd edn. Wiley (1967)
2. Taylor, D.W.: Fundamentals of Soil Mechanics. Wiley, Chapman & Hall Limited, pp. 522–524 (1948)
3. Bolton, M.D.: A Guide to Soil Mechanics. Universities Press (India) Private Ltd (1991)
4. Bjerrum, L.: Embankments on soft ground. In: Proceedings of the Specialty Conference on Performance of Earth and Earth Supported Structures, vol. 2, pp. 1–54. ASCE (1972)
5. Reese, L.C., Isenhower, W.M., Wang, S.T.: Analysis and Design of Shallow and Deep Foundations, Section 11.6.3. Wiley (2006)
6. Reese, L.C., O'Neill, M.W.: Drilled Shafts: Construction Procedures and Design Methods. U.S. Department of Transportation, Federal Highway Administration, Office of Implementation, McLean, VA (1998)
7. O'Neill, M.W., Reese, L.C.: Drilled Shafts: Construction Procedures and Design Methods. Report No: FHWA-IF-99-025, HIBT, Washington DC (1999)

8. Radhakrishnan, R., Anirudhan, I.V.: Ground improvement with pre-fabricated V drains and pre-load—a case study. In: Proceedings of the Symposium on Advances in Geotechnical Engineering, SAGE 2003, IIT Kanpur, pp. 426–430 (2003)
9. Anirudhan, I.V., Ramaswamy, S.V.: Revisiting a full raft foundation constructed on soft clay. In: Proceedings of the TC 207 Workshop on Soil-Structure Interaction and Retaining Walls, Paris, France, September 2013, pp. 38–50 (2013)
10. Ramaswamy, S.V., Anirudhan, I.V.: Case history as aid to teaching foundations on expansive soils. In: Proceedings of the Indian Geotechnical Conference IGC 2012 at Mumbai, vol. 2, pp. 1122–1125 (2012)
11. IS 8009 Part 1: Code of Practice for Calculation of Settlements of Foundations Part I Shallow Foundations Subjected to Symmetrical Static Vertical Loads (Fifth Reprint March 1999) (1976)
12. Anirudhan, I.V., Balakumar, V.: Pile foundation as settlement reducer for large MS storage tanks. In: Proceedings of the Indian Geotechnical Conference IGC 2010, Mumbai, pp. 995–998 (2010)
13. Kee, C.F.: The inverse slope as prediction of ultimate bearing capacity of piles. In: Proceedings of the 3rd South East Asian Conference on Soil Engineering, pp. 83–91 (1972)
14. Kota, V.K., Madhav, M.R., Bandi, V.: Estimation of shaft and base responses from pile load test. In: Proceedings of the 7th Annual Conference on Deep Foundation Technologies for Infrastructure Development in India, Chennai (2017)
15. IS 2911 (Part 1 Section 2): Design and Construction of Pile Foundations—Code of Practice, Part 1 Concrete Piles, Section 2 Bored Cast In-situ Concrete Piles (2010)
16. Penman, A.D.M., Watson, G.H.: The improvement of the tank foundation by the weight of its own test load. In: 6th Proceedings of ICSMFE, vol. 2, Montreal (1965)
17. Darragh, R.D.: Controlled water tests on preload tank foundation. J. Soil Mech. Found. Div. Proc. ASCE (SM5) 90, 303–329 (1964)
18. Bjerrum, L., Overland, A.: Foundation failure of an oil tank in Fredrikstad, Norway. In: Proceedings of the 4th International Conference on Soil Mechanics and Foundation Engineering, vol. 1, pp. 287–290 (1957)
19. Kalaiarasi, V., Aniudhan, I.V.: Controlling post construction settlement of storage tank foundation by pre-loading—a field study. In: Proceedings of the Indian Geotechnical Conference 2018, Chennai, vol. 1, pp. 459–462 (2006)
20. Chellis, R.D.: Pile foundations. In: Leonard, G.A. (ed.) Foundation Engineering, Chapter 7, pp. 653–661
21. IS 2911 (Part 1 Section 1): Design and Construction of Pile Foundations—Code of Practice, Part 1 Concrete Piles, Section 1Driven Cast In-situ Concrete Piles (1979)
22. IS 2911 (Part 4): Design and Construction of Pile Foundations—Code of Practice, Part 4 Load Test on Piles (2013)
23. Whitaker, T.: The Design of Piled Foundations, 2nd edn, pp. 26–47. Pergamon Press (1976)
24. Gandhi, S.R., Raju, V.S., Rawat, P.C.: Load transfer in piles—analysis and field test on instrumented pile. In: Proceedings of the Soil Properties—Problems and Practices, Indian Geotechnical Conference 1987, vol. 1, pp. 151–154 (1987)
25. Tomlinson, M.J.: Foundation Design and Construction, 5th edn., pp. 27–29, 408–410, 419–421, 474–480. ELBS (1986)
26. Kruizinga, J.: State of Art Report on SPT-CPT Correlation. II European Symposium on Penetration Testing (1982)

Chapter 21
Role of Professional Vigilance in Design and Construction—A Case Study of Solar Projects

N. Ranganath, Debasis Sarkar, Surendra Singh Kachuwaha, Vinaykumar S. Mathad and Saurav Kumar Ghosh

21.1 Introduction

Rapid urbanization, advent of global market, industrial growth, and subsequent infrastructural development have led to exponential energy needs to be met. In the present scenario, use of fossil fuels or nonrenewable sources for energy generation will have adverse effect on environment and mankind as a whole. Climate change is a derivative of unsustainable use of fossil fuels. To minimize the adverse alteration of weather phenomenon and related effects, it is imperative to reduce the use of fossils fuels by sourcing alternate and suitable renewable energy. Tapping freely available abundant divine solar energy, a nonconventional energy source is one such endeavor to mitigate the process of climate change. Solar plants minimize greenhouse gas emissions compared to conventional fossil fuel based power plants. However, solar power generation requires large tracts of contiguous wastelands with optimal climatic conditions.

Realizing the need for nonconventional energy and in order to mitigate the threats of global warming, in the recent past proactive steps have been taken in setting up solar power plants across the globe and even in India.

Setting up of solar parks involves in-depth study of suitability of land, environmental impact, availability of water, and electricity at the site. Accordingly, both macro- and micro-level planning for various infrastructures of the solar park are developed including resource allocation and quality assurance. It is to be noted that, during the course of implementation, i.e., at the time of site selection, reconnaissance survey,

---

N. Ranganath (✉) · V. S. Mathad
E I Technologies Pvt. Ltd., Bangalore, India
e mail: n.ranganath@eitech.in

D. Sarkar · S. S. Kachuwaha
Pandit Deendayal Petroleum University, Gandhinagar, Gujarat, India

S. K. Ghosh
HKBK College of Engineering, Bangalore, India

M. Latha G. (ed.), *Frontiers in Geotechnical Engineering*, Developments in Geotechnical Engineering, https://doi.org/10.1007/978-981-13-5871-5_21

and survey and investigations including design and construction, various factors have to be considered simultaneously and cautiously to execute the solar projects.

21.2 Major Components for Setting of Solar Plants

The solar plant shall comprise minimum components [1, 2] as listed below apart from the panels used in generating solar energy:

(a) Suitable location,
(b) Availability of land at identified location,
(c) Accessibility to location,
(d) Internal road network within plant,
(e) Arrangement of structure with foundation for placing panels,
(f) Drainage facility,
(g) Source of water for plant operation,
(h) Fire station,
(i) Electrical network substation for supply of auxiliary power,
(j) Developed office space for operation,
(k) Residential accommodation for plant staff,
(l) Substation to evacuate power from each subunits in the plant,
(m) Medical assistance,
(n) Telecom network,
(o) Operation and maintenance facility,
(p) Hotels/guest house,
(q) Helpline/assistance for 24/7, and
(r) Provision of green belt.

It is evident that the infrastructure required for the implementation of a solar plant needs a thorough study of the area comprising various surveys, investigations, economic design (at the same time, these designs shall be stable enough for various adverse conditions), and availability of good construction materials for execution. Apart from this, performance of the project depends on the quality of materials and construction as obvious for any construction project.

21.3 Role of Professionals in Setting up of Solar Projects

The following aspects of the project need to be addressed to avoid any kind of effect on solar plant during operations [3, 4].

- Project Performance
  For any project to be commissioned and performed reliably within the stipulated time, effective project management techniques are required to be in place. There are simple and effective tools that have proven to be fruitful in the past.
  - Deployment of qualified and experienced team in handling the projects of multidisciplinary nature.
  - True and accurate survey and investigation of the site.
  - Taking preventive actions during the course of the project with regard to non-availability of good quality construction materials, timely approvals, local and administrative issues, etc., and advising the project authority to take preventive actions at appropriate time.
  - Project activities are planned up to the micro-level and are executed from part to whole, thereby avoiding any duplication and repetition.
  - Dedicated and experienced manpower resources for obtaining necessary approvals/permits well in time.
- Project Integration Management
  Project management integration is considerably a complex matter/issue, during the implementation of the solar park. The role of individual and team as a whole and timely advice during the course of the project will help integration management to a great extent. The following narration gives briefly, the management of overall changes, including controlling various aspects during the project execution.
  - A detailed scope including the approach and required documents need to be prepared and discussed with the project authorities. Consequent to the acceptance of the same, a comprehensive and detailed master plan shall be prepared covering all the aspects envisaged in the project document. The master plan shall then be reviewed, agreed, and accepted by the project authorities, which forms the subsequent basis for the remaining activities.
  - All the details pertaining to topographical survey drawing, geotechnical, and geological investigation report along with levels for drainage planning for proper integration with the master plan shall be made available.
  - During construction, as far as possible, deviation/changes shall be avoided by the statutory authority and the approving authorities shall be available for interaction, consultation on 24/7 basis, especially during commissioning stage.
  - The activities shall be monitored by the top management with system of reporting procedure at all levels, besides carrying out surprise and periodical inspection to the site. All the above aspects will ensure timely commissioning of the project with emphasis on quality.
- Project Scope Management
  The infrastructure development includes acquisition of government and private lands, development of approach and internal roads, creation of reservoir for storing rainwater, raw water supply, treatment and distribution, smart power evacuation facility including auxiliary electrical and telecom network, etc., online tracking

system, development of local manpower by providing in-house training, community development, etc.

- Project Time/Schedule Management
  A comprehensive planning is required to ensure completion of project in all aspects without time overrun. This is achieved by having work breakdown structure and assigning proper time to each activity taking into account the likely factors contributing to delay resulting in time overrun beyond the schedule leading to additional cost. Hence, a detailed project schedule should be in place to monitor all the critical activities for ensuring timely completion of project.
- Project Cost/Resources Management
  Project cost/resource management shall be undertaken to ensure there is no major deviation in the estimated cost and, generally, the project is executed within the budget. The sources for project funding needs to be identified.
- Project Quality Management
  The following concepts need to be practiced in order to achieve total quality management:

  – Every person involved must be trained to understand the purpose of the project and measures in place to achieve the target and quality.
  – Improvise implementation practices.
  – Identify core team of reliable and competent contractors and award the work based on competency.
  – Have regular meetings with the contractors and the project management team to address technical and nontechnical issues and try to build a good relationship between all stakeholders.
  – Develop quality standards at a very early stage of the project.
  – Address and provide necessary technical assistance to the workforce.

- Project Human Resource Management
  Project human resource management includes processes required to make most effective use of people involved with the project to complete it within the stipulated time limit. The organization of human resource employed for the successful completion of the project shall be planned well in advance.
- Project Communications Management
  An effective communication system shall be established between the parties involved in the project to avoid any sort of communication gaps leading to delay in project execution.
- Project Risk Management
  Risk management pertaining to external and internal factors likely to affect the progress of the work shall be dealt carefully. External factors, generally, are unpredictable and thereby not under control but precautions required for such factors based on the previous experience/data shall be in place.
- Project Contract/Procurement Management
  In order to achieve timely completion of the project, the total project shall be divided into smaller subprojects like internal roads, site development, treatment

plants, auxiliary power plants, telecom, etc. These subprojects then shall be awarded to different contractors having expertise in their respective fields.

21.4 Data Collection and Analysis

The solar power plants established in Rajasthan had a major setback. Within 2 years from the date of commissioning, due to reasons related to investigation, earthwork filling, lack of supervision, limiting in the codes, etc., leading to a major disaster in the power plant within a short span of 12–18 months which could have been avoided with timely vigilance of field engineers and professionals, design engineers, and the project execution authorities.

21.5 A Case Study on Defective Design, Poor Execution, and Project Management [5, 6]

21.5.1 Project Location

A State-of-the-art 5.83 MW (DC) solar power project in Bikaner District, Rajasthan was developed by an entrepreneur/agency who had entrusted the work of entire project involving engineering, procurement, construction, commissioning including operation and maintenance to a single Agency (EPC).

21.5.2 Project Condition

The power plant is in operation since the last 3 years. However, during the recent past (June to October 2015), due to heavy monsoon rains along with gale/heavy wind (60–90 Kmph), extensive damage to some of the solar panels including the structure has occurred, resulting in considerable generation loss. Out of the total structures generating 5.83 MW, approximately structures generating 1.05 MW has been damaged extensively and most of the remaining structures are in the distress conditions. The structure is M.S. supporting column fixed on a single pile foundation [7].

On May 19 and May 27, 2015, the site witnessed high-speed winds. These resulted in structures to be uprooted from their foundation and were damaged significantly. The purlins and rafters were damaged beyond repair at certain locations. At certain areas where the purlins and rafters could not be mended, different ones were used. Identifying the scale of damage, the owner decided to carry out a detailed investigation into the failure of structures and utilized the structure consultant's ser-

Fig. 21.1 Photos indicating the extent of damage witnessed due to heavy wind

vices to carry out the revised design of foundation and support the module mounting structures (Fig. 21.1).

21.5.3 Findings and Solutions for Damage

(1) *Field Observations after Damage due to Heavy Rain and Wind*:

- Terrain of the project area is flat. In order to provide effective drainage, part of the project site has been filled up and the ground level is raised. The extent of filling varies between 0.4 and 0.6 m from the original ground level.
- Structures are erected on pile foundation of 300 mm diameter, with a pile cap of 375 × 375 × 300 mm. The depth of the pile is 1.6 m.
- The effective length of the pile in the original soil (where the land is filled with borrowed material by 0.4–0.6 m) is only 0.9–1.2 m in the natural ground.
- Soil exploration details furnished reveals the SBC of $7/m^2$ under dry condition.
- Anchorage between the pile and the pile cap was poor and the reinforcement was not concentric within the pile.
- Length of the foundation bolts was insufficient and MS bolts have been used instead of E8 bolts.

- One of the common failures was the shearing of the pile cap at the foundation bolt level.
- Quality of the concrete work was poor and the cover for the reinforcement was less.

(2) *Modes of Failure*:

- Considerable number of panels was uprooted due to the wind force pulling along with it the supporting steel section, pile cap, and the pile (Fig. 21.1).
- The steel sections were also twisted and the pile cap was separated around the base of the foundation bolt.
- In majority of the cases, there was no bond between the pile and the pile cap.
- The pile and pile cap were not cast monolithically.
- The depth of the pile is indicated as 1.7 m below formed ground level (FGL).
- The diameter of the pile is 300 mm and provided with 10 mm diameter reinforcement.
- The wind speed considered is 44.4 m/s and the design wind speed considered is 160 kmph. This is based on risk coefficient, terrain factor, and topography factor which are considered as 1. According to wind map given in IS 875, the Bikaner district of Rajasthan falls under green zone for which the basic wind speed is 47 m/s. Since the solar park producing solar energy is a power plant, the risk coefficient for a wind speed of 47 m/s for power plant structures is 1.07 as per IS 875 (Part-3). Similarly, the terrain coefficient and the topography factor considered in the modified design are in accordance with IS 875. Considering the above factors, the design wind speed works out to 177.42 kmph.
- The main reinforcement for a circular pile should be 6 numbers, 12 mm diameter as per IS 2911 (Part-2) for bored cast-in-situ pile and as per the same code, the minimum grade of concrete shall be M25 for pile foundation. In reality, the foundation bolts provided were 4 numbers MS bolts, less than 300 mm long and majority of the pile caps have suffered failure by separation at this point.
- As per IS 2911 (Part-2), the overhang of the pile cap beyond the pile face shall be a minimum of 150 mm. Hence, the dimension of the pile cap should be 300 mm more than the diameter of the pile.

(3) *Contributing Factors for the Failure*:

- Insufficient embedding of pile in natural soil.
- Lack of integrity in the structure due to absence of monolithic action between pile and pile cap.
- Lesser value of wind design speed considered than what is applicable for the area.
- Noncompliance with codal provisions regarding minimum reinforcement in pile, minimum grade of concrete, and size of pile cap.

- Diameter, quality, and length of the foundation bolts were found not up to the standards.
- The client had discussions with representatives of contractor and designer, wherein the rectification methods were discussed and one of the main criteria to be satisfied during the process of rectification was to minimize the loss of power generation and minimizing the time of construction and simultaneously ensuring the safety and durability of the rectified structure. It was also decided to utilize the existing materials to the maximum extent possible.
- Poor quality of construction and lack of supervision.
- In view of the bitter experience of the failure and for ensuring safety and durability of the rectified structure, the following criteria was decided for the design and execution of rectification proposals duly deliberating all the points in several meetings.
  - Adopting wind load coefficients as per ASCE.
  - Adopting single pile structure. Later as per site conditions, decision is taken to go with single pile foundations in blocks C, D, and E and twin pile foundation in blocks A, B, and F.
  - Providing EN8 bolts instead of M.S. bolts.
  - Adopting strict quality control measures during construction.
  - Rectification works were carried out accordingly and completed within the stipulated time satisfying the code requirements and simultaneously ensuring that the structure is safe against any probable high wind forces.
- The rectification work carried out involved only essential components required for the safety and durability of the structure besides ensuring minimum generation loss.

(4) *Reasons for Failures*:

- Inadequate investigation and insufficient data.
- Design not confirming to provisions mentioned in relevant code.
- Poor construction.
- Lack of supervision.
- Insufficient anchorage of the pile.
- Location of some of the piles in filled up soil.
- Inadequate designs and detailing.

(5) *Corrective Actions*:

- Designs have been carried out considering the depth of foundation to reach natural ground level.
- Considering the code provision and also as per client suggestions, twin pile structures have been adopted for 170 structures. For remaining structures, single pile of 400 mm diameter, 3.5 m deep, and pile cap of 400 mm is considered.
- Minimum embedment depth of 2.5 m in natural soil is provided.

Fig. 21.2 Photos indicating the repair, rectification, and maintenance of foundation and module mounting structure

- Structures have been designed considering site condition and also effect of moisture in the soil as per code.
- Additional inner reinforcement is provided along with cover blocks to ensure concentricity of reinforcement.
- EN8 bolts of 6.8 grade having a length of 800 mm have been used.
- Main pile reinforcement has been extended up to top of pile cap with additional anchorage length of 150 mm. Both pile and pile cap have been casted monolithically.
- Mix design was carried out and accordingly M25 grade concrete was used. Required cover was provided using cover block. Strict supervision was undertaken during execution.

(6) *Proposed Corrective and Preventive Designs*: Upon review of the site conditions and field investigations, it was planned to suggest remedial measures to ensure restoration of plant for operations. Preventive and corrective designs proposed depending upon the extent of damage witnessed for the module mounting structures. Two types of designs as detailed below were proposed (Fig. 21.2).

(a) Type A: twin pile corrective foundation and
(b) Type B: single pile preventive foundation.

The comparison of original design which was executed at site and modified design after witness of damage due to reason stated above have been presented in Table 21.1.

21.5.4 Review of Proposed Corrective and Preventive Design by Independent Agency [3]

(1) *Review of Type A: Twin Pile Corrective Foundation*.

The design forces and design calculations for the foundation were reviewed and following are the discussions presented by Excelsior Engineering Solution as an independent agency.

Overview of the Design of Type A Twin Pile:

Table 21.1 Comparison of design parameters

No.	Item (Parameters)	As per original designer	As per modified design
1	No. of supports per structure	4	4
2	Reference ground level	F.G.L (formed ground level)	N.G.L (natural ground level)
3	Embedment of pile in natural soil	0.7–1.2 m	2.50 m and above
4	No. of piles per support	1	2
5	Diameter of pile	300 mm	350 mm for twin pile configuration and 400 mm for single pile configuration
6	Grade of concrete	Less than M20	M25
7	Grade of steel	Fe-250	Fe-500
8	Reinforcement	6 nos. 10 dia	6 nos. 12 dia
9	Pile cap	375 × 375 × 300 mm	Twin pile of 2600 × 600 × 350 mm configuration and 400 mm diameter pile with a pile cap of 700 mm diameter and 600 mm thick for single pile configuration
10	Foundation bolt	M.S., 300 mm long, 12 dia	EN-8, 400 mm long, 16 diameter
11	Basis for analysis	One structure having 3 rows X × 4 columns module as space frame	One structure having 3 rows X × 4 columns module as space frame
12	Inclination	25	25
13	Wind velocity	44.4 km/s	47 km/s
14	Design wind speed	160 kmph	177.42 kmph
15	Analysis	STAAD. Pro	Analytical
16	Wind load coefficients	As per IS 875	As per ASCE-7
17	SBC	7 tons/m^2	7 tons/m^2

(a) Grade of concrete—M25; grade of reinforcement—Fe500.
(b) For each steel column, two piles of 350 mm diameter each at a center-to-center spacing of 2400 mm; total length of pile in the soil is 3500 mm.
(c) Pile cap 650 mm width, 3050 mm long; 600 mm projected above NGL.
(d) Pile reinforcement: 6 numbers 12 mm diameter bars.
(e) Pile cap reinforcement: 4 numbers 12 mm and 2 numbers 10 mm diameter bars, both at top and bottom.
(f) Anchor bolts: 6 numbers 16 mm diameter, 450 mm long 6.8 grade anchor bolts with 400 mm embedment in the concrete.

(g) Base plate 400 × 400 × 10 mm dimension.

Independent Review of Type A Twin Pile:

An independent design check was carried out based on IS 875-3 and the design was submitted based on which the following inferences are made. As a combined effect of the multiple discrepancies in the design calculations, e.g., overestimation of force calculations and some overestimation of the capacity of piles, the final recommended twin pile of 350 mm diameter with 3500 mm length inside the soil is found to be adequate based on the checks carried out. Table 21.2 summarizes the outcome of the checks for utility ratio of the various actions that the foundation would be subjected to various actions mentioned.

It is observed that the Type A foundation design is safe, and the margins of safety are adequate. Prima facie, it might seem that the margins in bending moment are significantly higher (a utility ratio of 0.13). However, optimizing the pile foundation for bending moment by reducing either the diameter or depth of the pile would result in reduction of its uplift capacity, which would result in failure against uplift as well as affect its lateral deflection as they are correlated.

(2) *Review of Type B: Single Pile Preventive Foundation*:

The design forces and design calculations for the foundation were reviewed and following are the discussions presented by Excelsior Engineering Solution.

Overview of the Design Type B Single Pile:

(a) Grade of concrete—M25; grade of reinforcement—Fe500.
(b) 450 mm diameter pile; total length of pile in the soil is 3800 mm.
(c) Pile cap 750 mm diameter; 300 mm in the soil, 300 mm projected above NGL.
(d) Pile reinforcement: 5 numbers 12 mm diameter + 1 number 16 mm diameter bars.

Table 21.2 Utility ration for Type A foundation

Sl. no	Action	Demand oneach pile	Capacity of one pile in corrective—twin pile design[a]	Utility Ratio
1	Vertical Bearing (kN)	12.33	23.97	0.51
2	Uplift (kN)	13.85	25.55	0.54
3	Lateral Deflection (mm)	4.46	5	OK
4	Bending Moment (kN-m)	15.24	120.74	0.13
5	Pile reinforcement (mm2)	384.8	678.6	0.57
6	Anchor bolts		Individual	0.89

[a]After applying factor of safety

Table 21.3 Utility ration for Type B foundation

Sl. no.	Action	Demand	Capacity of preventive single pile design[a]	Utility ratio
1	Vertical bearing (kN)	14.88	58.25	0.26
2	Uplift (kN)	17.93	28.19	0.64
3	Lateral deflection (mm)	4.38	5	OK
4	Bending moment (kN-m)	30.40	197.98	0.15
5	Pile reinforcement (mm^2)	636.2	766.5	0.83
6	Anchor bolts		Individual	0.51
			Combined	0.26

[a]After applying factor of safety

(e) Pile cap reinforcement: 6 numbers 16 mm diameter bars.
(f) Anchor bolts: 6 numbers 16 mm diameter, 900 mm long 6.8 grade anchor bolts with 700 mm embedment in the concrete and 100 mm L-bend beyond that.
(g) Base plate 350 × 350 × 8 mm dimension.

Independent Review of Type B Single Pile:

An independent design check based on IS 875-3, IS 456 (200), and IS 2911 Part-1 Sect.-2 and considering the missing criteria from the submitted design, the following can be concluded. As a combined effect of the multiple discrepancies in the design calculations, e.g., overestimation of force calculations, some overestimation of the capacity of piles, and some underestimation of the capacity, the final recommended single pile of 450 mm diameter with 3800 mm length inside the soil is found to be adequate based on the checks carried out. Table 21.3 summarizes the outcome of the checks for utility ratio of the various actions that the foundation would be subjected to utility ratio for various actions of the foundation.

It is observed that the Type B foundation design is safe and the margin of safety considered is adequate. It may seem that the margins in bending moment are very high (a utility ratio of 0.15). However, optimizing the pile foundation for bending moment by reducing either the diameter or depth of the pile would result in reduction of its uplift capacity, which would result in failure against uplift as they are correlated.

(3) *Outcome of Independent Review*:

- A detailed review of the foundation Type A: Twin pile corrective as per the ASCE and several IS codes as indicated, revealed that the designed foundations are safe. The margins of safety considered for the design of foundation are adequate for a safe operational life of the project. The anchor bolts utilized are adequate to withhold as well as transmit the forces acting on them as per IS-456.
- A detailed review of the foundation Type B: Single pile preventive as per the ASCE and several IS codes as indicated, revealed that the designed foundations are safe. The margins of safety considered for the design of foundation

are adequate for a safe operational life of the project. The anchor bolts utilized are adequate to withhold as well as transmit the forces acting on them as per IS-456.

21.6 Conclusion

A detailed study of available data is carried out along with a thorough investigation and identification of the causes for failure of the structure (foundation). The validation and simulation of actual data with respect to the site condition across and verification of the codes and practices followed by international standards, strict vigilance at the time of project execution has led to successful re-commissioning of the project in a record time. The remedy measures though effective overshoots the initial budget as the improvisation and unforeseen scenarios like the heavy winds in Bikaner led to more cost. Proper planning and insight about the issues would have helped prevent the disaster in the first place which is now repaired at an additional cost. Furthermore, the project was monitored for one complete year including the peak disaster time and found that the alternative design and execution suggested by the professional is effective and has substantially validated their forecasting and the adopted approach is found to be beneficial.

References

1. Alagh, Y.K., Shah, J., Shah, V.K.: Power Economics in Gujarat. Har-Anand Publications (1998)
2. Shukla, P., Biswas, D., Nag, T., Yajnik, A., Heller, T., Victor, D.G.: Impact of power sector reforms on technology, efficiency and emissions: case study of Gujarat, India. In: Program on Energy and Sustainable Development Working Paper, vol. 21 (2004)
3. Sharma, N.K., Tiwari, P.K., Sood, Y.R.: Solar energy in India: strategies, policies, perspectives and future potential. Renew. Sustain. Energy Rev. **16**(1), 933–941 (2012)
4. Sharma, A.: A comprehensive study of solar power in India and World. Renew. Sustain. Energy Rev. **15**(4), 1767–1776 (2011)
5. Sahu, B.K.: Solar energy developments, policies and future prospectus in the state of Odisha, India. Renew. Sustain. Energy Rev. **61**, 526–536 (2016)
6. Shrimali, G., Rohra, S.: Indias solar mission: a review. Renew. Sustain. Energy Rev. **16**(8), 6317–6332 (2012)
7. Khan, M.A.: Policy review and recommendations on the promotion of renewable energy and energy efficiency (2016)

Printed by Printforce, the Netherlands